Fundamentals of Transport Processes

The study of transport phenomena is an essential part of chemical engineering, as well as other disciplines concerned with material transformations such as biomedical engineering, mechanical engineering and materials engineering. Material transformations require the motion of constituents relative to each other, the transfer of heat across materials and fluid flow.

This lucid textbook introduces the student to the fundamentals and applications of transport phenomena in a single volume, and explains how the outcomes of transformation processes depend on fluid flow and heat/mass transfer. It demonstrates the progression from physical concepts to the mathematical formulation, followed by the solution techniques for predicting outcomes in industrial applications. The ordering of the topics, gradual build-up of complexity and easy to read language make it a vital resource for anyone looking for an introduction to the domain. It also provides a foundation for advanced courses in fluid mechanics, multiphase flows and turbulence. The author explains the book in a series of video lectures in the supplements package.

V. Kumaran is Professor in the Department of Chemical Engineering, Indian Institute of Science, Bangalore. He received his B. Tech from the Indian Institute of Technology Madras in 1987, and his PhD from Cornell University in 1992. His research is in the areas of statistical mechanics, fluid mechanics and the dynamics of complex fluids. He has received the Bhatnagar Prize for Engineering Science in 2000, the The World Academy of Sciences (TWAS) Prize for Engineering Sciences in 2014, and the Infosys Prize for Engineering and Computer Science in 2016.

CAMBRIDGE–IISc SERIES

Cambridge–IISc Series aims to publish the best research and scholarly work in different areas of science and technology with emphasis on cutting-edge research.

The books aim at a wide audience including students, researchers, academicians and professionals and are being published under three categories: research monographs, centenary lectures and lecture notes.

The editorial board has been constituted with experts from a range of disciplines in diverse fields of engineering, science and technology from the Indian Institute of Science, Bangalore.

IISc Press Editorial Board:

Amaresh Chakrabarti, *Professor, Centre for Product Design and Manufacturing*
Diptiman Sen, Professor, *Centre for High Energy Physics*
Prabal Kumar Maiti, *Professor, Department of Physics*
S. P. Arun, Associate Professor, *Centre for Neuroscience*

Titles in print in this series:

- *Continuum Mechanics: Foundations and Applications of Mechanics* by C. S. Jog

- *Fluid Mechanics: Foundations and Applications of Mechanics* by C. S. Jog

- *Noncommutative Mathematics for Quantum Systems* by Uwe Franz and Adam Skalski

- *Mechanics, Waves and Thermodynamics* by Sudhir Ranjan Jain

- *Finite Elements: Theory and Algorithms* by Sashikumaar Ganesan and Lutz Tobiska

- *Ordinary Differential Equations: Principles and Applications* by A. K. Nandakumaran, P. S. Datti and Raju K. George

- *Lectures on von Neumann Algebras, 2nd Edition* by Serban Valentin Strătilă and László Zsidó

- *Biomaterials Science and Tissue Engineering: Principles and Methods* by Bikramjit Basu

- *Knowledge Driven Development: Bridging Waterfall and Agile Methodologies* by Manoj Kumar Lal

- *Partial Differential Equations: Classical Theory with a Modern Touch* by A. K. Nandakumaran and P. S. Datti

- *Modular Theory in Operator Algebras, 2nd Edition* by Şerban Valentin Strătilă

- *Notes on the Brown-Douglas-Fillmore Theorem* by Sameer Chavan and Gadadhar Misra

Cambridge IISc Series

Fundamentals of Transport Processes

V. Kumaran

CAMBRIDGE
UNIVERSITY PRESS

CAMBRIDGE
UNIVERSITY PRESS

University Printing House, Cambridge CB2 8BS, United Kingdom

One Liberty Plaza, 20th Floor, New York, NY 10006, USA

477 Williamstown Road, Port Melbourne, vic 3207, Australia

314 to 321, 3rd Floor, Plot No. 3, Splendor Forum, Jasola District Centre, New Delhi 110025, India

103 Penang Road, #05-06/07, Visioncrest Commercial, Singapore 238467

Cambridge University Press is part of the University of Cambridge.

It furthers the University's mission by disseminating knowledge in the pursuit of education, learning and research at the highest international levels of excellence.

www.cambridge.org
Information on this title: www.cambridge.org/9781009005333

First published 2022

Printed in India by India Binding House, Noida

A catalogue record for this publication is available from the British Library

ISBN 978-1-009-00533-3 Paperback

Dedicated to the memory of my father,

whose belief in my abilities was the 'driving force'

that 'transported' me to where I am.

Contents

Preface

An anecdote about Prof. P. K. Kelkar, founding director of IIT Kanpur and former director of IIT Bombay, was narrated to me by Prof. M. S. Ananth, my teacher and former director of IIT Madras. A distraught young assistant professor at IIT Kanpur approached the director and complained that 'the syllabus for the course is too long, and I am will not be able to cover everything'. Prof. Kelkar replied, 'You do not have to cover everything, you should try to uncover a few things.' In this book, my objective is to uncover a few things regarding transport processes.

The classic books on transport processes, notably the standard text *Transport Phenomena* by Bird, Stewart and Lightfoot written about 60 years ago, provided a comprehensive overview of the subject organised into different subject areas. At that time, engineers were required to do design calculations and modeling for different unit operations, and for the sequencing of these operations in process design. This required expertise in laboratory and pilot scale experiments on unit operations and scaling up of these operations using correlations. Proficiency in developing, understanding and using design handbooks and correlations was also needed. In this context, the study of transport processes at the microscopic level, and its implications for design for unit operations, was a pioneering advance that has since become an essential part of the chemical engineering curriculum.

In the last half century, sophisticated computational tools have been developed for detailed flow modeling within unit operations, and for the selection and concatenation of unit operations for achieving the required material transformations. The ease of search for information and data today was inconceivable half a century ago. Routine calculations have been automated, and there is little need for routine tasks such as unit conversion, graphical construction and interpreting engineering tables. There is now a greater need for understanding physical phenomena and processes and their mathematical description.

Using a rigorous understanding of transport processes, an engineer usually contributes to process design in one of two ways. The first is the development and enhancement of models and computational tools for modeling of flows and transformations in unit operations; these result in higher resolution, better representation of the essential physics and inclusion of new phenomena. The second is the use of these tools for design of unit operations, and the coupling between them. In the present context, the objective of this text is to assist the student in internalising some common conceptual frameworks for transport processes.

(1) At the level of unit operations, the use of dimensional analysis is enhanced to provide a physical understanding of the dimensionless groups for common internal and external flows in chapters 1–2. Justification is provided for the forms of the correlations in different parameter regimes.

(2) The interplay between convective and diffusive transport is one of the central ideas. This is emphasised in chapter 2, and the molecular origins of diffusion are discussed in chapter 3. Approximate methods for estimating diffusion coefficients in gases and liquids are developed.

(3) The progression from the balance condition for a differential volume, the inclusion of the constitutive relation for transport across surfaces, to the derivation of the partial differential equation for the densities of mass/momentum/energy is shown in chapter 4 for a Cartesian co-ordinate system and chapter 5 for curvilinear co-ordinate systems. This demonstrates how the mathematical formulation arises from the physical description.

(4) The procedure for deriving conservation equations for a general co-ordinate system is the subject of chapter 7. A compact representation in terms of vector differential operators is obtained, and the operators are determined for different co-ordinate systems. This enables a student to comprehend the meaning of the vector notation, and the form of the vector operators in different co-ordinate systems.

(5) The use of physical insight in reducing partial differential equations to one or more ordinary differential equations is demonstrated in chapters 4 and 5 for unidirectional transport. Two procedures are emphasised, similarity transforms and separation of variables.

(6) The similarity transform procedure is then developed into the concept of a boundary layer in chapter 9 for forced convection, and chapter 10 for natural convection. This is an important conceptual basis for transport under strong convection, and it provides a means for understanding the form of the correlations for laminar flows.

(7) The separation of variables procedure is used for solving the diffusion equation in Cartesian and spherical co-ordinates in chapter 8. This is further developed

to the spherical harmonic analysis in a spherical co-ordinate system, and the resulting fields are interpreted in terms of sources and sinks.

(8) The role of pressure in momentum transport is the subject of chapter 6. After examining potential flows, the laminar and turbulent flows in a pipe are discussed in detail. The discussion of a turbulent flow is more detailed than that in most courses on the same subject, because it is important for a student to have a physical picture of the structure of a turbulent flow. The Navier–Stokes equations for a Newtonian fluid are presented without derivation in chapter 7, along with reasonably advanced treatments of the stress and rate of deformation tensor in a fluid.

Some important topics that do not fit into the set of common frameworks listed above are not included in this study. Multicomponent diffusion is discussed at a very basic level in chapter 4, only for binary mixtures. The derivation of the diffusion flux using non-equilibrium thermodynamics is a specialised topic which is not discussed here. Radiation heat transfer is also not examined here, since it is not possible to do justice to this without a basic coverage of electromagnetic waves.

Computational methods are also not covered in this course. There have been significant advances in computational methods in the last few decades, and one or more specialised courses are required to cover this area to a reasonable extent.

Some of the content in the book requires only a rudimentary knowledge of mathematics, while there is more advanced material requiring calculus, special functions and solution procedures for ordinary differential equations. All of the course material is suitable for students who have taken undergraduate courses in calculus, linear algebra and ordinary differential equations. Where necessary, additional background for the mathematical formulation has been provided in the appendices.

For students without adequate exposure to mathematical methods, an elementary course could be designed by selecting topics from chapters 1, 2, 3, 4.1–4.3, 4.8, 5.1.1–5.1.2, 5.2.1–5.2.3 and 6.1–6.3. The remaining chapters and sections could constitute an advanced course which requires proficiency in advanced mathematical techniques at the undergraduate level.

The important concepts in each section are emphasised in the summary at the end of the section. The important equations in the section are enclosed in boxes, and these are also highlighted in the summary section. There are numerical examples throughout the text, which enable the students to acquire a feel for the numbers and magnitudes involved in practical situations. An adequate number of exercise problems are provided at the end of each chapter, which can be solved using procedures discussed in the respective chapter.

At the end of the course, it is hoped that the conceptual understanding provided here enables the prospective engineer to develop a 'feel' for the phenomena involved when faced with a new problem, assess the dominant forces and the flow regime from dimensionless groups, and carry out quick estimates of the magnitudes of the transport rates which can be used to evaluate feasibility. These estimates can also be used to check the consistency of solutions from more rigorous analytical or numerical procedures. When using computational tools, it is important to know the operating flow regimes so that the appropriate models can be used, and to validate the consistency of the numerical scheme with known analytical solutions in limiting parameter regimes. An understanding of the meaning of the conservation equations in vector notation, and their representation as partial differential equations in different co-ordinate systems, is essential for those using computational tools to determine the field variables within unit operations. The book is also a useful starting point for prospective researchers who will develop models for new phenomena, incorporate transport processes into novel applications and develop better analytical/numerical solution procedures for the transport equations.

I learnt the subject of transport processes as a student at IIT Madras and at Cornell University, but I began to understand the subject only after teaching for many years, and after research discussions with students who often had a better understanding than me. For that, I am grateful to the generations of students who have helped me in this journey from learning to understanding.

I would like to thank Prof. Kesava Rao for his detailed review and critical comments on the section on binary diffusion. My colleagues in the Department of Chemical Engineering at the Indian Institute of Science welcomed me into this department many years ago, and guided and supported me as I progressed in my career. They provided a congenial atmosphere and freedom to think and explore, and they did not burden me with even a fair share of administrative responsibilities. To them, and to the administrative staff who have assisted me many times beyond the call of duty, I am grateful. I would like to thank my family, Nandini, Radha and Sudha, for their support and patience while I was writing the book, and it appeared the process would never end.

1

Introduction and Dimensional Analysis

1.1 Significance of Transport Processes

The conversion of raw materials into useful products in a predictable, efficient, economical and environment-friendly manner is an essential part of many branches of engineering. There are two types of transformations: chemical transformations (involving chemical reactions) and physical transformations (melting, evaporation, filtering, mixing, etc.). Both of these transformations involve the motion of constituents relative to each other, and they often involve the transfer of energy in the form of heat. In operations involving fluid flow and mixing, there are forces exerted on the fluid due to pumps, impellers, etc. (input of mechanical energy), in order to overcome the frictional resistance generated by the flow. The subject of this text is the transport of the components in materials relative to each other, the transport of heat energy and the transport of momentum due to applied forces.

This text is limited to operations carried out in the fluid phase. Although solids transport and mixing does form an important part of material transformation processes, fluid-phase operations are the preferred mode for conversion because the transport is enabled by the two fundamental processes: convection and diffusion. Convection is the transport of mass, momentum and energy along with the flowing fluid. Diffusion is transport due to the fluctuating motion of the molecules in a fluid, which takes place even in the absence of fluid flow. Convection does not take place in solids since they do not flow, and diffusion in solids due to vacancy or interstitial migration is a very slow process, which makes it infeasible to effect material transformations over industrial timescales.

Fluids are of two types: liquids and gases. In liquids, the molecules are closely packed, and the distance between molecules is comparable to the molecular diameter.

1

In contrast, in gases, the distance between molecules is about 10 times larger than the molecular diameter under conditions of standard temperature and pressure (STP). Due to this, the density of a liquid is about 10^3 times that of a gas. In a gas, the molecules interact through discrete collisions, and the period of a collision is much smaller than the average time between collisions. In contrast, a molecule in a liquid is always under the influence of forces exerted by neighbouring molecules. Despite these differences, the conservation equations for liquids are identical in form to those for gases, but the coefficients in the equations (density, viscosity, diffusion coefficient, etc.) are very different. Therefore, the procedure for solving the equations is identical for liquids and gases. In this text, we will discuss the mechanism of diffusion separately for liquids and gases, but after that, there will be no distinction between liquids and gases when the equations are formulated and solved.

Let us consider some illustrative examples, in order to understand the importance of transport processes.

1.1.1   Stirred-tank Reactor with Solid Catalyst

Consider a continuous stirred-tank reactor in which reactant A is converted into product B, as shown in Fig. 1.1. The reactant and product are dissolved in the liquid phase. The reaction requires a solid catalyst which is in the form of pellets, and the conversion occurs on the surface of the catalyst. An impeller is necessary for mixing and to enhance the reaction rate.

In order to ensure that the consumption of A takes place at the desired rate, it is not sufficient to feed in the reactant A at the reactor inlet at this rate. Since the reaction is taking place at the solid surface of the catalyst, conditions have to be created such that the transfer of A from the liquid phase to the catalyst surface occurs at the desired rate. This is a mass transfer problem. At the catalyst surface, reactant A is converted into product B. The reaction will proceed only if product B is also transported from the solid surface to the liquid phase at the rate at which it is produced. Otherwise, the products will occupy the catalyst surface and prevent further reaction from taking place. Therefore, it is necessary to engineer the system in such a way that the mass transfer rates of both the reactants to and products from the catalyst surface are sufficient to obtain the desired conversion rate. In our study of transport processes, we will examine how the rate of transport of materials from/to a solid surface depends on the difference in concentration between the catalyst surface and the bulk of the fluid, and how this transport is affected by convection and diffusion.

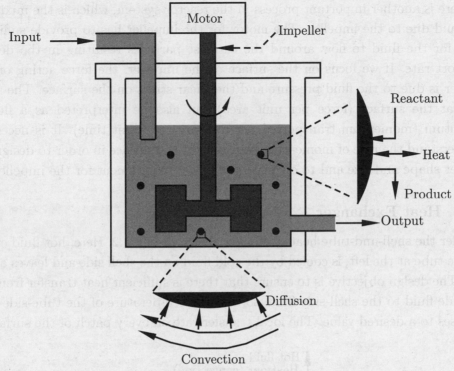

FIGURE 1.1. Stirred-tank reactor for conversion of reactant A into product B in the presence of a solid catalyst.

In addition, reactions are usually exothermic or endothermic, and heat transfer is a vital component of the reaction process. Heat is usually added/removed by heating/cooling coils immersed in the reactor. However, it is not sufficient to design the coils to add/remove heat at the desired rate. Conditions have to be created to ensure that the heat transfer rate from the catalyst surface to the fluid is sufficiently fast so that the specified temperature is maintained. If the reaction is exothermic and if the heat transfer rate is too low, there will be a rise in temperature at the catalyst surface, and this could lead to hot spots and runaway reactions. If the reaction is endothermic, conditions have to be created to ensure the desired heat transfer rate from the fluid to the catalyst surface; otherwise, there will be a decrease in the temperature and reaction rate. Thus, it is necessary to engineer the system in such a way that there is sufficient heat transfer from/to the catalyst surface to ensure that a steady state is reached for an exo/endothermic reaction. In the study of heat transfer processes, we will examine how the transfer of heat between the catalyst surface and the fluid depends on the difference between the surface and bulk temperatures, with attention to the specific roles of convection and diffusion in determining the transport rates.

There is another important process in the reactor system, which is the mixing of the liquid due to the impeller. The motor for the impeller has to provide sufficient power for the fluid to flow around the catalyst particles resulting in the desired transport rate. If we focus on the surface of the impeller, the force acting on the impeller is due to the fluid pressure and the shear stress on the surface. The shear stress at the surface (force per unit area) can also be interpreted as a flux of momentum (momentum transported per unit area per unit time). It is necessary to understand the rate of momentum transport at the surface in order to design the impeller shape and size and to estimate the power requirement for the impeller.

1.1.2　Heat Exchanger

Consider the shell-and-tube heat exchanger shown in Fig. 1.2. Here, hot fluid enters into the tube at the left, is cooled by the cold fluid on the shell side and leaves at the right. The design objective is to ensure that there is sufficient heat transfer from the tube-side fluid to the shell-side fluid, so that the temperature of the tube-side fluid decreases to a desired value. The local transfer rate on every patch of the surface of

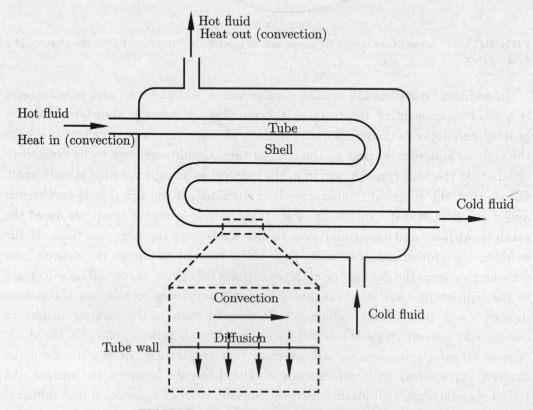

FIGURE 1.2.　Shell-and-tube heat exchanger.

the tube determines the overall performance of the entire heat exchanger, and the local transfer rates could vary since the fluid temperature changes with downstream distance along the tube. In order to design the system, it is necessary to predict the overall heat transfer rate based on the temperatures and flow rates on the two sides. To design the heat exchanger, it is necessary to understand the dependence of the local rate of transport on the local conditions (flow rate, temperature difference), and to determine the overall transport rate by adding the local transport rates.

In a heat exchanger, heat and momentum transfer are intricately linked. For a given temperature difference and flow rate, we may naively think that the maximum possible area (maximum possible tube length) will be desirable, since it will result in a larger rate of heat transfer. However, as the tube length increases, the pressure difference required to drive the flow increases (proportional to the length of the tube), as shown in Fig. 1.3. When there is fluid flow in a tube, the wall of the tube exerts a shear stress in the direction opposite to the flow. A pressure difference across the ends of the tube is required to compensate for the shear stress exerted by the pipe wall on the fluid. It is necessary to know how the shear stress depends on the fluid flow rate to optimise between the additional heat transfer gain and the energy cost for pumping the fluid through the extra tube length. As mentioned earlier, the shear stress can also be interpreted as the momentum flux (momentum transfer per unit area per unit time), and so it is necessary to understand the momentum transfer at the surface.

1.1.3   Spray Drier

This is an instance where fluid flow is coupled with heat and mass transfer. In the spray drier, the liquid flows through nozzles and breaks up into small droplets, and the water in these droplets evaporates due to the heat transferred from the hot air, as shown in Fig. 1.4. The droplets in nozzle spray driers have sizes up to 100 μm,

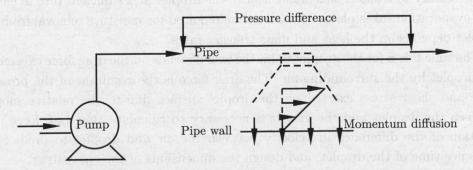

FIGURE 1.3.   Momentum transfer at the surface of a pipe.

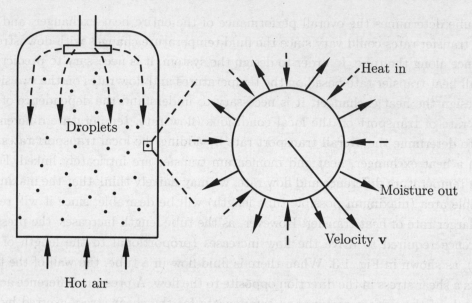

FIGURE 1.4. Schematic of a spray drier.

the velocities could be as large as large as 0.1–1 m/s, and so the time required for drying the droplets is usually of the order of seconds or less.

After the droplet is ejected into the drier, it is necessary to ensure that all the water has evaporated before it hits the wall of the drier, to prevent agglomeration. The drying occurs due to dry hot air blown from below. The drier should be designed such that the residence time of the droplet in the air is long enough for all the moisture to evaporate from the surface. This involves mass transport of water vapour from the surface to the air; the transport rate depends on the speed of the droplet and the humidity of the air. The evaporation of water also requires latent heat, and it is necessary to transfer this latent heat to the droplet at a sufficient rate to ensure that evaporation takes place. Thus, the time required for moisture removal from the droplet depends on the heat and mass transfer rates.

The time taken for the droplet to hit the wall depends on the drag force exerted on the droplet by the surrounding air. The drag force is the resultant of the pressure force and shear stress exerted on the droplet surface due to the relative motion between the droplet and the gas. It is necessary to calculate this drag force as a function of the difference in velocity between the air and droplet to predict the residence time of the droplet, and design the dimensions of the spray drier.

1.2 Unit Operation

In the design of process units such as the reactor in Section 1.1.1, it is necessary to be able to predict the overall mass transfer rate as function of the average difference in the concentration between the catalyst surface and the bulk of the fluid. Similarly, in the design of the heat exchanger discussed in Section 1.1.2, the overall heat transfer rate has to be predicted as a function of the average temperature difference between the fluids on the shell and tube sides. Instead of the overall transfer rate, which is an extensive quantity, predictions are made for the *average* flux (mass/heat/momentum transfer rate divided by the total area), which is an intensive quantity. The average fluxes are related to the *average* concentration/temperature/velocity difference driving the transport.

Correlations for the average fluxes are expressed in the form of relations between dimensionless groups. In Sections 1.5 and 1.6, dimensional analysis is used to obtain the dimensionless groups for different problems in mass, heat and momentum transfer, including those discussed in Sections 1.1.1, 1.1.2 and 1.1.3. Dimensionless groups are classified into different types, such as dimensionless fluxes, ratios of convection and diffusion and ratios of different types of diffusion. Two different regimes of operation, the diffusion-dominated and strong-convection regimes, are identified based on the values of the dimensionless groups, and the nature of the correlations is discussed in each of these regimes. The dependence of the correlations on the flow pattern (laminar or turbulent) is also discussed.

The correlations were originally obtained by experimentation. In this text, correlations are derived using mathematical analysis and approximations in the diffusion-dominated and the strong-convection regimes. It is important to make a distinction between the local values of the flux and the driving force on each patch of, for example, the catalyst surface in Section 1.1.1, or the pipe surface in Section 1.1.2, and average fluxes and driving forces in the correlations; the latter are determined by averaging the local values over the entire surface. In this text, our objective is to determine the local values at each patch of surface, and then to integrate these to obtain the average fluxes as a function of the average difference in concentration/temperature/velocity. Using this procedure, we not only derive standard correlations, but also develop the methodology to derive correlations for any system of interest.

1.3 Methodology

In our analysis of transport processes, we will treat the concentration, temperature and velocity as spatially varying fields, which are continuous functions of space and time. If a catalyst pellet is generating heat, the temperature decreases continuously from the catalyst surface to the bulk of the fluid. In the flow through the pipe, the velocity increases continuously from the wall to the centre. Similarly, in a heat exchanger where the heat is transferred from the tube side to the shell side, the temperature in the tube decreases continuously from the centre to the wall of the tube. These variations generate local fluxes and stresses at every point within the fluid, which depend on the concentration/temperature/velocity variations and the nature of the fluid flow. The purpose of our analysis is to obtain the concentration/temperature/velocity field over the entire domain of interest, and from this, calculate the local flux at the surface. The local flux is then averaged over the surface to determine the average flux.

The solution procedure involves two steps. In the first step, conservation equations are derived for mass/momentum/energy which are differential equations valid locally at each point in the flow. In the second step, these equations are 'integrated', subject to boundary conditions at the bounding surfaces, to calculate the mass/momentum/energy field over the entire domain.

1.3.1 Governing Equations

The first objective is to obtain a set of governing equations that determine how the concentration, temperature and velocity fields vary in space and time. Since there are three spatial co-ordinates and one time co-ordinate, these equations are 'partial differential equations' which contain derivatives with respect to the spatial co-ordinates and time derivatives. There are two inputs used in deriving these equations.

The first input is the conservation principle. The mass/energy conservation principle states that the change in the mass/energy in a differential volume over a differential time interval is equal to the net mass/energy entering the differential volume and the mass/energy generated within the volume. The mass generation could be due to chemical reactions, while the energy generation could be due to reactions, phase change, fluid friction or other reasons. The momentum conservation principle states that the rate of change of the momentum in a differential volume is the net rate of transport of momentum into the volume plus the sum of all forces exerted on the volume. The forces are of two types—body forces (such as gravitational force, centrifugal force, etc.) whose magnitude is proportional to the volume, and surface forces (such as pressure and stress) which act on the bounding surfaces.

In the conservation equations, it is necessary to specify the flux of mass/energy entering or leaving the volume through the bounding surfaces, or the surface stresses acting on the bounding surfaces for momentum transfer. The transfer across surfaces takes place by two mechanisms: convection and diffusion.

Convection is the transport due to the fluid flow. In the continuous stirred-tank reactor in Section 1.1.1, the reactant is transported into the reactor by convection, since it is carried in by the fluid. Similarly, the product at the outlet is carried out of the reactor by convection. Convection is also responsible for the circulation of the reactant and product in the tank, due to the fluid motion caused by the impeller. In the heat exchanger discussed in Section 1.1.2, heat is carried into the heat exchanger at the inlet and conveyed along the length of the tube by convection. In our continuum description of the concentration/temperature/velocity fields, fluid flow causes convection through each differential volume in the fluid, and the convective flux can be determined if the fluid velocity field is known.

In the example of the reactor discussed in Section 1.1.1, transport of reactant/product to/from the catalyst surface cannot take place due to convection, since there is no fluid motion perpendicular to the surface of the catalyst particle. Therefore, the transport of mass and heat to the catalyst surface has to take place due to diffusion. In a similar manner, in the heat exchanger in Section 1.1.2, the transport across the wall of the tube cannot be caused by convection, because there is no fluid flow perpendicular to the wall of the heat exchanger. The transport across the surface can take place only due to diffusion. Diffusion forms an essential mechanism for mass/heat/energy transfer both within the fluid and to the boundaries and suspended particles.

Diffusion is caused by the fluctuating motion of the molecules. The molecular origin of diffusion is discussed in detail in Chapter 3. In a fluid, all molecules are in a state of constant motion due to thermal fluctuations. At standard temperature and pressures the fluctuating velocity of the molecules varies in the range 100–1000 m/s. The motion of the molecules is random and isotropic—that is, molecular motion takes place with equal probability in all directions. This fluctuating motion will not result in mass transfer if the concentration field is uniform, because the motion of the molecules is not biased in any direction. However, when there is a variation in concentration, the random motion will result in the net transport of molecules from regions of higher concentration to regions of lower concentration. Thus, a variation in concentration results in a flux of mass. In a similar manner, a variation in temperature results in a flux of energy from regions of higher temperature to regions of lower temperature. A spatial variation in the fluid velocity results in the flux of momentum. The constitutive relations which relate the fluxes of mass/momentum/energy to the variations in the concentration/velocity/temperature provide the second input into the governing equations.

1.3.2   Solution Procedure

The second objective is to solve the partial differential equations for simple situations and derive the correlations for the average fluxes that can be used for the design of unit operations. Unlike linear ordinary differential equations, there is no general procedure for solving partial differential equations. In order to solve these equations, it is necessary to have a physical understanding of the underlying processes represented by the terms in the equations. Different approximations are made in different parameter regimes to reduce the partial differential equation to a set of ordinary differential equations, and these are then solved to determine the concentration/temperature/velocity fields. The two important regimes in transport processes are the the diffusion-dominated regime and the strong-convection regime.

In the 'diffusion-dominated' regime, transport due to diffusion is much faster than that due to convection. Convection is neglected, and the 'diffusion equation' is solved to obtain the concentration or temperature field. The mathematical procedure for solving the diffusion equation, called the method of 'separation of variables', is demonstrated for different transport problems in Chapter 8.

In the 'strong-convection' regime, it might naively be assumed that diffusive transport can be neglected. However, transport across surfaces (such as the catalyst pellet surface in Fig. 1.1 and the surface of the tube in the heat exchanger in Fig. 1.2) cannot take place due to convection. This is because there is no fluid flow across these surfaces. The transport across the surfaces has to take place due to diffusion. Therefore, even when there is strong convection, diffusion is important near flow boundaries or particle surfaces. Mass/heat diffusing from a surface is rapidly swept downstream by the strong convection, and therefore, the effect of diffusion from the surface is restricted to concentration/temperature 'boundary layers' of thickness much smaller than the characteristic system size. The 'boundary layer' theory is used to solve problems involving strong convection in Chapters 9 and 10.

1.4 Outline

The use of dimensional analysis for simplifying a problem and expressing the problem in terms of a set of dimensionless groups is discussed, and illustrated using archetypal examples, in Chapter 2. The 'unit operations' approach is discussed in this chapter, where the relevant dimensionless groups are identified, and then experimental results are used to formulate correlations between these dimensionless groups. Going beyond the unit operations approach, the physical interpretation of dimensionless groups as the ratio of two different types of forces is an important concept explained in this chapter. The correlations for heat and mass transfer depend on the geometry and the flow pattern in cases where convective transport is important, and the typical flow patterns encountered in the flow through conduits and the flow past objects are presented. Correlations for different types of flows in representative geometries are listed; many of these correlations are derived using mathematical analysis in subsequent chapters.

The molecular picture of diffusion is discussed in Chapter 3, in order to provide a physical understanding of the diffusion of heat, mass and momentum. Kinetic theory of gases is used to obtain approximate expressions for the diffusion coefficients in gases in terms of molecular properties such as the molecular diameter, number density and temperature. Quantitative estimates are obtained for the diffusion coefficients in gases. Diffusion in liquids is discussed from a molecular perspective, with an explanation of the different mechanisms of mass, momentum and energy diffusion. Estimates are also provided for the diffusion coefficients in liquids.

One-dimensional unsteady transport in Cartesian and curvilinear co-ordinates are discussed in Chapters 4 and 5 respectively. Here, there are variations in the concentration, temperature or velocity fields in one spatial dimension and in time. The two parts of the solution procedure discussed in Section 1.3, the derivation of governing equations valid at each point in the flow, and the procedure for solving the resulting partial differential equations, are explained for simple geometries. The concentration/temperature/velocity fields are determined by integrating the governing equations over the entire domain, subject to boundary conditions, using different techniques. For unbounded domains, a 'similarity transform' is used to convert the partial differential equation to an ordinary differential equation. For bounded domains, the separation of variables procedure is used for separating the partial differential equation into a series of ordinary differential equations.

The equations for mass and heat transfer are very similar, since the quantities being transported are scalars. The equations for momentum transfer are different from those for mass/heat transfer because the quantity being transported is a vector.

The transport of momentum due to fluid pressure does not have an analogue in mass and heat transfer. Fluid flow due to pressure variations is the subject of Chapter 6.

The general procedure for deriving balance equations in different three dimensional co-ordinate systems is presented in Chapter 7. These equations are expressed in terms of vector fields, such as the velocity and flux vectors, and vector operators, such as the gradient and divergence operators. It is shown that the vector form of the conservation equations is universal, but the equation has different forms when these vector operators are expressed in different co-ordinate systems. The form of the Navier–Stokes equations for momentum transfer is explained, and the equations are presented without a detailed derivation.

Diffusive transport is the subject of Chapter 8. The diffusion equation is solved using the method of separation of variables in different co-ordinate systems for problems of practical importance, such as the temperature field around a heated sphere and the effective conductivity of a composite. A physical interpretation of the solution in the spherical co-ordinate system as the superposition of heat/mass sources and sinks is discussed.

Transport in forced convection, where external forcing generates the flow through conduits or past objects, is the subject of Chapter 9. The fluid velocity field is specified, and it is necessary to determine the rates of mass/energy transport.[1] When there is strong convection, one might naively attempt to neglect diffusion altogether. However, as explained in Section 1.3.2, convective transport occurs only along the flow velocity. At surfaces bounding the fluid, such as the surface of conduits and suspended particles, there is no velocity perpendicular to the surface. Transport perpendicular to the surface necessarily occurs due to diffusion, so diffusive transport cannot be neglected even in the case of strong convection. There are thin concentration/temperature 'boundary layers' close to the surface, of thickness much smaller than the characteristic flow dimension (tube diameter or particle size) where the downstream transport due to convection is balanced by the transport due to diffusion. The solution for the concentration/temperature fields within these boundary layers is derived using similarity transforms. From the concentration/temperature fields, the local flux is obtained at each point on the surface of the conduit or the suspended particle. The average fluxes are obtained by averaging over the surface, and these are used to formulate correlations for heat/mass transfer.

---

[1]The fluid velocity field can be obtained by solving the fluid momentum equations subject to boundary conditions, but this is not within the scope of this text.

Natural convection is considered in Chapter 10. When a heated object is placed in a cold fluid, the temperature of the fluid near the object increases. The hot fluid becomes lighter and moves upwards carrying the heat from the object. Cold ambient fluid moves in towards the object to replace the hot fluid. This mechanism could result in a significant increase in the heat transfer rate in comparison to heat conduction. In contrast to forced convection where the fluid flow is forced by external forcing, the flow here is generated due to buoyancy effects caused by a hot object. Thus, the flow velocity has to be determined as a part of the solution. The fluid equations are scaled to determine the boundary layer thickness as a function of the dimensionless parameters that characterise the flow, and the estimate of the boundary layer thickness is used to derive correlations for the average fluxes for heat transfer.

There are several excellent textbooks which treat, in further detail, the subjects covered here; a few are listed here as a starting point for further reference. The classic book by Bird, Stewart and Lightfoot [1] has a more comprehensive coverage and is organised on the basis of different kinds of transport. The analytical solutions and asymptotic techniques for convective transport in Chapters 9 and 10 are covered more rigorously in Leal [2]. A more in-depth analysis of complex phenomena in mass transport can be found in Cussler [3]. The techniques for solving the diffusion equation (Chapter 8) are discussed in further detail in the context of electrodynamics in Griffiths [4]. A more advanced treatment of fluid mechanics and incompressible laminar flows can be found in Batchelor [5] and Panton [6], and an introduction to turbulence is covered in Tennekes and Lumley [7].

1.5 Dimensions and Units

The measure of a physical quantity is called its *dimension*, and the standard for measurement is called the *unit*. For example, the speed is the distance moved in unit time, so the dimension of speed is length/time, while the unit for measurement could be m/s, km/hr, etc. Though it is necessary to have only one unit for a quantity, for historical reasons and for convenience, there are often many different units for the same quantity. Examples are feet and meters for length, or pounds and kilograms for mass. In this text, we will exclusively use the SI (Systeme International) system of units (meters, kilograms, seconds) where it is necessary to specify units. To the extent possible, we will work with dimensionless groups, whose value is independent of units.

Dimensions are divided into two categories—fundamental dimensions and derived dimensions. The dimensions of all known physical quantities can be expressed in terms of a set of six fundamental dimensions. Though the choice of fundamental dimensions is arbitrary, by international convention, the following are considered to be fundamental dimensions and their corresponding units in the SI system.

1. Mass ($\mathcal{M}$) with unit kilogram (kg).

2. Length ($\mathcal{L}$) with unit meter (m).

3. Time ($\mathcal{T}$) with unit second (s).

4. Temperature (Θ) with unit Kelvin (K).[2]

5. Current ($\mathcal{A}$) with unit ampere (A).

6. Light intensity with unit candela.

The quantities of relevance in this text contain the mass $\mathcal{M}$, length $\mathcal{L}$, time $\mathcal{T}$ and temperature Θ dimensions. Some of the important quantities used in transport processes and their units in the SI system are given in Table 1.1.

The dimensions of all derived quantities are determined from relations that involve these quantities. For example,

1. The dimension of **velocity** is $\mathcal{L}\mathcal{T}^{-1}$, since the velocity is the distance travelled per unit time.

2. The **angular velocity** is the ratio of the linear velocity and the distance from the axis of rotation, and the dimension of angular velocity is $\mathcal{T}^{-1}$.

3. **Acceleration** is defined as the rate of change of velocity, and the dimension of acceleration is $\mathcal{L}\mathcal{T}^{-2}$.

4. The dimension of **force** is determined from Newton's second law, which states that the force on an object is mass times acceleration. Therefore, the dimension of force, $\mathcal{M}\mathcal{L}\mathcal{T}^{-2}$, is the dimension of mass times $\mathcal{M}$ times the dimension of acceleration $\mathcal{L}\mathcal{T}^{-2}$. The SI unit of force is Newton (N), which is 1 kg m/s^2.

---

[2]The unit 'Kelvin' (K) is used for temperature where thermodynamic relations such as the ideal gas law is involved. The 'degree Celsius' (°C) is used for heat transfer problems where the relevant quantity is the temperature difference.

TABLE 1.1. Dimensions and units of some commonly used quantities in transport processes.

Quantity	Dimension	Unit (SI)
Mass (m)	$\mathcal{M}$	kilogram (kg)
Length (l)	$\mathcal{L}$	meter (m)
Time (t)	$\mathcal{T}$	second (s)
Temperature (T)	Θ	K, °C
Velocity (v)	$\mathcal{L}\mathcal{T}^{-1}$	(m/s)
Angular velocity (Ω)	$\mathcal{T}^{-1}$	s^{-1}
Acceleration (a)	$\mathcal{L}\mathcal{T}^{-2}$	(m/s^2)
Force (F)	$\mathcal{M}\mathcal{L}\mathcal{T}^{-2}$	(kg m/s^2) (Newton, N)
Torque (T)	$\mathcal{M}\mathcal{L}^2\mathcal{T}^{-2}$	(kg m^2/s^2) (Newton m, N m)
Energy (E)	$\mathcal{M}\mathcal{L}^2\mathcal{T}^{-2}$	(kg m^2/s^2) (Joule, J)
Work (W)	$\mathcal{M}\mathcal{L}^2\mathcal{T}^{-2}$	(kg m^2/s^2) (Joule, J)
Power (P)	$\mathcal{M}\mathcal{L}^2\mathcal{T}^{-3}$	(kg m^2/s^3) (Watt, W)
Density (ρ)	$\mathcal{M}\mathcal{L}^{-3}$	(kg/m^3)
Concentration (c)	$\mathcal{M}\mathcal{L}^{-3}$	(kg/m^3)
Mass flux (j_z)	$\mathcal{M}\mathcal{L}^{-2}\mathcal{T}^{-1}$	(kg/m^2/s)
Diffusivity $(\mathcal{D})$	$\mathcal{L}^2\mathcal{T}^{-1}$	(m^2 s^{-1})
Heat energy (E)	$\mathcal{M}\mathcal{L}^2\mathcal{T}^{-2}$	(kg m^2/s^2) (Joule, J)
Heat flux (q)	$\mathcal{M}\mathcal{T}^{-3}$	(kg/s^3) (Joules/m^2/s, J/m^2/s)
Thermal conductivity (k)	$\mathcal{M}\mathcal{L}\mathcal{T}^{-3}\Theta^{-1}$	(kg m s^{-3} K^{-1})
Specific heat (C_p, C_v)	$\mathcal{L}^2\mathcal{T}^{-2}\Theta^{-1}$	(m^2 s^{-2} K^{-1})
Pressure (p)	$\mathcal{M}\mathcal{L}^{-1}\mathcal{T}^{-2}$	(kg/m/s^2) (Pascal, Pa)
Stress (τ)	$\mathcal{M}\mathcal{L}^{-1}\mathcal{T}^{-2}$	(kg/m/s^2) (Pascal, Pa)
Viscosity (μ)	$\mathcal{M}\mathcal{L}^{-1}\mathcal{T}^{-1}$	(kg/m/s) (Pascal s, Pa s)
Surface tension (γ)	$\mathcal{M}\mathcal{T}^{-2}$	(kg/s^2)
Elasticity (G)	$\mathcal{M}\mathcal{L}^{-1}\mathcal{T}^{-2}$	(kg/m/s^2) (Pascal, Pa)
Boltzmann constant (k_B)	$\mathcal{M}\mathcal{L}^2\mathcal{T}^{-2}\Theta^{-1}$	(kg m^2 s^{-2} K^{-1}) (J/K)

5. The **torque** is the product of the force and the perpendicular distance from the axis of rotation. The dimension of torque is $\mathcal{M}\mathcal{L}^2\mathcal{T}^{-2}$, and the unit is N m.

6. The **work** done on an object is the product of the force exerted on the object and the displacement in the direction of the force. Therefore, the dimension of work is the dimension of force times that of displacement, $\mathcal{M}\mathcal{L}^2\mathcal{T}^{-2}$. The SI unit of work is Joule (J), which is 1 kg m^2/s^2.

7. **Energy** has the same dimension as work, $\mathcal{M}\mathcal{L}^2\mathcal{T}^{-2}$. In thermodynamics, enthalpy, Gibbs free energy and Helmholtz free energy have the same dimension as energy. The SI unit of energy is also Joule (J).

8. **Power** is the rate at which work is done—that is, the work done per unit time—and the dimension of power is $\mathcal{ML}^2\mathcal{T}^{-3}$. The SI unit of power is Watt (W), which is 1 kg m^2/s^3.

9. **Pressure** is the force acting per unit area of the surface. The dimension of force is $\mathcal{MLT}^{-2}$, and the dimension of area is $\mathcal{L}^2$, and so the dimension of pressure is $\mathcal{ML}^{-1}\mathcal{T}^{-2}$. The SI unit of pressure is Pascal, which is 1 kg/m/s^2.

10. The **density** is the mass of a unit volume of the material, so the density has dimension $\mathcal{ML}^{-3}$.

11. **Concentration** is the mass of solute per unit volume of solution, so it also has dimension $\mathcal{ML}^{-3}$.

12. The **mass flux** j is the mass transported across a surface per unit area per unit time, with dimension $\mathcal{ML}^{-2}\mathcal{T}^{-1}$.

13. The **diffusion coefficient** or **mass diffusivity** is the proportionality constant in Fick's law for diffusion. Consider a cuboidal volume of solution with area of cross section A in the $x-y$ plane and height Δz in the z direction, as shown in Fig. 1.5(a). If the opposite faces in the z direction are maintained at different concentrations $c|_{z+\Delta z}$ and $c|_z$, and the concentration is uniform in the $x-y$ plane, there is a mass flux from the surface at higher concentration to the surface at lower concentration. Fick's law for mass diffusion states that the flux j_z in the z direction is directly proportional to the difference in concentration, and inversely proportional to the separation between the two surfaces,

$$j_z = -\mathcal{D}\frac{\Delta c}{\Delta z}. \qquad (1.5.1)$$

Here, $\Delta c = (c|_{z+\Delta z} - c|_z)$ is the difference in the concentrations at the locations $(z+\Delta z)$ and z. The negative sign on the right in Eq. 1.5.1 indicates that the flux is directed from the surface at higher concentration to the surface at lower concentration. The constant $\mathcal{D}$ is the diffusion coefficient. The mass flux has dimension $\mathcal{ML}^{-2}\mathcal{T}^{-1}$, and the concentration has dimension $\mathcal{ML}^{-3}$; therefore, the diffusion coefficient has dimension $\mathcal{L}^2\mathcal{T}^{-1}$.

14. The **specific heat** C_p at constant pressure is the proportionality constant in the relation between the change in enthalpy ΔH and the change in temperature of an object ΔT,

$$\Delta H = mC_p\Delta T, \qquad (1.5.2)$$

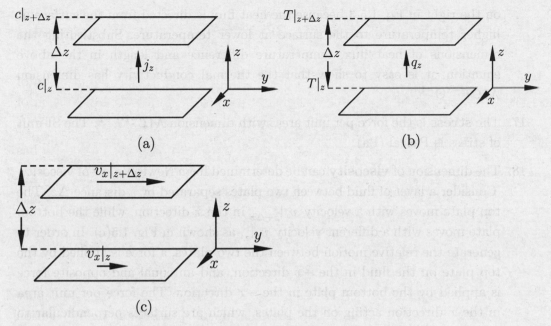

FIGURE 1.5. The mass flux due to a concentration difference (a), the heat flux due to a temperature difference (b), and the shear stress due to a velocity difference (c).

where m is the mass. In the above equation, ΔH has dimension $\mathcal{M}\mathcal{L}^2\mathcal{T}^{-2}$, mass has dimension $\mathcal{M}$ and temperature has dimension Θ. Therefore, specific heat has dimension $\mathcal{L}^2\mathcal{T}^{-2}\Theta^{-1}$. The SI unit of specific heat is J/kg/K, which is the same as m^2/(s^2 K).

The specific heat at constant volume C_v is related to the change in internal energy ΔU for a process at constant volume. This has the same dimension as the specific heat at constant pressure.

15. The **heat flux** is the thermal energy transferred per unit area per unit time. The dimension of energy is $\mathcal{M}\mathcal{L}^2\mathcal{T}^{-2}$, and so the dimension of heat flux is $\mathcal{M}\mathcal{T}^{-3}$. The SI unit of heat flux is often written as J/m^2/s, which is the same as kg/s^3.

16. The **thermal conductivity** is defined in Fourier's law for heat conduction. If we consider a slab of material of thickness Δz subjected to a temperature difference ΔT, as shown in Fig. 1.5(b), the heat flux q_z in the z direction is,

$$q_z = -k\frac{\Delta T}{\Delta z}, \qquad (1.5.3)$$

where k is the thermal conductivity, and $\Delta T = T|_{z+\Delta z} - T|_z$ is the difference in temperature between the locations $z + \Delta z$ and z. There is a negative sign on the right in Eq. 1.5.3 because the heat flux is directed from the surface at higher temperature to the surface at lower temperature. Substituting the dimensions of heat flux, temperature difference and length in the above equation, it is easy to show that the thermal conductivity has dimension $\mathcal{MLT}^{-3}\Theta^{-1}$.

17. The **stress** is the force per unit area, with dimension $\mathcal{ML}^{-1}\mathcal{T}^{-2}$. The SI unit of stress is Pascal (Pa).

18. The dimension of **viscosity** can be determined from Newton's law of viscosity. Consider a layer of fluid between two plates separated by a distance Δz. The top plate moves with a velocity $v_x|_{z+\Delta z}$ in the x direction, while the bottom plate moves with a different velocity $v_x|_z$, as shown in Fig. 1.5(c). In order to generate the relative motion between the two plates, a force is applied by the top plate on the fluid in the $+x$ direction, and an equal and opposite force is applied by the bottom plate in the $-x$ direction. The force per unit area in the x direction acting on the plates, which are surfaces perpendicular to the z direction, is called the shear stress τ_{xz}. Newton's law of viscosity states that the stress is directly proportional to the velocity difference and inversely proportional to the separation between the plates,

$$\tau_{xz} = \mu \frac{\Delta v_x}{\Delta z}. \tag{1.5.4}$$

Here, $\Delta v_x = (v_x|_{z+\Delta z} - v_x|_z)$ is the difference in the velocity at the locations $z + \Delta z$ and z.[3] Substituting the dimensions for the stress, velocity and length in Eq. 1.5.4, we find that the viscosity has dimension $\mathcal{ML}^{-1}\mathcal{T}^{-1}$. The SI unit of viscosity is Pascal × second (Pa s), which is the same as kg/m/s.

19. The **surface tension** is the force per unit length at a surface, with dimension $\mathcal{MT}^{-2}$. The SI unit of surface tension is often written as (N/m) or (J/m^2), which is the same as kg/s^2.

20. When a force is applied on an object, either extensional or compressional, there is a change in the length ΔL of the object. The **elasticity** is the proportionality constant in the relation between the stress τ, which is a force per unit area, and the strain $(\Delta L/L_0)$, where L_0 is the original length of the

---

[3]It is important to note that velocity is a vector, and it is necessary to take the vector difference of the velocities on the two surfaces. When the top plate moves in the $+x$ direction and the bottom plate in the $-x$ direction, the velocity difference is the sum of the magnitudes of the two velocities.

object. Since the strain is dimensionless, elasticity has the same dimension as stress, $\mathcal{ML}^{-1}\mathcal{T}^{-2}$. The SI unit of elasticity is Pascal (Pa).

21. The thermal energy at absolute temperature T is given by $k_B T$, where k_B is the Boltzmann constant. The Boltzmann constant has dimension energy divided by temperature, $\mathcal{ML}^2\mathcal{T}^{-2}\Theta^{-1}$. The ideal gas law can be expressed in terms of either the Boltzmann constant, or the ideal gas constant R. These are related, $R = $ Avogadro number $\times k_B$. The ratio of the Boltzmann constant and the mass of a molecule is equal to the ratio of the ideal gas constant and mass of one mole,

$$\frac{k_B}{\text{Mass of one molecule}} = \frac{R}{\text{Mass of one mole}}. \tag{1.5.5}$$

Here, the Boltzmann constant and the mass of a molecule in kilograms (SI units) is usually used; this avoids confusion between the ideal gas constant expressed in terms of moles and kilomoles.

The notation $[X]$ is used to represent the dimension of a quantity X. For example, the dimension of pressure is written as $[p] = \mathcal{ML}^{-1}\mathcal{T}^{-2}$, and that of thermal conductivity as $[k] = \mathcal{MLT}^{-3}\Theta^{-1}$.

The dimensions and units of commonly used quantities in the study of electromagnetic phenomena are given in Table 1.2. Some of these are derived in the solved examples, while others are included in the exercises at the end of the chapter.

EXAMPLE 1.5.1: For a power-law fluid, the relation between the stress τ_{xz} and the strain rate $(\Delta v_x/\Delta z)$ is,

$$\tau_{xz} = \kappa \frac{\Delta v_x}{\Delta z}\left|\frac{\Delta v_x}{\Delta z}\right|^{n-1}, \tag{1.5.6}$$

where the velocity v_x and the cross-stream distance z are shown in Fig. 1.5(c), n is the 'power-law index', and κ is the 'flow consistency index'. What is the dimension of κ?

Solution: The dimension of stress $[\tau_{xz}] = \mathcal{ML}^{-1}\mathcal{T}^{-2}$, and that of the strain rate $[(\Delta v_x/\Delta z)] = \mathcal{T}^{-1}$. The dimension of the flow consistency index is

$$[\kappa] = [\tau_{xz}][(\Delta v_x/\Delta z)]^{-n} = (\mathcal{ML}^{-1}\mathcal{T}^{-2}) \times \mathcal{T}^n = \mathcal{ML}^{-1}\mathcal{T}^{n-2}. \tag{1.5.7}$$

$\square$

TABLE 1.2. Dimensions and units of some commonly used quantities in electromagnetic phenomena.

Quantity	Dimension	Unit (SI)	Derivation
Current (I)	$\mathcal{A}$	Ampere (A)	
Charge (Q)	$\mathcal{AT}$	A s	Example 1.5.2
Voltage, Electric potential (V)	$\mathcal{ML}^2\mathcal{T}^{-3}\mathcal{A}^{-1}$	Volt (V)	Example 1.5.2
Electric field (E)	$\mathcal{MLT}^{-3}\mathcal{A}^{-1}$	V/m	Voltage difference/Length
Current density (J)	$\mathcal{AL}^{-2}$	A/m^2	Current/Area
Electrical resistance (R)	$\mathcal{ML}^2\mathcal{T}^{-3}\mathcal{A}^{-2}$	Ohm	Example 1.5.2
Electrical conductance (G)	$\mathcal{M}^{-1}\mathcal{L}^{-2}\mathcal{T}^3\mathcal{A}^2$	Mho	R^{-1}
Capacitance (C)	$\mathcal{M}^{-1}\mathcal{L}^{-2}\mathcal{T}^4\mathcal{A}^2$	Farad	Example 1.5.2
Inductance (L)	$\mathcal{ML}^2\mathcal{T}^{-2}\mathcal{A}^{-2}$	Henry	Example 1.5.2
Electrical resistivity (ϱ)	$\mathcal{ML}^3\mathcal{T}^{-3}\mathcal{A}^{-2}$	Ohm m	(E/J)
Electrical conductivity (κ)	$\mathcal{M}^{-1}\mathcal{L}^{-3}\mathcal{T}^3\mathcal{A}^2$	Mho/m	ϱ^{-1}
Electric permittivity (ϵ)	$\mathcal{M}^{-1}\mathcal{L}^{-3}\mathcal{T}^4\mathcal{A}^2$	Farad/m	Exercise 1.6
Vacuum permittivity (ϵ_0)	$\mathcal{M}^{-1}\mathcal{L}^{-3}\mathcal{T}^4\mathcal{A}^2$	Farad/m	Exercise 1.6
Relative permittivity (ϵ_r)	Dimensionless		(ϵ/ϵ_0)
Magnetic flux density (B)	$\mathcal{MT}^{-2}\mathcal{A}^{-1}$	Tesla	Exercise 1.7
Magnetic permeability (μ)	$\mathcal{MLT}^{-2}\mathcal{A}^{-2}$	Newton/Ampere^2	Exercise 1.8
Vacuum permeability (μ_0)	$\mathcal{MLT}^{-2}\mathcal{A}^{-2}$	Newton/Ampere^2	Exercise 1.8
Relative permeability (μ_r)	Dimensionless		(μ/μ_0)
Magnetic field (H)	$\mathcal{AL}^{-1}$	Ampere/m	(B/μ)
Magnetic moment ($\boldsymbol{m}$)	$\mathcal{AL}^2$	Ampere m^2	Exercise 1.13

EXAMPLE 1.5.2: Determine the dimension of electric charge Q, voltage V, electrical resistance R, capacitance C and inductance L in terms of the fundamental dimensions.

Solution: Electric current is the flow of electric charge per unit time. Therefore the dimension of electric charge $[Q]$ is the product of the dimension of current and time, $\mathcal{AT}$. The SI unit of charge is Coulomb, which is 1 Ampere × 1 second.

The dimension of voltage V is determined from the relation,

$$\text{Power} = VI. \tag{1.5.8}$$

From Table 1.1, the dimension of power is [Power] $= \mathcal{ML}^2\mathcal{T}^{-3}$. Therefore, the dimension of voltage is the ratio of the dimensions of power and current, [Power]/[Current] $= \mathcal{ML}^2\mathcal{T}^{-3}\mathcal{A}^{-1}$. The SI unit of voltage is Volt, which is 1 kg m^2 s^{-3} A^{-1}.

The dimension of electrical resistance R is determined from Ohm's law,

$$R = \frac{\Delta V}{I}, \qquad (1.5.9)$$

where ΔV is the difference in voltage across a conductor and I is the current flowing through it. The dimension of resistance is the ratio of the dimensions of voltage and current, $[V]/[I] = \mathcal{ML}^2\mathcal{T}^{-3}\mathcal{A}^{-2}$. The SI unit of resistance is Ohm, which is $1 \text{ kg m}^2\text{s}^{-3}\text{A}^{-2}$. The inverse of the resistance is the conductance, with SI unit Mho.

The dimension of capacitance C is determined from the relation,

$$Q = C\Delta V, \qquad (1.5.10)$$

where Q is the charge held by the capacitor and ΔV is the voltage difference across the capacitor. Therefore, the dimension of C is $[Q]/[\Delta V] = \mathcal{M}^{-1}\mathcal{L}^{-2}\mathcal{T}^4\mathcal{A}^2$. The SI unit of capacitance is Farad, which is $1 \text{ kg}^{-1}\text{m}^{-2}\text{s}^4\text{A}^2$.

The dimension of inductance L is determined from the relation

$$\Delta V = L\frac{\mathrm{d}I}{\mathrm{d}t}, \qquad (1.5.11)$$

where ΔV is the voltage difference across an inductor, I is the instantaneous current, and the time derivative $(\mathrm{d}/\mathrm{d}t)$ has dimension $\mathcal{T}^{-1}$. Therefore, the dimension of inductance is $[V]\mathcal{T}/[I]$, which is $\mathcal{ML}^2\mathcal{T}^{-2}\mathcal{A}^{-2}$. The SI unit of inductance is Henry, which is $1 \text{ kg m}^2\text{s}^{-2}\text{A}^{-2}$. $\qquad\square$

1.5.1 Dimension of an Equation

An equation containing multiple terms has to be dimensionally consistent. The equation has to be valid for any system of units used, provided the same system is used for expressing all the quantities. Therefore, the dimension of every additive term in the equation is required to be the same. If the dimension were not the same, then different terms would transform differently when the system of units is changed. This requirement can be used to verify the dimensional consistency of an equation and determine the dimension of unknown quantities in the equation. Equations can appear complicated and a little intimidating because they contain unfamiliar symbols and operators. However, to determine dimensional consistency, it is sufficient to understand that each symbol has dimension, and then work out the dimension of each term in the equation.

An example is the Bernoulli equation for the pressure p for an inviscid flow, discussed in Section 6.1 (Chapter 6),

$$p + \tfrac{1}{2}\rho v^2 + \rho g h = p_0, \qquad (1.5.12)$$

where ρ and v are the fluid density and velocity, g is the acceleration due to gravity, z is the co-ordinate in the direction opposite to gravity, and p_0 is a constant. The dimension of the different terms in the above equation are,

$$[p] = \mathcal{ML}^{-1}\mathcal{T}^{-2},$$
$$[\rho][v^2] = (\mathcal{ML}^{-3})(\mathcal{LT}^{-1})^2 = \mathcal{ML}^{-1}\mathcal{T}^{-2},$$
$$[\rho][g][z] = (\mathcal{ML}^{-3})(\mathcal{LT}^{-2})\mathcal{L} = \mathcal{ML}^{-1}\mathcal{T}^{-2}. \qquad (1.5.13)$$

Therefore, the dimensions of all the terms in the equation are the same, and this is the dimension of the equation.

The mass, momentum and energy equations that we will consider later in the text contain some quantities which are vectors. For example, the velocity v is a vector expressed as a boldface letter. A vector has both magnitude and direction, and is written as the sum of the component times the unit vector in three co-ordinate directions. The dimension of a vector v, its magnitude v and the dimension of each of its components is the same. The unit vector has no dimension. For example, the dimension of the vector velocity v is $\mathcal{LT}^{-1}$. Section 7.3.2 (Chapter 7) explains that the stress is a tensor, which contains nine components. Each component has the same dimension, $\mathcal{ML}^{-1}\mathcal{T}^{-2}$.

The conservation equations contain derivatives with respect to position and time, and these derivatives have dimension. The time derivatives, $(\mathrm{d}/\mathrm{d}t)$ and $(\partial/\partial t)$, have dimension $\mathcal{T}^{-1}$. Similarly, the spatial derivatives $(\mathrm{d}/\mathrm{d}x)$ and $(\partial/\partial x)$, have dimension $\mathcal{L}^{-1}$, and the second derivatives, $(\mathrm{d}^2/\mathrm{d}x^2)$ and $(\partial^2/\partial x^2)$, have dimension $\mathcal{L}^{-2}$. A vector spatial gradient operator, ∇, also appears in these equations. This has dimension $\mathcal{L}^{-1}$. Similarly, integrals also have dimension. For example, the time integral $\int \mathrm{d}t$ has dimension $\mathcal{T}$, the spatial integral $\int \mathrm{d}x$ has dimension $\mathcal{L}$, and the volume integral $\int \mathrm{d}V$ has the dimension of volume $\mathcal{L}^3$.

Consider the Navier–Stokes momentum conservation equation for an incompressible flow,

$$\rho \left(\frac{\partial v}{\partial t} + v \cdot \nabla v \right) = -\nabla p + \mu \nabla^2 v, \qquad (1.5.14)$$

where ρ is the fluid density, μ is the fluid viscosity and ∇ is the gradient operator. The dimension is determined in the following steps. The brackets/braces in the equation are first expanded, so that the equation is the additive sum of terms,

$$\rho\frac{\partial \boldsymbol{v}}{\partial t} + \rho\boldsymbol{v}\cdot\boldsymbol{\nabla}\boldsymbol{v} = -\boldsymbol{\nabla}p + \mu\boldsymbol{\nabla}^2\boldsymbol{v}. \tag{1.5.15}$$

The dimensional consistency of the equation can now be verified. The dimensions of the terms from left to right are,

$$[\rho][(\partial/\partial t)][\boldsymbol{v}] = (\mathcal{ML}^{-3})(\mathcal{T}^{-1})(\mathcal{LT}^{-1}) = \mathcal{ML}^{-2}\mathcal{T}^{-2},$$

$$[\rho][\boldsymbol{v}][\boldsymbol{\nabla}][\boldsymbol{v}] = (\mathcal{ML}^{-3})(\mathcal{LT}^{-1})(\mathcal{L}^{-1})(\mathcal{LT}^{-1}) = \mathcal{ML}^{-2}\mathcal{T}^{-2},$$

$$[\boldsymbol{\nabla}][p] = (\mathcal{L}^{-1})(\mathcal{ML}^{-1}\mathcal{T}^{-2}) = \mathcal{ML}^{-2}\mathcal{T}^{-2},$$

$$[\mu][\boldsymbol{\nabla}]^2[\boldsymbol{v}] = (\mathcal{ML}^{-1}\mathcal{T}^{-1})(\mathcal{L}^{-1})^2(\mathcal{LT}^{-1}) = \mathcal{ML}^{-2}\mathcal{T}^{-2}. \tag{1.5.16}$$

Thus, each term in Eq. 1.5.15 has the same dimension, $\mathcal{ML}^{-2}\mathcal{T}^{-2}$, and the equation is dimensionally consistent.

EXAMPLE 1.5.3: The 'Model-H' equations for the concentration field c and the velocity field $\boldsymbol{v}$ for a binary fluid mixture are,

$$\frac{\partial c}{\partial t} + \boldsymbol{\nabla}\cdot(\boldsymbol{v}c) = \Gamma\boldsymbol{\nabla}^2\Xi, \tag{1.5.17}$$

$$\rho\left(\frac{\partial \boldsymbol{v}}{\partial t} + \boldsymbol{v}\cdot\boldsymbol{\nabla}\boldsymbol{v}\right) = -\boldsymbol{\nabla}p + \mu\boldsymbol{\nabla}^2\boldsymbol{v} + c\boldsymbol{\nabla}\Xi, \tag{1.5.18}$$

where p is the pressure, μ is the viscosity, Γ is the Onsager coefficient and Ξ is the chemical potential. Determine the dimensions of the chemical potential Ξ and the Onsager coefficient Γ.

Solution: The dimension of all the terms in Eq. 1.5.18 for the velocity field, with the exception of the last term on the right, was shown to be $\mathcal{ML}^{-2}\mathcal{T}^{-2}$ in Eq. 1.5.16. The dimension of the last term on the right is $[c][\boldsymbol{\nabla}][\Xi] = (\mathcal{ML}^{-3})(\mathcal{L}^{-1})[\Xi]$. Therefore, the dimension of the chemical potential is $[\Xi] = \mathcal{L}^2\mathcal{T}^{-2}$.

The dimensions of the terms in the the concentration Eq. 1.5.17, from left to right, are

$$[(\partial/\partial t)][c] = \mathcal{T}^{-1}(\mathcal{ML}^{-3}),$$

$$[\boldsymbol{v}][\boldsymbol{\nabla}][c] = (\mathcal{LT}^{-1})\mathcal{L}^{-1}(\mathcal{ML}^{-3}) = \mathcal{ML}^{-3}\mathcal{T}^{-1},$$

$$[\Gamma][\boldsymbol{\nabla}^2][\Xi] = [\Gamma](\mathcal{L}^{-2})(\mathcal{L}^2\mathcal{T}^{-2}). \tag{1.5.19}$$

For dimensional consistency, the dimension of the Onsager coefficient Γ is $\mathcal{ML}^{-3}\mathcal{T}$. $\qquad\square$

Summary (1.5)

---

1. There are six fundamental dimensions: Mass ($\mathcal{M}$), Length ($\mathcal{L}$), Time ($\mathcal{T}$), Temperature (Θ), Current ($\mathcal{A}$) and Light intensity.

2. The dimension of any quantity can be expressed in terms of these fundamental dimensions.

3. The dimension of all terms in an equation have to be the same, and this is the dimension of the equation. Vectors, their magnitude and their components all have the same dimension.

4. In the equations, derivatives and integrals have dimension. The dimension of the nth spatial derivative is $\mathcal{L}^{-n}$, and the nth time derivative is $\mathcal{T}^{-n}$. A spatial integral has dimension $\mathcal{L}$, and a time integral has dimension $\mathcal{T}$.

1.6 Dimensional Analysis

The fundamental principle in dimensional analysis, the Buckingham Pi theorem, states that if a problem contains n dimensional parameters, and these parameters involve a total of m fundamental dimensions, then the problem can be equivalently formulated using $n-m$ dimensionless groups. The choice of the dimensionless groups is subjective. For example, if Φ is a dimensionless group, then $\Phi^2, \Phi^{-3/2}$, etc. are also dimensionless group. Similarly, if Φ and Ψ are two dimensionless groups, then the product $\Phi^a \times \Psi^b$ is also a dimensionless group for any values of a and b. Though there is subjectivity in selecting the dimensionless groups, the Buckingham Pi theorem guarantees that the problem can be equivalently formulated in terms of $n - m$ dimensionless groups.

The subjectivity in the choice of dimensionless groups can be reduced by classification on the basis of physical understanding. In any problem involving n quantities, it is necessary to predict the value of one quantity when the values of $n - 1$ quantities are specified. The quantity to be predicted is the *dependent* quantity, and the other $(n - 1)$ quantities that are specified are the *independent* quantities. The objective of dimensional analysis is to obtain the relationship between dependent and independent quantities in a more compact form, involving $n - m$ dimensionless groups, if m is the number of dimensions. The *dependent dimensionless group* is obtained by non-dimensionalising the dependent quantity with a combination of the independent quantities. Dimensionless groups formed by combination of the independent quantities are the *independent dimensionless groups*. If there are $n - 1$ independent parameters and m dimensions, the number of independent dimensionless groups is $n - m - 1$. Therefore, when expressed in terms of dimensionless groups, a relationship is obtained between the dependent dimensionless group and the $n - m - 1$ independent dimensionless groups.

The independent dimensionless groups can further be interpreted as the ratio of different types of forces which cause the transport of mass, momentum or energy. From the numerical values of these dimensionless groups, the dominant forces can be identified. In addition, the number of dimensions can be augmented if, for example, a distinction can be made between thermal and mechanical energy as different types of energies, as shown in the analysis of a heat exchanger in Section 1.6.3, or between solute mass and total mass, as shown in the analysis of diffusion from a particle in Section 1.6.2. The classification and interpretation of dimensionless groups is illustrated in the following sub-sections.

1.6.1 Sphere Settling in a Fluid

Consider a sphere of diameter d settling in a tank containing fluid of density ρ and viscosity μ shown in Fig. 1.6. For simplicity, we assume that the width of the tank L is much larger than the sphere diameter, so that the flow around the sphere is modelled as the flow in a fluid of infinite extent. The equation of motion for the sphere velocity v in the vertical direction is,

$$m\frac{\mathrm{d}v}{\mathrm{d}t} = mg - F_B - F_D, \qquad (1.6.1)$$

where m is the mass of the sphere, g is the acceleration due to gravity, F_B is the buoyancy force and F_D is the drag force. It is desired to obtain an expression for the drag force F_D as a function of the sphere velocity and the fluid properties.

The physical reason for the drag force is as follows. The motion of the sphere causes flow and shearing in the surrounding fluid. There is frictional resistance to flow due to the fluid viscosity, resulting in a drag force. Since the frictional resistance to flow is due to viscous stresses, the viscosity μ and the sphere velocity v are relevant parameters for determining the force. There could also be an asymmetry in the flow pattern around the upstream and downstream hemispheres, resulting in a larger pressure on the front (upstream) side in comparison to the rear (downstream) side. The pressure difference between the front and the rear could also result in a net force. This pressure force, which is due to fluid inertia, depends on the flow velocity v and the mass density of the fluid ρ.

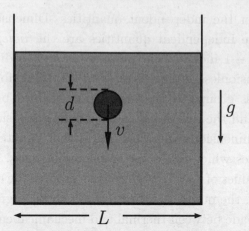

FIGURE 1.6.   A sphere of diameter d settling under gravity in a tank of size L containing a fluid with density ρ and viscosity μ.

It is important to note that the mass of the sphere or its mass density is not directly relevant for determining the flow, once the sphere velocity is specified. This is because the fluid velocity around the sphere, which results in the frictional and pressure forces, is determined by the velocity with which the sphere is moving. Similarly, the acceleration due to gravity determines the gravitational and buoyancy forces on the sphere, but does not directly affect the fluid velocity around the sphere. The velocity of the fluid around the sphere is entirely specified from the fluid equations of motion, and the condition that the fluid velocity at the surface of the sphere is equal to the velocity of the sphere. Thus, the drag force directly depends only on the sphere's velocity, diameter and the fluid properties.

Based on the above discussion, the dependent parameter, the drag force F_D with dimension $\mathcal{MLT}^{-2}$, is a function of the independent parameters:

1. diameter of the sphere d, with dimension $\mathcal{L}$,

2. velocity of the sphere v, with dimension $\mathcal{LT}^{-1}$,

3. viscosity of the fluid μ, with dimension $\mathcal{ML}^{-1}\mathcal{T}^{-1}$, and

4. density of the fluid ρ, with dimension $\mathcal{ML}^{-3}$.

In addition to the dependent dimensional quantity (drag force), there are four independent dimensional quantities and three dimensions. Therefore, there is one independent and one dependent dimensionless group.

The dependent dimensionless group is the drag force non-dimensionalised by a combination of the independent parameters. The two independent parameters that contain the mass dimension are the density and viscosity, so either of these can be chosen for non-dimensionalising the drag force. As we will see later, the choice depends on whether the inertial or viscous forces are dominant. For now, we will non-dimensionalise the force by the viscosity, velocity and the particle radius. The scaled force F_{DV}^* (the subscript V indicates that the force is scaled by the viscosity), is a combination of the force, viscosity, velocity and particle diameter,

$$F_{DV}^* = F_D \mu^\alpha v^\beta d^\gamma, \tag{1.6.2}$$

where α, β and γ are the exponents which render F_{DV}^* dimensionless. The dimensions of the left and right sides of the equation are

$$\mathcal{M}^0 \mathcal{L}^0 \mathcal{T}^0 = (\mathcal{M}\mathcal{L}\mathcal{T}^{-2})(\mathcal{M}\mathcal{L}^{-1}\mathcal{T}^{-1})^\alpha (\mathcal{L}\mathcal{T}^{-1})^\beta \mathcal{L}^\gamma. \tag{1.6.3}$$

For the dimensions on two sides to be equal,

$$\mathcal{M} : 0 = 1 + \alpha,$$
$$\mathcal{L} : 0 = 1 - \alpha + \beta + \gamma,$$
$$\mathcal{T} : 0 = -2 - \alpha - \beta. \tag{1.6.4}$$

These equations can be solved simultaneously, to obtain $\alpha = -1, \beta = -1, \gamma = -1$. Thus, the dependent dimensionless group is,

$$F_{DV}^* = \frac{F}{\mu d v}. \tag{1.6.5}$$

The independent dimensionless group, which we call Re, is a combination of the density, diameter, sphere velocity and viscosity,

$$\mathrm{Re} = \rho v^\alpha d^\beta \mu^\gamma. \tag{1.6.6}$$

In the dimensionless group, the exponent of any one of the quantities can be set equal to 1 without loss of generality. In Eq. 1.6.6, we have chosen to set the exponent of the density equal to 1, to make it consistent with the conventional definition of the

Reynolds number. The dimensions of the quantities in Eq. 1.6.6 are,

$$\mathcal{M}^0\mathcal{L}^0\mathcal{T}^0 = (\mathcal{M}\mathcal{L}^{-3})(\mathcal{L}\mathcal{T}^{-1})^\alpha\mathcal{L}^\beta(\mathcal{M}\mathcal{L}^{-1}\mathcal{T}^{-1})^\gamma. \qquad (1.6.7)$$

The following relations between the exponents have to be satisfied to render Re dimensionless,

$$\mathcal{M} : 1 + \gamma = 0,$$

$$\mathcal{L} : -3 + \alpha + \beta - \gamma = 0,$$

$$\mathcal{T} : -\alpha - \gamma = 0. \qquad (1.6.8)$$

These equations are solved to obtain $\alpha = 1$, $\beta = 1$ and $\gamma = -1$. Therefore, the dimensionless group Re, referred to as the Reynolds number, is

$$\text{Re} = \frac{\rho v d}{\mu}. \qquad (1.6.9)$$

The relation between force and velocity can also be written as

$$F^*_{DV} = F/(\mu dv) = \Phi(\text{Re}), \qquad (1.6.10)$$

where $\Phi(\text{Re})$ is a function of the Reynolds number, $\text{Re} = (\rho v d/\mu)$, which is the ratio of inertial and viscous effects.

The expression for the drag force, Eq. 1.6.10, can be simplified in the limits of small and large Reynolds number. When the Reynolds number is small, $\text{Re} \ll 1$, the fluid flow is dominated by viscous effects. In this case, the drag force does not depend on the fluid density, and from dimensional analysis, the drag force scaled by the viscous force scale F^*_{DV} (Eq. 1.6.5), is a constant. The expression for the drag force is,

$$F_D = \text{Constant} \times \mu v d. \qquad (1.6.11)$$

Eq. 1.6.11 is called the 'Stokes' drag law' for a settling particle in a viscous fluid. The value of the constant in the Stokes' drag law cannot be determined from dimensional analysis. It is necessary to solve the fluid mass and momentum conservation equations subject to appropriate boundary conditions to determine the value of this constant. This calculation, beyond the scope of this text, reveals that the value of the constant is 3π for solid sphere, and 2π for a spherical gas bubbles. However, to within a dimensionless constant factor, the expression for the drag force has been determined using dimensional analysis.

In the opposite limit of high Reynolds number, Re $\gg 1$, inertial effects are large compared to viscous effects. In this case, it is necessary to non-dimensionalise the mass dimension in the drag force by the density instead of using the viscosity. This leads to the dimensionless group

$$F_{DI}^* = \frac{F_D}{\rho v^2 d^2},$$
(1.6.12)

where the subscript $_I$ in F_{DI}^* indicates that the force is scaled by the 'inertial force scale'. In the high Reynolds number limit, it is expected that the drag force does not depend on the viscosity, and so F_{DI}^* is a constant. It is easily verified that the dimensionless forces scaled in these two different ways are related by,

$$F_{DI}^* = \text{Re}^{-1} F_{DV}^*.$$
(1.6.13)

The 'drag coefficient' c_D, which is the dimensionless group for momentum transfer (discussed in Section 2.1.3), is defined as the ratio of the average pressure on the object and the fluid kinetic energy per unit volume. It is easily verified that the dimensions of pressure and kinetic energy per unit volume are the same, so that the ratio is dimensionless. The average pressure is defined as the ratio of the drag force F_D and the projected area of the object perpendicular to the flow, which is $(\pi d^2/4)$. A measure of the kinetic energy per unit volume of the fluid is $(\rho v^2/2)$. Therefore, the drag coefficient is defined as,

$$c_D = \frac{(F_D/(\pi d^2/4))}{\rho v^2/2}.$$
(1.6.14)

The drag coefficient is proportional to the force scaled by inertial scales in Eq. 1.6.12 based on dimensional analysis, $c_D = (8/\pi) F_{DI}^*$.

If the size of the tank L in Fig. 1.6 is comparable to the particle diameter, the force will depend on the dimensionless ratio (L/d) in addition to the Reynolds number.

EXAMPLE 1.6.1: Colloidal particles of sub-micron size, such as dust and pollen grains, in a liquid undergo random motion called Brownian motion due to the force exerted by the liquid molecules on the particle. Brownian motion is characterised by the Brownian diffusivity, D_B, which is the mean square of the distance travelled by the particle in unit time. The Brownian diffusivity has the same dimension, $\mathcal{L}^2 \mathcal{T}^{-1}$, as the mass diffusion coefficient. The Brownian diffusivity depends on the thermal fluctuating energy $k_B T$ of the liquid molecules, the liquid viscosity and the particle diameter. Here, k_B is the Boltzmann constant with dimension $\mathcal{ML}^2\mathcal{T}^{-2}\Theta^{-1}$, and T is the absolute temperature. Obtain an expression for the Brownian diffusivity using dimensional analysis.

Solution: There is one dependent quantity, the Brownian diffusivity D_B, and four independent quantities, the Boltzmann constant k_B, temperature T, particle diameter d and viscosity μ. There are four dimensions, $\mathcal{M}, \mathcal{L}, \mathcal{T}$ and Θ. Therefore, there are no independent dimensionless groups, and there is one dependent dimensionless group,

$$\Phi = D_B k_B^\alpha T^\beta d^\gamma \mu^\delta. \tag{1.6.15}$$

Substituting the dimensions on the left and right side of the above equation, we obtain,

$$\mathcal{M}^0\mathcal{L}^0\mathcal{T}^0\Theta^0 = (\mathcal{L}^2\mathcal{T}^{-1})(\mathcal{ML}^2\mathcal{T}^{-2}\Theta^{-1})^\alpha\Theta^\beta\mathcal{L}^\gamma(\mathcal{ML}^{-1}\mathcal{T}^{-1})^\delta. \tag{1.6.16}$$

Equating the exponents of the fundamental dimensions on both sides, we obtain,

$$\mathcal{M}: 0 = \alpha + \delta,$$
$$\mathcal{L}: 0 = 2 + 2\alpha + \gamma - \delta,$$
$$\mathcal{T}: 0 = -1 - 2\alpha - \delta,$$
$$\Theta: 0 = -\alpha + \beta. \tag{1.6.17}$$

These equations can be solved to obtain $\alpha = -1, \beta = -1, \gamma = 1$ and $\delta = 1$. Therefore, the dependent dimensionless group is,

$$\frac{D_B\mu d}{k_B T} = \text{Constant}. \tag{1.6.18}$$

The Einstein calculation for the Brownian diffusivity shows that the constant is $(3\pi)^{-1}$, and the Brownian diffusivity is,

$$D_B = \frac{k_B T}{3\pi\mu d}. \tag{1.6.19}$$

□

EXAMPLE 1.6.2: The torque T on a spherical particle of diameter d rotating in a quiescent fluid with angular velocity Ω depends on the particle diameter, angular velocity and fluid properties. Obtain an expression for the torque using dimensional analysis.

Solution: There is one dependent quantity, the torque on the particle T, which depends on the angular velocity Ω, diameter d, the fluid density ρ and viscosity μ. These contain three dimensions: mass, length and time. There is one dependent variable, torque, and four independent variables, ρ, Ω, d and μ. Therefore, there is one independent and one dependent dimensionless group. The mass dimension in the torque can be non-dimensionalised by either the viscosity or the density. Here, we use the viscosity for non-dimensionalisation, and define the scaled torque as

$$T^* = T\mu^\alpha d^\beta \Omega^\gamma. \tag{1.6.20}$$

Substituting the dimensions for torque, viscosity and angular velocity provided in Table 1.1, the dimensions of the quantities on the left and right sides of Eq. 1.6.20 are,

$$\mathcal{M}^0\mathcal{L}^0\mathcal{T}^0 = (\mathcal{M}\mathcal{L}^2\mathcal{T}^{-2})(\mathcal{M}\mathcal{L}^{-1}\mathcal{T}^{-1})^\alpha \mathcal{L}^\beta (\mathcal{T}^{-1})^\gamma. \tag{1.6.21}$$

The relations between the exponents are,

$$\mathcal{M}: 1 + \alpha = 0,$$
$$\mathcal{L}: 2 - \alpha + \beta = 0,$$
$$\mathcal{T}: -2 - \alpha - \gamma = 0. \tag{1.6.22}$$

The above equations are solved to obtain $\alpha = -1, \beta = -3, \gamma = -1$, and the independent dimensionless groups is,

$$T^* = \frac{T}{\mu d^3 \Omega}. \tag{1.6.23}$$

The independent dimensionless group depends on the density ρ, angular velocity Ω, diameter d and viscosity μ. It is easy to infer that this independent dimensionless group, the Reynolds number for the rotating sphere, is

$$\text{Re} = \frac{\rho \Omega d^2}{\mu}. \tag{1.6.24}$$

Therefore, the expression for the torque is,

$$T^* = \text{Function(Re)}. \tag{1.6.25}$$

In the limit of low Reynolds number, the torque does not depend on the fluid density. The scaled torque is a constant; therefore, the expression for the torque is,

$$T = \text{Constant} \times \mu d^3 \Omega. \tag{1.6.26}$$

The constant in the above equation, which has to be determined by a calculation which is outside the scope of this text, turns out to be π. ☐

EXAMPLE 1.6.3: The torque T on a charged particle rotating in a magnetic field depends on the magnetic flux density B, the charge Q, the angular velocity Ω and the particle diameter d. Use dimensional analysis to determine the torque.

Solution: There is one dependent quantity T with dimension $\mathcal{ML}^2\mathcal{T}^{-2}$, and four independent quantities, the charge Q with dimension $\mathcal{AT}$, the magnetic flux density B with dimension $\mathcal{MT}^{-2}\mathcal{A}^{-1}$, the angular velocity Ω with dimension $\mathcal{T}^{-1}$ and the particle diameter d with dimension $\mathcal{L}$. There are four dimensions, $\mathcal{M}, \mathcal{L}, \mathcal{T}$ and $\mathcal{A}$. Therefore, there is one dimensionless group, which is the scaled torque,

$$T^* = TQ^\alpha B^\beta \Omega^\gamma d^\delta. \tag{1.6.27}$$

The relation between the dimensions on the left and right sides is,

$$\mathcal{M}^0 \mathcal{L}^0 \mathcal{T}^0 \mathcal{A}^0 = (\mathcal{ML}^2\mathcal{T}^{-2})(\mathcal{AT})^\alpha (\mathcal{MT}^{-2}\mathcal{A}^{-1})^\beta \mathcal{T}^{-\gamma}\mathcal{L}^\delta. \tag{1.6.28}$$

The relations between the exponents are,

$$\mathcal{M} : 1 + \beta = 0,$$
$$\mathcal{L} : 2 + \delta = 0,$$
$$\mathcal{T} : -2 + \alpha - 2\beta - \gamma = 0,$$
$$\mathcal{A} : \alpha - \beta = 0. \tag{1.6.29}$$

These four equations are solved to provide $\alpha = -1$, $\beta = -1$, $\gamma = -1$ and $\delta = -2$. Therefore, the dimensionless torque is,

$$T^* = TQ^{-1}B^{-1}\Omega^{-1}d^{-2}. \tag{1.6.30}$$

☐

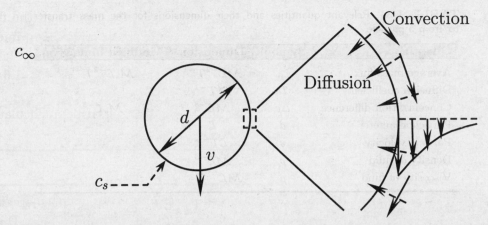

FIGURE 1.7. Mass transfer to a particle due to the concentration difference between the bulk concentration c_∞ and the surface concentration c_s. In the figure at the right, the fluid velocity is equal to the particle velocity at the surface, and it decreases to zero far from the particle. The solid lines show the direction of the convective transport along the flow, and the dashed lines show that diffusion is necessary for transporting mass perpendicular to the surface, since there is no flow perpendicular to the surface.

1.6.2 Mass Transfer to a Suspended Particle

Consider the stirred-tank reactor for heterogeneous catalysis shown in Fig. 1.1, where the reactants and products are in solution and the catalyst is in the form of solid particles. The reactant has to be transported from the bulk of the fluid to the catalyst surface, and then react on the catalyst surface. The product is then transported from the catalyst surface to the bulk of the fluid. The driving force for this transport is the concentration difference between the bulk fluid far from the particle and the fluid at the surface of the catalyst.

Here, we focus on determining the rate of transport of the reactant/product to/from the catalyst surface as a function of the concentration difference between the bulk and the catalyst surface, as shown in Fig. 1.7. Instead of the average transport rate, correlations are written for the average flux, which is the mass transported per unit surface area per unit time. The rate of transport is then equal to the product of the average flux and the total surface area of the catalyst particles. It should be emphasised that correlations are written for the average flux, and not for the local flux at a particular position on the catalyst surface depicted in Fig. 1.7.

The dependent quantity, the average mass flux to/from the surface j_{av} depends on the particle diameter d, the diffusion coefficient $\mathcal{D}$ and the difference in concentration $\Delta c = (c_\infty - c_s)$, where c_∞ is the concentration in the bulk and c_s is the concentration in the fluid at the surface of the particle. Mass transfer is also affected by the motion of the particle relative to the fluid caused by stirring. The flow pattern around the

TABLE 1.3. Relevant quantities and their dimensions for the mass transfer to/from a particle.

Quantity	Symbol	Dimension	Modified dimension
Average mass flux	j_{av}	$\mathcal{ML}^{-2}\mathcal{T}^{-1}$	$\mathcal{M}_s\mathcal{L}^{-2}\mathcal{T}^{-1}$
Diffusion coefficient	$\mathcal{D}$	$\mathcal{L}^2\mathcal{T}^{-1}$	
Concentration difference	Δc	$\mathcal{ML}^{-3}$	$\mathcal{M}_s\mathcal{L}^{-3}$
Particle diameter	d	$\mathcal{L}$	
Particle velocity	v	$\mathcal{LT}^{-1}$	
Density of fluid	ρ	$\mathcal{ML}^{-3}$	
Viscosity of fluid	μ	$\mathcal{ML}^{-1}\mathcal{T}^{-1}$	

particle results in the convective transport of mass to/from the particle, and thereby enhances mass transfer. Thus, the average flux is affected by the flow pattern, which depends on the fluid density ρ and viscosity μ, the particle velocity v relative to the fluid and the particle diameter. The different dimensional quantities of relevance are shown in Table 1.3.

In Table 1.3, there is one dependent dimensional quantity (average flux), six independent dimensional quantities and three dimensions. On this basis, there should be one dependent and three independent dimensionless groups. However, a further simplification can be made by distinguishing between the mass dimension in mass transport (which involves the solute mass) and that in the flow dynamics (which involves the total mass of the fluid). The flux and the concentration difference involve the mass of the solute, whereas the flow pattern around the particle is affected by the total mass density (solute plus solvent). If the solute concentration does not affect the fluid density and viscosity, we can make a distinction between the mass dimension for the solute, $\mathcal{M}_s$, and the mass dimension for the fluid, $\mathcal{M}$. In the fourth column in Table 1.3, the dimensions are modified so that the mass flux and the concentration difference depend on the solute mass $\mathcal{M}_s$, while the density and viscosity depend on the fluid mass $\mathcal{M}$. Now, there are four dimensions, $\mathcal{M}$, $\mathcal{M}_s$, $\mathcal{L}$ and $\mathcal{T}$, and consequently there are two independent and one dependent dimensionless groups.

The dependent dimensionless group, called the Sherwood number, can be formulated by non-dimensionalising the flux by the diffusion coefficient, the concentration difference and the particle diameter,

$$\mathrm{Sh} = j_{av}(\Delta c)^\alpha \mathcal{D}^\beta d^\gamma. \qquad (1.6.31)$$

Though indices α, β and γ can be determined from the powers of the fundamental dimensions, it is easier to deduce the dimensionless group from the Fick's law for diffusion, Eq. 1.5.1. It should be noted that Fick's law is written for the local flux at each point in the fluid, and not for the average flux. Nevertheless, since the right and left sides of Fick's law have the same dimension. Therefore, the dependent dimensionless group can be written as the ratio of the average flux j_{av} and $(\mathcal{D}\Delta c/d)$, which has the same dimension as the right side of Fick's law,

$$\mathrm{Sh} = \frac{j_{av}d}{\mathcal{D}\Delta c}. \tag{1.6.32}$$

There are two independent dimensionless groups. We have already determined one of these, the Reynolds number $\mathrm{Re} = (\rho v d/\mu)$, in Section 1.6.1. The second dimensionless group can be defined in two ways. Since the diffusion coefficient $\mathcal{D}$ has the dimension $\mathcal{L}^2\mathcal{T}^{-1}$, a dimensionless group can be defined as the Peclet number $\mathrm{Pe} = (vd/\mathcal{D})$. The Peclet number can be understood as the ratio of mass convection and diffusion. Thus, the relation between the dependent and independent dimensionless groups for mass transfer is,

$$\mathrm{Sh} = \mathrm{Function}(\mathrm{Re}, \mathrm{Pe}). \tag{1.6.33}$$

The second, more common, formulation of the second independent dimensionless group is the Schmidt number $\mathrm{Sc} = (\mu/\rho\mathcal{D}) = (\mathrm{Pe}/\mathrm{Re})$. The Sherwood number correlation in terms of the Reynolds and Schmidt numbers is obtained by substituting $\mathrm{Pe} = \mathrm{Re}\,\mathrm{Sc}$ in Eq. 1.6.33.

The Sherwood number correlation, Eq. 1.6.33, can be simplified in the limits of low and high Peclet numbers. In the limit of low Peclet number, $(vd/\mathcal{D}) \ll 1$, mass transport due to diffusion is much larger than that due to convection. In this case, the fluid flow pattern and the velocity v do not affect the mass flux, and so the Sherwood number does not depend on either the Reynolds or Peclet numbers. Therefore, the Sherwood number tends to a constant value in the limit of low Peclet number.

In the limit of high Peclet number $(vd/\mathcal{D}) \gg 1$, the rate of convection of mass is much larger than the rate of mass diffusion. It might be simplistically inferred that the average mass flux does not depend on the mass diffusion coefficient. This inference is not correct for the following reason. Convection results in transport along the direction of the local fluid velocity. There is no convective transport across the flow, and so cross-stream transport is necessarily due to diffusion. If we focus on the flow close to the particle surface in Fig. 1.7, the fluid flow relative to the particle is along the surface, and there is no fluid velocity perpendicular to the surface due to the no-penetration condition. There is no convective mass transport perpendicular to the surface, and transport of mass to the surface is necessarily due to diffusion.

Therefore, the mass diffusion coefficient is a relevant parameter even in the limit of high Peclet number, and the Sherwood number is a function of the Reynolds and Peclet numbers.

EXAMPLE 1.6.4: For the drying of a wet particle of diameter d in air, the difference between the vapour pressure of water at the particle surface p_w^{sat} and the partial pressure of water in air p_w provides the driving force. What is the definition of the Sherwood number in this case?

Solution: If the partial pressure of water is p_w, the number density of water molecules in air can be estimated from the ideal gas law, $N = (p_w/k_BT)$, where k_B is the Boltzmann constant and T is the absolute temperature. The mass concentration is the number density times the mass of one molecule of water M_w, $c = (pM_w/k_BT)$. The concentration difference between the surface of the particle and the ambient air is $\Delta c = ((p_w^{sat} - p_w)M_w/k_BT)$. Therefore, the Sherwood number $(j_{av}/(\mathcal{D}\Delta c/d))$, is

$$\text{Sh} = \frac{j_{av}}{\mathcal{D}(p_w^{sat} - p_w)M_w/(k_BTd)}. \qquad (1.6.34)$$

$\square$

1.6.3   Heat Transfer in a Heat Exchanger

A heat exchanger is used to transport heat from one fluid to another for heating/cooling the feed/waste streams in chemical processes. The shell-and-tube heat exchanger, shown in Fig. 1.2 in Section 1.1.2, consists of one fluid flowing inside the tube, and the second fluid pumped through the shell on the outside of the tube. Heat transfer takes place across the wall of the tube due to the average temperature difference ΔT between the fluids in the shell and tube. The simpler problem of the rate of heat transfer from the tube-side fluid to the tube wall due to the temperature difference ΔT is considered here. Instead of the total heat transfer rate, correlations are written for the average flux, which is the ratio of the total heat transfer rate and the surface area of the tube.

The dependent quantity in this problem is the average flux q_{av}. The independent parameters for heat transfer are the dimensions of tube—the diameter d and length L, and the thermal properties of the fluid, the thermal conductivity k, the specific heat C_p, and the average difference in temperature between the fluid and the wall ΔT. Heat is also transported by convection of fluid along the tube, and so the average flux depends on the nature of the flow in the tube. The flow pattern depends on the

TABLE 1.4. Relevant quantities and their dimensions for the heat transfer to a fluid flowing in a pipe.

Quantity	Symbol	Dimension	Modified Dimension
Average heat flux	q_{av}	$\mathcal{M}\mathcal{T}^{-3}$	$\mathcal{H}\mathcal{L}^{-2}\mathcal{T}^{-1}$
Diameter of pipe	d	$\mathcal{L}$	$\mathcal{L}$
Length of pipe	L	$\mathcal{L}$	$\mathcal{L}$
Average fluid velocity	v_{av}	$\mathcal{L}\mathcal{T}^{-1}$	$\mathcal{L}\mathcal{T}^{-1}$
Density of fluid	ρ	$\mathcal{M}\mathcal{L}^{-3}$	$\mathcal{M}\mathcal{L}^{-3}$
Viscosity of fluid	μ	$\mathcal{M}\mathcal{L}^{-1}\mathcal{T}^{-1}$	$\mathcal{M}\mathcal{L}^{-1}\mathcal{T}^{-1}$
Specific heat of fluid	C_p	$\mathcal{L}^2\mathcal{T}^{-2}\Theta^{-1}$	$\mathcal{H}\mathcal{M}^{-1}\Theta^{-1}$
Thermal conductivity	k	$\mathcal{M}\mathcal{L}\mathcal{T}^{-3}\Theta^{-1}$	$\mathcal{H}\mathcal{L}^{-1}\mathcal{T}^{-1}\Theta^{-1}$
Temperature difference	ΔT	Θ	Θ

average flow velocity v_{av}, the density ρ and viscosity μ. The dependent dimensional quantity and eight independent dimensional quantities, listed in Table 1.4, contain four fundamental dimensions, $\mathcal{M}$, $\mathcal{L}$, $\mathcal{T}$ and temperature Θ. On this basis, it is expected that there is one dependent dimensionless group and four independent dimensionless groups.

A further reduction is possible when there is no interconversion between heat energy (which is being transferred) and mechanical energy (which is driving the flow). In this case, it is possible to consider heat energy as a separate dimension $\mathcal{H}$ which is different from mechanical energy. There are now five dimensions, $\mathcal{H}$, $\mathcal{M}$, $\mathcal{L}$, $\mathcal{T}$ and Θ, and a total of nine dimensional quantities, as shown in the fourth column in Table 1.4. The problem now contains one dependent dimensionless group and three independent dimensionless groups.

The dependent dimensionless group, which is a scaled average heat flux, can be deduced from the Fourier's law for heat conduction, Eq. 1.5.3. It is important to note that Fourier's law defines the local heat flux at each point in the heat exchanger, and not the average flux. Nevertheless, the left and right sides of Fourier's law have the same dimension. Therefore the dependent dimensionless group, called the Nusselt number, can be defined by scaling the average flux by a term with the same dimension as the right side of Eq. 1.5.3,

$$\mathrm{Nu} = \frac{q_{av}d}{k\Delta T}. \qquad (1.6.35)$$

Of the three independent dimensionless groups, the simplest is the ratio (L/d) of the length and diameter of the tube. A second dimensionless group is the Reynolds number, $\mathrm{Re} = (\rho v_{av}d/\mu)$, which is the ratio of fluid inertia and viscosity. The third dimensionless group contains the specific heat C_p and the thermal conductivity k, which have not been included so far. Since C_p contains both the thermal energy and mass dimensions, the dimensionless group has to contain the thermal conductivity k, as well as the viscosity μ or the density ρ. The dimensionless group constructed with the specific heat, viscosity and conductivity is the Prandtl number,

$$\mathrm{Pr} = \frac{C_p \mu}{k}. \tag{1.6.36}$$

The third dimensionless group can, alternatively, be expressed as the Peclet number for heat transfer Pe, which is the product of the Reynolds and Prandtl numbers,

$$\mathrm{Pe} = \mathrm{Re}\,\mathrm{Pr} = \frac{\rho v_{av} d C_p}{k}. \tag{1.6.37}$$

As we shall see in the next chapter, this is the ratio of heat convection and diffusion. Based on the above dimensional analysis, the correlation is of the form,

$$\mathrm{Nu} = \mathrm{Function}\left(\frac{L}{d}, \mathrm{Re}, \mathrm{Pe}\right). \tag{1.6.38}$$

EXAMPLE 1.6.5: The dimensionless groups for the heat flux in a heat exchanger were determined assuming that there is no interconversion between mechanical and thermal energy. If mechanical energy can be converted to thermal energy due to the viscous dissipation of energy, then the Nusselt number would depend on one additional dimensionless group. What is this dimensionless group, and what is its significance?

Solution: To obtain the dimensionless group, we determine the dependence of the rate of viscous dissipation of energy per unit length on the flow parameters, and the rate of conduction of energy from the tube per unit length due to the difference between the fluid temperature and the wall temperature. The ratio of these two provides the dimensionless group.

The rate of viscous dissipation of energy per unit length, which has dimension Energy/length/time, $\mathcal{MLT}^{-3}$, depends on the viscosity μ, the flow velocity v_{av} and the pipe diameter d. From dimensional analysis, it is easily inferred that the rate of dissipation of energy per unit length is proportional to μv_{av}^2.

The rate of transfer of energy across the wall of the tube per unit length depends on the thermal conductivity k, the temperature difference between the fluid and the wall, ΔT, and the tube diameter d. From dimensional analysis, the rate of transfer per unit length is proportional to $k\Delta T$, which has dimension $\mathcal{MLT}^{-3}$.

The ratio of these two is the dimensionless Brinkman number,

$$\text{Br} = \frac{\mu v_{av}^2}{k \Delta T}. \tag{1.6.39}$$

When the Brinkman number is small, the rate of viscous dissipation of energy can be neglected in comparison to the rate of conduction in the energy balance equation. When the Brinkman number is comparable to 1, a significant fraction of the heat energy transferred across the wall of the pipe is generated due to the viscous dissipation. □

EXAMPLE 1.6.6: A particle of diameter 1 mm and temperature 40^oC is cooled in a fluid at temperature 0^oC. It takes 60 s for the temperature to decrease from 40^oC to 10^oC. If a particle of diameter 2 mm made of the same material is placed in the same fluid, how long will it take for the particle temperature to decrease from 40^oC to 10^oC? The rate of cooling is determined by the conduction of heat in the fluid surrounding the particle.

Solution: The time required for cooling depends on the initial and final temperatures, the temperature of the ambient fluid, the thermal properties of the fluid, the thermal conductivity k, the density ρ and the specific heat C_p, and the particle diameter d. Since the initial and final temperatures and the ambient temperature are the same for the two particles, the dependent dimensional parameter is the time of drying t, and the independent parameters are k, ρ, C_p and d. There are five dimensional parameters and four dimensions, $\mathcal{M}, \mathcal{L}, \mathcal{T}$ and Θ. There is only one dimensionless group, which is easily determined as $(tk/\rho C_p d^2)$. This dimensionless group is the same for the small and large particles, and the time required for cooling is proportional to d^2 if the fluid properties are unchanged. Therefore, the time required for cooling the 2 mm particle is 240 s. □

Another important consideration in the design of a heat exchanger is the pressure difference across the pipe that is required to generate the desired flow rate. The pressure difference across the ends is required to compensate for the shear stress exerted by the walls on the fluid due to fluid friction, as shown in Fig. 1.8. The pressure difference and the wall shear stress are related by a force balance across a pipe of length L. The net force applied at the two ends, which is the pressure difference (Δp) times the area of cross section ($\pi d^2/4$), is equal and opposite to the force exerted by the walls which is the wall shear stress τ_w times the curved surface

TABLE 1.5. Relevant quantities and their dimensions for the pressure gradient required for fluid flow in a pipe.

Quantity	Symbol	Dimension
Pressure gradient	$-(\Delta p/L)$	$\mathcal{ML}^{-2}\mathcal{T}^{-2}$
Diameter of pipe	d	$\mathcal{L}$
Average fluid velocity	v_{av}	$\mathcal{LT}^{-1}$
Density of fluid	ρ	$\mathcal{ML}^{-3}$
Viscosity of fluid	μ	$\mathcal{ML}^{-1}\mathcal{T}^{-1}$

area of the pipe, (πdL),

$$(\Delta p)(\pi d^2/4) = -\tau_w \pi dL. \tag{1.6.40}$$

Therefore, the pressure gradient $-(\Delta p/L)$, which is the pressure drop per unit length, is equal to $(4\tau_w/d)$.[4]

The average wall shear stress depends on the pipe diameter d, the average flow velocity v_{av}, the fluid density ρ and viscosity μ. These quantities and their dimensions are listed in Table 1.5. There are four independent dimensional quantities, and one dependent dimensional quantity, and these quantities contain three dimensions. Therefore, there is one independent dimensionless group and one dependent dimensionless group. The wall shear stress could be non-dimensionalised by either viscous or inertial scales. It is conventional to define the friction factor f as the wall shear stress scaled by the kinetic energy per unit volume, $(\rho v_{av}^2/2)$. The

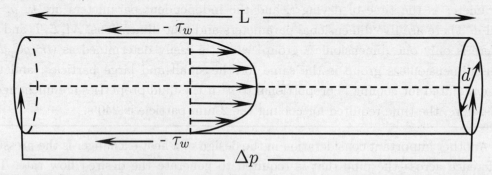

FIGURE 1.8. For the pressure-driven flow in a pipe of diameter d and length L, the pressure difference Δp times the area of cross section $(\pi d^2/4)$ is equal to the negative of the wall shear stress τ_w times the curved surface area πdL.

---

[4]Note that the pressure difference Δp, defined as the difference between the outlet and inlet pressures, is negative in order to pump fluid from the inlet to the outlet. The wall shear stress is positive if it is in the flow direction.

Fanning friction factor is defined as,

$$f = \frac{\tau_w}{(\rho v_{av}^2/2)} = -\frac{(\Delta p/L)}{2\rho v_{av}^2/d}. \tag{1.6.41}$$

The second relation above follows from the force balance equation, Eq. 1.6.40. The independent dimensionless group is the Reynolds number, $(\rho v_{av} d/\mu)$. Therefore, for the flow in a pipe, the friction factor is a function of the Reynolds number,

$$f = \text{Function(Re)}. \tag{1.6.42}$$

Summary (1.6)

1. Buckingham Pi theorem: If there are n dimensional quantities and m dimensions, the problem can be reduced to a relation between $(n - m)$ dimensionless groups.

2. Dimensional and dimensionless quantities can be divided into independent quantities, which are specified, and a dependent quantity which is to be determined.

3. When there is no interconversion between heat and mechanical energy in heat transfer, or solute and solvent mass in mass transfer, these can be treated as independent fundamental dimensions, to reduce the number of independent dimensionless groups.

Exercises

EXERCISE 1.1 Consider an alternate convention, where the fundamental dimensions are force $\mathcal{F}$, length $\mathcal{L}$ and time $\mathcal{T}$, instead of mass, length and time. What is the dimension of mass in this alternate convention?

EXERCISE 1.2 Consider an alternate convention, where the fundamental dimensions are voltage $\mathcal{V}$, current $\mathcal{A}$, length $\mathcal{L}$ and time $\mathcal{T}$, instead of mass, current, length and time. What is the dimension of mass in this alternate convention?

EXERCISE 1.3 The relaxation time of a polymer molecule t_R, with dimension $\mathcal{T}$, depends on the solvent viscosity μ_s, the radius of gyration of the molecule R_g with dimension length, the Boltzmann constant k_B and the absolute temperature T. Find an expression for the relaxation time using dimensional analysis.

EXERCISE 1.4 When a pressure difference Δp is applied across a porous medium of thickness h, the average velocity v_{av} of the fluid through the porous medium is given by the Darcy–Forchheimer equation,

$$\frac{\Delta p}{h} = -\frac{\mu}{\kappa}v_{av} - \beta\rho v_{av}^2,$$

where ρ and μ is the density and viscosity of the fluid. Determine the dimension of the Darcy constant κ and the Forchheimer constant β.

EXERCISE 1.5 In the Hertz model for the interaction force between two smooth particles such as sand grains or glass beads, the force F between the particles acting along the line joining their centers is expressed as,

$$F = -k_H\delta^{3/2},$$

where k_H is the Hertz spring constant, and δ is the linear compression of the particle surfaces at the contact point with dimension $\mathcal{L}$. What is the dimension of k_H? Suggest an expression for k_H if it depends only on the shear modulus of elasticity G and the particle diameter d.

EXERCISE 1.6 When two point charges Q_1 and Q_2 are separated by a distance r in vacuum, the electrostatic energy is given by,

$$E = \frac{Q_1Q_2}{4\pi\epsilon_0 r}.$$

Determine the dimension of ϵ_0, the vacuum permittivity.

EXERCISE 1.7 The Lorentz force F on a charged particle with charge Q moving with velocity v perpendicular to a magnetic field with magnetic flux density B is given by the relation,

$$F = QvB.$$

Determine the dimension of the magnetic flux density B.

EXERCISE 1.8 Ampere's law states that the magnetic flux density B at a distance r from a current-carrying wire is

$$B = \frac{\mu I}{2\pi r},$$

where I is the current in the wire. Determine the dimension of the magnetic permeability μ.

EXERCISE 1.9 The Maxwell equations in electrodynamics are,

$$\boldsymbol{\nabla} \cdot \boldsymbol{E} = (\rho_s/\epsilon_0),$$

$$\boldsymbol{\nabla} \cdot \boldsymbol{B} = 0,$$

$$\boldsymbol{\nabla} \times \boldsymbol{E} = -\frac{\partial \boldsymbol{B}}{\partial t},$$

$$\boldsymbol{\nabla} \times \boldsymbol{B} = \mu_0 \boldsymbol{J} + \mu_0 \epsilon_0 \frac{\partial \boldsymbol{E}}{\partial t},$$

where $\boldsymbol{E}$ is the electric field (voltage difference per unit length), $\boldsymbol{B}$ is the magnetic flux density whose dimension was calculated in Exercise 1.7, ρ_s is the charge density (charge per unit volume), ϵ_0 and μ_0 are the electric permittivity and magnetic permeability in vacuum, and $\boldsymbol{J}$ is the current density (current per unit area). Determine the dimensions of the equations, and verify that all terms in each equation have the same dimension.

EXERCISE 1.10 The Schrödinger equation in quantum mechanics is written for the wave function ψ, which is defined such that the probability of finding a particle within a differential volume dV is $\psi^*\psi\, dV$, where ψ^* is the complex conjugate of ψ with the same dimension as ψ. The total probability of finding a particle in all space is 1, so that the wave function satisfies the equation,

$$\int_{\text{All space}} dV\, \psi^*\psi = 1.$$

The Schroedinger equation is,

$$\imath\hbar\frac{\partial \psi}{\partial t} = -\frac{\hbar^2}{2m}\boldsymbol{\nabla}^2\psi + P\psi,$$

where $\imath = \sqrt{-1}$, P is the potential energy, m is the particle mass, $\boldsymbol{\nabla}$ is the gradient operator and t is time. Determine the dimension of the wave function ψ and Planck's constant $\hbar$.

EXERCISE 1.11 Slender objects such as an airfoil or an airplane wing, while moving relative to air with velocity v, experience a lift force. The schematic of the cross section of the airfoil is shown in Fig. 1.9. The lift force per unit length perpendicular to the plane of the paper, F_L, depends on the shape, a linear dimension L usually chosen as the chord of the airfoil, the velocity v, the density ρ of the fluid and the 'angle of attack' θ, which is the angle made by the chord of the airfoil with respect to the incident flow velocity. For a given shape of the airfoil, use dimensional analysis to obtain an expression for the lift force.

EXERCISE 1.12 A charged particle of diameter d and charge Q in an electric field E suspended in a viscous liquid with viscosity μ moves with velocity v due to the electrostatic forces acting on it. Use dimensional analysis to determine the dependence of v on Q, E, d and μ.

EXERCISE 1.13 When a conducting particle rotates in a magnetic field, eddy currents are induced due to Ampere's circutal law. A current loop in a conductor results in a magnetic

FIGURE 1.9. The lift force on an airfoil of characteristic length L moving through air with angle of attack θ.

moment due to Faraday's law of induction. The interaction between the particle magnetic moment m and the magnetic flux density B results in a torque $\mathbf{T}$ on the particle,

$$\mathbf{T} = m \times B,$$

where $\times$ is the vector cross product which is dimensionless. Determine the dimension of the magnetic moment m. The magnetic moment depends on the magnetic flux density B, the vacuum permeability μ_0, the electrical resistivity of the particle ϱ, the particle angular velocity Ω and the particle diameter d. Express the dependence of the magnetic moment on the angular velocity in dimensionless form.

EXERCISE 1.14 For the flow of water on the tube side of a heat exchanger, with average temperature difference of 10°C between the fluid and wall, what should be the velocity for the Brinkman number to be 1? Can the Brinkman number be large for a steady flow? The viscosity of water is 10^{-3}kg/m/s, and the thermal conductivity of water is 0.6 W/m/°C.

2

Dimensionless Groups and Correlations

The number of independent parameters in a problem is reduced when the dependent and independent parameters are expressed in dimensionless form. In the problem of the settling sphere in Section 1.6.1 and the flow through a pipe in Section 1.6.3, the original problem contained one dependent and four independent dimensional quantities. Using dimensional analysis, this was reduced to one independent and one dependent dimensionless groups. The mass transfer problem in Section 1.6.2 contained one dependent and six independent quantities. The problem was reduced to a relationship between one dependent and two independent dimensionless groups, using dimensional analysis and the assumption that the solute mass and total mass can be considered as different dimensions. In the heat transfer problem in Section 1.6.3, there were one dependent and eight independent dimensional quantities. This was reduced to a relationship between one dependent and three independent dimensionless groups, using dimensional analysis and the assumption that the thermal and mechanical energy can be considered as different dimensions. Thus, dimensional analysis has significantly reduced the number of parameters in the problem.

It is not possible to further simplify the problem using dimensional analysis. In order to progress further, experiments can be carried out to obtain empirical correlations between the dimensionless groups. Another option, pursued in this text, is to do analytical calculations based on a mathematical description of transport processes. Before proceeding to develop the methodology for the analytical calculations, a physical interpretation of the different dimensionless groups is provided in this chapter.

In dimensional analysis, there is ambiguity in the selection of the dimensional parameters for forming the dimensionless groups. This ambiguity is reduced by a physical understanding of the dimensionless groups as the ratio of different types of forces. Here, a broad framework is established for understanding the different dimensionless groups and the relations between them. The forms of the correlations depend on several factors, such as the flow regime, flow patterns and the boundary conditions.

It is important to note that the correlations listed here are indicative, but not exhaustive. Some commonly used correlations are presented to obtain a physical understanding of the terms in the correlation, and to illustrate their application. More accurate correlations applicable in specific domains can be found in specialised handbooks/technical reports.

2.1 Dimensionless Groups

Dimensionless groups can be classified into three broad categories: the dependent dimensionless groups which are dimensionless fluxes, the ratios of convection and diffusion, and the ratios of different types of diffusion. Before proceeding to discuss dimensionless groups and correlations, it is necessary to first define convective and diffusive fluxes, and the diffusion coefficients for mass, momentum and energy.

2.1.1 Convective Flux

The convective flux is the transport of a quantity (mass, momentum or energy) across a surface per unit area per unit time due to the fluid flow. Consider a surface ΔS within the fluid, with fluid velocity v_n perpendicular to the surface, as shown in Fig. 2.1. If we consider a time interval Δt, the volume of fluid transported across the surface is $(v_n \Delta t \Delta S)$, where v_n is the fluid velocity in the direction perpendicular to the surface. (Note that the component of the fluid velocity parallel to the surface does not result in transport across the surface.) The mass flux, which is the mass transported per unit area per unit time, is

$$\text{Convective mass flux} = c v_n. \tag{2.1.1}$$

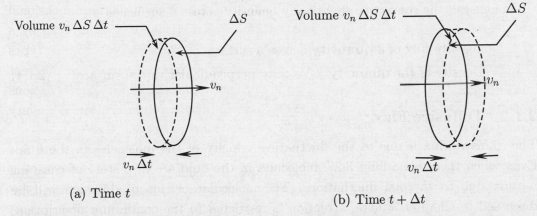

Volume $v_n \Delta S \Delta t$ — ΔS

v_n

$v_n \Delta t$

(a) Time t

Volume $v_n \Delta S \Delta t$ — ΔS

v_n

$v_n \Delta t$

(b) Time $t + \Delta t$

FIGURE 2.1. For a surface ΔS with fluid velocity v_n across the surface from left to right, the fluid volume $v_n \Delta t \Delta S$ which was on the upstream side of the surface at time t (a) moves across the surface to the downstream side in the direction of the flow velocity at time $t + \Delta t$ (b).

In a similar manner, the heat flux across the surface is the product of the thermal energy density and the fluid velocity perpendicular to the surface,

$$\text{Convective heat flux} = (\rho C_p T) v_n, \qquad (2.1.2)$$

where $\rho C_p T$ is the energy density (energy per unit volume).[1]

While defining the momentum flux, it is important to be aware of the distinction between the direction of the momentum and the direction of transport. Momentum is a vector that has both magnitude and direction, while the transport across the surface is also directional along the perpendicular to the surface. If we consider the transport of momentum in the x direction across a surface with velocity v_n perpendicular to the surface, the convective momentum flux is,

$$\text{Convective flux of } x \text{ momentum} = (\rho v_x) v_n, \qquad (2.1.3)$$

where the momentum density in the x direction, ρv_x, is the product of the mass density and the flow velocity. Similar expressions can be written for the momentum density in the y and z directions.

---

[1]Formally, the change in $\rho C_p T$ is the change in the thermodynamic enthalpy density; the change in the internal energy density is the change in $\rho C_v T$, where C_v is the specific heat at constant volume. Here, we will refer to the change in $\rho C_p T$ as the change in the thermal energy density. For liquids, the specific heats at constant pressure volume are nearly equal, so there is little distinction between the internal energy and enthalpy densities. For gases, the formulation here applies to open systems where processes occur at constant thermodynamic pressure.

In general, the convective flux of any quantity across a surface is

Convective flux of a **quantity** across a surface

$=$ Density of the **quantity** $\times$ Velocity perpendicular to the surface. (2.1.4)

2.1.2 Diffusive Flux

The diffusive flux is due to the fluctuating velocity of the molecules in the fluid. Even when there is no fluid flow, molecules in the fluid are in a state of constant motion due to thermal fluctuations. The molecular origins of diffusion will be discussed in Chapter 3; here, attention is restricted to the continuum description. The constitutive relations, Fick's law for mass diffusion (Eq. 1.5.1), Fourier's law for heat conduction (Eq. 1.5.3) and Newton's law for viscosity (Eq. 1.5.4) all have the general form,

$$\left(\begin{array}{c} \textbf{Flux of a quantity} \\ \text{(Transport per unit area} \\ \text{per unit time)} \end{array} \right) = - \left(\begin{array}{c} \text{Diffusion} \\ \text{coefficient} \end{array} \right) \times \dfrac{\left(\begin{array}{c} \text{Difference in \textbf{density}} \\ \text{of the \textbf{quantity}} \end{array} \right)}{\text{Thickness}},$$

(2.1.5)

where the quantity could be mass, momentum or energy. There is a negative sign in Eq. 2.1.5 because the flux is directed from a higher to lower density. The diffusion coefficients or diffusivities are the proportionality constants in the relationship between the flux of a quantity (amount of the quantity transferred per unit area per unit time), and the difference in the density (quantity per unit volume) divided by the thickness of the fluid across which transport is taking place. If $\mathcal{Q}$ is the dimension of the quantity, the flux has dimension $\mathcal{Q}\mathcal{L}^{-2}\mathcal{T}^{-1}$ and the density of the quantity has dimension $\mathcal{Q}\mathcal{L}^{-3}$. Therefore, all diffusion coefficients have the dimension $\mathcal{L}^2\mathcal{T}^{-1}$.

1. For mass transport, the quantity is mass, the flux is the mass flux, and the density of the quantity is the concentration (mass of solute per unit volume). For the configuration shown in Fig. 1.5(a), the flux is,

$$j_z = -\mathcal{D}\frac{\Delta c}{\Delta z},$$

(2.1.6)

where $\mathcal{D}$ is the diffusion coefficient or mass diffusivity.

2. The heat flux is related to the temperature difference by Fourier's law, Eq. 1.5.3 (Fig. 1.5(b)). The heat flux q_z in Eq. 1.5.3 can be equivalently expressed in

terms of the thermal energy density $\rho C_p T$, which is the thermal energy per unit volume of the fluid,

$$q_z = -\frac{k}{\rho C_p}\frac{\Delta(\rho C_p T)}{\Delta z} = -\alpha\frac{\Delta(\rho C_p T)}{\Delta z}. \qquad (2.1.7)$$

The proportionality constant $\alpha = (k/\rho C_p)$ is the *thermal diffusivity*, with dimension $\mathcal{L}^2 \mathcal{T}^{-1}$.

3. The Newton's law of viscosity, Eq. 1.5.4, relates the shear stress τ_{xz} (force per unit area in the x direction on the wall) to the strain rate (change in velocity per unit length across the flow) for the flow shown in Fig. 1.5(c). Eq. 1.5.4 can be expressed in terms of the difference in the x momentum density across the distance Δz, where the momentum density is ρv_x, the product of the mass density and the velocity difference,

$$\tau_{xz} = \frac{\mu}{\rho}\frac{\Delta(\rho v_x)}{\Delta z} = \nu\frac{\Delta(\rho v_x)}{\Delta z}. \qquad (2.1.8)$$

The kinematic viscosity $\nu = (\mu/\rho)$ is the *momentum diffusivity* with dimension $\mathcal{L}^2 \mathcal{T}^{-1}$. [2]

2.1.3 Dimensionless Flux

Characteristic Length and Velocity

The dimensionless groups involve a characteristic velocity and a characteristic length scale. The definitions of these are obvious for regular shapes such as a cylindrical pipe or a spherical particle, but these need to be defined carefully for irregular shapes. The characteristic velocity for an internal flow is the average velocity, which is the ratio of the volumetric flow rate $\dot{V}$ and the area of cross section A_{cs}. The characteristic length for the flow in a cylindrical pipe is the pipe diameter. For irregular shapes, the characteristic length is defined based on the area of cross section and the 'wetted

---

[2]It should be noted that there is no negative sign in Newton's law, Eq. 1.5.4, in contrast to Fick's law and Fourier's law. This is due to the difference in the convention for the definition of stress in fluid mechanics, and the definition of fluxes in heat/mass transfer. The shear stress τ_{xz} in Newton's law is defined as the force per unit area at a surface in the x direction whose outward unit normal is in the z direction. In contrast, the fluxes are defined as positive if they are directed into the volume. Therefore, the shear stress is actually the negative of the momentum flux. As we will see later in this course, this difference in convention will not affect the balance equation for the rate of change of momentum.

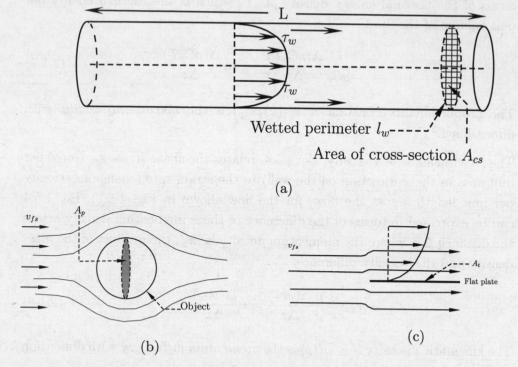

FIGURE 2.2. The area of cross-section A_{cs}, the wetted perimeter l_w and length L for the internal flow in a conduit (a); the projected area of an immersed object which is the area of cross section perpendicular to the flow A_p (b), and tangential surface area A_t for a slender body (c).

perimeter' l_w, which is the perimeter of the cross-section wetted by the fluid, as shown in Fig. 2.2(a). The hydraulic radius r_H is defined as

$$r_H = \frac{A_{cs}}{l_w},$$

(2.1.9)

and the characteristic length is the 'equivalent diameter',

$$d_e = 4r_H.$$

(2.1.10)

For a cylindrical pipe, the area of cross-section is $A_{cs} = (\pi d^2/4)$, the wetted perimeter is $l_w = \pi d$, the hydraulic radius is $r_H = (d/4)$, and the equivalent diameter is equal to the pipe diameter, $d_e = d$. The average flux is the ratio of the total transfer rate of heat or mass per unit time or the tangential force exerted on the walls, and the surface area Ll_w, the product of the length and the wetted perimeter,

$$q_{av} \text{ or } j_{av} \text{ or } \tau_{av} = \frac{\text{Total transfer rate or tangential force}}{Ll_w}, \qquad (2.1.11)$$

where L is the length of the conduit.

For an external flow around an object, the characteristic velocity is the 'free stream' velocity v_{fs}, which is the difference between the velocity of the object and the velocity far from the object. The hydraulic radius is defined as,

$$r_H = \frac{V}{A}, \qquad (2.1.12)$$

where V and A are the volume and the surface area of the object. The equivalent diameter is six times the hydraulic radius,

$$d_e = 6r_H. \qquad (2.1.13)$$

For a spherical object of diameter d, $V = (\pi d^3/6)$, $A = \pi d^2$, $r_H = (d/6)$, and the equivalent diameter is equal to the diameter of the object, $d_e = d$.

For heat/mass transfer around an object, the average flux is the ratio of the total heat/mass transfer rate and the surface area A,

$$q_{av} \text{ or } j_{av} = \frac{\text{Total transfer rate}}{A}. \qquad (2.1.14)$$

The definition of the dimensionless momentum flux is discussed in the following section on Momentum Transfer.

Heat and Mass Transfer

The dimensionless fluxes for mass and heat transfer are defined based on Fick's law (Eq. 1.5.1) and Fourier's law (Eq. 1.5.3) respectively. The driving force for mass diffusion from a catalyst particle is Δc, the difference between the concentration at the surface and in the bulk of the fluid far from the surface. Similarly, the driving force for the heat flux in the flow through a pipe is the temperature difference ΔT between the pipe wall and fluid. On this basis, the Sherwood number for mass transfer is defined as,

$$\mathrm{Sh} = \frac{j_{av}}{(\mathcal{D}\Delta c/d_e)}, \tag{2.1.15}$$

and the Nusselt number for heat transfer is defined as,

$$\mathrm{Nu} = \frac{q_{av}}{(k\Delta T/d_e)}. \tag{2.1.16}$$

The average heat and mass flux non-dimensionalised by the convection scale, $\rho C_p v_{av}\Delta T$ and $v_{av}\Delta c$, are called the Stanton number. The dimensionless fluxes are summarised in Table 2.1.

The ratio $(q_{av}/\Delta T)$ and $(j_{av}/\Delta c)$ are called the 'heat transfer coefficient' h and 'mass transfer coefficient' k_c, respectively. The Nusselt and Sherwood numbers are expressed in terms of the heat/mass transfer coefficients,

$$\mathrm{Nu} = (hd_e/k), \ \ \mathrm{Sh} = (k_c d_e/\mathcal{D}). \tag{2.1.17}$$

It is important to note that heat and mass transfer coefficients are not material properties. They do depend on the system geometry and flow conditions, and can be calculated only after solving the specific transport problem. Therefore, in this text, we shall use the Nusselt and Sherwood numbers to avoid notational complexity. In case the heat and mass transfer coefficients are necessary, they can be calculated from the Nusselt and Sherwood numbers using Eq. 2.1.17.

Momentum Transfer

The non-dimensional flux for mass/heat transfer was defined as the ratio of average flux and the characteristic diffusive flux. If the same procedure were used, the non-dimensional momentum flux would be defined by scaling the average pressure or shear stress by $(\mu v_c/l_c)$, where μ is the viscosity, and v_c and l_c are the characteristic velocity and length. Historically, the non-dimensional fluxes have been defined by scaling the average momentum flux (force per unit area) by the kinetic energy density (kinetic energy per unit volume), as discussed in Sections 1.6.1 and 1.6.3.

The definition of the dimensionless flux depends on the flow geometry. For internal flows in conduits, there is a force exerted at the walls due to fluid friction. The friction factor is defined as the ratio of the average stress, τ_{av}, and the average kinetic energy per unit volume, $(\rho v_{av}^2/2)$, where v_{av} is the average flow velocity,

$$f = \frac{\tau_{av}}{\rho v_{av}^2/2}. \tag{2.1.18}$$

Here, the average shear stress τ_{av} is defined in Eq. 2.1.11.

TABLE 2.1. Dimensionless ratios of convection and diffusion, different types of diffusion and dimensionless fluxes. Here, v_c is the characteristic velocity (average velocity for internal flows and free stream velocity for external flows) and l_c is the characteristic length scale, which is the equivalent diameter for irregular shaped objects or conduits of irregular cross section.

Dimensionless group	Expression	Ratio

Ratio of convection & diffusion

Reynolds number	$\frac{\rho v_c l_c}{\mu}$	$\dfrac{\text{Momentum convection}}{\text{Momentum diffusion}}$
Peclet number (heat)	$\frac{v_c l_c}{\alpha}$	$\dfrac{\text{Heat convection}}{\text{Heat diffusion}}$
Peclet number (mass)	$\frac{v_c l_c}{\mathcal{D}}$	$\dfrac{\text{Mass convection}}{\text{Mass diffusion}}$

Ratio of different kinds of diffusion

Prandtl number	$\frac{\nu}{\alpha}$	$\dfrac{\text{Momentum diffusion}}{\text{Thermal diffusion}}$
Schmidt number	$\frac{\nu}{\mathcal{D}}$	$\dfrac{\text{Momentum diffusion}}{\text{Mass diffusion}}$
Lewis number	$\frac{\alpha}{\mathcal{D}}$	$\dfrac{\text{Thermal diffusion}}{\text{Mass diffusion}}$

Dimensionless flux

Sherwood number	$\frac{j_{av}}{(\mathcal{D}\Delta c/l_c)}$	$\dfrac{\text{Mass flux}}{\text{Diffusive flux scale}}$
Nusselt number	$\frac{q_{av}}{(k\Delta T/l_c)}$	$\dfrac{\text{Mass flux}}{\text{Diffusive flux scale}}$
Stanton number (heat)	$\frac{q_{av}}{\rho C_p v_c \Delta T}$	$\dfrac{\text{Heat flux}}{\text{Convective flux scale}}$
Stanton number (mass)	$\frac{j_{av}}{v_c \Delta c}$	$\dfrac{\text{Mass flux}}{\text{Convective flux scale}}$
Friction factor	$\frac{\tau_{av}}{(\rho v_{av}^2/2)}$	$\dfrac{\text{Average wall stress}}{\text{Kinetic energy density}}$
Drag coefficient	$\frac{(F_D/A_p)}{(\rho v_{fs}^2/2)}$	$\dfrac{\text{Force/Projected area}}{\text{Kinetic energy density}}$
Skin friction coefficient	$\frac{(F_D/A_t)}{(\rho v_{fs}^2/2)}$	$\dfrac{\text{Force/Tangential area}}{\text{Kinetic energy density}}$

An alternate definition is in terms of the pressure difference applied across the ends to drive the flow, which is related to the wall shear stress by a force balance condition. The force on the fluid due to the shear stress at the wall is balanced by the force due to the difference in pressure at the two ends of the conduit (see Fig. 2.2(a)),

$$\tau_{av} L l_w = (-\Delta p) A_{cs}, \tag{2.1.19}$$

where L is the length of the conduit, l_w is the wetted perimeter and A_{cs} is the area of cross section. From Eq. 2.1.19, the average shear stress and pressure difference are related by,

$$\tau_{av} = -\frac{\Delta p}{L} \frac{A_{cs}}{l_w} = -\frac{\Delta p}{L} \frac{d_e}{4}, \tag{2.1.20}$$

where d_e, the equivalent diameter, was defined in Eqs. 2.1.9 and 2.1.10. Substituting the above expression for τ_{av} in Eq. 2.1.18, we obtain the friction factor in terms of the pressure gradient $(\Delta p / L)$,

$$f = -\frac{\Delta p}{L} \frac{d_e}{2\rho v_{av}^2}. \tag{2.1.21}$$

The friction factor defined in Eqs. 2.1.18 and 2.1.21 is the 'Fanning friction factor'. The Darcy friction factor f_D, which is defined as

$$f_D = -\frac{\Delta p}{L} \frac{d_e}{(\rho v_{av}^2 / 2)}, \tag{2.1.22}$$

is four times the Fanning friction factor. In this text, the Fanning friction factor defined in Eq. 2.1.18 will be used.

In the case of flow past suspended objects such as spherical particles, there is substantial curvature of flow streamlines around the object as shown in Fig. 2.2(b). The drag force is due to the pressure difference between the upstream and downstream halves of the object, and due to the viscous drag exerted by the fluid at the surface of the object. Correlations are written for the 'drag coefficient',

$$\boxed{c_D = \frac{(F_D / A_p)}{(\rho v_{fs}^2 / 2)}.} \tag{2.1.23}$$

Here, F_D is the drag force, A_p is the projected area of the object perpendicular to the flow shown in Fig. 2.2(b), the 'free stream' velocity v_{fs} is the velocity of the object relative to the fluid, and $(\rho v_{fs}^2 / 2)$ is the fluid kinetic energy per unit volume based on v_{fs}.

For 'slender' objects, such as a flat plate of infinitesimal thickness shown in Fig. 2.2(c), there is very little obstruction of the flow due to the object, and very little deformation of the fluid streamlines. However, there is a tangential stress exerted at the surface, and a resultant drag force. The average tangential stress is the ratio of the drag force F_D and the surface area of the plate parallel to the flow streamlines, A_t. The skin friction coefficient is,

$$c_f = \frac{(F_D/A_t)}{(\rho v_{fs}^2/2)}, \tag{2.1.24}$$

where the free stream velocity v_{fs} is the relative velocity between the fluid and the object. The dimensionless momentum fluxes for different configurations are summarised in Table 2.1, and the correlations for the friction factor, the drag coefficient and skin friction coefficient are discussed in Section 2.2.

2.1.4   Independent Dimensionless Groups

Based on the discussion in Section 2.1.1 and 2.1.2, there are two types of dimensionless groups: the ratio of convection and diffusion and the ratio of two different types of diffusion. The dimensionless ratio of convection and diffusion is the ratio of $(v_c l_c)$ divided by the corresponding diffusivity, where v_c and l_c are the characteristic velocity and length. The different ratios of convection and diffusion and two different types of diffusion are listed in Table 2.1.

2.1.5   Other Dimensionless Groups

The ratio of inertial and gravitational forces is given by the Froude number,

$$\mathrm{Fr} = \sqrt{\frac{v_c^2}{g l_c}}, \tag{2.1.25}$$

where v_c and l_c are the characteristic velocity and length, and g is the gravitational acceleration. In applications involving rotation of fluids, the Froude number is the ratio of centrifugal and gravitational forces.

There are two important dimensionless groups involving surface tension γ (dimension $\mathcal{M}\mathcal{T}^{-2}$), which are the Weber number and the Capillary number. The Weber number is the ratio of inertial and surface tension forces,

$$\mathrm{We} = \frac{\rho v_c^2 l_c}{\gamma}, \tag{2.1.26}$$

and the Capillary number is the ratio of viscous and surface tension forces,

$$\mathrm{Ca} = \frac{\mu v_c}{\gamma}. \tag{2.1.27}$$

When a fluid interface is curved in a gravitational field, the ratio of gravitational and surface tension forces is given by the Bond number,

$$\text{Bo} = \frac{\rho g l_c^2}{\gamma}. \tag{2.1.28}$$

Summary (2.1)

1. The convective flux of a quantity (mass, momentum, energy) across a surface is v_n times the density of that quantity, where v_n is the velocity perpendicular to the surface.

2. Eq. 2.1.5 provides the general form for the diffusive flux across a surface, where the diffusion coefficients are the mass diffusivity $\mathcal{D}$ for mass transfer, the thermal diffusivity $\alpha = (k/\rho C_p)$ for heat transfer, and the kinematic viscosity $\nu = (\mu/\rho)$ for momentum transfer. Here, ρ is the mass density, μ is the viscosity, k is the thermal conductivity and C_p is the specific heat.

3. For an internal flow in a conduit, the characteristic velocity is the average flow velocity, the ratio of the flow rate and the area of cross section. The characteristic length is the equivalent diameter based on the area of cross section and the wetted perimeter, Eqs. 2.1.9–2.1.10. The average flux is ratio of the transfer rate and the wetted surface area, Eq. 2.1.11.

4. For an external flow around an object, the characteristic velocity is the free-stream velocity, the difference in velocity between the object and the fluid far from the object. The characteristic length is the equivalent diameter, Eqs. 2.1.12–2.1.13. The average flux for heat/mass transfer is the ratio of the transfer rate and the surface area of the object, Eq. 2.1.14.

5. The dimensionless ratio of convection and diffusion are the Peclet number for mass/heat transfer and the Reynolds number for momentum transfer listed in Table 2.1.

6. The dimensionless ratios of diffusivities are the Prandtl number (momentum/ thermal), Schmidt number (momentum/mass) and Lewis number (thermal/ mass) listed in Table 2.1.

7. The dimensionless mass and heat flux, the Sherwood number, Eq. 2.1.15, and Nusselt number, Eq. 2.1.16, respectively, are defined by scaling the average flux by the characteristic diffusive flux $(\mathcal{D}\Delta c/d_e)$ and $(k\Delta T/d_e)$. Here, d_e is the

equivalent diameter, and Δc and ΔT are the concentration and temperature differences that drive the fluxes.

8. The dimensionless momentum flux for internal flows, the friction factor f, Eq. 2.1.18, is the ratio of the average wall shear stress and kinetic energy per unit volume.

9. For the flow past a particle, the drag coefficient c_D (Eq. 2.1.23) is the ratio of the drag force per unit area perpendicular to the flow, (F_D/A_p), divided by the kinetic energy per unit volume. The skin friction coefficient c_f (Eq. 2.1.24) for slender objects is the ratio of the drag force per unit area tangential to the surface, (F_D/A_t) and the kinetic energy per unit volume.

2.2 Momentum Transfer

2.2.1 Flow Characteristics

Fluid flows occur in one of two disparate forms: laminar and turbulent flows. The distinction between laminar and turbulent flows is an important one in the context of transport processes, because the flow properties and the transport rates in laminar and turbulent flows are very different. There is a transition from laminar to turbulent flow when the Reynolds number increases above a critical value, called the transition Reynolds number.

A laminar flow is characterised by smooth streamlines, and transport across the streamlines occurs due to molecular diffusion. There are large instantaneous velocity fluctuations in all directions in a turbulent flow. The flow contains 'eddies' which are parcels of fluids in correlated motion moving in all directions. Here, the cross-stream transport of mass, momentum and energy occurs due to the convective transport by turbulent eddies moving across the flow. Since the eddy transport mechanism is significantly more efficient than the molecular mechanism, the heat and mass transfer coefficients in a turbulent flow are much larger than those in a laminar flow. For the same flow rate, the fluid energy dissipation rate and the power required to drive a turbulent flow are also much higher than those for a laminar flow.

The patterns in four archetypal geometries—that is, the flow through conduits, flow past a flat plate, flow past a suspended particle and flow through a

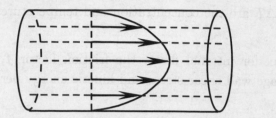 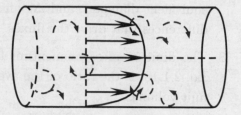

FIGURE 2.3. Laminar (left) and turbulent (right) flows in a pipe.

packed column—are discussed, along with a summary of the momentum transfer correlations at low and high Reynolds number.

2.2.2 Flow through a Pipe

There is a sharp transition from laminar to turbulent flow when the Reynolds number, $(\rho v_{av} d / \mu)$, exceeds about 2100 in a pipe. It is shown in Section 6.2 (Chapter 6) that the velocity profile for the laminar flow in a cylindrical pipe is parabolic. In the turbulent flow, the profile of the average velocity is more plug-like, as shown in Fig. 2.3, with a smaller curvature at the centre and a larger slope near the wall. Though the mean velocity is along the axis, the instantaneous velocity exhibits large fluctuations in all directions in a turbulent flow. The transition and turbulence in a pipe are discussed in further detail in Section 6.3 (Chapter 6).

For conduits of non-circular cross section, the transition Reynolds number depends on the shape of the conduit, but the transition process is qualitatively similar to that in a cylindrical pipe. The transition is abrupt, and the friction factor changes discontinuously at the transition Reynolds number. The mean velocity profile for the turbulent flow is flatter near the centre and steeper at the walls, and the instantaneous velocity exhibits large fluctuations in all directions. The velocity profile for a laminar flow in a channel of rectangular cross section is calculated using the method of separation of variables in Example 8.1.1 (Chapter 8).

Friction Factor: Low Reynolds number

The wall shear stress in the flow through a pipe is independent of the fluid density when the Reynolds number is below the transition Reynolds number. From dimensional analysis, the average wall shear stress is necessarily of the form $\tau_{av} \propto (\mu v_{av}/d)$. The friction factor, Eq. 2.1.18, is,

$$f \propto \frac{\mu v_{av}/d}{\rho v_{av}^2/2} = \frac{\text{Constant}}{\text{Re}}. \qquad (2.2.1)$$

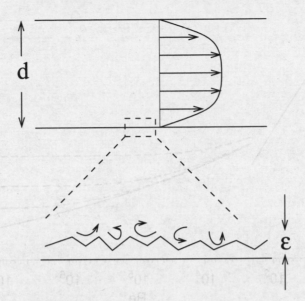

FIGURE 2.4. The wall roughness in a pipe.

Thus, the friction factor is proportional to the inverse of the Reynolds number in the limit of low Reynolds number. It is shown in Section 6.2 (Chapter 6) that the friction factor is (16/Re) for the flow in a cylindrical pipe. For the flow in a channel of height d and infinite extent perpendicular to the plane of the flow, the friction factor is (12/Re). The value of the constant is different for conduits of different cross sections, but the friction factor is proportional to the inverse of the Reynolds number for a laminar flow in all cases.

Friction Factor: High Reynolds number

Correlations for the friction factor for turbulent flows in channels and tubes have so far been obtained by fitting the data from experiments, and have not yet been derived from the flow equations. The friction factor in a turbulent flow does depend on the Reynolds number as well as the wall roughness. This is because rough walls generate complicated patterns, which result in greater energy dissipation in comparison to a smooth wall, as shown in Fig. 2.4. The wall roughness is quantified by a wall roughness coefficient ε which is the characteristic feature size of the rough wall.

The Moody diagram, which is a plot of the friction factor vs. the Reynolds number on logarithmic axes, can be used to determine the friction factor for a given Reynolds number and pipe roughness. The Moody diagram for the Fanning friction factor is given in Fig. 2.5. The low Reynolds number relation $f = (16/\text{Re})$ is a straight

Fundamentals of Transport Processes

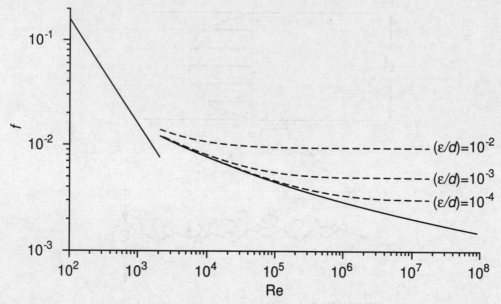

FIGURE 2.5. The Moody diagram[8] for the Fanning friction factor vs. Reynolds number for a pipe flow for different wall roughness coefficients. The solid line on the left is the friction factor for a laminar flow $f = (16/\text{Re})$, the solid line on the right is the friction factor for a smooth pipe, and the dashed lines are for different different values of the wall roughness.

line of slope -1 in the Moody diagram[8][3], when the Reynolds number is below the transition Reynolds number of 2100. At high Reynolds numbers, the friction factor exhibits a much slower decrease with Reynolds number, and the friction factor increases as the wall roughness is increased.

A useful, though approximate, correlation for the friction factor in a pipe is the Colebrook correlation,

$$\frac{1}{\sqrt{f}} = -1.737 \ln\left(\frac{1.25}{\text{Re}\sqrt{f}} + \frac{\varepsilon}{3.7d}\right). \qquad (2.2.2)$$

The logarithm in Eq. 2.2.2 is the natural logarithm; the logarithm referenced to the base 10 is usually used for writing the Colebrook correlation, in which case the prefactor of the logarithm on the right side is -4.0. The equation for a smooth pipe is obtained by setting $\varepsilon = 0$ in Eq. 2.2.2. The Colebrook correlation is applicable for a turbulent flow in a cylindrical pipe when the Reynolds number is greater than about 4000. Though the Colebrook correlation appears complicated, it is actually an explicit relation for the average flow velocity if the pressure drop is specified. If

---

[3]Note that the Moody diagram is usually plotted for the Darcy friction factor, whereas Fig. 2.5 is plotted for the Fanning friction factor from the Colebrook correlation.

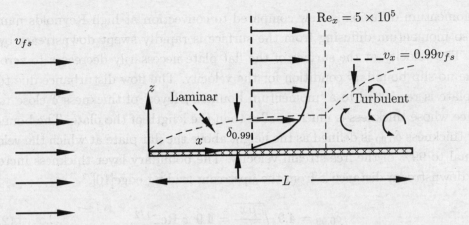

FIGURE 2.6. Flow past a flat plate.

Eq. 2.1.21 and the definition $\mathrm{Re} = (\rho v_{av} d/\mu)$ are substituted for the friction factor and the Reynolds number, the Colebrook correlation[9] is,

$$v_{av}\sqrt{\frac{2\rho}{d(\Delta p/L)}} = -1.737 \ln\left(\frac{1.77\mu}{\sqrt{\rho d^3(\Delta p/L)}} + \frac{\varepsilon}{3.7d}\right). \tag{2.2.3}$$

The Reynolds number based on the average velocity is,

$$\mathrm{Re} = \frac{\rho v_{av} d}{\mu} = -\frac{1.228\sqrt{\rho d^3(\Delta p/L)}}{\mu}\ln\left(\frac{1.77\mu}{\sqrt{\rho d^3(\Delta p/L)}} + \frac{\varepsilon}{3.7d}\right). \tag{2.2.4}$$

2.2.3 Flow past a Flat Plate

The flow past a thin flat plate is an example where the area projected by a suspended object perpendicular to the flow is small, and the force on the object is due to the shear stress exerted on surface parallel to the flow direction. For the uniform flow past a flat plate, shown in Fig. 2.6, the characteristic length is the total length of the plate L, while the characteristic velocity is the constant free-stream velocity far from the plate v_{fs}.

The Reynolds and Peclet numbers are defined in two ways. For calculating the total drag force and transport rates, the average Reynolds and Peclet numbers are based on the total length L and the free stream velocity v_{fs}. The Reynolds number is $(\rho v_{fs}L/\mu)$, the Peclet number for heat transfer is $(v_{fs}L/\alpha)$, and that for mass transfer is $(v_{fs}L/\mathcal{D})$. At a specific downstream location x from the leading edge of the plate, the 'local Reynolds number' is defined based on the distance x, $\mathrm{Re}_x = (\rho v_{fs}x/\mu)$.

Momentum diffusion is slow compared to convection at high Reynolds number, and so momentum diffusing from the surface is rapidly swept downstream by the flow. The velocity at the surface of the flat plate necessarily decreases to zero due to the no-slip boundary condition for the velocity. The flow disturbance due to the flat plate is restricted to a 'momentum boundary layer' of thickness δ close to the surface whose thickness is much smaller than the length of the plate. The boundary layer thickness $\delta_{0.99}$ is defined as the height above the flat plate at which the velocity is equal to 99% of the free-stream velocity. The boundary layer thickness increases with downstream distance x from the upstream leading edge[10],

$$\delta_{0.99} = 4.9 \sqrt{\frac{\mu x}{\rho v_{fs}}} = 4.9 \ x \ \mathrm{Re}_x^{-1/2}, \tag{2.2.5}$$

for a laminar flow. There is a transition from a laminar to a turbulent flow at a downstream location x where the Reynolds number $(\rho v_{fs} x / \mu)$ is 5×10^5. After transition, the turbulent boundary layer thickness increases as,

$$\delta_{0.99} = 0.37 \ x \ \mathrm{Re}_x^{-1/5}. \tag{2.2.6}$$

Correlations

For a plate of length L along the flow direction, width W perpendicular to the flow plane and infinitesimal thickness, the average stress is the total force exerted by the fluid divided by the area of the plate $A_t = (LW)$ in Eq. 2.1.24. For a laminar flow, the correlation for the skin friction coefficient (Eq. 2.1.24) can be calculated using boundary layer theory[10],

$$\boxed{c_f = \frac{1.328}{\mathrm{Re}^{1/2}}.} \tag{2.2.7}$$

If the plate sufficiently long that the flow is turbulent over most of the plate surface, the empirical correlation for the skin friction coefficient is,

$$\boxed{c_f = \frac{0.074}{\mathrm{Re}^{1/5}},} \tag{2.2.8}$$

for $\mathrm{Re} < 10^7$.

EXAMPLE 2.2.1: Consider the flow of air of density 1.25 kg/m^3 and viscosity 1.8×10^{-5} kg/m/s over a flat plate with free stream velocity $v_{fs} = 2$ m/s (Fig. 2.6). At what downstream distance L_t does the flow become turbulent? What is the boundary layer thickness at this downstream distance? What is the drag force per unit width of the plate perpendicular to the flow direction if the plate length is much smaller and much larger than L_t?

Solution: The flow becomes turbulent when the Reynolds number $(\rho v_{fs} L_t / \mu) = 5 \times 10^5$,

$$\frac{1.25 \text{ kg/m}^3 \times 2 \text{ m/s} \times L_t}{1.8 \times 10^{-5} \text{ kg/m/s}} = 5 \times 10^5 \Rightarrow L_t = 3.6 \text{ m}. \tag{2.2.9}$$

The boundary layer thickness $\delta_{0.99}$ at which the velocity reaches 99% of its free-stream value is

$$\delta_{0.99} = 4.9 \times L_t \times \text{Re}^{-1/2} = 4.9 \times (3.6 \text{ m}) \times (5 \times 10^5)^{-1/2} = 2.5 \times 10^{-2} \text{m}. \tag{2.2.10}$$

Thus, the boundary layer thickness is approximately 2.5 cm at a distance 3.6 m from the leading edge of the plate.

When the length is much smaller than 3.6 m, the flow is laminar. The skin friction coefficient is given by Eq. 2.2.7. Substituting the definition $(\rho v_{fs} L / \mu)$ for the Reynolds number and Eq. 2.1.24 for the skin friction coefficient, we obtain,

$$\frac{F_D}{W} = \frac{\rho v_{fs}^2 L}{2} \times \frac{1.328 \mu^{1/2}}{(\rho v_{fs} L)^{1/2}} = 8.9 \times 10^{-3} L^{1/2} \text{ N/m}, \tag{2.2.11}$$

where L is the length in meters. The force per unit width perpendicular to the flow direction increases proportional to $L^{1/2}$ for a laminar flow.

When the length is much larger than 3.6 m, the flow is turbulent. The skin friction coefficient is given by Eq. 2.2.8. The equation for the force is,

$$\frac{F_D}{W} = \frac{\rho v_{fs}^2 L}{2} \times \frac{0.074 \mu^{1/5}}{(\rho v_{fs} L)^{1/5}} = 1.73 \times 10^{-2} L^{4/5} \text{ N/m}, \tag{2.2.12}$$

where L is the length in meters. The force per unit width perpendicular to the plane of flow increases proportional to $L^{4/5}$ in this case. □

2.2.4 Flow past a Sphere

The flow pattern around a sphere has a complicated dependence on the Reynolds number[11, 12]. For Reynolds number in the range 0–1.5, the streamlines of the flow around the sphere are smooth and symmetric about the mid-plane that separates the upstream and downstream hemispheres, as shown in Fig. 2.7(a). There are

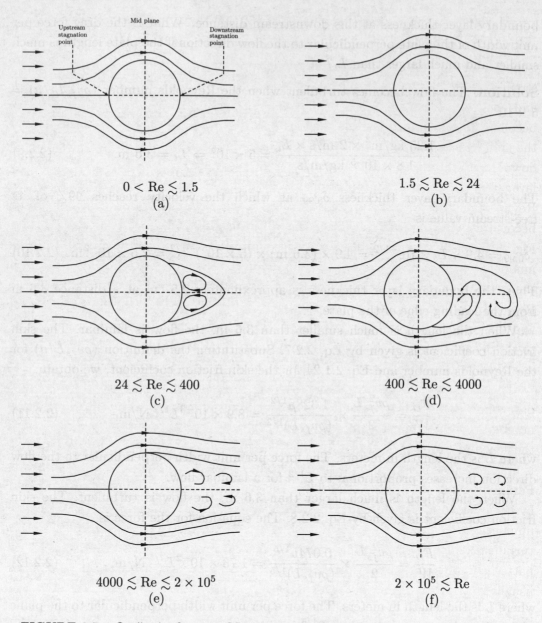

FIGURE 2.7. Qualitative features of flow past a sphere for different Reynolds number ranges.

two 'stagnation' points at the upstream and downstream ends of the sphere where the velocity relative to the sphere is zero. In this regime, the drag force is due to viscous stresses at the surface, and is well approximated by Stokes' drag law, Eq. 1.6.11. For $1.5 \lesssim \text{Re} \lesssim 24$, the symmetry about the mid-plane between upstream and downstream hemispheres is broken, but the streamlines encircle the sphere, as shown in Fig. 2.7(b). As the Reynolds number is increased in the range 24–400, the external flow 'detaches' from the sphere surface at a location slightly downstream of the plane bisecting the upstream and downstream hemispheres, and there is a 'separation bubble' on the downstream side within which there are closed circulating streamlines, as shown in Fig. 2.7(c). The separation bubble is steady for Re $\lesssim$ 130, and small oscillations are observed at 130 $\lesssim$ Re $\lesssim$ 400. For Re > 400, the flow becomes unsteady, and there is the cyclic shedding of 'vortices' in a 'wake' behind the sphere, as shown in Fig. 2.7(d). On the upstream side, the incident flow slows down due to the viscous friction at the sphere surface within a 'momentum boundary layer' whose thickness is proportional to $Re^{-1/2}$ times the sphere diameter. As the Reynolds number is further increased in the range 4000–2×10^5, the vortex shedding becomes less regular, and the flow in the wake region becomes turbulent, as shown in Fig. 2.7(e). The transition to a turbulent wake is a gradual process as the Reynolds number is increased to a few thousands. For Re $\lesssim 2 \times 10^5$, the boundary layer on the upstream hemispherical surface is laminar, as shown in Fig. 2.7(e). When the Reynolds number increases beyond 2×10^5, there is a laminar-turbulent transition and the flow in the boundary layer is also turbulent. The attachment point moves to the downstream hemisphere resulting in a decrease in the size of the wake region, as shown in Fig. 2.7(f).

The characteristic features of the flow past a sphere are also observed for the flow past 'bluff' bodies, such as the flow past a cylinder or other shapes. At low Reynolds number, there is symmetry about the mid-plane through the object perpendicular to the flow direction if the object is symmetric. As the Reynolds number increases, there is first a steady asymmetric flow with a separation bubble on the downstream side. Periodic vortex shedding occurs when the Reynolds number is increased, followed by irregular vortex shedding and a transition to a turbulent wake. As the Reynolds number is further increased, the boundary layer at the surface of the object becomes turbulent. Though the Reynolds number ranges of the different characteristic flow patterns do depend on the specific configurations, the sequence of flow patterns is usually observed.

Low Reynolds number

The drag coefficient for an immersed particle moving in a fluid was defined in Eq. 2.1.23. For a particle settling in a fluid under conditions of low Reynolds number, it was shown in Section 1.6.1 that the drag force is proportional to $\mu v_{fs} d$, where v_{fs} is the free stream velocity (particle velocity relative to the fluid). Substituting this into Eq. 2.1.23 for the drag coefficient, we obtain,

$$c_D \propto \frac{(\mu v_{fs} d)}{(\pi d^2/4)(\rho v_{fs}^2/2)} = \frac{\text{Constant}}{\text{Re}}. \qquad (2.2.13)$$

As indicated in Section 1.6.1, detailed calculations show that the constant in the Stokes' drag law (Eq. 1.6.15) is 3π for a spherical particle. Therefore, the correlation for the drag coefficient is

$$c_D = \frac{24}{\text{Re}}. \qquad (2.2.14)$$

High Reynolds number

The drag coefficient is shown as a function of the Reynolds number in Fig. 2.8. The correlation, Eq. 2.2.14, for low Reynolds number is applicable for $\text{Re} \lesssim 1.5$, where the flow is symmetric between the upstream and downstream hemispheres as shown in Fig. 2.7(a). The drag coefficient departs from Eq. 2.2.14 when the flow becomes asymmetric, as shown in Fig. 2.7(b). An accurate empirical correlation for the drag coefficient for low-to-moderate values of the Reynolds number is the Schiller–Naumann correlation[13] shown by the dot-dash line in Fig. 2.8,

$$c_D = \frac{24(1 + 0.15\,\text{Re}^{0.687})}{\text{Re}}. \qquad (2.2.15)$$

This correlation is accurate for $\text{Re} \lesssim 400$, where there is an attached separation bubble at the rear of the sphere, as shown in Fig. 2.7(c).

Fig. 2.8 shows that the drag coefficient does approach a constant value of about 0.4 for Reynolds number in the range $4 \times 10^3 \lesssim \text{Re} \lesssim 2 \times 10^5$, where there is turbulence in the wake and a laminar flow in the upstream boundary layer. Here, the pressure in the wake region is much smaller than the pressure in the upstream hemisphere, and the drag is due to the pressure difference between the upstream and downstream hemispheres. There is a sharp decrease in the drag coefficient at $\text{Re} = 2 \times 10^5$, called the 'drag crisis', due to the transition from a laminar boundary layer in Fig. 2.7(e) to a turbulent boundary layer in Fig. 2.7(f). The boundary layer separation, which took place in the upstream hemisphere for a laminar boundary layer in Fig. 2.7(e), abruptly shifts to the downstream hemisphere as shown in Fig. 2.7(f). Since there is a smaller low pressure area in the rear, the average pressure on the downstream side increases relative to that on the upstream side, and the net force on the sphere decreases. This results in a sharp decrease in the drag force when there is a transition from a laminar to a turbulent boundary layer.

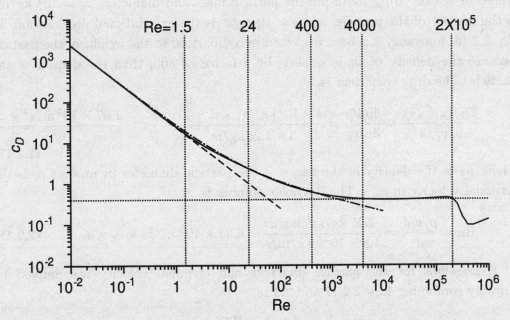

FIGURE 2.8. The drag coefficient as a function of Reynolds number for the flow around a sphere. The solid line is the correlation of Morrison[14] which fits the data for Reynolds number up to 10^6. The vertical dotted lines are the boundaries between different regimes in Fig. 2.7. The dashed line is $c_D = (24/Re)$ (Eq. 2.2.14), the horizontal dotted line is $c_D = 0.4$, and the dot-dash line is the Schiller–Naumann correlation (Eq. 2.2.15).

EXAMPLE 2.2.2: For a particle of mass density 10^3 kg/m^3 settling in air of mass density 1.25 kg/m^3 and viscosity 1.8×10^{-5} kg/m/s, estimate the diameter at the boundaries of the different regimes in Fig. 2.7, that is, Re = 1.5, 24, 400, 4000, 2 $\times$ 10^5.

Solution: At the terminal settling velocity, the drag force on the particle is equal to the weight,

$$F_D = mg = \rho_p(\pi d^3/6)g, \qquad (2.2.16)$$

where $m = (\pi d^3/6)\rho_p$ and d are the particle mass and diameter, $\rho_p = 10^3$ kg/m^3 is the density of the particle, and $g = 10$ m/s^2 is the gravitational acceleration. In Eq. 2.2.16, buoyancy has been neglected in comparison to the weight of the particle because the density of air is smaller, by a factor of 800, than the density of the particle. The drag coefficient is,

$$c_D = \frac{F_D/(\pi d^2/4)}{\rho_f v_t^2/2} = \frac{4\rho_p dg}{3\rho_f v_t^2} = \frac{4 \times 10^3 \, \text{kg/m}^3 \times d \times 10 \, \text{m/s}^2}{3 \times 1.25 \, \text{kg/m}^3 \times v_t^2} = \frac{1.07 \times 10^4 \, \text{m/s}^2 \times d}{v_t^2},$$
$$(2.2.17)$$

where ρ_f is the density of the gas, d is the particle diameter in m, and v_t is the terminal velocity in m/s. The Reynolds number is,

$$\text{Re} = \frac{\rho_f v_t d}{\mu} = \frac{1.25 \, \text{kg/m}^3 \times v_t d}{1.8 \times 10^{-5} \, \text{kg/m/s}} = 6.94 \times 10^4 \, \text{m}^{-2}\text{s} \times v_t \times d. \qquad (2.2.18)$$

Expressions for the particle diameter and terminal velocity are derived by suitably combining Eqs. 2.2.17 and 2.2.18,

$$c_D \times \text{Re}^2 = 5.14 \times 10^{13} \text{m}^{-3} \times d^3, \qquad \frac{\text{Re}}{c_D} = 6.51(\text{m}^{-3}\text{s}^3) \times v_t^3. \qquad (2.2.19)$$

The particle diameter and terminal velocity for different Reynolds numbers are provided in Table 2.2.   □

2.2.5   Flow past a Bubble

For the flow past a bubble, the flow pattern depends on the nature of the bubble surface. If there are adsorbed surfactants, these restrict the fluid mobility at the surface, and the flow pattern is similar to that in the flow past a particle. If the surface is clean, the zero stress condition is applicable at the surface instead of a no-slip condition. In this case, there is no boundary layer separation and wake formation at high Reynolds number.

TABLE 2.2. The particle diameter and settling velocity corresponding to the Reynolds numbers demarcating different regimes in Fig. 2.7.

Re	Drag correlation	c_D	d (m)	v_t (m/s)
1.5	Eq. 2.2.14	16.00	8.88×10^{-5}	0.24
24	Eq. 2.2.15	2.33	2.97×10^{-4}	1.17
400	Eq. 2.2.15	0.61	1.24×10^{-3}	4.65
4000	$c_D = 0.4$	0.40	4.99×10^{-3}	11.54
2×10^5	$c_D = 0.4$	0.40	6.77×10^{-2}	42.51

The sequence of flow patterns in Fig. 2.7 is not observed for a bubble in a pure liquid in the absence of surfactants. This is because of a difference in the boundary conditions between a solid–liquid interface and a liquid–gas interface discussed later in Section 2.3.2. The boundary condition at the surface of the bubble, which is the zero stress condition (Fig. 2.10(c)), is different from the zero velocity condition (Fig. 2.10(b)) for the flow around a particle. For a bubble, there is no boundary layer separation and wake formation of the type shown in Fig. 2.7; the streamlines are smooth and exhibit symmetry about the mid-plane even at high Reynolds number. The drag coefficient for a spherical bubble at low Reynolds number is,

$$c_D = \frac{16}{\text{Re}}, \qquad (2.2.20)$$

where Re is the Reynolds number based on the liquid density and viscosity and the bubble diameter and velocity.[4] The drag coefficient for a spherical bubble at high Reynolds number in a pure liquid in the absence of surfactants is,

$$c_D = \frac{48}{\text{Re}}. \qquad (2.2.21)$$

It should be emphasised that the correlations Eqs. 2.2.20–2.2.21 only apply to bubbles rising in pure liquids without surfactants where a zero stress condition (Fig. 2.10(c)) is applicable at the interface; if there are surfactants, the interface behaves like a rigid surface and the drag coefficient could be higher.

EXAMPLE 2.2.3: For a spherical bubble of diameter d rising due to buoyancy at low Reynolds number, express the drag coefficient and the Capillary number in

---

[4]The gas density is usually neglected in comparison to the liquid density because it is smaller by a factor of $\sim 10^{-3}$.

terms of the Archimedes and Bond numbers. For what range of diameters is the Reynolds number small, and for what range of diameters is the Capillary number small, for bubbles in water? The density and viscosity of water are 10^3kg/m^3 and 10^{-3}kg/m/s, and the air-water surface tension is 0.072 kg/s^2.

Solution: The drag force on a bubble of diameter d rising at its terminal velocity v_t in a liquid with density ρ and viscosity μ is determined from the drag coefficient, Eq. 2.2.20,

$$c_D = \frac{F_D/(\pi d^2/4)}{(\rho v_t^2/2)} = \frac{16\mu}{\rho v_t d} \Rightarrow F_D = 2\pi\mu dv_t. \tag{2.2.22}$$

The density of air is neglected in comparison to the density of water in the expression for the buoyancy force, which is the product of the mass of water displaced by the bubble, $\rho(\pi d^3/6)$ times the gravitational acceleration g. The force balance equation for a rising bubble is solved to determine the terminal velocity,

$$g\rho \times (\pi d^3)/6 - F_D = 0 \Rightarrow v_t = \frac{\rho g d^2}{12\mu}. \tag{2.2.23}$$

The condition $\text{Re} \ll 1$ is equivalent to

$$\frac{\rho d}{\mu}\frac{\rho g d^2}{12\mu} \ll 1 \Rightarrow \frac{\text{Ar}}{12} \ll 1, \tag{2.2.24}$$

where the Archimedes number $\text{Ar} = (\rho^2 g d^3/\mu^2)$. For bubbles in water, the diameter range is,

$$d \ll \left[\left(\frac{12\mu^2}{\rho^2 g} \right)^{1/3} = \left(\frac{12 \times (10^{-3} \text{ kg/m/s})^2}{(10^3 \text{ kg/m}^3)^2 \times 10 \text{ m/s}^2} \right)^{1/3} = 1.06 \times 10^{-4} \text{ m} \right]. \tag{2.2.25}$$

Thus, the Reynolds number is low for bubbles of diameter less than about 100 microns. From Eqs. 2.2.22, and 2.2.23, the drag coefficient is,

$$c_D = \frac{16\mu}{\rho d}\frac{12\mu}{\rho g d^2} = \frac{192}{\text{Ar}}. \tag{2.2.26}$$

The Capillary number, Eq. 2.1.27, is,

$$\text{Ca} = \frac{\mu v_t}{\gamma} = \frac{\mu}{\gamma}\frac{\rho g d^2}{12\mu} = \frac{\text{Bo}}{12}, \tag{2.2.27}$$

where the Bond number is defined in Eq. 2.1.28. Bubble deformation is small when the Capillary number is small,

$$d \ll \left[\sqrt{\frac{12\gamma}{\rho g}} = \sqrt{\frac{12 \times 0.072 \text{ kg/s}^2}{10^3 \text{ kg/m}^3 \times 10 \text{ m/s}^2}} = 9.29 \times 10^{-3} \text{ m} \right]. \tag{2.2.28}$$

The limit on the diameter for low Reynolds number, Eq. 2.2.25, is more restrictive than Eq. 2.2.28 for low Capillary number. It can be concluded that the deformation is always small in the limit of low Reynolds number for bubbles with diameter less than about 100 microns. $\qquad\qquad\square$

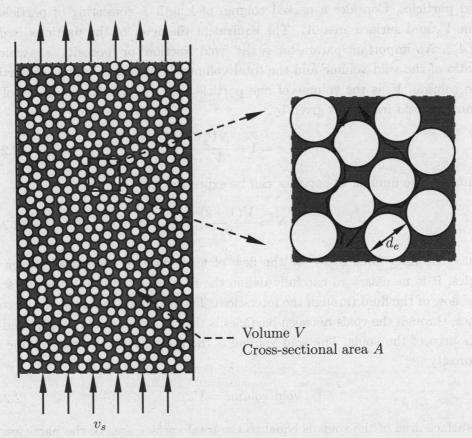

Volume V
Cross-sectional area A

v_s

FIGURE 2.9. Flow through a packed column of volume V and area of cross section A and porosity (ratio of void volume and total volume) ε consisting of particles of equivalent diameter d_e. The fluid superficial velocity (flow rate divided by area of cross section) is v_s.

2.2.6 Packed Column

Unit operations for processes such as adsorption, filtration and chromatography involve the flow through a densely packed column of particles, as shown in Fig. 2.9. The particle equivalent diameter d_e is typically much smaller than the diameter of the column. The volume fraction for the particles could range from 0.64 for randomly packed spherical particles to as low as 0.2 for odd shaped particles. The large area of contact between the fluid and densely packed particles enhances processes such as adsorption and mass/heat transport. However, there is also a significant increase in the pressure drop required to drive the flow, because the larger area of contact with the particles increases the frictional resistance to flow.

Friction Factor

The friction factor is defined for the flow of the fluid in the gaps between densely packed particles. Consider a packed column of length L consisting of particles of volume V_p and surface area A_p. The equivalent diameter of the particles is $d_e = (6V_p/A_p)$. An important parameter is the 'void fraction' or 'porosity' ε, which is the ratio of the void volume and the total volume. If N is the number of particles in the column, V_p is the volume of one particle and V is the total volume of the column, the void fraction is given by,

$$\varepsilon = 1 - \frac{NV_p}{V}. \tag{2.2.29}$$

Alternately, the number of particles can be expressed as,

$$N = \frac{V(1-\varepsilon)}{V_p}. \tag{2.2.30}$$

Since the drag force is due to the flow of fluid through the gaps between the particles, it is necessary to carefully define the relevant length and velocity scales for the flow of the fluid through the interstices. The hydraulic radius of the tortuous channels through the voids between particles is the ratio of the void volume and the surface area of the voids. The void volume is the product of the total volume and the porosity,

$$\text{Void volume} = V\varepsilon. \tag{2.2.31}$$

The surface area of the voids is equal to the total surface area of the particles,

$$\text{Void surface area} = NA_p = \frac{V(1-\varepsilon)A_p}{V_p}. \tag{2.2.32}$$

The above equation has been simplified using Eq. 2.2.30 for N. The hydraulic radius for the voids is the ratio of the volume and surface area,

$$r_H^{void} = \frac{\text{Void volume}}{\text{Void surface area}} = \frac{\varepsilon V_p}{(1-\varepsilon)A_p}. \tag{2.2.33}$$

The equivalent diameter for the flow through the voids between the particles is defined as $d_e^{void} = 6r_H^{void}$,

$$d_e^{void} = \frac{6\varepsilon V_p}{(1-\varepsilon)A_p} = \frac{\varepsilon d_e}{1-\varepsilon}. \tag{2.2.34}$$

Note that d_e^{void} is the equivalent diameter for the void spaces between the particle, and $d_e = (6V_p/A_p)$ (Eq. 2.1.13) is the equivalent diameter of the particles.

The characteristic velocity is the velocity of the fluid through the voids between the particles. This is usually expressed in terms the 'superficial velocity' v_s, which is the hypothetical velocity through a column with the same area of cross section if there were no particles present. Since the area of cross section with particles present is ε times the area of cross section if there were no particles, the average flow velocity through the column is higher than the superficial velocity by a factor $(1/\varepsilon)$ due to volume conservation,

$$v^{void} = (v_s/\varepsilon). \tag{2.2.35}$$

The Reynolds number is defined using the equivalent diameter d_e^{void} and the characteristic velocity v^{void} of the voids, $\text{Re} = (\rho v^{void} d_e^{void}/\mu)$. Using Eq. 2.2.34 for d_e^{void} and Eq. 2.2.35 for v^{void}, the Reynolds number is,

$$\text{Re} = \frac{\rho v^{void} d_e^{void}}{\mu} = \frac{\rho v_s d_e}{(1-\varepsilon)\mu}. \tag{2.2.36}$$

The friction factor for packed columns is defined by scaling the pressure drop per unit length, $(-\Delta p/L)$, by the inertial scale $(\rho(v^{void})^2/d_e^{void})$.[5] Substituting Eq. 2.2.34 for d_e^{void} and Eq. 2.2.35 for v^{void}, the definition of the friction factor is,

$$f = -\frac{\Delta p}{L}\frac{d_e^{void}}{\rho(v^{void})^2} = -\frac{\Delta p}{L}\frac{\varepsilon^3 d_e}{\rho v_s^2(1-\varepsilon)}. \tag{2.2.37}$$

---

[5]The pressure gradient is divided by $(\rho(v^{void})^2/d_e^{void})$ in the case of packed columns, in contrast to $(2\rho v_{av}^2/d)$ for flow through conduits.

Correlation

The Ergun correlation[15] relates the friction factor for the packed column, Eq. 2.2.37, to the Reynolds number Eq. 2.2.36,

$$f = \frac{150}{\mathrm{Re}} + 1.75. \tag{2.2.38}$$

The first term on the right in Eq. 2.2.38 captures the viscous resistance to flow at low Reynolds number; the pressure difference is independent of the fluid density and proportional to the fluid viscosity in this limit. The correlation $f = (150/\mathrm{Re})$ in the low Reynolds number limit is called the Kozeny–Carman correlation. At high Reynolds number, the difference in the pressure between the upstream and downstream hemispheres results in a drag force that is proportional to $(\rho(v^{void})^2/2)$, the kinetic energy per unit volume. The friction factor tends to a constant value of 1.75 in this limit. The correlation $f = 1.75$ in the limit of high Reynolds number is called the Burke–Plummer correlation.

EXAMPLE 2.2.4: Water of density $10^3 \mathrm{kg/m}^3$ and viscosity $10^{-3}\mathrm{kg/m/s}$ is pumped upwards through a packed column consisting of spherical particles of density $1.2 \times 10^3\mathrm{kg/m}^3$, diameter 1 cm and void fraction 0.4. Estimate the minimum superficial velocity above which the particles are fluidised. What is the Reynolds number?

Solution: The particles are fluidised when the upward force due to the pressure difference is equal to the difference between the weight of the particles and the buoyancy force due to the liquid displaced. The force balance is,

$$-\Delta p\, A = (1 - \varepsilon)\, \Delta\rho\, AgL, \tag{2.2.39}$$

where Δp is the pressure difference between the outlet and inlet, A is the area of cross section of the bed, $-\Delta p\, A$ is the upward force on the particles, $\Delta\rho$ is the difference in density between the particles and liquid, L is the length of the bed, ε is the void fraction, g is the gravitational acceleration, and $((1 - \varepsilon)\,\Delta\rho gAL)$ is difference between the weight of the particles and the buoyancy force. Eq. 2.2.39 is simplified as,

$$-\frac{\Delta p}{L} = (1 - \varepsilon)\Delta\rho g. \tag{2.2.40}$$

Eq. 2.2.37 is used to express the left side of the above equation in terms of the friction factor,

$$\frac{f\rho v_s^2(1-\varepsilon)}{\varepsilon^3 d_p} = (1-\varepsilon)\Delta\rho\, g. \tag{2.2.41}$$

The above equation is simplified, and the correlation, Eq. 2.2.38, is substituted for the friction factor,

$$\left(\frac{150(1-\varepsilon)\mu v_s}{\rho d_p^2 \varepsilon^3} + \frac{1.75 v_s^2}{\varepsilon^3 d_p}\right) = \frac{\Delta\rho g}{\rho}. \tag{2.2.42}$$

Numerical values of the void fraction, particle diameter and fluid and particle properties are substituted in Eq. 2.2.42,

$$\frac{150 \times 0.6 \times (10^{-3}\ \text{kg/m/s}) \times v_s}{10^3\ \text{kg/m}^3 \times (10^{-2}\ \text{m})^2 \times (0.4)^3} + \frac{1.75 v_s^2}{(0.4)^3 \times 10^{-2}\ \text{m}} = \frac{2 \times 10^2\ \text{kg/m}^3 \times 10\ \text{m/s}^2}{10^3\ \text{kg/m}^3}. \tag{2.2.43}$$

The above quadratic equation is solved to determine the maximum superficial velocity, $v_s = 0.0246$ m/s. The Reynolds number, Eq. 2.2.36, is

$$\text{Re} = \frac{10^3\ \text{kg/m}^3 \times 0.0246\ \text{m/s} \times 0.01\ \text{m}}{0.6 \times 10^{-3}\ \text{kg/m/s}} = 410. \tag{2.2.44}$$

$\square$

2.2.7 Mixing

Dimensional analysis plays an important role in industrial scale-up. While scaling up a process from the laboratory scale to the pilot plant and industrial scale, it is essential to ensure that the important dimensionless groups are kept a constant. This is illustrated using the example of the impeller in a stirred-tank reactor, of the type discussed in Section 1.1.1. Here, the liquid is stirred using an impeller of a certain shape, and the impeller is to be designed such that efficient mixing is achieved with minimum power requirement. For a given impeller shape, it is necessary to estimate the power consumption for stirring at a frequency of rotation f_r. The power consumption will, in general, depend on the shape and dimension of the impeller and the vessel, as well as other details such as baffles, etc. If we keep the ratios of the sizes of the impeller, vessel, baffles, etc. a constant between the actual reactor and the scale model, then there is only one length scale in the problem, which is considered to be the impeller diameter d. In addition, the power

TABLE 2.3. Relevant quantities and their dimensions for the calculation of the power required for the impeller in a reactor.

Parameter	Dimension
Power (P)	$\mathcal{ML}^2\mathcal{T}^{-3}$
Frequency (f_r)	$\mathcal{T}^{-1}$
Diameter (d)	$\mathcal{L}$
Density (ρ)	$\mathcal{ML}^{-3}$
Viscosity (μ)	$\mathcal{ML}^{-1}\mathcal{T}^{-1}$
Gravity (g)	$\mathcal{LT}^{-2}$
Surface tension (γ)	$\mathcal{MT}^{-2}$

can also depend on the density of the liquid, ρ, the viscosity μ, and the frequency of rotation f_r. The acceleration due to gravity g and the surface tension γ are also relevant parameters. This is because during stirring, the height of the liquid surface increases at the wall and decreases at the centre due to the centrifugal force, and the interface shape could depend on the gravity and the surface tension. The relevant parameters and their dimensions are shown in Table 2.3.

Dimensionless groups are most easily derived by considering ($f_r d$) as a characteristic velocity for the flow. The dependent dimensionless group is the 'Power number' ($P/f_r^3 d^5 \rho$). The independent dimensionless groups are the Reynolds number based on the diameter and the frequency of rotation, ($\rho d^2 f_r/\mu$), the Froude number $\sqrt{f_r^2 d/g}$, the ratio of centrifugal and gravitational forces, and the Weber number ($\rho f_r^2 d^3/\gamma$) the ratio of centrifugal and surface tension forces. The correlation for the power number has the form,

$$\frac{P}{\rho f_r^3 d^5} = \text{Function}\left(\frac{\rho d^2 f_r}{\mu}, \sqrt{\frac{f_r^2 d}{g}}, \frac{\rho f_r^2 d^3}{\gamma}\right). \qquad (2.2.45)$$

The above scaling of the power is appropriate when the Reynolds number is large and inertial forces are dominant. At low Reynolds number when viscous forces dominate, the power required depends on the fluid viscosity and is independent of the density. The scaled power is ($P/(\mu f_r^2 d^3)$), which is the product of the power number and the Reynolds number $\text{Re} = (\rho d^2 f_r/\mu)$.

EXAMPLE 2.2.5: Consider a reactor with an impeller of diameter 1 m designed to rotate with frequency of 1 rev/s, in which a liquid of density 10^3 kg/m^3 and viscosity 1 kg/m/s. The surface tension γ of the liquid–air interface is 0.1 kg/s^2. In order to estimate the power requirement, we design a smaller model reactor with an impeller of size 10 cm. What is the fluid that should be used, and what is the speed at which the model reactor should operate, so that all relevant dimensionless groups are the same for the actual and model reactor? How is the power requirement of the model reactor related to that of the actual reactor?

Solution: First, we determine the dimensionless groups that need to be maintained constant in the model and actual reactor. For the actual reactor, the relevant dimensionless groups are,

$$\text{Re} = \frac{\rho d^2 f_r}{\mu} = \frac{10^3 \text{ kg/m}^3 \times (1 \text{ m})^2 \times 1 \text{ s}^{-1}}{1 \text{ kg/m/s}} = 10^3,$$

$$\text{Fr} = \sqrt{\frac{d f_r^2}{g}} = \sqrt{\frac{1\text{m} \times (1\text{s}^{-1})^2}{10 \text{ m/s}^2}} = 0.316,$$

$$\text{We} = \frac{\rho d^3 f_r^2}{\gamma} = \frac{10^3 \text{kg/m}^3 \times (1 \text{ m})^3 \times (1 \text{ s}^{-1})^2}{0.1 \text{ kg/s}^2} = 10^4.$$

The Reynolds number is large, and it might simplistically be assumed that viscous effects can be neglected. However, if there is a significant contribution to the force on the impeller due to the tangential stress at the impeller surface, the Reynolds number could still be a relevant parameter. The Froude number is not too small, so gravity is a relevant parameter. The Weber number is large, so inertial effects are much larger than surface tension effects, and surface tension is not a relevant parameter. The power number is a function of the Reynolds and Froude numbers.

The speed of rotation can be determined from the consideration that the Froude number has to be the same for the model and actual reactors. The quantities for the actual reactor are denoted by the subscript a and those for the model reactor are denoted by the subscript m. The constant Froude number condition is,

$$\frac{f_{ra}^2 d_a}{g} = \frac{f_{rm}^2 d_m}{g}. \tag{2.2.46}$$

Substituting the dimensions and the frequency of the actual reactor, we obtain the impeller frequency of the model reactor as $f_{rm} = 3.16$ rev/s. The choice of fluid to be used in the model reactor is determined by the condition that the Reynolds number has to be a constant,

$$\frac{\rho_a f_{ra} d_a^2}{\mu_a} = \frac{\rho_m f_{rm} d_m^2}{\mu_m}. \tag{2.2.47}$$

Relating the frequency and diameters of the two reactors, we obtain,

$$\frac{\rho_m}{\mu_m} = 31.6 \frac{\rho_a}{\mu_a}. \tag{2.2.48}$$

The density of the liquid for the model reactor cannot be changed very much, because the density of most liquids is of the order of 10^3 kg/m^3. However, the viscosity of liquids varies over a wide range from 10^{-3}–1 kg/m/s. If the density of the fluid in the model reactor is maintained as 10^3 kg/m^3, the viscosity ratio is $(\mu_m/\mu_a) = 0.0316$.

For the model reactor, it is important to confirm that the Weber number is large, so that the surface tension is not a significant parameter. Based on the frequency $f_{rm} = 3.16$ rev/s, diameter $d_m = 0.1$ m, the Weber number $(\rho f_{rm}^2 d_m^3/\gamma)$ is 10^2. This is large, and the effect of surface tension is small for the model reactor.

Since the Reynolds and Froude numbers are the same in the two configurations, the Power number is also the same,

$$\frac{P_a}{f_{ra}^3 d_a^5} = \frac{P_m}{f_{rm}^3 d_m^5}. \tag{2.2.49}$$

Therefore, the ratio of the power required in the two configurations is

$$\frac{P_a}{P_m} = 3162. \tag{2.2.50}$$

□

Impellers are used to break up drops and bubbles in foams, emulsions and suspensions. The diameter of the droplet is usually much smaller than the dimension of the mixer or impeller, and the break-up depends only on the local conditions at the droplet scale and not the details of the large-scale flow. The droplet diameter depends on the energy dissipation rate per unit mass designated ϵ with dimension $\mathcal{L}^2\mathcal{T}^{-3}$. The droplet diameter also depends on the surface tension γ, the fluid density ρ and viscosity μ. If the Reynolds number based on the droplet diameter and velocity is large, the droplet break-up does not depend on the viscosity; it depends only on ϵ, γ and the density ρ. From dimensional analysis, the dimensionless group

$$\frac{\rho^{3/5}\epsilon^{2/5}d_d}{\gamma^{3/5}} = \text{Constant}, \tag{2.2.51}$$

where d_d is the diameter of the droplet. Experimentally, the constant on the right side of Eq. 2.2.51 is found to be around 1.1[16].

EXAMPLE 2.2.6: A mixer of characteristic dimension d_a and frequency f_{ra} is to be designed for the break-up of oil droplets in water down to a desired maximum size d_{max}. A scale model of dimension $d_m = (d_a/10)$ is used to model the break-up process. If scale model provides the correct droplet diameter d_{max} for frequency f_{rm}, what is the frequency of the actual mixer f_{ra} to obtain the same droplet diameter? What is the ratio of the power of the actual mixer P_a and the model P_m? Assume inertial forces are dominant.

Solution: If the droplet diameter d_{max} is the same between the scale model and the actual mixer, then the energy dissipation rate per unit mass ϵ in Eq. 2.2.51 is the same. The energy dissipation rate per unit mass is proportional to the ratio of the power and the mass of the reactor. The mass of the reactor is the product of the density ρ and the volume; the latter is proportional to d^3, where d is the characteristic dimension, if the ratios of all the dimensions are the same for the actual and model mixer. Therefore, the rate of dissipation of energy per unit mass is,

$$\epsilon \propto \frac{\text{Power}}{\rho d^3} \propto d^2 f_r^3, \qquad (2.2.52)$$

where f_r is the frequency of the impeller. It should be noted that the rate of dissipation of energy in the reactor is non-uniform, and it could vary depending on the location and the shape of the impeller. However, if all relevant dimensionless groups are kept constant, the relation, Eq. 2.2.52, is valid on average. Therefore, the relation between the diameter and frequency is,

$$d_m^2 f_{rm}^3 = d_a^2 f_{ra}^3 \Rightarrow \frac{f_{ra}}{f_{rm}} = \left(\frac{d_m}{d_a}\right)^{2/3} = 0.215. \qquad (2.2.53)$$

The relation between the power of the actual reactor and scale model is,

$$\frac{P_a}{P_m} = \frac{d_a^5 f_{ra}^3}{d_m^5 f_{rm}^3} = 10^3. \qquad (2.2.54)$$

□

A summary of the appropriate length and velocity scales used for the Reynolds number, and the expressions for the drag coefficient/friction factor for different types of flows is provided in Table 2.4.

Summary (2.2)

1. The definitions of the characteristic length and velocity, Reynolds number and the non-dimensional momentum flux are summarised in Table 2.4.

2. For laminar flows, the non-dimensional momentum flux is inversely proportional to the Reynolds number in all cases.

3. At high Reynolds number, the correlation depends on the flow geometry, and these are not often available in analytical form.

2.3   Heat/Mass Transfer

2.3.1   Equivalence of Mass and Heat Transfer

Empirical correlations for mass/heat transfer relate the dimensionless fluxes (Sherwood/Nusselt number) to the Reynolds number (ratio of convection and momentum diffusion) and the Schmidt/Prandtl number (ratio of momentum and mass/heat diffusion). As we shall see in this discussion, the correlations are more compactly expressed in terms of the Peclet number (ratio of mass/heat convection and diffusion) instead of the Schmidt/Prandtl number both at low Peclet number and at high Peclet number for laminar flows.

The correlations do depend on the geometry and flow configuration, and are different for flow through conduits and flow around suspended objects. A central idea in heat/mass transfer is the Chilton–Colburn analogy[17], that for a given geometry and flow pattern, the correlation is the same for heat and mass transfer. The correlation for mass transfer can be obtained by replacing the Nusselt and Prandtl numbers by the Sherwood and Schmidt numbers, respectively, in the correlation for heat transfer. This is because of the equivalence of the local constitutive relations and the the dimensionless groups for mass and heat transfer. The Fourier's law for heat conduction, Eq. 2.1.7, can be obtained from the Fick's law for diffusion, Eq. 2.1.6, by substituting the heat flux for the mass flux, the energy density for the mass density and the thermal diffusivity for the mass diffusivity. Similarly, the Nusselt number, Eq. 2.1.16, is obtained from the Sherwood number, Eq. 2.1.15, by substituting the average heat flux for the average mass flux, the thermal diffusion coefficient for the mass diffusion coefficient and the difference in energy density for the difference in concentration. The Peclet number for heat transfer can be obtained from the Peclet number for mass transfer by substituting the thermal diffusivity for the mass diffusivity. Therefore, for the same geometry and flow pattern, the correlation for heat transfer can be obtained from that for mass transfer by substituting the Nusselt number for the Sherwood number, and heat transfer Peclet number for the mass transfer Peclet number.

TABLE 2.4. The characteristic length and velocity used in the Reynolds number, the average stress/pressure and the definition of the friction factor/drag coefficient for flows in different geometries.

Type of flow	Length l_c	Velocity v_c	Average stress/pressure	Non-dimensional momentum flux	Correlations Laminar	Correlations Turbulent
Flow in a pipe	Pipe diameter d	Average velocity v_{av}	Wall shear stress τ_{av}	$f = \dfrac{\tau_{av}}{(\rho v_{av}^2/2)}$	$Re \lesssim 2100$ $f = \dfrac{16}{Re}$	$Re \gtrsim 2100$ Moody diagram Fig. 2.5
Flow in a conduit	Equivalent diameter d_e $\left(4 \times \dfrac{\text{Cross-sectional area}}{\text{Wetted perimeter}}\right)$	Average velocity v_{av} $\left(\dfrac{\text{Flow rate}}{\text{Cross-sectional area}}\right)$	Average wall shear stress τ_{av}	$f = \dfrac{\tau_{av}}{(\rho v_{av}^2/2)}$	$f \propto \dfrac{1}{Re}$	
Flow past flat plate	Plate length L	Free stream velocity v_{fs}	Average wall shear stress τ_{av}	$c_f = \dfrac{\tau_{av}}{\rho v_{fs}^2/2}$	$(Re \lesssim 5 \times 10^5)$ $c_f = 1.328\,Re^{-\frac{1}{2}}$	$Re \gtrsim 5 \times 10^5$ $c_f = 0.074\,Re^{-\frac{1}{5}}$
Flow past a spherical particle	Sphere diameter d	Sphere velocity relative to fluid v_{fs}	Drag force / Projected area $\dfrac{F_D}{A_p}$	$c_D = \dfrac{(F_D/A_p)}{(\rho v_{fs}^2/2)}$	$(Re \lesssim 1.5)$ $c_D = \dfrac{24}{Re}$	$(Re \gtrsim 1.5)$ Fig. 2.8
Flow past a particle	Equivalent diameter d_e $\left(\dfrac{6 \times \text{Volume}}{\text{Surface area}}\right)$	Particle velocity relative to fluid v_{fs}	Drag force / Projected area $\dfrac{F_D}{A_p}$	$c_D = \dfrac{(F_D/A_p)}{(\rho v_{fs}^2/2)}$	$c_D \propto \dfrac{1}{Re}$	
Flow past a spherical bubble	Bubble diameter d	Bubble velocity relative to fluid v_{fs}	Drag force / Projected area $\dfrac{F_D}{A_p}$	$c_D = \dfrac{(F_D/A_p)}{(\rho v_{fs}^2/2)}$	$c_D = \dfrac{16}{Re}$ $(Re \ll 1)$ $c_D = \dfrac{48}{Re}$ $(Re \gg 1)$	
Packed column, Length L, Void fraction ε, Superficial velocity v_s, Particle eq. dia. d_e	Void equivalent diameter $d_e^{void} = \dfrac{d_e \varepsilon}{1-\varepsilon}$	Fluid velocity in voids $v^{void} = \dfrac{v_s}{\varepsilon}$	Pressure difference Length $-\dfrac{\Delta p}{L}$	$f = -\dfrac{(\Delta p/L)d^{void}}{\rho (v^{void})^2}$ $= -\dfrac{\Delta p}{L}\dfrac{d_e}{\rho v_s^2}\dfrac{\varepsilon^3}{1-\varepsilon}$		$f = \dfrac{150}{Re} + 1.75$

The Chilton–Colburn $\jmath$-factor for heat and mass transfer are defined as,

$$\jmath_{heat} = \frac{\mathrm{Nu}}{\mathrm{Re}\,\mathrm{Pr}^{1/3}}, \quad \jmath_{mass} = \frac{\mathrm{Sh}}{\mathrm{Re}\,\mathrm{Sc}^{1/3}}. \tag{2.3.1}$$

For heat transfer, the variation in viscosity between the wall of the tube and the bulk of the fluid is incorporated in the definition of the $\jmath$ factor. In the Chilton–Colburn analogy, it is postulated that $\jmath$-factors are functions of the Reynolds number only.

It is important to note that the Chilton–Colburn analogy is valid only when the transport coefficients are constants, independent of temperature and concentration. That is, the variations in the temperature or concentration do not affect the flow or transport properties. This is a strong assumption which is violated in many cases, especially in heat transfer. The thermal conductivity and viscosity are temperature-dependent, and it is essential to incorporate this temperature dependence in the calculation of fluxes and flow. The Chilton–Colburn analogy can be used only in situations where the variation in the transport properties is small compared to their absolute values over the range of temperatures considered. For dilute gases, the condition is that the variation in temperature should be much smaller than the absolute temperature. For liquids, the viscosity is often a strong function of the temperature, and this has to be incorporated in the calculation of the velocity field.

The equivalence does not extend to momentum transfer. In the case of mass and heat transfer, the quantity being transported (mass or heat) is different from the mechanism of transport, which is the flow velocity. In the case of momentum transfer, the quantity transported, the momentum density, is proportional to the mechanism of convective transport, the flow velocity. The flow patterns are affected by the transport of momentum, and therefore the correlations for momentum transfer are different from those for mass and heat transfer. The Reynolds analogy is a relation between the Chilton–Colburn factors and the friction factor,

$$\jmath_{mass} = \jmath_{heat} = \frac{f}{2}. \tag{2.3.2}$$

The conditions for the applicability of this analogy are more restrictive than those for the Chilton–Colburn analogy. This is applicable only for gases where the Schmidt and Prandtl numbers are close to 1 and in the flow past slender bodies such as flat plates where the drag force is due to skin friction. This equivalence is observed, for example, in the momentum transfer correlations Eqs. 2.2.7–2.2.8 and the heat/mass transfer correlations Eqs. 2.3.11–2.3.12 if the Prandtl/Schmidt numbers are close to 1. The Reynolds analogy is not applicable for liquids where the Prandtl/Schmidt

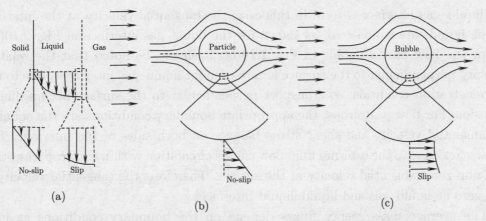

FIGURE 2.10. The velocity on the liquid side at the solid–liquid and gas–liquid interfaces (a), for the flow around a particle (b) and the flow around a bubble (c).

numbers could be large or small, and for flows in conduits and around suspended objects.

2.3.2 Surface Conditions

Transport due to convection depends on the flow characteristics discussed in Sections 2.2.2 through to 2.2.5, and the velocity boundary condition at the surface across which transport takes place. The boundary conditions are of two types, the no-slip boundary condition where the tangential velocity relative to the surface is zero, and the slip boundary condition where there is a non-zero tangential velocity relative to the surface, as shown in Fig. 2.10.   In both cases, the velocity perpendicular to the surface is equal to the surface velocity due to the no-penetration condition.

The no-slip velocity condition is appropriate for flow past a solid surface, as shown at the solid–liquid interface in Fig. 2.10(a), or for the flow around solid particles in Fig. 2.10(b). The solid is considered to be rigid and non-deformable under the application of fluid stresses. Due to the condition that the solid and fluid velocities have to be equal at the interface, the fluid velocity at the surface is zero in a reference frame moving and rotating with the solid. Consequently, the fluid tangential velocity relative to the surface decreases to zero at the interface, as shown in Figs. 2.10(a) and (b).

In the flow at a liquid–gas interface or the flow past a gas bubble, the viscosity of the gas is much smaller than that in the liquid. If the rates of deformation in the gas and liquid are comparable, it is appropriate to use the zero-stress boundary condition on the liquid side of the interface—that is, the shear stress in the liquid at the liquid–gas interface is zero. In this case, the tangential velocity at the interface in the liquid side is non-zero, as shown for the liquid–gas interface in Fig. 2.10(a), or for the flow past a bubble in Fig. 2.10(c). It should be noted that the relative velocity perpendicular to the surface is zero even for liquid–gas interfaces, due to the no-penetration condition, so transport perpendicular to the surface occurs due to diffusion. For flow past drops, the appropriate boundary conditions are the equality of tangential velocity and shear stress balance on both sides of the interface. The shear stress due to the external fluid flow causes circulation within the drop, resulting in a non-zero tangential velocity at the surface. Therefore, the tangential velocity is non-zero at liquid–gas and liquid–liquid interfaces.

The average mass/energy fluxes depend on the boundary conditions at high Peclet number. The strength of stream-wise convection relative to the surface is determined by the flow pattern close to the surface, which depends on the boundary condition. For surfaces with a no-slip boundary condition, the strength of convection decreases to zero as the surface is approached; in contrast, when there is a slip condition at the surface, the strength of convection is non-zero at the surface. Due to this, there is a difference in the Nusselt/Sherwood number correlations for flows past rigid and mobile surfaces.

2.3.3   Low Peclet number Heat/Mass Transfer

In the limit of low Peclet number, mass/heat transfer is primarily due to diffusion, and transport due to convection can be neglected in comparison to that due to diffusion. The Sherwood/Nusselt number does not depend on the flow velocity, and is therefore independent of the Reynolds or Peclet numbers. This implies that the Sherwood/Nusselt number tends to a constant in the limit of low Peclet number. Due to the equivalence of heat and mass transfer, the Sherwood and Nusselt numbers assume the same constant value in the low Peclet number limit for the same configuration,

$$(\mathrm{Sh}, \mathrm{Nu}) = \text{Constant for Pe} \ll 1. \qquad (2.3.3)$$

It is shown in Section 5.2.2 (Chapter 5) that this constant is 2 for transport from/to a spherical particle.

EXAMPLE 2.3.1: A wet particle of diameter d saturated with water is dried in quiescent dry air. It takes 10 s for the water content in the particle to decrease to 5% of its initial value. For a wet particle of diameter $10\,d$ made of the same material in quiescent dry air, how long does it take for the water content to decrease to 5% of its initial value? Assume that there is no convection, and the rate limiting step is the diffusion of water vapour from the particle surface into the surrounding air.

Solution: Since the rate limiting step is the diffusion of water in air, the time required for drying depends on the driving force, diffusion coefficient, the particle diameter, and the extent of drying. The driving force is the difference in the water concentration between the particle surface, which is the saturation concentration of the water in air, and the concentration far from the particle, which is zero for dry air. When the driving force and the extent of drying are same, the time required depends only on two dimensional parameters, the diffusion coefficient $\mathcal{D}$ and the particle diameter d. Based on dimensional analysis, the time require is $t \propto (d^2/\mathcal{D})$. Therefore, if a particle of diameter d takes 10 s to lose 95% of its moisture, a particle of diameter $10d$ will take 1000 s to lose 95% of its moisture. $\qquad\square$

The low Peclet number regime is not of practical importance in the flow through conduits for the following reason. The time for mass/heat diffusion across the diameter d is $(d^2/\mathcal{D})$ or (d^2/α), where $\mathcal{D}$ and α are the mass and thermal diffusivities. The distance moved by the fluid within this time is $(v_{av}d^2/\mathcal{D})$ or $(v_{av}d^2/\alpha)$. This distance is small compared to the pipe diameter d at low Peclet number where $(v_{av}d/\mathcal{D})$ or $(v_{av}d/\alpha)$ is small. Therefore, the mass or energy diffuses across the entire cross section over a distance smaller than the pipe diameter, and there is no advantage in pumping a fluid through a pipe with length larger than the diameter.

2.3.4 High Peclet number Heat/Mass Transfer

In the strong-convection limit at high Peclet number, the dimensionless fluxes do depend on the Reynolds and Peclet number. It might naively be assumed that diffusion can be neglected at high Peclet number because the rate of transport due to convection is much higher than that for diffusion. Such a conclusion is misleading, due to an important difference in the directionality of transport by the two mechanisms. Convection is directional and the transport is along the direction of the fluid flow, while diffusive transport is along the direction of concentration/temperature variation. There is no fluid flow perpendicular to the solid surfaces, such as the surface of the catalyst particle in Section 1.6.2, or the wall of the pipe in the heat exchanger in Section 1.6.3. Therefore, transport from/to the surface is necessarily due to diffusion.

In the strong convection limit, mass/heat is transported perpendicular to the surface due to diffusion, and is rapidly swept downstream due to the strong convection. Due to this, the concentration/temperature variation is restricted to a thin layer near the surface, called the 'boundary layer', and the concentration/temperature is almost uniform in the fluid outside the boundary layer. The thickness of the boundary layer decreases as the Peclet number increases, but diffusion in the boundary layer is comparable to convection in the limit of high Peclet number. The fluxes depend on the flow characteristics discussed in Section 2.2.1, as well as the boundary conditions discussed in Section 2.3.2 which determine the strength of convection at the surface.

It is useful to make a distinction between the low and high Reynolds number flows for convection-dominated transport. If the momentum diffusivity is large compared to mass/thermal diffusivity—that is, the Schmidt/Prandtl number is large—the system could simultaneously be in the low Reynolds number and high Peclet number regime. As we shall see in the discussion of diffusion in Chapter 3, the mass, momentum and thermal diffusivities in gases are comparable, and so the Reynolds and Peclet numbers are also comparable. However, in liquids, the mass diffusion coefficient is usually much smaller than the momentum diffusion coefficient, the Schmidt number is large, and so the low Reynolds number high Peclet number parameter regime is of practical importance. Similarly, for heat transfer, the thermal diffusion coefficient could be much smaller than the momentum diffusion coefficient for substances such as organic liquids, which have large Prandtl number.

At low Reynolds number, the correlations are simplified due to the 'linearity' of the velocity profiles. In this regime, for Reynolds number less than 2100 in a pipe and less than about 1.5 in the flow past a particle, the velocity at every point in the fluid is a linear function of the pressure gradient for the flow in conduits, or the free stream velocity in the flow around suspended particles. If the pressure gradient or the free stream velocity is increased by a constant factor C, the magnitude of the velocity at every point in the fluid increases by the same factor C, and the direction remains unchanged. Therefore, the flow pattern does not change, only the characteristic velocity changes.

For example, the velocity profile in a pipe is parabolic when the Reynolds number is less than about 2100. The flow velocity does change when the pressure gradient is increased, but the parabolic shape of the velocity profile does not change. A consequence is that the ratio of the local velocity at a point and the average flow velocity is independent of the pressure gradient or the Reynolds number for a laminar flow. The velocity profile for the flow around a spherical particle is laminar, and the flow is symmetric about the mid-plane perpendicular to the flow when the Reynolds number is less than about 1.5. The velocity in the fluid does increase as the particle velocity increases, but the ratio of the velocity at a point and the particle velocity remains unchanged as the particle velocity increases. Since the flow pattern does not change, the effect of convection is captured entirely by the Peclet number, $(v_c l_c / \mathcal{D})$ or $(v_c l_c / \alpha)$, and geometric parameters for the low Reynolds number laminar flow. Here, v_c and l_c are the characteristic velocity and length scale.

In contrast, at high Reynolds number, the flow pattern does change as the Reynolds number changes, and therefore, the correlations for heat and mass transfer depend on the Peclet number and the Reynolds number. Cross-stream transport in turbulent flow stakes place primarily by turbulent eddies, and so the transport rates are much higher than those in laminar flows. However, the fluctuations decrease as the surface of a particle or the wall of the tube is approached, because the fluid cannot penetrate the solid surface. Therefore, transport very close to the surface takes place by molecular diffusion. Due to the combination of eddy diffusion and molecular diffusion, the Nusselt and Sherwood numbers depend on both the Peclet and the Reynolds numbers in turbulent flows.

Flow in a Pipe

For a laminar flow in a pipe with Reynolds number less than 2100, the correlation for the Nusselt/Sherwood number reported by Sieder and Tate[18] is,

$$(\mathrm{Nu}, \mathrm{Sh}) = 1.86 \, \mathrm{Pe}^{1/3} (d/L)^{1/3}, \qquad (2.3.4)$$

where d and L are the pipe diameter and length respectively.

The Peclet number in the above correlation is often expressed as Re Pr, but we choose to express this in terms of the Peclet number because the non-dimensional flux depends only on the ratio of the mass/heat convection and diffusion, and does not depend on the ratio of momentum convection and diffusion when the flow is laminar. The correlation, Eq. 2.3.4, is derived to within a multiplicative constant in Section 9.1.2 (Chapter 9); it is also shown that the correlation is restricted to geometries with the ratio (d/L) below a threshold due to the requirement that the 'boundary layer thickness' should be smaller than the pipe diameter for this correlation to be valid. When there is a difference between the temperature of the wall of the pipe and the average temperature of the fluid in heat transfer, there is a correction factor due to the variation of viscosity with temperature, and the correlation is of the form,

$$\mathrm{Nu} = 1.86 \, \mathrm{Pe}^{1/3} (d/L)^{1/3} (\mu/\mu_w)^{0.14}, \qquad (2.3.5)$$

where μ_w is the viscosity at the wall temperature and μ is the viscosity at the average fluid temperature in the pipe. For the turbulent flow through a pipe, the following correlation is commonly used for the Nusselt and Sherwood numbers,

$$(\text{Nu}, \text{Sh}) = 0.023 \, \text{Re}^{0.8}(\text{Pr}, \text{Sc})^{1/3}(\mu/\mu_w)^{0.14} \qquad (2.3.6)$$

EXAMPLE 2.3.2: A heavy oil with density $\rho = 10^3$ kg/m^3, viscosity $\mu = 0.1$ kg/m/s, thermal conductivity $k = 0.15$ W/m/$^\circ$C, and specific heat $C_p = 2 \times 10^3$ J/kg/$^\circ$C flows through the tube side of the heat exchanger of diameter 2 cm and length 4 m. If the flow rate is 100 l/hr and the average temperature difference between the shell and tube side is 50°C, what is the heat transfer rate?

Solution: The flow rate is 100 l/hr = 100×10^{-3} m^3/3600 s = 2.78×10^{-5} m^3/s. The average velocity of the flow through the pipe is,

$$v_{av} = \frac{2.78 \times 10^{-5} \text{ m}^3/\text{s}}{\pi (0.02 \text{ m})^2/4} = 8.85 \times 10^{-2} \text{ m/s}. \qquad (2.3.7)$$

The Reynolds and Prandtl numbers are,

$$\text{Re} = \frac{10^3 \text{ kg/m}^3 \times 8.85 \times 10^{-2} \text{ m/s} \times 0.02 \text{ m}}{0.1 \text{ kg/m/s}} = 17.70,$$

$$\text{Pr} = \frac{2 \times 10^3 \text{ J/kg/}^\circ\text{C} \times 0.1 \text{ kg/m/s}}{0.15 \text{ W/m/}^\circ\text{C}} = 1.33 \times 10^3. \qquad (2.3.8)$$

The flow is laminar because the Reynolds number is less than 2100. The correlation Eq. 2.3.4 for the Nusselt number is,

$$\text{Nu} = 1.86 \times (17.70)^{1/3} \times (1.33 \times 10^3)^{1/3} \times (0.02 \text{ m/4 m})^{1/3} = 9.12. \qquad (2.3.9)$$

The heat transfer rate is,

$$Q = \pi dL \times q_{av} = \pi dL \times \frac{k\Delta T}{d} \times \text{Nu} = \pi L k \Delta T \text{Nu}$$
$$= \pi \times 4 \text{ m} \times 0.15 \text{ W/m/}^\circ\text{C} \times 50^\circ\text{C} \times 9.12 = 860 \text{ W}. \qquad (2.3.10)$$

$\square$

Flow past a Flat Plate

The average fluxes are defined per unit surface area of the plate. For mass/heat transfer, the driving force in the Sherwood/Nusselt number is the concentration/temperature difference between the surface of the plate and the free stream far from the plate.

The laminar flow past a flat plate, discussed in Section 2.2.3, is characterised by a 'momentum boundary layer' of thickness $\mathrm{Re}^{-1/2}$ times the plate length where the friction due to the presence of the plate affects the flow. The correlation for the Nusselt and Sherwood number are of the form

$$(\mathrm{Nu}, \mathrm{Sh}) = 0.664 \, \mathrm{Re}^{1/2}(\mathrm{Pr}, \mathrm{Sc})^{1/3}(\mu/\mu_w)^{0.14}, \qquad (2.3.11)$$

at high Peclet number for a laminar flow with Reynolds number less than 5×10^5. For a turbulent flow with Reynolds number greater than 5×10^5, correlations similar to Eq. 2.3.6 have been proposed for the flow past a flat plate.

$$(\mathrm{Nu}, \mathrm{Sh}) = 0.037 \, \mathrm{Re}^{0.8}(\mathrm{Pr}, \mathrm{Sc})^{1/3}(\mu/\mu_w)^{0.14}. \qquad (2.3.12)$$

For mass transfer, the factor (μ/μ_w) is 1 if the viscosity is independent of the concentration.

Flow around a Particle

For the high Peclet number heat/mass transfer to a spherical particle in laminar flow with a no-slip condition at the surface, we shall see in Section 9.1.3 (Chapter 9) that the Nusselt/Sherwood numbers are proportional to $\mathrm{Pe}^{1/3}$. The correlation for the flow past a spherical particle, with a no-slip boundary condition at the surface, is

$$(\mathrm{Nu}, \mathrm{Sh}) = 0.992 \, \mathrm{Pe}^{1/3}. \qquad (2.3.13)$$

For particles of other shapes, the Nusselt and Sherwood numbers are proportional to the one-third power of the Peclet number, but the constant is different from that in Eq. 2.3.13. In general, it is shown in Chapter 9 that for the laminar flow past solid surfaces, the dimensionless fluxes are proportional to $\mathrm{Pe}^{1/3}$.

For Reynolds number greater than 1.5, correlations for the heat/mass flux are difficult to derive analytically because of the complexity of the flow patterns in

Fig. 2.7. The Ranz–Marshall correlation[19] for the Nusselt/Sherwood number,

$$(\text{Nu}, \text{Sh}) - 2 = 0.6 \, \text{Re}^{1/2}(\text{Pr}, \text{Sc})^{1/3}, \tag{2.3.14}$$

was developed for Reynolds number up to about 200 and Prandtl number in the range 0.7 to 380. The Whitaker correlation,

$$(\text{Nu}, \text{Sh}) - 2 = (0.4 \, \text{Re}^{1/2} + 0.06 \, \text{Re}^{2/3})(\text{Pr}, \text{Sc})^{1/3}, \tag{2.3.15}$$

is reported to provide results accurate to within 30% for $0.7 \leq \text{Pr} \leq 380$ and $3.5 \leq \text{Re} \leq 7.8 \times 10^4$ for the flow around a sphere. The term on the left in Eq. 2.3.15 is the difference between the Nusselt/Sherwood number and its value for diffusive transport. In the original formulation of Eq. 2.3.15 [20], the exponent of the Prandtl number was reported as 0.4. Here, the exponent $\frac{1}{3}$ is used because it is consistent with the correlation for a packed column, Eq. 2.3.34, and the analytical result, Eq. 2.3.15, for the flow around a particle at low Reynolds number. This modification makes very little difference to the predicted Nusselt/Sherwood numbers, in comparison to the 30% variation reported between the correlation and experimental results.

The correlations for flows around other blunt objects, such as a cylinder perpendicular to the flow, have the same power-law relationships between the Nusselt/Sherwood, Reynolds and Peclet numbers as Eqs. 2.3.13 and 2.3.15, but the constants in the correlations do depend on the geometry.

In combined heat and mass transfer problems, it is necessary to determine the mass and heat fluxes, and then determine the time required for heat and mass transfer, as shown in the following example for the drying of a droplet in a spray drier.

EXAMPLE 2.3.3: In a spray drier, droplets of diameter 0.1 mm and density 10^3 kg/m^3, containing 80% by weight of water, are dried into porous particles of the same diameter. The droplets at a temperature of 20°C are ejected at a velocity of 1 m/s from a nozzle into an upward stream of hot air. What is the time taken for a droplet to dry completely? What should be the air temperature so that the time required for heat transfer of the required latent heat to the droplet is equal to the time for mass transfer? The vapour pressure of water at 20°C is 2.3 kPa, and the latent heat of vapourisation of water is 2.3×10^6 J/kg. Assume approximate values of $\rho = 1.095$ kg/m^3, $\mu = 1.97 \times 10^{-5}$ kg/m/s, $k = 2.8 \times 10^{-2}$ W/m/°C, $C_p = 10^3$ J/kg/°C and $\mathcal{D} = 3 \times 10^{-5}$ m^2/s for the density, viscosity, thermal conductivity, specific heat of air and diffusivity of water vapour in air in the range 20°–80°C.

Solution:

The dimensionless groups are,

$$\text{Re} = \frac{1.095 \text{ kg/m}^3 \times 1 \text{ m/s} \times 10^{-4} \text{ m}}{1.97 \times 10^{-5} \text{ kg/m/s}} = 5.6, \qquad (2.3.16)$$

$$\text{Pr} = \frac{10^3 \text{ J/kg/}^\circ\text{C} \times 1.97 \times 10^{-5} \text{ kg/m/s}}{2.8 \times 10^{-2} \text{ W/m/}^\circ\text{C}} = 0.70, \qquad (2.3.17)$$

$$\text{Sc} = \frac{1.97 \times 10^{-5} \text{ kg/m/s}}{1.095 \text{ kg/m}^3 \times 3 \times 10^{-5} \text{ m}^2/\text{s}} = 0.60. \qquad (2.3.18)$$

From the correlation, Eq. 2.3.15, the Nusselt and Sherwood numbers are,

$$\text{Nu} = 2 + (0.4(5.6)^{1/2} + 0.06(5.6)^{2/3})(0.70)^{1/3} = 3.01,$$
$$\text{Sh} = 2 + (0.4(5.6)^{1/2} + 0.06(5.6)^{2/3})(0.60)^{1/3} = 2.96. \qquad (2.3.19)$$

The total mass of water to be evaporated is the water density times the volume of water, the latter is $0.8(\pi d^3/6)$, 80% of the volume of the particle with diameter d.

$$\text{Mass of water} = 10^3 \text{ kg/m}^3 \times 0.8 \times \frac{\pi(10^{-4} \text{ m})^3}{6} = 4.2 \times 10^{-10} \text{ kg}. \qquad (2.3.20)$$

The saturation concentration of gas at the surface is determined from the vapour pressure $p_v = 2.3$ kPa,

$$c_s = \frac{p_v M_w}{k_B T} = \frac{2.3 \times 10^3 \text{ Pa} \times (0.018 \text{ kg/}(6.023 \times 10^{23}))}{1.38 \times 10^{-23} \text{ J/K} \times 293 \text{ K}} = 1.7 \times 10^{-2} \text{ kg/m}^3. \qquad (2.3.21)$$

The mass transfer rate is the product of the surface area πd^2 and the average flux j_{av}, the latter equal to $(\mathcal{D}c_s/d) \times \text{Sh}$,

$$\begin{aligned}
\text{Mass transfer rate} &= \pi d^2 j_{av} = \pi d \mathcal{D}c_s \times \text{Sh} \\
&= \pi \times 10^{-4} \text{ m} \times 3 \times 10^{-5} \text{ m}^2/\text{s} \times 1.7 \times 10^{-2} \text{ kg/m}^3 \times 2.96 \\
&= 4.73 \times 10^{-10} \text{ kg/s}. \qquad (2.3.22)
\end{aligned}$$

The time required for mass transfer of the water in the droplet is,

$$\text{Time for mass transfer} = \frac{\text{Mass}}{\text{Mass transfer rate}} = \frac{4.2 \times 10^{-10} \text{ kg}}{4.73 \times 10^{-10} \text{ kg/s}} = 0.89 \text{ s}. \qquad (2.3.23)$$

The time required for the transfer of the heat for vapourisation from the gas to the particle is now calculated. The total heat of vapourisation is the product of the mass and the latent heat,

$$\text{Heat of vapourisation} = 4.2 \times 10^{-10} \text{ kg} \times 2.3 \times 10^6 \text{ J/kg} = 9.63 \times 10^{-4} \text{ J}. \quad (2.3.24)$$

The heat transfer rate is the product of the surface area πd^2 and the average heat flux q_{av}, the latter related to the Nusselt number $q_{av} = (\text{Nu}k\Delta T/d)$,

$$\text{Heat transfer rate} = \pi d^2 q_{av} = \pi dk\Delta T\text{Nu}$$
$$= \pi \times 10^{-4} \text{ m} \times 2.8 \times 10^{-2} \text{ W/m/}^\circ\text{C} \times \Delta T \times 3.01$$
$$= 2.64 \times 10^{-5} \text{ W/}^\circ\text{C} \times \Delta T. \quad (2.3.25)$$

The time for heat transfer is,

$$\text{Time for heat transfer} = \frac{\text{Heat of vapourisation}}{\text{Heat transfer rate}} = \frac{9.63 \times 10^{-4} \text{ J}}{2.64 \times 10^{-5} \text{ W/}^\circ\text{C} \times \Delta T}$$
$$= \frac{36.43(\text{s} \times^\circ \text{C})}{\Delta T}. \quad (2.3.26)$$

If the time for heat transfer is equal to that for mass transfer, Eq. 2.3.23, the temperature difference between the droplet and air is,

$$\Delta T = \frac{36.43(\text{s} \times^\circ \text{C})}{0.89\text{s}} = 40.9^\circ\text{C}. \quad (2.3.27)$$

Thus, the latent heat times mass transfer rate is equal to the heat transfer rate when the air temperature is 60.9°C.   □

Falling Film

For transport to a falling liquid film from the gas phase, where there is a slip velocity condition at the surface, we show in Section 4.4.1 (Chapter 4) that the Nusselt/Sherwood numbers are,

$$(\text{Nu}, \text{Sh}) = 1.38 \text{ Pe}^{1/2} \left(\frac{h}{L}\right)^{1/2}, \quad (2.3.28)$$

where the Peclet number $\text{Pe} = (v_{av}h/\mathcal{D})$, v_{av} is the average velocity, h is the film thickness and L is the length of the film in the stream-wise direction. The correlation, Eq. 2.3.28, is valid only for $(L/h) \ll \text{Pe}$.

Flow around a Bubble

In the case of transport to a spherical bubble, for which there is a non-zero tangential velocity at the surface, it is shown in Section 9.2.1 (Chapter 9) that the Nusselt/Sherwood number is proportional to the square root of the Peclet number for low Reynolds number,

$$(\text{Nu}, \text{Sh}) = 0.652 \, \text{Pe}^{1/2} = 0.652 \, \text{Re}^{1/2} (\text{Pr}, \text{Sc})^{1/2}. \tag{2.3.29}$$

The Nusselt and Sherwood numbers are proportional to the square root of the Peclet number for non-spherical bubbles as well, but the constant in the correlation is different from that in Eq. 2.3.29. In general, it is shown that in laminar flow when there is a slip velocity at the surface, the dimensionless flux is proportional to $\text{Pe}^{1/2}$.

At high Reynolds number, the potential flow solution can be used for the flow around a bubble if there is a zero shear stress condition at the bubble surface. In this case, there is no boundary layer separation at the rear of the bubble, and the thickness of the wake region is much smaller than the size of the bubble. The correlation,

$$(\text{Nu}, \text{Sh}) = 1.13 \, \text{Pe}^{1/2} = 1.13 \, \text{Re}^{1/2} (\text{Pr}, \text{Sc})^{1/2}, \tag{2.3.30}$$

can be derived in Exercise 9.9 in Chapter 9.

Packed Column

Since a packed column consists of a collection of particles, the fluxes at the surface depend on the fluid flow pattern around a particle. These flow patterns are similar to those around a single particle discussed in Section 2.3.4, but these are modified by the presence of the neighbouring particles. Therefore, the correlations are expected to have the same power-law relationship between the Nusselt, Reynolds and Peclet numbers as Eqs. 2.3.13 and 2.3.15, but with different coefficients to account the presence of particles in close proximity. The characteristic length is $(d_e \varepsilon / (1 - \varepsilon))$ (Eq. 2.2.34), where d_e is the equivalent diameter of the particle and ε is the porosity, while the characteristic velocity is (v_s / ε) (Eq. 2.2.35), where v_s is the superficial velocity.

The average mass/heat fluxes are defined as the ratio of the total transport rates and the packing surface area, which is the total surface area of all the particles. The total surface area is,

$$\text{Total surface area} = N A_p = \frac{V(1-\varepsilon)A_p}{V_p} = \frac{6V(1-\varepsilon)}{d_e}, \tag{2.3.31}$$

where V is the total volume of the packed column, A_p, V_p and d_e are the surface area, volume and equivalent diameter of a particle respectively, N is the number of particles, and Eq. 2.2.30 is substituted for N. The average fluxes are defined as the ratio of the total transport rates and the total surface area,

$$(q_{av}, j_{av}) = \frac{\text{Transport rates}}{\text{Total surface area}} = \frac{\text{Transport rates}}{6V(1-\varepsilon)/d_e}. \tag{2.3.32}$$

The Reynolds number is defined in Eq. 2.2.36, and the Nusselt and Sherwood numbers are defined based on the characteristic length,

$$\text{Nu} = \frac{q_{av}}{k\Delta T((1-\varepsilon)/d_e\varepsilon)}, \quad \text{Sh} = \frac{j_{av}}{\mathcal{D}\Delta c((1-\varepsilon)/d_e\varepsilon)}. \tag{2.3.33}$$

A correlation of the form [20],

$$(\text{Nu}, \text{Sh}) = (0.5\text{Re}^{1/2} + 0.2\text{Re}^{2/3})(\text{Pr}, \text{Sc})^{1/3}, \tag{2.3.34}$$

has been found to fit the data for $0.7 \leq \text{Pr} \leq 380$ and $20 \leq \text{Re} \leq 10^4$. The derivation of the correlations such as Eq. 2.3.34 is difficult, since the geometry and velocity field cannot be precisely specified, and several idealisations have to be made.

Summary (2.3)

1. If the diffusivities are constants, the correlations for heat and mass transfer have the same form. For the same configuration and Reynolds number, the correlations for mass transfer can be obtained by the substitution $\text{Nu} \to \text{Sh}$, $\text{Pr} \to \text{Sc}$ in the correlations for heat transfer.

2. For low Peclet number, transport is due to diffusion. If convection is neglected, the Nusselt/Sherwood numbers are a constant for a specified configuration.

3. For high Peclet number, some examples of correlations for different configurations are summarised in Table 2.5; some of these are derived in Chapter 9.

TABLE 2.5. The definitions of the Nusselt and Peclet numbers and the correlations for high Peclet number heat transfer. The Reynolds number was defined in Table 2.4. The mass transfer correlations are obtained using the substitutions Nu $\to$ Sh, Pr $\to$ Sc, $\alpha \to \mathcal{D}$, $\Delta T \to \Delta c$, $k \to \mathcal{D}$ and $q_{av} \to j_{av}$. Here, q_{av} is the average heat flux, j_{av} is the average mass flux, d is the particle/bubble/pipe diameter, L is the length of a pipe, flat plate or falling film, v_{fs} is the free-stream velocity past a flat plate/particle/bubble, and h is the thickness of a falling film. For a packed column, d_e is the particle equivalent diameter, v_s is the superficial velocity and ε is the porosity.

Type of flow	Pe	Nu	Parameter regime	Interface
Flow in a pipe	$(v_{av}d/\alpha)$	$q_{av}/(k\Delta T/d)$ $\mathrm{Nu} = 1.86\mathrm{Re}^{1/3}\mathrm{Pr}^{1/3}(d/L)^{1/3}(\mu/\mu_w)^{0.14}$	Laminar Re < 2100	Rigid
Flow in a pipe	$(v_{av}d/\alpha)$	$q_{av}/(k\Delta T/d)$ $\mathrm{Nu} = 0.023\mathrm{Re}^{0.8}\mathrm{Pr}^{1/3}(d/L)^{1/3}(\mu/\mu_w)^{0.14}$	Turbulent Re $> 10{,}000$	Rigid
Flow past flat plate	$(v_{fs}L/\alpha)$	$q_{av}/(k\Delta T/L)$ $\mathrm{Nu} = 0.664\mathrm{Re}^{1/2}\mathrm{Pr}^{1/3}(\mu/\mu_w)^{0.14}$	Laminar	Rigid
Flow past flat plate	$(v_{fs}L/\alpha)$	$q_{av}/(k\Delta T/L)$ $\mathrm{Nu} = 0.037\mathrm{Re}^{0.8}\mathrm{Pr}^{1/3}(\mu/\mu_w)^{0.14}$	Turbulent	Rigid
Falling film	$(v_{av}h/\alpha)$	$(q_{av}/(k\Delta T/h))$ $\mathrm{Nu} = 1.38\mathrm{Pe}^{1/2}(h/L)^{1/2}$	Laminar $(L/h) \ll \mathrm{Pe}$	Mobile
Spherical particle	$(v_{fs}d/\alpha)$	$q_{av}/(k\Delta T/d)$ $\mathrm{Nu} = 0.992\mathrm{Pe}^{1/3}$	Re $\ll 1$	Rigid
Spherical particle	$(v_{fs}d/\alpha)$	$q_{av}/(k\Delta T/d)$ $\mathrm{Nu} - 2 = (0.4\mathrm{Re}^{1/2} + 0.06\mathrm{Re}^{2/3})\mathrm{Pr}^{1/3}$	$0.7 \leq \mathrm{Pr} \leq 380$, $3.5 \leq \mathrm{Re} \leq 7.8 \times 10^4$	Rigid
Spherical bubble	$(v_{fs}d/\alpha)$	$q_{av}/(k\Delta T/d)$ $\mathrm{Nu} = 0.652\mathrm{Pe}^{1/2}$	Re $\ll 1$	Mobile
Spherical bubble	$(v_{fs}d/\alpha)$	$q_{av}/(k\Delta T/d)$ $\mathrm{Nu} = 1.13\mathrm{Pe}^{1/2}$	Re $\gg 1$	Mobile
Packed column	$(v_s d_e/(1-\varepsilon)\alpha)$	$q_{av}\varepsilon d_e/(k\Delta T(1-\varepsilon)\alpha)$ $\mathrm{Nu} = (0.5\mathrm{Re}^{1/2} + 0.2\mathrm{Re}^{2/3})\mathrm{Pr}^{1/3}$	$0.7 \leq \mathrm{Pr} \leq 380$, $20 \leq \mathrm{Re} \leq 10^4$	Mobile

2.4   Natural Convection

When the temperature of an object is higher than the ambient, the hotter fluid close to the object has a a lower density in comparison to the fluid further away. The buoyancy force due to the lower density causes the fluid near the object to rise, and the rising fluid carries heat from the surface due to convection. The surrounding colder fluid from further away is drawn in to replace the rising fluid, as shown in Fig. 2.11. This mechanism, called 'natural convection', results in significantly higher heat transfer rates in comparison to conduction.

In natural convection, the driving force for convection is the body force caused by a variation in the density of the fluid, which is in turn caused by variation in temperature. This is in contrast to the 'forced convection' problems such as the flow through conduits or flow past objects, where the flow velocity is specified. In natural convection, the flow velocity is determined by a balance between the buoyancy force and the larger of the inertial or viscous forces.

The force density (force per unit volume) due to buoyancy can be estimated as $(\Delta \rho \, g)$, where $\Delta \rho$ is the density difference between the fluid at the surface of the object and the ambient fluid far away. The density difference is expressed as $\Delta \rho = \rho \beta \Delta T$, where ρ is the density of the ambient fluid, ΔT $(T_0 - T_\infty$ in Fig. 2.11) is the temperature difference between the object and the ambient fluid, and β is the coefficient of thermal expansion with dimension Θ^{-1}. Therefore, the buoyancy force density is expressed as $\rho \, \beta \, \Delta T \, g$. If the characteristic velocity is v_c, the inertial force density (force per unit volume) is $(\rho v_c^2 / l_c)$ where l_c is the characteristic size of the object shown in Fig. 2.11. Balancing the buoyancy and inertial force densities, the characteristic velocity is

$$v_c = \sqrt{l_c g \beta \Delta T}. \tag{2.4.1}$$

The dependent dimensionless group is the Nusselt number, which is the ratio of the average heat flux q_{av} and $(k\Delta T / l_c)$, where ΔT is the temperature difference between the surface of the object and the ambient fluid. There are two independent dimensionless groups. One independent dimensionless group is the ratio of momentum convection and diffusion based on the characteristic velocity, $(v_c l_c / \nu)$. Historically, the 'Grashof number' is defined as the square of the Reynolds number based upon this convection velocity,

$$\text{Gr} = \left(\frac{v_c l_c}{\nu} \right)^2 = \frac{\rho^2 l_c^3 g \beta \Delta T}{\mu^2}, \tag{2.4.2}$$

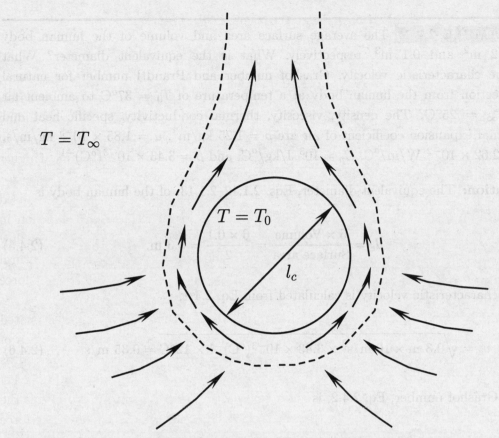

FIGURE 2.11. Natural convection from a heated object of characteristic size l_c with surface temperature T_0 placed in a colder fluid with temperature $T_\infty < T_0$.

where the kinematic viscosity $\nu = (\mu/\rho)$. The other independent dimensionless group is the Prandtl number, which is the ratio of momentum and thermal diffusion, Eq. 1.6.36.

An alternative dimensionless group which is sometimes used is the Rayleigh number, which is the product of the Grashof and Prandtl numbers,

$$\mathrm{Ra} = \mathrm{Gr}\,\mathrm{Pr} = \left(\frac{v_c l_c}{\nu}\right)^2 \frac{\nu}{\alpha} = \frac{l_c^3 g \beta \Delta T}{\nu \alpha}, \qquad (2.4.3)$$

where $\alpha = (k/\rho C_p)$ is the thermal diffusivity. Correlations for natural convection are of the form,

$$\mathrm{Nu} = \mathrm{Function}(\mathrm{Gr}, \mathrm{Pr}). \qquad (2.4.4)$$

EXAMPLE 2.4.1: The average surface area and volume of the human body are 2 m^2 and 0.1 m^3, respectively. What is the equivalent diameter? What is the characteristic velocity, Grashof number and Prandtl number for natural convection from the human body at a temperature of $T_0 = 37°C$ to ambient air at $T_\infty = 25°C$? The density, viscosity, thermal conductivity, specific heat and thermal expansion coefficient of air are $\rho = 1.25$ kg/m^3, $\mu = 1.85 \times 10^{-5}$ kg/m/s, $k = 2.62 \times 10^{-2}$ W/m/°C, $C_p = 10^3$ J/kg/°C, and $\beta = 3.43 \times 10^{-3}$(°C)$^{-1}$.

Solution: The equivalent diameter, Eqs. 2.1.12–2.1.13, of the human body is

$$d_e = \frac{6 \times \text{Volume}}{\text{Surface area}} = \frac{6 \times 0.1}{2} = 0.3 \text{ m.} \tag{2.4.5}$$

The characteristic velocity is calculated from Eq. 2.4.1,

$$v_c = \sqrt{0.3 \text{ m} \times 10 \text{ m/s}^2 \times 3.43 \times 10^{-3}(°C)^{-1} \times 12°C} = 0.35 \text{ m/s.} \tag{2.4.6}$$

The Grashof number, Eq. 2.4.2, is

$$Gr = \frac{(1.25 \text{ kg/m}^3)^2 \times 10 \text{ m/s}^2 \times 3.43 \times 10^{-3}(°C)^{-1} \times 12°C \times (0.3 \text{ m})^3}{(1.85 \times 10^{-5} \text{ kg/m/s})^2} = 5 \times 10^7.$$
$$\tag{2.4.7}$$

The Prandtl number, Eq. 1.6.36, is

$$Pr = \frac{10^3 \text{ J/kg/°C} \times 1.85 \times 10^{-5} \text{ kg/m/s}}{2.62 \times 10^{-2} \text{ W/m/°C}} = 0.71. \tag{2.4.8}$$

□

EXAMPLE 2.4.2: What is the characteristic velocity and the Grashof number for natural convection into water from a horizontal cylinder of diameter 1 cm heated to 35°C in ambient water at 25°C. The density, viscosity, thermal conductivity, specific heat and thermal expansion coefficient of water are $\rho = 10^3$ kg/m^3, $\mu = 10^{-3}$ kg/m/s, $k = 0.6$ W/m/°C, $C_p = 4.2 \times 10^3$ J/kg/°C, and $\beta = 2 \times 10^{-4}$(°C)$^{-1}$.

Solution: The characteristic velocity, Eq. 2.4.1, is

$$v_c = \sqrt{0.01 \text{ m} \times 10 \text{ m/s}^2 \times 2 \times 10^{-4}(^o \text{ C})^{-1} \times 10^o\text{C}} = 1.4 \times 10^{-2} \text{ m/s}. \qquad (2.4.9)$$

The Grashof number, Eq. 2.4.2, is

$$\text{Gr} = \frac{(10^3 \text{ kg/m}^3)^2 \times 10 \text{ m/s}^2 \times 2 \times 10^{-4}(^o\text{C})^{-1} \times 10^o\text{C} \times (0.01 \text{ m})^3}{(10^{-3} \text{ kg/m/s})^2} = 2 \times 10^4. \qquad (2.4.10)$$

The Prandtl number, Eq. 1.6.36, is

$$\text{Pr} = \frac{4.2 \times 10^3 \text{ J/kg/}^o C \times 10^{-3} \text{ kg/m/s}}{0.6 \text{ W/m/}^oC} = 7.0. \qquad (2.4.11)$$

□

The examples illustrate that the the characteristic velocity due to natural convection is in the range cm/s to m/s, and the Grashof number is usually much larger than 1, in the range 10^4–10^{11} in practical applications. The flow due to natural convection is laminar when the product (Gr Pr) is less than about 10^9, whereas the flow becomes turbulent for (Gr Pr) $> 10^9$. The correlations for a laminar flow are derived in Chapter 10, to within a proportionality constant, by scaling the fluid momentum and temperature equations. When the Grashof number is high, momentum convection due to buoyancy is larger than diffusion, and the flow is restricted to a 'boundary layer' of thickness much smaller than the characteristic size l_c, as shown in Fig. 2.11. Thermal convection due to buoyancy is also much stronger than thermal diffusion, and the increase in temperature due to the presence of the object is restricted to a thermal boundary layer whose thickness is much smaller than the characteristic size. Nusselt number correlations are derived based on the variation of the thickness of the momentum and thermal boundary layers with Grashof and Prandtl numbers. The general form of the correlation for natural convection due to a laminar flow, derived in Section 10.2 (Chapter 10) is,

$$\boxed{\text{Nu} = C(\text{Pr}) \text{ Gr}^{1/4},} \qquad (2.4.12)$$

where $C(\text{Pr})$ depends on the Prandtl number.

For high Prandtl number, momentum diffusion is fast compared to thermal diffusion, and it is shown in Section 10.3.2 (Chapter 10) that the Nusselt number correlation is of the form,

$$\boxed{\text{Nu} \propto \text{Pr}^{1/4}\text{Gr}^{1/4}.} \qquad (2.4.13)$$

The proportionality constant depends on the specific configuration. The proportionality constant is reported to be 0.51 for a horizontal plate, and 0.59 for a sphere. Momentum diffusion is slow compared to thermal diffusion in the limit of low Prandtl number, and so the fluid flow is confined to a thinner region when compared to the temperature disturbance. In this case, it is shown in Section 10.3.1 (Chapter 10) that the correlation for the Nusselt number is,

$$\boxed{\text{Nu} \propto \text{Pr}^{1/2}\text{Gr}^{1/4},} \qquad (2.4.14)$$

where the proportionality constant depends on the shape of the object.

Summary (2.4)

1. When a hotter object is placed in a colder ambient fluid, the fluid close to the object rises because it is hotter and less dense. Fluid from further away is drawn in. This mechanism of heat transfer is called natural convection.

2. The characteristic flow velocity due to buoyancy, Eq. 2.4.1, is in the range of a few centimeters to a few meters per second in practical applications.

3. The Grashof number, Eq. 2.4.2, is the square of the Reynolds number based on the characteristic flow velocity due to buoyancy. The Grashof number is in the range 10^4–10^9 in practical applications.

4. The Nusselt number correlation for natural convection in laminar flow has the general form, Eq. 2.4.12, and the correlations for the special cases of low and high Prandtl number are Eqs. 2.4.14 and 2.4.13, respectively.

Exercises

EXERCISE 2.1 Using dimensional analysis, determine the exponents in the following dimensionless groups. Here, L is the characteristic length, T and ΔT are the temperature and temperature difference, ρ is the density, μ is the viscosity, $\mathcal{D}$ is the mass diffusivity, α is the thermal diffusivity, γ is the surface tension and g is the acceleration due to gravity.

1. Galilei number—ratio of gravitational and viscous forces,
   $\text{Ga} = gL^{\square}\mu^{\square}\rho^{\square}$.

2. Marangoni number—relates stress due to variation of surface tension with temperature and viscous stress,
   $\text{Mg} = \left(-\dfrac{\mathrm{d}\gamma}{\mathrm{d}T}\right)L^{\square}(\Delta T)^{\square}\mu^{\square}\alpha^{\square}$.

3. Laplace number—ratio of surface tension and viscous forces,
   $La = \gamma L^{\Box} \rho^{\Box} \mu^{\Box}$.

4. Taylor number—ratio of centrifugal and viscous forces,
   $Ta = (\Omega L^{\Box} \rho^{\Box} \mu^{\Box})^2$,
   where Ω is the angular velocity with dimension of $\mathcal{T}^{-1}$.

5. Damkohler number—ratio of reaction and diffusion rates,
   $Da = (\mathcal{K} \mathcal{D}^{\Box} L^{\Box})$,
   where $\mathcal{K}$ is the first-order reaction rate constant.

6. Elasticity number—ratio of elastic and viscous forces,
   $El = t_r \rho^{\Box} \mu^{\Box} L^{\Box}$,
   where t_r is the relaxation time of the fluid with dimensions of time.

7. Savage number—ratio of shear stress and pressure,
   $Sa = (\dot{\gamma} \rho^{\Box} L^{\Box} p^{\Box})^2$, where p is the pressure, and $\dot{\gamma}$ is the strain rate with dimension $\mathcal{T}^{-1}$.

EXERCISE 2.2 Consider the flow of water of density 10^3 kg/m^3 and viscosity 10^{-3} kg/m/s in a pipe of length 2 m and diameter 2 cm. What is the maximum pressure difference across the ends of the pipe for which the flow is laminar.

EXERCISE 2.3 The flow of water with density 10^3 kg/m^3 and viscosity 10^{-3} kg/m/s in a pipe of diameter 2.5 cm and length 5 m is driven by a pressure difference of 50 kPa across the ends of the pipe. What is the flow rate if the pipe is smooth, and if the pipe has wall roughness $(\epsilon/d) = 0.01$? What is the Reynolds number in each case?

EXERCISE 2.4 Determine the pressure drop per unit length required for the flow of air of density 1.25 kg/m^3 and viscosity 1.8×10^{-5} kg/m/s in a smooth pipe of diameter 2 cm with a velocity of 3 m/s. Use the Colebrook correlation, Eq. 2.2.4, for the friction factor.

EXERCISE 2.5 The constitutive relation for the flow of a power-law fluid is Eq. 1.5.6. What is the Reynolds number for the pipe flow of a power-law fluid?

EXERCISE 2.6 An airplane wing of length 1 m is moving with velocity 300 km/hr through air with density 1.25 kg/m^3 and viscosity 1.8×10^{-5} kg/m/s. If the wing is treated as a flat plate, what is the downstream distance at which there is a transition? What is the drag force per unit length of the wing?

EXERCISE 2.7 If the Reynolds number for the flow past a flat plate shown in Fig. 2.6 is expressed in terms of the boundary layer thickness for a laminar flow $\delta_{0.99}$ (Eq. 2.2.5) instead of the downstream length x, what is the Reynolds number for the transition from laminar to turbulent flow?

EXERCISE 2.8 Instead of a particle settling in air, solve Example 2.2.2 for a particle of density 1.5×10^3 kg/m^3 settling in water of density 10^3 kg/m^3 and viscosity 10^{-3} kg/m/s.

EXERCISE 2.9 A deformable spherical elastic particle with diameter d and elasticity modulus G is settling in a quiescent fluid of density ρ and viscosity μ. What are the dimensionless ratios of (viscosity/elasticity) and (inertia/elasticity)? For an elastic particle of mass density 1.2×10^3 kg/m^3 and elasticity modulus 10^3 kg/m/s^2 settling at its terminal velocity in air of density 1.25 kg/m^3 and viscosity 1.8×10^{-5} kg/m/s, what is the restriction on the diameter of the particle for small deformation?

EXERCISE 2.10 If air of density 1.2 kg/m^3 and viscosity 1.8×10^{-5} kg/m/s is passed through the packed column in Example 2.2.4 instead of water, what is the minimum superficial velocity for fluidisation?

EXERCISE 2.11 A water dispenser consists of a tank of height 1 m below which is a filter of height 20 cm and diameter 2 cm connected to an outlet. The filter consists of activated carbon particles of diameter 1 mm and void fraction 0.45. Determine the flow rate of water when the outlet is opened. The density and viscosity of water are 10^3 kg/m^3 and 10^{-3} kg/m/s, respectively.

EXERCISE 2.12 The mixing of granular materials could occur in one of two regimes, the 'slow flow' regime where the stress is approximately equal to the yield stress τ_Y, and the 'rapid flow' regime where the constitutive relation is $\tau_{xz} = B(\Delta v_x/\Delta z)^2$, where B is called the Bagnold coefficient. What is the definition of the power number for the mixing by an impeller with frequency f_r in a tank with characteristic dimension d for slow and rapid flow regimes? What is the equivalent of the Reynolds number based on the impeller diameter and frequency for rapid flow?

EXERCISE 2.13 An extruder consisting of a helical screw in a cylinder is used for extruding a highly viscous liquid into moulds of different shapes. An extruder of length L with a screw of pitch p, which is used for extruding a polymeric liquid of density ρ and viscosity μ with volumetric flow rate Q, requires power P. If all linear dimensions of the extruder are increased by a factor of 10, and the flow rate is increased by a factor of 10, what is the power required?

EXERCISE 2.14 Solve the Example 2.2.6 for a mixer at low Reynolds number, where the droplet diameter depends only on the energy dissipation rate per unit mass ϵ, the viscosity μ and the surface tension γ. What is the equivalent of Eq. 2.2.51? What are the ratios (f_a/f_m) and P_a/P_m?

EXERCISE 2.15 A spherical iodine crystal of diameter 0.5 cm sublimes in air. The molecular weight of iodine is 254 g/mol, the density is 5×10^3 kg/m^3, and the vapour pressure at 25°C is 7.58×10^{-2} atm. The mass of iodine sublimed is 10 micrograms in one minute. What is the diffusion coefficient for iodine in air?

EXERCISE 2.16 Polymer microbeads are prepared by freezing polymer droplets in cold air under conditions of low Reynolds number and low Peclet number. If t is the time required for freezing a droplet of diameter d, what is the time required for freezing a droplet of the same material of diameter $2d$ under the same conditions? Assume that the latent heat of freezing is much larger than the sensible heat for reducing the temperature to freezing point.

EXERCISE 2.17 Water at average temperature 70°C is pumped at a pressure of 10 kPa into the tube side of a heat exchanger of diameter 2 cm and length 2 m. The temperature on the shell side is 30°C. What is the heat transfer rate? What is the power required for pumping the fluid, $-\Delta p \times \dot{V}$, where $\dot{V}$ is the volumetric flow rate? The density, viscosity, thermal conductivity and specific heat of water are $\rho = 10^3$ kg/m^3, $\mu = 10^{-3}$ kg/m/s, $k = 0.6$ W/m/°C and $C_p = 4.2 \times 10^3$ J/kg/°C.

EXERCISE 2.18 Consider the laminar flow in a pipe of length L and diameter d, with pressure difference Δp across the ends of the pipe. The average temperature difference between the wall and the fluid in the pipe is ΔT. The power required for driving the flow, Power $= -\Delta p \times \dot{V}$, where $\dot{V}$ is the volumetric flow rate. Express the power required for driving the flow and the heat transfer rate per unit time as a function of the pressure drop, temperature difference, pipe dimensions and the fluid properties. What is the optimum pressure drop where the difference between the heat transfer rate and the power for driving the flow is a maximum?

EXERCISE 2.19 Molten aluminium droplets of diameter 1 mm, density 2.7×10^3 kg/m^3 and temperature 660°C are dropped into a pool of viscous oil of density 850 kg/m^3, viscosity 3×10^{-2} kg/m/s, thermal conductivity 0.15 W/m/°C and specific heat 2×10^3 J/kg/°C maintained at a temperature of 100 °C. What should be the depth of the pool of oil so that the aluminium droplets freeze before they reach the bottom? The latent heat of melting of aluminium is 4×10^5 J/kg.

EXERCISE 2.20 Polymer particles of diameter 0.1 mm, density 600 kg/m^3 and temperature 0°C are dried as the settle in an upward current of air to remove the solvent ethanol. The mass fraction of ethanol in the polymer is 20%, the vapour pressure of ethanol in air at 0°C is 1.5 kPa, the molecular weight of ethanol is 46 g/mol, the latent heat of evaporation is 8.46×10^5 J/kg and the diffusion coefficient of ethanol in air is 1.1×10^{-5} m^2/s. What should be the temperature of air so that the time taken for heat and mass transfer are equal? What is the distance travelled before the ethanol is completely evaporated from the particle? The density, viscosity, thermal conductivity and specific heat of air are 1.28 kg/m^3, 1.73×10^{-5} kg/m/s, 2.43×10^{-2} W/m/°C and 10^3 J/kg/°C, respectively.

EXERCISE 2.21 The dimensions of the human body and the properties of air were presented in Example 2.4.1. Estimate the average heat loss per unit time due to forced convection if the wind speed is 10 m/s, the body temperature is 37°C and the ambient temperature is 25°C.

EXERCISE 2.22 Liquid sodium is used as a coolant in nuclear reactors at temperatures in the range 1000–1200°C. Determine the characteristic velocity, Grashof and Prandtl numbers for natural convection of liquid sodium around a tube of diameter 1 cm if the tube wall is at a temperature of 1200°C, and the ambient temperature is 1000°C. The properties of liquid sodium are as follows: density $\rho = 800$ kg/m^3, viscosity $\mu = 1.81 \times 10^{-4}$ kg/m/s, specific heat $C_p = 1.25 \times 10^3$J/kg/°C, thermal conductivity $k = 52$ W/m/°C and thermal expansion coefficient $\beta = 4 \times 10^{-4}(°C)^{-1}$.

EXERCISE 2.23 Using the parameters in Example 2.4.1, estimate the average daily heat loss from the body in air due to natural convection. Use the correlation, Eq. 2.4.14, with proportionality constant 0.5. What would be the heat loss if the mechanism of transfer is conduction, and the Nusselt number is 2?

3

Diffusion and Dispersion

The two transport mechanism considered in this text are convection and diffusion. Convection is transport due to the flow. It is directional, and takes place only along the flow streamlines. Transport across streamlines, and transport across surfaces (where there is no fluid velocity perpendicular to the surface) necessarily takes place due to diffusion.

Diffusion is the process by which material is transported by the random thermal motion of the molecules within the fluid, even in the absence of fluid flow. The random velocity fluctuations of the molecules are isotropic, and they have no preferred direction. The characteristic velocity and length for the thermal motion are the molecular velocity and the microscopic length scale, which is the molecular size in a liquid or the mean free path (distance between intermolecular collisions) in a gas. While random molecular motion is always present in fluids, when the concentration/temperature/velocity fields are uniform, there is no net transport due to the random motion. Diffusion takes place only when there is a spatial variation, and transport is along direction of variation.

The molecular mechanisms of mass, momentum and thermal diffusion, are discussed in this chapter. Constitutive relations for the fluxes are derived from a molecular description, and the diffusion coefficients are estimated.

The gas diffusivities are estimated using kinetic theory for an ideal gas made of hard spheres, which undergo instantaneous collisions when the surfaces are in contact, but which do not exert any intermolecular force when not in contact. Real gas molecules do not interact like hard spheres—the interaction force between molecules is repulsive at small separations and attractive at larger separations. Diatomic and polyatomic molecules are also not spherically symmetric, and their interaction depends on the relative orientation of the molecules. The diffusion coefficients in the hard sphere model are proportional to $\sqrt{T}$, where T is the absolute temperature. For molecules with continuous intermolecular potential, the diffusion coefficients are proportional to a power of the temperature which higher than $\frac{1}{2}$. The pressure-density relationship for real gases is also more complicated than that for an ideal gas, and the virial corrections need to be included for dense gases. The hard sphere model is used here because of its simplicity and ease of visualisation, but the diffusion coefficients calculated here should be considered estimates, and are not quantitatively accurate. Detailed calculations with accurate models are required for quantitative predictions; these are the subject of current research and are not within the scope of this text.

The 'dispersion' of mass, momentum and energy in turbulent flows, in packed columns and the axial dispersion in the flow through pipes are examined in Section 3.2. Dispersion is the transport due to the fluid velocity fluctuations, and not the molecular velocity fluctuations. The distinction between laminar and turbulent flows was discussed in Section 2.2.1 (Chapter 2). In a turbulent flow, there are large instantaneous fluid velocity fluctuations over length scales much larger than the molecular diameter or mean free path. This is in contrast to a laminar flow, where the streamlines are smooth and cross-stream transport occurs due to molecular diffusion. Turbulent flows contain 'eddies' which are parcels of fluid in correlated motion, and these eddies transport mass, momentum and energy at a rate that is significantly higher than that due to diffusion. The transport due to eddies in a turbulent flow, termed turbulent dispersion, is discussed in Section 3.2.1.

In a packed column, a fluid flows through a densely packed assembly of particles. While the flow velocity is, on average, along the axis of the column, the fluid follows a tortuous path in the voids between the particles at the particle scale. The velocity field is expressed as the sum of an average velocity along the axis and a spatially fluctuating velocity around the particles. The transport of mass and energy due to the spatial velocity fluctuations is the subject of Section 3.2.2. An interesting case of axial dispersion in a pipe flow, called 'Taylor dispersion', is described in Section 3.2.3.

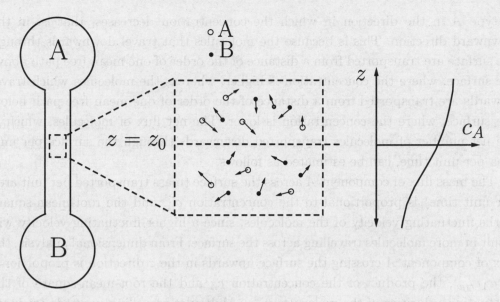

FIGURE 3.1. Mass diffusion due to a concentration variation in a gas.

3.1 Diffusion

3.1.1 Mass Diffusion

Mass Diffusion in Gases

Consider two gas chambers connected by a tube, as shown in Fig. 3.1. One of the chambers contains component A, while the other contains component B. For simplicity, A and B are considered to have equal molecular mass and diameter, and the initial pressures and temperatures in the two chambers are equal. When the barrier between the two is opened, there is no net transfer of mass between the two chambers, since the pressures and temperatures are equal. However, there is transport of component A from the chamber above to the chamber below, and transport of B in the opposite direction, until the concentrations in the two chambers are equal in the long time limit.

Consider a horizontal surface across which there is a variation of c_A, the concentration of component A, shown on the right in Fig. 3.1. There is a transport of molecules in both directions across this surface due to the fluctuating velocity. The flux of molecules at the surface is defined as the number of molecules crossing the surface per unit area per unit time. If the concentration and temperature are uniform in the z direction perpendicular to the surface, the total downward flux and the total upward flux are equal. However, if there is a variation of concentration across the surface, as shown in Fig. 3.1, then there is a net transport of molecules of type A in the direction in which the concentration decreases, that is, in the downward direction. This is because the molecules that travel downwards through the surface are transported from a distance of the order of one mean free path above the surface, where the concentration is higher, whereas the molecules which travel upwards are transported from a distance of the order of one mean free path below the surface, where the concentration is lower. The net flux of molecules, which is the net number of molecules transported downwards through the surface per unit area per unit time, can be estimated as follows.

The mass flux of component A across the surface (mass transported per unit area per unit time) is proportional to the concentration c_A, and the root-mean-square of the fluctuating velocity of the molecules, since a higher fluctuating velocity will result in more molecules travelling across the surface. From dimensional analysis, the flux of component A crossing the surface upwards in the z direction is proportional to $c_A v_{rms}$, the product of the concentration c_A and the root-mean-square of the fluctuating velocities of the molecules v_{rms}. Molecules travelling upwards originate at a location approximately $b\lambda$ below the surface, whereas those crossing the surface downwards originate at a distance which is on average $b\lambda$ above the surface. Here, b is a constant and λ is the mean free path. The concentration c_A is different at the starting locations of the molecules travelling downwards and upwards. Due to this, there is a net flux across the surface.

Since the molecules travelling downwards across the surface $z = z_0$ originate at a location approximately one mean free path above the surface, the downward flux of molecules at the location z_0, $j_{z\downarrow}$, is $a_j v_{rms}$ times the concentration $c_A(z_0 + b\lambda)$ at $z = z_0 + b\lambda$,

$$j_{z\downarrow} = a_j v_{rms} c_A(z_0 + b\lambda). \tag{3.1.1}$$

The factor a_j incorporates the ratio of the average z velocity of the downward travelling molecules and the root-mean-square fluctuating velocity, and the correlations between the concentration of species A and the downward fluctuating velocity. The detailed calculation of the correlation between the fluctuating velocity and the concentration of the molecules is beyond the scope of this text, and is not required for an estimate of the diffusion coefficient.

The molecules travelling upwards at the location $z = z_0$ originate at, on average, a distance $b\lambda$ below the surface, and so the upward flux of molecules is,

$$j_{z\uparrow} = a_j v_{rms} c_A(z_0 - b\lambda). \tag{3.1.2}$$

The net flux in the z direction is,

$$j_z = j_{z\uparrow} - j_{z\downarrow} = a_j v_{rms} [c_A(z_0 - b\lambda) - c_A(z_0 + b\lambda)]. \tag{3.1.3}$$

The concentration is expressed as a Taylor series expansion in z,

$$c_A(z_0 \pm b\lambda) = c_A(z_0) \pm b\lambda \left.\frac{dc_A}{dz}\right|_{z=z_0} + \frac{(b\lambda)^2}{2}\left.\frac{d^2c_A}{dz^2}\right|_{z=z_0} + \ldots \quad (3.1.4)$$

If L is the length of the tube connecting the two chambers in Fig. 3.1, the second term on the right scales as $(\lambda/L)\Delta c_A$, the third term on the right scales as $(\lambda/L)^2\Delta c_A$, and the nth term scales as $(\lambda/L)^n\Delta c_A$, where Δc_A is the difference in concentration of A between the two chambers. In the continuum limit where the macroscopic length scale L is much larger than the mean free path λ, each successive term in the expansion is smaller than the previous term by a factor (λ/L). Therefore, the series in Eq. 3.1.4 is approximated by the first two terms. When this truncated expansion is substituted into Eq. 3.1.3 for the flux, we obtain,

$$j_z|_{z=z_0} = -2a_j v_{rms} b\lambda \left.\frac{dc_A}{dz}\right|_{z=z_0}. \quad (3.1.5)$$

The diffusion coefficient is determined by comparing this equation with Fick's law for diffusion, Eq. 1.5.1 in differential form,

$$\boxed{\mathcal{D} = 2a_j b v_{rms} \lambda.} \quad (3.1.6)$$

The dimensionless constant $a_j b$ can not be obtained from the present calculation. But to within a dimensionless constant, Eq. 3.1.6 shows that the mass diffusion coefficient in a gas is proportional to the product of the molecular velocity and the mean free path.

The mean free path and the molecular velocity are estimated as follows. In a dilute gas at equilibrium, the root-mean-square of the molecular velocity is obtained from the equipartition condition, that the kinetic energy of the molecules $\frac{1}{2}Mv_{rms}^2$ is equal to the thermal energy $\frac{3}{2}k_BT$ for the translational degrees of freedom, where M is the molecular mass, k_B is the Boltzmann constant and T is the absolute temperature. The root-mean-square velocity is,

$$\boxed{v_{rms} = \sqrt{\frac{3k_BT}{M}}.} \quad (3.1.7)$$

EXAMPLE 3.1.1: Estimate the root-mean-square fluctuating velocity for hydrogen, nitrogen and oxygen at 300 K and atmospheric pressure.

Solution: The mass of one gram mole (6.023×10^{23} molecules) of hydrogen is 2 g (2×10^{-3}kg). Therefore, the mass of one molecule is

$$M = \frac{2 \times 10^{-3} \text{ kg}}{6.023 \times 10^{23}} = 3.32 \times 10^{-27} \text{ kg.} \qquad (3.1.8)$$

The Boltzmann constant $k_B = 1.38 \times 10^{-23}$ J/K. At $T = 300$ K, the root-mean-square of the fluctuating velocity is,

$$v_{rms} = \sqrt{\frac{3k_BT}{M}} = \sqrt{\frac{3 \times 1.38 \times 10^{-23} \text{ J/K} \times 300 \text{ K}}{3.32 \times 10^{-27} \text{ kg}}} = 1934 \text{ m/s.} \qquad (3.1.9)$$

The molecular mass of oxygen is 32 g/mol, which is 16 times that of hydrogen. From Eq. 3.1.7, the root-mean-square velocity of oxygen molecules is $\frac{1}{4}$ that of hydrogen, or about 483 m/s. The molecular velocity in nitrogen is slightly higher than oxygen, because its molecular mass (28 g/mol) is slightly less than oxygen. The root-mean-square molecular velocity in nitrogen is 517 m/s. The molecular velocity is comparable to the speed of sound, which is about 1290 m/s in hydrogen, and about 330 m/s in air. □

The mean free path λ is the distance travelled by a molecule between successive intermolecular collisions. The mean free path in a gas of hard sphere molecules, in which the molecule is modelled as a rigid sphere of diameter d_m travelling in a straight line between successive collisions, is schematically shown in Fig. 3.2. As it travels, the molecule sweeps out a cylindrical 'exclusion' volume of radius equal to the molecular diameter along the path. The volume of the cylinder swept out is $\pi d_m^2 L$, where L is the distance travelled by the molecule. The molecule will collide with a second molecule if the centre of the second molecule is located within the cylinder. The total number of second molecules within this volume is $(\pi d_m^2 L) \times N$, where N is the number density of the gas molecules, that is, the number of molecules per unit volume. On average, a collision will take place when the number of second molecules in the volume swept by the first molecule is approximately 1. Thus, the average distance between collisions, which is the mean free path λ, can be estimated from the condition $(N\pi d_m^2 \lambda) \sim 1$. The result from a more detailed calculation based on kinetic theory of gases is,

$$\boxed{\lambda = \frac{1}{\sqrt{2}\pi N d_m^2}.} \qquad (3.1.10)$$

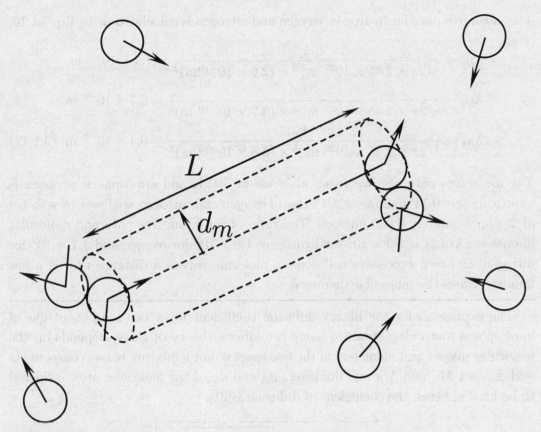

FIGURE 3.2. Schematic for estimating the distance travelled by a molecule between successive collisions in a hard sphere gas of molecular diameter d_m and number density N.

EXAMPLE 3.1.2: Estimate the mean free path of hydrogen, oxygen and nitrogen molecules, with molecular diameters 2.6×10^{-10} m, 3.7×10^{-10} m and 3.8×10^{-10} m, respectively, at 300 K and atmospheric pressure.

Solution: The number of gas molecules per unit volume N is determined from the ideal gas law,

$$N = \frac{p}{k_B T} = \frac{1.013 \times 10^5 \, \text{Pa}}{1.38 \times 10^{-23} \, \text{J/K} \times 300 \, \text{K}} = 2.45 \times 10^{25} \, \text{m}^{-3}. \qquad (3.1.11)$$

The mean free path for hydrogen, oxygen and nitrogen is calculated using Eq. 3.1.10,

$$\lambda_{H_2} = \frac{1}{\sqrt{2}\pi \times 2.45 \times 10^{25} \text{ m}^{-3} \times (2.6 \times 10^{-10} \text{m})^2} = 1.4 \times 10^{-7} \text{ m},$$

$$\lambda_{O_2} = \frac{1}{\sqrt{2}\pi \times 2.45 \times 10^{25} \text{ m}^{-3} \times (3.7 \times 10^{-10} \text{ m})^2} = 6.7 \times 10^{-8} \text{ m},$$

$$\lambda_{N_2} = \frac{1}{\sqrt{2}\pi \times 2.45 \times 10^{25} \text{ m}^{-3} \times (3.8 \times 10^{-10} \text{ m})^2} = 6.4 \times 10^{-8} \text{ m}. \quad (3.1.12)$$

The mean free path of a hydrogen molecule at 300 K and atmospheric pressure is approximately 0.14 microns, while those of oxygen and nitrogen are lower by a factor of 2, approximately 0.065 microns. The ratio of the mean free path and molecular diameter, (λ/d_m) is 5.2×10^2 for hydrogen, 1.8×10^2 for oxygen and 1.7×10^2 for nitrogen. Between successive collisions, a molecule travels a distance that is a few hundred times the molecular diameter. □

The expression for the binary diffusion coefficient for a two-component gas of hard sphere molecules, obtained using the kinetic theory of gases, depends on the molecular masses and diameters of the two species. For a mixture of two components with masses M_A and M_B and diameters d_A and d_B, if the molecules are considered to be hard spheres, the coefficient of diffusion is[21],

$$\mathcal{D}_{AB} = \frac{3}{8Nd_{AB}^2}\sqrt{\frac{k_B T(M_A + M_B)}{2\pi M_A M_B}}, \quad (3.1.13)$$

where $d_{AB} = (d_A + d_B)/2$ and $N = N_A + N_B$. The coefficient of 'self diffusion', which is the diffusion of a molecule in a gas composed of molecules of the same type, can be obtained by setting $d_A = d_B$ and $M_A = M_B$ in Eq. 3.1.13,

$$\mathcal{D}_{AA} = \frac{3}{8N_A d_A^2}\sqrt{\frac{k_B T}{\pi M_A}}. \quad (3.1.14)$$

Similar expressions can be derived for more complicated molecular models based on the interaction potential between the molecules. In all cases, the diffusion coefficient is proportional to the product of the mean free path and the root-mean-square of the fluctuating velocity.

EXAMPLE 3.1.3: Determine the mass diffusion coefficient for water vapour in air at atmospheric pressure and 300 K. Consider air as a single component gas with molecular weight 29 g/mol and molecular diameter 3.7 Å, and water has molecular weight 18 g/mol and diameter 2.6 Å.

Solution: The average diameter $d_{AB} = (d_{air} + d_{H_2O})/2 = 3.15$ Å. The molecular masses of air and water are,

$$M_{air} = \frac{29 \times 10^{-3} \text{ kg}}{6.023 \times 10^{23}} = 4.82 \times 10^{-26} \text{ kg}, \quad M_{H_2O} = \frac{18 \times 10^{-3} \text{ kg}}{6.023 \times 10^{23}} = 2.99 \times 10^{-26} \text{ kg}.$$
(3.1.15)

From Example 3.1.2, the number density is 2.45×10^{25} m^{-3}. The diffusion coefficient, Eq. 3.1.13, is

$$\mathcal{D}_{AB} = \frac{3}{8 \times 2.45 \times 10^{25} \text{ m}^{-3} \times (3.15 \times 10^{-10}\text{m})^2}$$
$$\times \sqrt{\frac{1.38 \times 10^{-23} \text{ J/K} \times 300 \text{ K} \times 7.81 \times 10^{-26} \text{ kg}}{2\pi \times 4.82 \times 10^{-26} \text{ kg} \times 2.99 \times 10^{-26} \text{ kg}}}$$
$$= 2.9 \times 10^{-5} \text{ m}^2/\text{s}.$$
(3.1.16)

This is close to the reported diffusion coefficient in the range $2.6 \times 10^{-5} - 2.8 \times 10^{-5}$ m^2/s[3]. □

Mass Diffusion in Liquids

If the diffusion coefficient for a liquid is calculated in a manner similar to that for a gas, the estimate turns out to be higher than the actual diffusion coefficient. In liquids, the molecules are densely packed, and the distance between molecules is comparable to the molecular diameter. The root-mean-square molecular velocity v_{rms} in liquids and gases are equal at the same temperature, because v_{rms} depends only on the temperature and molecular mass. The molecular diameters for liquids of small molecules are in the range 1–10 Å (10^{-10}–10^{-9} m), which is smaller than our estimate of the mean free path in a gas (10^{-7}–10^{-8} m) by a factor of 10^{-2}–10^{-3}. On this basis, the diffusion coefficient in liquids is expected to to be smaller than that in gases by a factor of 10^{-2}–10^{-3}. However, the diffusion coefficient of small molecules in a liquid such as water is smaller than that in a gas by a factor of 10^{-4}. For example, the diffusion coefficient of nitrogen in water is 1.9×10^{-9} m^2/s, while that of hydrogen in water is 4.5×10^{-9} m^2/s. The diffusion coefficient of larger molecules, such as polymers, in water is smaller still; the diffusion coefficient of haemoglobin in water is 6.9×10^{-11} m^2/s.

Eq. 3.1.6 cannot be used to estimate the diffusion coefficients in liquids because 'co-operative motion' is necessary for diffusion in a liquid. The molecules in a liquid

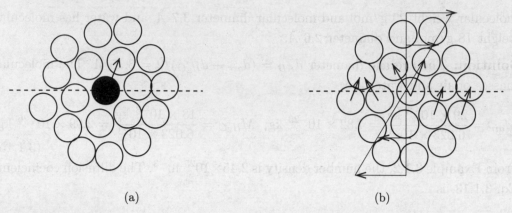

FIGURE 3.3. The requirement for cooperative motion for the transport of mass across a surface in a liquid (a) and the transport of momentum in a shear flow due to the intermolecular forces exerted by molecules on one side of the surface on molecules on the other side of the surface (b).

are closely packed, as shown in Fig. 3.3. The translation of one molecule requires the collective motion of surrounding molecules. For example, the motion of the dark molecule in Fig. 3.3(a) is confined by a 'cage' formed by the surrounding molecules, and translation of this molecule takes place due to rare cage breakage events which requires co-operative motion of the neighbouring molecules. Due to this, the diffusion coefficient in a liquid is 10^{-1}–10^{-2} times smaller than that estimated by Eq. 3.1.6.

An estimate for the diffusion coefficient can be obtained using the Stokes–Einstein equation,

$$\mathcal{D} = \frac{k_B T}{3\pi \mu d_m},$$
(3.1.17)

where d_m is the diameter of the molecule that is diffusing, and μ is the viscosity of the liquid. This formula is strictly applicable only for colloidal particles in a liquid where the particle diameter is large compared to the diameter of the liquid molecules, but it is also used for estimating the diffusivity of small molecules in a liquid.

EXAMPLE 3.1.4: Determine the diffusion coefficient of nitrogen molecules in water at 300 K. The molecular diameter of nitrogen is 3.8×10^{-10} m, and the viscosity of water is 10^{-3} kg/m/s.

Solution: The diffusion coefficient, Eq. 3.1.17, is

$$\mathcal{D} = \frac{1.38 \times 10^{-23} \text{ J/K} \times 300 \text{ K}}{3\pi \times 10^{-3} \text{ kg/m/s} \times 3.8 \times 10^{-10} \text{ m}} = 1.16 \times 10^{-9} \text{ m}^2/\text{s}.$$
(3.1.18)

This result is not in quantitative agreement with the experimental value of 1.9×10^{-9} m^2/s[3], though the order of magnitude is correct. This is because Eq. 3.1.17 is not strictly valid when the size of the diffusing molecule is comparable to the size of the liquid molecules. $\square$

The diffusion coefficient in a liquid increases with an increase in temperature. This is because an increase in temperature increases the molecular fluctuating velocity, and thereby increases the frequency of rearrangements and the co-operative motion. Eq. 3.1.17 indicates that the diffusion coefficient increases proportional to the ratio of the absolute temperature and the viscosity. The viscosity decreases as temperature increases for reasons discussed in Section 3.1.3, resulting in an increase in the diffusion coefficient with temperature.

3.1.2 Thermal Diffusion

Thermal diffusion is the transfer of energy due to the random motion of molecules when there is a variation in the temperature in the system. In a gas, thermal diffusion takes place due to the physical motion of molecules across a surface. Transport processes usually occur under conditions of constant pressure, where the heat transferred per unit volume to a material is equal to the change in the enthalpy density (enthalpy per unit volume) h.[1] Consider a surface at $z = z_0$ across which there is a variation in the temperature, and therefore a variation in the enthalpy density h. The flux of heat downwards through the surface is the result of molecules travelling downwards from a distance approximately $b\lambda$ above the surface. Analogous to Eq. 3.1.1 for the mass flux, the downward heat flux is,

$$q_{z\downarrow} = a_e v_{rms}(z_0 + b\lambda) h(z_0 + b\lambda), \tag{3.1.19}$$

where a_e is a constant that incorporates the correlations between the heat energy and the fluctuating velocity of the molecules. Note that we have considered the value of v_{rms} at the location $z_0 + b\lambda$; this is because a variation in the temperature results in a variation in the root-mean-square of the fluctuating velocity from Eq. 3.1.7. The average rate of transport of heat per unit area upwards through the surface,

---

[1]Enthalpy is defined as $H = U + pV$, where U is the internal energy, p is the pressure and V is the volume. The change in enthalpy is $\Delta H = \Delta U + p\Delta V + V\Delta p$. The change in internal energy is $\Delta U = \Delta Q - p\Delta V$ from the first law of thermodynamics, where ΔQ is the heat transferred to the system. Therefore, the change in enthalpy is $\Delta H = \Delta Q + V\Delta p$. For a constant pressure system, $\Delta p = 0$, and the change in enthalpy is equal to the heat transferred to the system.

analogous to Eq. 3.1.2, is

$$q_{z\uparrow} = a_e v_{rms}(z_0 - b\lambda)h(z_0 - b\lambda). \tag{3.1.20}$$

The total energy flux is

$$
\begin{aligned}
q_z|_{z=z_0} &= q_{z\uparrow} - q_{z\downarrow} \\
&= a_e v_{rms}(z_0 - b\lambda)h(z_0 - b\lambda) - a_e v_{rms}(z_0 + b\lambda)h(z_0 + b\lambda) \\
&= \left(a_e(v_{rms}h)|_{z=z_0} - a_e b\lambda \left.\frac{\mathrm{d}(v_{rms}h)}{\mathrm{d}z}\right|_{z=z_0} \right) \\
&\quad - \left(a_e v_{rms}h|_{z=z_0} + a_e b\lambda \left.\frac{\mathrm{d}(v_{rms}h)}{\mathrm{d}z}\right|_{z=z_0} \right) \\
&= -2a_e b\lambda \left.\frac{\mathrm{d}(v_{rms}h)}{\mathrm{d}z}\right|_{z=z_0}. \tag{3.1.21}
\end{aligned}
$$

The enthalpy density h is $\rho C_p T$, where C_p is the specific heat at constant pressure. Considering the variations in both h and v_{rms} with temperature in Eq. 3.1.21, the expression for the heat flux has the same form as Fourier's law,[2]

$$q_z|_{z=z_0} = -3a_e b v_{rms}\lambda\rho C_p \left.\frac{\mathrm{d}T}{\mathrm{d}z}\right|_{z=z_0} = -k \left.\frac{\mathrm{d}T}{\mathrm{d}z}\right|_{z=z_0}, \tag{3.1.22}$$

where the thermal conductivity is,

$$k = 3a^e b v_{rms}\lambda\rho C_p. \tag{3.1.23}$$

The thermal diffusivity is $(k/\rho C_p)$,

$$\boxed{\alpha = \frac{k}{\rho C_p} = 3a_e b v_{rms}\lambda.} \tag{3.1.24}$$

The thermal diffusivity in gases is also proportional the $v_{rms}\lambda$, the product of the root-mean-square fluctuating velocity and the mean free path. A calculation based on kinetic theory for hard sphere molecules results in the following expression for the thermal conductivity of a hard sphere gas[21],

$$\boxed{k = \frac{25}{32d_m^2}\frac{1}{\gamma - 1}\left(\frac{k_B^3 T}{\pi M}\right)^{1/2},} \tag{3.1.25}$$

where d_m is the molecular diameter, M is the molecular mass, and $\gamma = (C_p/C_v)$ is the ratio of specific heats.

---

[2]The derivative $(\mathrm{d}v_{rms}/\mathrm{d}z)$ is $\mathrm{d}(\sqrt{3k_B T/M}/\mathrm{d}z)$ from Eq. 3.1.7. This can be simplified as $\sqrt{3k_B/M}/(2\sqrt{T})(\mathrm{d}T/\mathrm{d}z) = (v_{rms}/2T)(\mathrm{d}T/\mathrm{d}z)$.

EXAMPLE 3.1.5: Determine the thermal conductivity of air if the molecules are considered diatomic molecules with molecular weight 29 g/mol and the molecular diameter 3.7 Å.

Solution: For a diatomic molecule, the ratio of specific heats $\gamma = \frac{7}{5}$. Eq. 3.1.25 for the thermal conductivity is,

$$k = \frac{25}{32 \times (3.7 \times 10^{-10} \text{ m})^2} \frac{5}{2} \left(\frac{(1.38 \times 10^{-23} \text{ J/K})^3 \times 300 \text{ K}}{\pi \times 4.82 \times 10^{-26} \text{ kg}} \right)^{1/2} = 0.032 \text{ W/m/K}.$$

(3.1.26)

The experimental values of the thermal conductivity are in the range $0.026 - 0.028$ W/m/K, slightly lower than the above estimate. $\square$

The thermal diffusivity for gases is of the same magnitude as the mass diffusivity, because the mechanism of energy transport (fluctuating motion of the gas molecules) is the same as that for mass transport. The Prandtl number $(C_p \mu/k)$, which is the ratio of the momentum and thermal diffusivity, is $(2/3)$ for monoatomic gases comprising hard sphere molecules, and experimentally observed values vary between 0.66 for unimolecular gases such as neon and argon, to approximately 1 for water at boiling point at atmospheric pressure. For polyatomic molecules, there is a transfer of energy between the translational and internal modes, and the Eucken formula $\text{Pr} = C_p/(C_p + 1.25(k_B/M))$ is found to provide good predictions, where k_B is the Boltzmann constant and M is the mass of a molecule.

In liquids, the mechanism of heat transport could be very different from that of mass and momentum transport. Whereas the transport of mass requires the physical transport of molecules across a surface, the transport of heat does not. One mechanism is the fluctuating motion of molecules transferring heat to neighbouring molecules. A second mechanism, in liquid metals, is the transport of heat through the electron cloud around the metal atoms. Transport of heat due to the latter mechanism is a very rapid process, resulting in high thermal diffusivity. Due to this, the Prandtl number of liquid metals is very low. For example, the Prandtl number of liquid mercury is 0.015 indicating that the thermal diffusivity is about 60 times higher than the momentum diffusivity. In contrast, for organic liquids, the thermal diffusivity is smaller than the mass diffusivity, and the Prandtl number varies between 10^2 and 10^4. The Prandtl number for water is about 7.

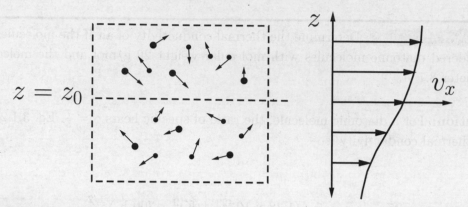

FIGURE 3.4. Flux of x momentum in the z direction due to the z variation of the x component of the velocity, v_x.

3.1.3 Momentum Diffusion

The shear stress τ_{xz} is defined as the force per unit area exerted in the x direction at the surface with outward unit normal in the z direction. The force per unit area can also be interpreted as the momentum flux (momentum transfer per unit area per unit time), with dimension $\mathcal{ML}^{-1}\mathcal{T}^{-2}$. It is important to note the physical meaning of the two subscripts in the stress τ_{xz}. Momentum is a vector with three components. The first subscript x in τ_{xz} is the direction of the momentum. The second subscript z is the direction in which the momentum is transported. The convention used here is that τ_{xz} is the rate of transport of the x momentum in the z direction. Alternately, τ_{xz} is the force per unit area exerted in the x direction on a surface perpendicular to the z direction.

The flux of momentum across a surface in a dilute gas can be estimated using arguments similar to that for mass diffusion. When the flow velocity v_x increases in the z direction, transport of x momentum takes place across a surface at the location z_0, as shown in Fig. 3.4. This is because molecules travel downwards from a distance approximately one mean free path above the surface, where v_x is higher, whereas the molecules travel upwards from a distance approximately one mean free path below the surface where v_x is lower. The flux of x momentum downwards through the surface is the product of the number of molecules travelling through the surface per unit area per unit time $a_m N v_{rms}$ and the average momentum of the molecules at the location $z_0 + b\lambda$,

$$\tau_{xz\downarrow} = (a_m N v_{rms})(M v_x(z_0 + b\lambda)). \tag{3.1.27}$$

The factor a_m includes the ratio of the average z velocity of the downward travelling molecules and the root-mean-square fluctuating velocity, and the correlations between the x momentum and the fluctuating velocity in the z direction in the presence of a mean shear flow. This could be different from a_j (Eq. 3.1.3) and a_e (Eq. 3.1.21) for mass and energy transport, respectively, because of the difference in the extents of correlations of mass, x momentum and energy with the fluctuating velocity. The upward flux of x momentum through the surface is $a_m N v_{rms}$ times the average momentum of the molecules at the location $z_0 - b\lambda$,

$$\tau_{xz\uparrow} = (a_m N v_{rms})(M v_x(z_0 - b\lambda)). \tag{3.1.28}$$

The sign convention for the shear stress τ_{xz} differs from that for the mass and energy fluxes, j_z and q_z. The mass and energy fluxes were defined to be positive if the the flux is in the $+z$ direction; therefore, the fluxes were calculated as $j_z = j_{z\uparrow} - j_{z\downarrow}$. A positive mass/energy flux increases the mass/energy of the volume above the surface at z_0, as shown in Fig. 3.1. In contrast, the shear stress is defined as positive if a force in the $+x$ direction is exerted at a surface with outward unit normal in the $+z$ direction. In Fig. 3.4, this implies that a positive force in the x direction is exerted on a the volume below the surface at z_0. This increases the x momentum of the volume below the surface at z_0. Therefore, the shear stress is defined as,

$$\tau_{xz} = \tau_{xz\downarrow} - \tau_{xz\uparrow}. \tag{3.1.29}$$

A Taylor series expansion is used for the fluid velocity v_x, similar to Eq. 3.1.4 for the concentration, and the series is approximated by the first two terms,

$$v_x(z_0 \pm b\lambda) = v_x(z) \pm b\lambda \left. \frac{dv_x}{dz} \right|_{z=z_0}. \tag{3.1.30}$$

This is substituted into Eqs. 3.1.27, 3.1.28 and 3.1.29 for the shear stress, to obtain,

$$\tau_{xz}|_{z=z_0} = 2a_m b N M v_{rms} \lambda \left. \frac{dv_x}{dz} \right|_{z=z_0} = 2a_m b \rho v_{rms} \lambda \left. \frac{dv_x}{dz} \right|_{z=z_0}, \tag{3.1.31}$$

where $\rho = NM$ is the mass density. Comparing the above equation the Newton's law of viscosity, the differential form of Eq. 1.5.4, the coefficient of viscosity is,

$$\boxed{\mu = 2a_m b \rho v_{rms} \lambda.} \tag{3.1.32}$$

The 'momentum diffusivity' can be obtained by recasting the right side of Eq. 3.1.31 in terms of the gradient in the momentum density, (ρv_x),

$$\tau_{xz}|_{z=z_0} = 2a_m b v_{rms} \lambda \left. \frac{d(\rho v_x)}{dz} \right|_{z=z_0} = \nu \left. \frac{d(\rho v_x)}{dz} \right|_{z=z_0}, \tag{3.1.33}$$

where $\nu = (\mu/\rho)$ is the 'kinematic viscosity' or the 'momentum diffusivity',

$$\nu = 2a_m b v_{rms} \lambda. \tag{3.1.34}$$

The expressions for μ (Eq. 3.1.32) and ν (Eq. 3.1.34) contain an unknown coefficient, a_m. The value of the coefficient can be obtained from a calculation based on kinetic theory of gases which is beyond the scope of this text. The viscosity for a dilute gas consisting of hard sphere molecules of mass M and molecular diameter d_m is[21],

$$\mu = \frac{5}{16 d_m^2} \sqrt{\frac{M k_B T}{\pi}}, \quad \nu = \frac{5}{16 N d_m^2} \sqrt{\frac{k_B T}{\pi M}}. \tag{3.1.35}$$

Eq. 3.1.35 indicates that the momentum diffusivity is proportional to $\sqrt{T}$, where T is the absolute temperature, because the root-mean-square of the molecular fluctuating velocity is proportional to $\sqrt{T}$. The momentum diffusivity is inversely proportional to the number density, because the mean free path is inversely proportional to the number density. The viscosity is proportional to $\sqrt{T}$ and independent of the number density in a dilute gas.

In liquids, the momentum diffusivity turns out to be much higher than the mass diffusivity, due to a difference in the mechanism of transport. In contrast to mass transport, momentum transport does not require the transport of molecules across a surface; momentum transport across a surface takes place due to intermolecular forces exerted by molecules with centres on one side of the surface on molecules with centres on the other side, as shown in Fig. 3.3(b). Due to the rapid transport of momentum by intermolecular forces, the momentum diffusivity in liquids is higher than the mass diffusivity by a factor of 10^2–10^3. For example, the self-diffusion coefficient for water molecules in liquid water is about 2.3×10^{-9} m^{2}s^{-1}, whereas the kinematic viscosity of water is 1×10^{-6} m^{2}s^{-1}. Consequently, the Schmidt number for liquids is in the range 10^2–10^3. The momentum diffusivity for liquids is smaller than that for gases by a factor of about 10; for example, the kinematic viscosity of water is 1×10^{-6} m^2 s^{-1} at 20°C and atmospheric pressure, whereas that for air is 1.5×10^{-5} m^2 s^{-1}.

Both the viscosity and the momentum diffusivity in a liquid decrease with an increase in temperature. It might naively be expected that the viscosity will increase with temperature due to an increase in the molecular fluctuating velocity, and the consequent increase in the rate of transport of momentum. However, there is another factor which affects momentum transport, which is the attractive forces between the molecules in the liquid. When the temperature increases, there is a slight increase in the distance between molecules, and a decrease in the attractive force between the molecules. This results in a decrease in the rate of momentum transfer, and therefore the viscosity decreases as the temperature increases in liquids.

Summary (3.1)

1. In gases, the distance moved by molecules is large compared to the molecular diameter. The mechanism of mass, momentum and energy diffusion is the physical transport of molecules across a surface from distances comparable to one mean free path on either side of the surface.

2. The mass diffusivity (Eq. 3.1.6), momentum diffusivity (Eq. 3.1.34) and thermal diffusivity (Eq. 3.1.24) are all proportional to $v_{rms}\lambda$, the product of the molecular fluctuating velocity, Eq. 3.1.7, and the mean free path, Eq. 3.1.10. For gases under conditions of standard temperature and pressure, the diffusivities are of the order of 10^{-5} m^2/s.

3. For specific molecular models, the transport coefficients in a dilute gases can be related to molecular parameters using kinetic theory of gases; for a gas of hard sphere molecules, the mass diffusion coefficient is given in Eqs. 3.1.13 and 3.1.14, the thermal conductivity is given in Eq. 3.1.25 and the viscosity is given in Eq. 3.1.35.

4. In liquids, mass diffusion is a slow process, because the transport of mass requires the physical transport of molecules. The molecules in a liquid are densely packed, as shown in Fig. 3.3, and molecules are confined within 'cages' formed by surrounding molecules. For a molecule to move a distance larger than its diameter in one direction, it is necessary for collective rearrangement and co-operative motion of the surrounding molecules. This results in mass diffusivities of the order 10^{-9} m^2/s for small molecules in a liquid such as water, which is much smaller than the $v_{rms}d_m$, the product of the root-mean-square velocity and the molecular diameter.

5. The diffusion coefficient in liquids can be estimated by the Stokes–Einstein relation, Eq. 3.1.17.

6. In liquids, momentum is transported across a surface due to the intermolecular forces between molecules with centres on either side, and momentum transport does not require the transport of a molecule across the surface. Due to this,

the momentum diffusivity could be a factor of 10^2–10^3 larger than the mass diffusivity in liquids, and the Schmidt number (ratio of momentum and mass diffusivity) is much larger than 1.

7. The value of the thermal diffusivity in liquids depends on the mechanism of energy transport. In liquid metals, there is rapid transport of heat due to the cloud of delocalised electrons, resulting in a very high thermal conductivity, and a low Prandtl number. In organic liquids, the thermal diffusivity is small compared to the momentum diffusivity, resulting in a high Prandtl number.

3.2 Dispersion

A distinction has hitherto been made between convection due to the fluid velocity and diffusion due to the molecular fluctuating velocity. In many flows, it is useful to separate the fluid velocity into the fluid mean velocity and the fluid fluctuating velocity. The fluid mean velocity is defined in different ways for different flows, such as the time average at a spatial location for turbulent flows, or the spatial average over a volume larger than a particle diameter for flows in packed columns filled with particles. The fluid fluctuating velocity is the difference between the instantaneous or local fluid velocity and the fluid mean velocity. In addition to the fluid velocity fluctuations, there are also fluctuations in the concentration/temperature in the case of mass/heat transport, and the fluid stress in the case of momentum transport. The concentration and temperature fields are also separated into the mean and fluctuating parts in a manner similar to the fluid velocity. The convective transport due to the correlations between fluid fluctuating velocity and the fluctuations in the quantities being transported is called dispersion.

In the commonly used models for dispersion, constitutive relations similar to Fick's law for diffusion, Fourier's law for heat conduction or Newton's law for viscosity relate the flux to the variations in the mean quantities. However, there is an important distinction between diffusion and dispersion. Dispersion is a flow property, which depends on the flow characteristics. In contrast, diffusion is due to the molecular fluctuating velocities, which is a material property of the fluid and is independent of the geometry or flow conditions.

3.2.1 Turbulent Dispersion

The distinction between laminar and turbulent flows was discussed in Section 2.2.1 (Chapter 2). A steady laminar flow consists of smooth streamlines, where the fluid velocity is independent of time, and transport perpendicular to the flow direction is necessarily by diffusion. Turbulent flows are characterised by the correlated motion of parcels of fluid called 'eddies', which fluctuate both in time and space. The eddies have a range of sizes, the largest eddies are comparable to the flow dimension, while the smallest eddies are much larger than the molecular size. Therefore, the motion of the eddies is governed by the continuum equations of fluid mechanics, and turbulence is a continuum phenomenon.

For example, for the turbulent flow in a pipe shown in Fig. 3.5, even when a steady pressure difference is applied across the ends, the velocity fluctuates in time and in all three directions. The components of the mean velocity $\bar{v}_x$, $\bar{v}_y$ and $\bar{v}_z$ at a location are defined as averages over a time t_{av},

$$\boxed{\bar{v}_x = \frac{1}{t_{av}} \int_0^{t_{av}} dt \, v_x(t), \ \bar{v}_y = \frac{1}{t_{av}} \int_0^{t_{av}} dt \, v_y(t) = 0, \ \bar{v}_z = \frac{1}{t_{av}} \int_0^{t_{av}} dt \, v_z(t) = 0.}$$
(3.2.1)

When the averaging time t_{av} is much larger than the correlation time for the eddies, the averages in Eq. 3.2.1 are independent of the time of averaging t_{av}. These represent the steady mean flow, upon which are superposed the fluid fluctuating velocities. In the example shown in Fig. 3.5, the flow is in the x direction on average, and so the averages $\bar{v}_y$ and $\bar{v}_z$ are zero. However, the instantaneous values of all three components of the velocity are non-zero in a turbulent flow.

The fluctuating velocity is defined as the difference between instantaneous velocity and the mean velocity. In the example shown in Fig. 3.5, the stream-wise fluctuating velocity is,

$$\boxed{v_x'(t) = v_x(t) - \bar{v}_x,}$$
(3.2.2)

while the other two components of the fluctuating velocity are equal to their instantaneous values. The time average of the fluctuating velocities is zero.

In turbulent flows involving mass and heat transfer, there are fluctuations in the concentration or temperature fields, as shown in Fig. 3.6. The mean concentration and temperature are defined in a manner similar to the mean velocity, Eq. 3.2.1,

$$\boxed{\bar{c} = \frac{1}{t_{av}} \int_0^{t_{av}} dt \, c(t), \ \bar{T} = \frac{1}{t_{av}} \int_0^{t_{av}} dt \, T(t).}$$
(3.2.3)

The fluctuating concentration/temperature is then defined as the difference between the instantaneous and mean concentration/temperature,

$$\boxed{c'(t) = c(t) - \bar{c}, \ T'(t) = T(t) - \bar{T}.}$$
(3.2.4)

Note that in Eqs. 3.2.2 and 3.2.4, the instantaneous and fluctuating quantities vary in time, but the mean quantities are independent of time.

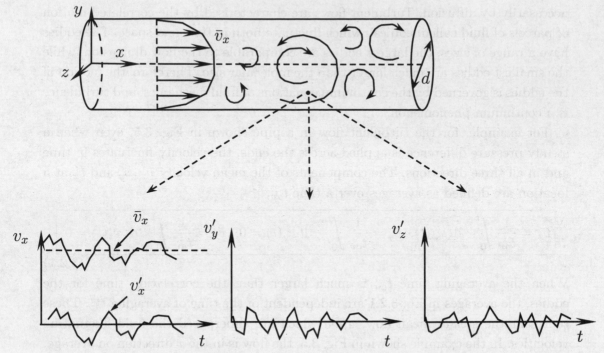

FIGURE 3.5. The turbulent flow in a pipe consists of a mean velocity in the x direction, $\bar{v}_x$, and fluctuations in all three directions, v'_x, v'_y, and v'_z at all points in the flow.

The physical reason for turbulent dispersion is as follows. The convective flux in the x direction at a location in the flow, shown in Fig. 3.5, is $j_x = cv_x$. If we take the time average of the convective flux, and express the concentration and velocity as the sum of the mean and fluctuating part, we obtain,

$$\bar{j}_x = \frac{1}{t_{av}} \int_0^{t_{av}} dt\, (\bar{c} + c'(t))(\bar{v}_x + v'_x(t)),$$

$$= \frac{1}{t_{av}} \int_0^{t_{av}} dt\, \bar{c}\bar{v}_x + \frac{1}{t_{av}} \int_0^{t_{av}} dt\, \bar{c}v'_x + \frac{1}{t_{av}} \int_0^{t_{av}} dt\, c'\bar{v}_x + \frac{1}{t_{av}} \int_0^{t_{av}} dt\, c'v'_x.$$

$$(3.2.5)$$

The first term on the right in the above expansion is $\bar{v}_x\bar{c}$, because the mean quantities are independent of time. The second and third terms on the right are products of one mean quantity and one fluctuating quantity; the former is independent of time and the latter is zero on average, so the second and third terms are both equal to zero. However, the fourth term on the right, which is the time average of two fluctuating quantities, is not zero in general. If there are correlations in the fluctuations of the concentration and velocity fields, then the transport due to these correlations could be non-zero. Therefore, the time average of the flux due to convection has two components, one due to the mean velocity and the second due to correlations between the fluctuating concentration and velocity fields,

$$\bar{j}_x = \bar{c}\bar{v}_x + \frac{1}{t_{av}} \int_0^{t_{av}} dt \, c' v_x'. \tag{3.2.6}$$

The first term on the right is the convective flux due to the mean velocity. The second term on the right, called the 'Reynolds flux' j_x^R, is the dispersion flux due to the correlation in the fluctuations of the concentration and the fluid velocity,

$$\boxed{j_x^R = \frac{1}{t_{av}} \int_0^{t_{av}} dt \, c'(t) v_x'(t) = \overline{c' v_x'}.} \tag{3.2.7}$$

It should be noted that the average of the fluctuations, $\overline{c' v_x'}$ could be non-zero, even though the averages $\overline{c'}$ and $\overline{v_x'}$ are both zero. To illustrate this, sequences of v_x' and c' which fluctuate in time are schematically shown in Fig. 3.6(a) and (b). Each of these sequences has zero average—that is, the area under the curve is zero for both of these sequences. However, when the two sequences are multiplied, the product clearly has a positive average, as shown in Fig. 3.6(c). This is because the two sequences for v_x' and c' are correlated—that is, when v_x' is positive, c' has a greater probability of being positive, and vice versa. This correlation results in a positive value of the Reynolds flux, $\overline{c' v_x'}$.

EXAMPLE 3.2.1: The fluctuations in the concentration and velocity at a point in the flow are,

$$c' = c_0' \sin(\omega t), \quad v_x' = v_{x0}' \sin(\omega t + \phi), \tag{3.2.8}$$

where c_0' and v_{x0}' are the amplitudes, ω is the frequency of fluctuations and ϕ is the phase shift. Determine the Reynolds flux.

Solution: The average of the fluctuations in the concentration is,

$$\bar{c} = \frac{c_0'}{t_{av}} \int_0^{t_{av}} dt \, \sin(\omega t) = \left. \frac{-c_0' \cos(\omega t)}{\omega t_{av}} \right|_0^{t_{av}}. \tag{3.2.9}$$

The averaging time is much larger than the characteristic time for the concentration fluctuations, which is $(2\pi/\omega)$, where ω is the frequency. Therefore, the above integral is evaluated in the limit $(t_{av}\omega) \gg 1$. Since the cosine function in the numerator of 3.2.9 is bounded, $-1 \leq \cos(\omega t) \leq 1$, the above integral tends to zero for $t_{av}\omega \gg 1$. In a similar manner, the average of the velocity fluctuation is also zero.

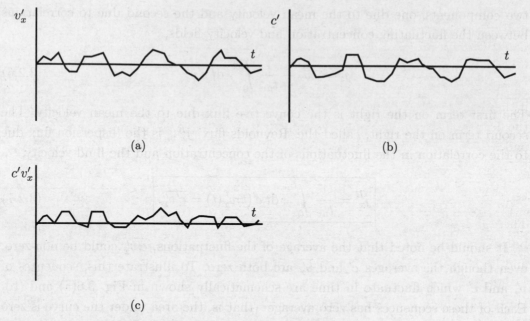

FIGURE 3.6. Fluctuations in the velocity v'_x (a) and concentration c' (b), each with zero time average, could result in a non-zero average for the product $(v'_x c')$ (c).

The Reynolds flux is,

$$\overline{c'v'_x} = \frac{c'_0 v'_{x0}}{t_{av}} \int_0^{t_{av}} dt \sin(\omega t) \sin(\omega t + \phi). \tag{3.2.10}$$

The integrand in the above equation is simplified using the trigonometric identity,

$$\sin(\omega t) \sin(\omega t + \phi) = \sin(\omega t)(\sin(\omega t) \cos(\phi) + \cos(\omega t) \sin(\phi))$$
$$= \tfrac{1}{2}[(1 - \cos(2\omega t)) \cos(\phi) + \sin(2\omega t) \sin(\phi)]. \tag{3.2.11}$$

This is substituted in Eq. 3.2.6, and integrated,

$$\overline{c'v'_x} = \frac{c'_0 v'_{x0}}{2t_{av}} \left[\left(t - \frac{\sin(2\omega t)}{2\omega} \right) \cos(\phi) - \frac{\cos(2\omega t)}{2\omega} \sin(\phi) \right]\Bigg|_0^{t_{av}}. \tag{3.2.12}$$

There is one term proportional to t in the square brackets, which increases linearly with time, while the other two terms are bounded. The contributions from the latter are proportional to $(t_{av}\omega)^{-1}$, and these tend to zero when the averaging time is much longer than the characteristic time of the fluctuations. Therefore, the Reynolds flux is,

$$\overline{c'v'_x} = \frac{c'_0 v'_{x0} t \cos(\phi)}{2t_{av}}\Bigg|_{t=0}^{t_{av}} = \frac{c'_0 v'_{x0} \cos(\phi)}{2}. \tag{3.2.13}$$

$\square$

A similar decomposition can be made for the average heat flux $\bar{q}_z$ in a turbulent flow, and the Reynolds heat flux $\bar{q}_z^R$ is,

$$\bar{q}_z^R = \rho C_p \overline{T' v_z'},$$

(3.2.14)

if the density and specific heat are constants. The stress in a turbulent flow due to the fluid velocity fluctuations is called the 'Reynolds stress'. Analogous to Eqs. 3.2.6 and 3.2.14, the Reynolds stress in a turbulent flow can be written as,

$$\bar{\tau}_{xz}^R = -\rho \overline{v_x' v_z'}.$$

(3.2.15)

There is a negative sign in the above equation because the stress τ_{xz}^R is defined to be positive if it acts at a surface whose outward perpendicular is in the z direction, and is therefore the negative of the momentum flux, as discussed in Section 3.1.3.

The Reynolds flux and stress are often modelled using equations similar to Fick's law for mass diffusion (Eq. 2.1.6), Fourier's law for thermal diffusion (Eq. 2.1.7), and Newton's law for momentum diffusion (Eq. 2.1.8), with the important difference that the gradients in the mean concentration, temperature or velocity are used in the expressions for the flux,

$$\bar{j}_z^R = -\mathcal{D}^R \frac{d\bar{c}}{dz},$$

(3.2.16)

$$\bar{q}_z^R = -\rho C_p \alpha^R \frac{d\bar{T}}{dz},$$

(3.2.17)

$$\bar{\tau}_{xz}^R = \rho \nu^R \frac{d\bar{v}_x}{dz},$$

(3.2.18)

where $\mathcal{D}^R$, α^R and ν^R are the turbulent or eddy diffusivities.[3]

The physical reason for the 'mixing length' models, Eqs. 3.2.16–3.2.18, is similar to that for molecular diffusion. Instead of the molecular length, the relevant length scale here is the 'mixing length' l which is the approximate length over which there is correlated motion of an eddy. Consider a variation in the mean concentration in a turbulent flow in the z direction, as shown in Fig. 3.7. When a turbulent eddy moves in the upward $(+z)$ direction across a surface in the flow, it carries with it fluid with mean concentration at a distance l below the surface. Similarly, a turbulent eddy moving in the downward $(-z)$ direction carries fluid with mean concentration at a distance l above the surface. This results in a transport of mass

---

[3]It would be more precise to refer to the coefficients $\mathcal{D}^R$, α^R and ν^R as 'turbulent dispersivities', but these have traditionally been called 'turbulent diffusivities', and so we employ the same terminology.

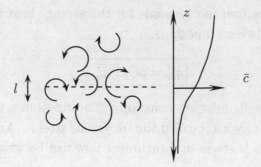

FIGURE 3.7. Transport of mass down a gradient of the mean concentration due to turbulent eddies.

from regions of higher concentration to regions of lower concentration. Based on this physical picture, the eddy diffusivities in Eqs. 3.2.16–3.2.18 are modelled as the product of the 'mixing length' l and the fluctuating velocity of the eddies v'. Since the mechanism of transport, the physical motion of turbulent eddies across a surface, is identical for mass, momentum and energy transport, the turbulent diffusivities for mass, momentum and energy are also comparable.

The expressions for the turbulent diffusivities appear similar to those for the molecular diffusivities in a gas. However, there are important distinctions that need to be emphasised. Firstly, the fluid fluctuating velocities and eddy sizes depend on the flow, and they could differ significantly at different locations in the flow. They are flow properties, in contrast to molecular diffusivities which are fluid properties and are independent of flow geometry. Secondly, Eqs. 3.2.16–3.2.18 are models, and they cannot be derived in a manner similar to molecular diffusivities. In the derivation of the molecular diffusivity, the Taylor series expansions for the concentration/temperature/velocity in Eqs. 3.1.4, 3.1.21 and 3.1.30 are expansions in powers of (λ/L), where λ is the mean free path and L is the macroscopic length scale. These series can be approximated by the first non-zero term, because λ is much smaller than the characteristic flow dimension. In contrast, the size of the largest eddies in a turbulent flow is comparable to the flow dimension. In the flow through a pipe in Fig. 3.5, for example, the largest eddies have the same size as the pipe diameter. The mixing length l is not significantly smaller than the pipe diameter d in turbulent flows, and series expansions in the ratio (l/d) can not be truncated after the first few terms. Therefore, the turbulent diffusivities have to be considered as flow-dependent model parameters obtained from transport rates in turbulent flows, and not as fluid properties. Though the mixing length models are approximate, they are widely used for modeling turbulent dispersion.

The turbulent diffusivities are typically many orders of magnitude larger than the molecular diffusivities. In Section 3.1.1, we had estimated that the diffusivities are typically 10^{-5} m^2/s in gases, and 10^{-6} m^2/s (for momentum diffusion) to 10^{-9} m^2/s (for mass diffusion) in liquids. Turbulent diffusivities depend on the flow velocity and length scales. If we take a flow with characteristic length in the range 0.1–1 m and fluctuating velocity in the range 0.1–1 m/s, the turbulent diffusivity is in the range 10^{-2}–1 m^2/s. This is 3–5 orders of magnitude higher than the molecular diffusivity in gases. Due to the efficient transport by turbulent eddies, the transport rates in a turbulent flow are much higher than those in a laminar flow for the same dimension and flow velocity.

3.2.2 Dispersion in a Packed Column

In a packed column, the fluid traverses a tortuous path through the voids between the particles. The 'superficial velocity' of the fluid through the column, v_s, is the velocity of the fluid in a column of the same diameter with out any particles. If the void fraction is ε, the upward velocity of the fluid through the column is (v_s/ε).

The actual fluid velocity fluctuates in magnitude and direction through the voids between the particles, though the fluid velocity is independent of time if the flow is steady. Since a packed column contains a very large number of particles, and the location of all the particles may not be specified, it is usually not feasible to resolve the local flow velocity within the voids between the particles. In a macroscopic description, the flow velocity is defined as an average over a volume larger than the particle size, but much smaller than the diameter of the packed column. Here, the cross-stream fluid motion at the particle scale is smoothed over, and the average velocity is along the axis of the column. However, the particle scale cross-stream flow

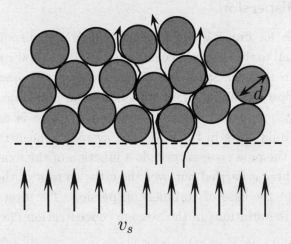

FIGURE 3.8. When fluid flows along tortuous paths in a packed column with particles of diameter d, nearby fluid elements spread by a distance approximately d when the fluid travels a distance d along the column.

does result in transport of mass/energy across the column, and this transport is not captured by the average flow. This cross-stream transport, due to the tortuosity of the fluid paths through the voids between particles, is termed dispersion.

Consider the flow through a column packed with particles of diameter d and fluid superficial velocity v_s, as shown in Fig. 3.8. As the fluid passes through the voids, two nearby fluid elements are separated by a distance comparable to the particle diameter d when they traverse an axial distance approximately d in the packed column. This spreading is modelled using equations similar to Fick's law for diffusion (Eq. 3.1.5) or Fourier's law for heat conduction (Eq. 3.1.22), but with 'dispersion coefficients' due to the fluid flow instead of molecular diffusion coefficients. As in the case of turbulent flows, the dispersion coefficients are flow properties, and not fluid properties. Since the velocity of the fluid through the voids is (v_s/ε), and the characteristic length scale for the separation of two nearby fluid elements is the particle diameter d, the dispersion coefficient is proportional to (dv_s/ε) from dimensional analysis. A microscopic model of a regular array of obstacles is used to derive the dispersion coefficient in a packed column in Section 9.3 (Chapter 9).

It should be emphasised that the dispersion is the result of processes taking place as the fluid travels through the particles over distances larger than the particle size. Therefore, dispersion coefficients can be used only for transport on the scale of the packed column, and these cannot be used for transport to/from the particles. For transport of mass/energy between the fluid and the particles at the particle scale, it is necessary to use correlations of the type discussed in Section 2.3.4 (Chapter 2).

3.2.3   Taylor Dispersion

Taylor dispersion is an unusual phenomenon where the dispersion coefficient is inversely proportional to the diffusion coefficient—that is, a lower molecular diffusion coefficient results in a faster spreading of mass/energy. A concentration pulse is injected into the pipe containing fluid with molecular diffusion coefficient $\mathcal{D}$ and average velocity v_{av} under laminar flow conditions. There is axial spread of the solute mass as the fluid flows in the pipe. The average concentration $\bar{c}(x)$, defined as the average over the pipe cross section, is a function of the axial distance x. Note that the averaging here is carried out over the cross section of the pipe, in contrast to time averaging in the case of turbulent dispersion. The average flux along the axial direction due to variations in the average concentration $\bar{c}(x)$ is found to be of the form,

$$\bar{j}_x = -\mathcal{D}_T \frac{\mathrm{d}\bar{c}}{\mathrm{d}x}. \tag{3.2.19}$$

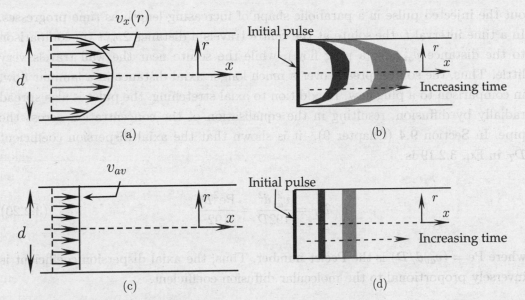

FIGURE 3.9. The laminar flow parabolic profile $v_x(r)$ in a pipe (a) and the resulting spreading of a pulse of solute injected into the pipe (b), compared with a plug flow with a constant velocity (c) and the resulting spreading of a pulse of solute injected into the pipe (d).

The dispersion coefficient $\mathcal{D}_T$ is derived in Chapter 9, and it is found that the dispersion coefficient is inversely proportional to $\mathcal{D}$, the molecular diffusion coefficient.

For a physical understanding of Taylor dispersion, consider the two velocity profiles in the pipe shown in Fig. 3.9. The momentum balance equation is solved in Chapter 5, and it is shown that the velocity profile for a laminar flow is a parabolic profile, qualitatively sketched in Fig. 3.9(a). Compare this with a hypothetical 'plug flow', shown in Fig. 3.9(c), which has a constant velocity, v_{av}. The latter is unphysical because it does not satisfy the momentum balance condition and the zero velocity condition at the wall, but it is a useful idealisation for comparison. In the laminar velocity profile shown in Fig. 3.9(a), the velocity at the pipe walls is zero, and the velocity at the centre is two times the average velocity, $2v_{av}$.

If a pulse of solute is injected into the plug flow profile shown in Fig. 3.9(c), it gets convected with the velocity v_{av}. Due to molecular diffusion, the pulse also spreads in the reference frame moving with the fluid; from dimensional analysis, an initial pulse of infinitesimal width has a width proportional to $\sqrt{\mathcal{D}t}$ after time t. The spreading of the pulse is shown in Fig. 3.9(d), and an exact relation for the spreading is derived in Chapter 4 using a similarity transform. Next, consider the axial dispersion in the laminar velocity profile shown in Fig. 3.9(a), with average velocity v_{av}. This relative velocity between the centre and the walls tends to stretch out the injected pulse in a parabolic shape of increasing length as time progresses. In a time interval t, the solute at the centre travels a distance $2v_{av}t$ (in comparison to the distance $v_{av}t$ for a plug flow), while the solute near the wall travels very little. Thus, the solute spreads over a much longer axial distance in a laminar flow, in comparison to a plug flow. In addition to axial stretching, the pulse is also spread radially by diffusion, resulting in the equalisation of the concentration across the pipe. In Section 9.4 (Chapter 9), it is shown that the axial dispersion coefficient $\mathcal{D}_T$ in Eq. 3.2.19 is

$$\mathcal{D}_T = \frac{v_{av}^2 d^2}{192\mathcal{D}} = \frac{\text{Pe}^2 \mathcal{D}}{192}, \qquad (3.2.20)$$

where $\text{Pe} = (v_{av}d/\mathcal{D})$ is the Peclet number. Thus, the axial dispersion coefficient is inversely proportional to the molecular diffusion coefficient.

Summary (3.2)

1. Dispersion is the transport of mass/momentum/energy due to the fluid velocity fluctuations, in contrast to diffusion which is due to the molecular velocity fluctuations. Therefore, dispersion is a flow property, in contrast to diffusion which is a fluid property.

2. In a turbulent flow, the quantities are separated into time-averaged mean quantities, Eqs. 3.2.1 and 3.2.3, and fluctuations, Eqs. 3.2.2 and 3.2.4. The correlation between the concentration/momentum/ temperature fluctuations and the velocity fluctuations results in a flux of mass/momentum/energy, Eqs. 3.2.7, 3.2.14 and 3.2.15.

3. The dispersion flux is modelled as the product of the dispersion coefficient and the (negative) of the gradient in the concentration/temperature/velocity, Eqs. 3.2.16–3.2.18. The dispersion coefficient is modelled as the product of the fluid velocity fluctuations and the 'mixing length' which is the approximate length over which the fluid is mixed due to turbulent eddies.

4. The turbulent diffusivities in Eqs. 3.2.16–3.2.18 are flow properties that depend on the flow configuration and the large-scale turbulence characteristics; these are not molecular properties, and they do not depend on the molecular diffusivities.

5. In the flow through packed columns, the flow is separated into the plug flow and the velocity fluctuations due to the tortuosity of the fluid paths through the densely packed particles. The dispersion coefficient is proportional to $(v_s d/\varepsilon)$, the product of the characteristic particle diameter d and the velocity (v_s/ε) of the fluid through the particles, where v_s is the superficial velocity (ratio of flow rate and area of cross section of the column without particles) and ε is the void fraction.

6. There is axial dispersion in the laminar flow in a pipe/channel because the velocity varies from zero at the wall to a maximum at the centre. The velocity is expressed as the sum of a plug flow and the spatial fluctuation which is the difference between the actual flow profile and the plug flow. The constitutive relation, Eq. 3.2.19, for dispersion in the axial direction has the same form as Fick's law. The dispersion coefficient $\mathcal{D}_T$, Eq. 3.2.20, is inversely proportional to the molecular diffusion coefficient.

Exercises

EXERCISE 3.1 Compute the mean free path and the root-mean-square molecular velocity of chlorine molecules (molecular weight 71 g/mol and molecular diameter 4.1 Å) at 300 K temperature and 10^5 Pa pressure. What is the ratio of the mean free path and the molecular diameter? Compute the viscosity from kinetic theory. Determine the binary diffusion coefficient of hydrogen in chlorine.

EXERCISE 3.2

a) Express the ratio of the mean free path and the molecular diameter, (λ/d_m), in terms of the fraction of the volume of the gas occupied by the molecules for a hard sphere gas.

b) The average distance between nearest neighbours in a random assembly of hard spheres is proportional to $N^{-1/3}$, where N is the number density. Express the ratio $(\lambda/N^{-1/3})$ in terms of the volume fraction.

c) What is the volume fraction of the molecules of hydrogen and oxygen, with molecular diameter 2.6 Å and 3.7 Å, at 300 K temperature and 1 atm pressure?

EXERCISE 3.3 Determine the thermal conductivity of helium gas at 300 K and 1 atm. The molecular mass and diameter of helium are 4 g/mol and 2.6 Å, respectively. Why is the thermal conductivity of helium much higher than that of air?

EXERCISE 3.4 In the kinetic theory of gases, there are two dimensionless numbers that relate the macroscopic flow properties to the molecular properties.

a) The Mach number is defined as (v_c/c_s), where v_c is the characteristic flow velocity and $c_s = \sqrt{\gamma k_B T/M}$ is the speed of sound. Here, γ is the ratio of specific heats, k_B is the Boltzmann constant, T is the absolute temperature and M is the molecular mass.

b) The Knudsen number is defined as (λ/l_c), where λ is the mean free path and l_c is the characteristic length.

How are the Reynolds and Peclet numbers related to the Mach and Knudsen numbers for an ideal gas of hard sphere molecules?

EXERCISE 3.5 A haemoglobin molecule has a diffusivity of 6.9×10^{-11} m²/s in water at 300 K. Using the Stokes–Einstein relation, estimate the diameter of this molecule. The viscosity of water is 10^{-3} kg/m/s.

EXERCISE 3.6 Use the Stokes–Einstein relation to determine the diffusion coefficient of hydrogen (molecular diameter 2.9 Å), oxygen (molecular diameter 3.43 Å) and benzene (molecular diameter 5.27 Å) in water at 300 K. Compare these with the measured values of 4.5×10^{-9} m²/s, 2.1×10^{-9} m²/s and 1.02×10^{-9} m²/s, respectively[3]. For which molecule would you expect the best and worst agreement with measured values?

EXERCISE 3.7 In non-equilibrium thermodynamics, the flux of a solute is proportional to the gradient of the chemical potential Ξ,

$$j_z = -\Gamma \frac{d\Xi}{dz},$$

where Γ is the Onsager coefficient. In a microscopic description, the chemical potential is the potential energy per unit mass of the molecules, and the force on the molecules per unit mass is $-(d\Xi/dz)$. Due to this force, the solute molecules accelerate between successive intermolecular collisions. This causes a drift velocity v_d of the solute molecules in a gas, which is the product of the acceleration and the time between collisions. Using this microscopic picture, estimate the Onsager coefficient in terms of the molecular parameters and the temperature. Verify that the dimensions of the chemical potential and Onsager coefficient are consistent with Example 1.5.3.

EXERCISE 3.8 In the classical theory of electrical conductivity, the nuclei of the atoms are at fixed locations, while the mobile electrons form a cloud around the nuclei. When there is an applied voltage difference across the material, there is a force exerted on the negatively charged electrons which causes a drift velocity. The electrons are considered to move as classical particles with mass M_e and charge e. The motion consists of two parts, acceleration of the electrons between the nuclei, and collisions with the nuclei which scatter electrons and decrease the the electron velocity.

Consider a conductor with area of cross-section A and length L, across which there is a potential difference ΔV. The force on an electron between two collisions is $e(\Delta V/L)$. At high temperature, the fluctuating velocity is, $v_{rms} = \sqrt{3k_B T/M_e}$. The spacing between the nuclei in the crystal is d_c.

a) What is the 'time of flight' t_f between two collisions of the electron with the nuclei?

b) What is the drift velocity v_d?

c) If the number of electrons in the material per unit volume is N, the flux (number of electrons transported per unit area of cross section) is $v_d \times N$. The electrical current density (I/A) (current per unit area) is the product of the electron flux and the charge on an electron. If equation for the current density is expressed as

$$(I/A) = \kappa(\Delta V/L),$$

how is the electrical conductivity κ related to the charge and mass of an electron, and the spacing of the nuclei?

d) The thermal conductivity k of the electron cloud is proportional $\rho C_p d_c v_{rms}$, where the mean free path is the distance between nuclei d_c, the density $\rho = NM_e$, and the specific heat $C_p = (5k_B/2M_e)$ for an electron. Show that the ratio of the thermal and electrical conductivities is independent of the lattice spacing or number density of electrons,

$$\frac{k}{\kappa} \propto \frac{k_B^2 T}{e^2}.$$

This is called the Wiedemann–Franz law for electrical conductors.

EXERCISE 3.9 The dispersion coefficient in a turbulent flow is a function of the kinetic energy of the fluid per unit mass K and the rate of dissipation of energy per unit mass ϵ. Based on dimensional analysis, what is the turbulent dispersion coefficient? For a pipe flow, determine the turbulent dispersion coefficient in terms of the pipe diameter, average velocity and the friction factor. The rate of dissipation of energy in a pipe flow is $-\Delta p \times \dot{V}$, where Δp is the pressure difference between the outlet and inlet and $\dot{V}$ is the volumetric flow rate.

EXERCISE 3.10 The fluctuations in the concentration and velocity fields in a turbulent flow can be expressed as a Fourier series,

$$c' = \sum_{n=1}^{n_{max}} c'_n \sin(n\omega t), \quad v'_x = \sum_{n=1}^{n_{max}} v'_{xn} \sin(n\omega t),$$

where ω is the fundamental frequency, c'_n and v'_{xn} are functions of position, n is an integer and n_{max} is the total number of Fourier modes. What is the Reynolds flux? Use the trigonometric identities,

$$\sin(n\omega t)\sin(m\omega t) = \tfrac{1}{2}[\cos((n-m)\omega t) - \cos((n+m)\omega t)], \quad \sin(n\omega t)^2 = \tfrac{1}{2}[1 - \cos(2n\omega t)].$$

4

Unidirectional Transport: Cartesian Co-ordinates

Problems involving mass, momentum and energy transport in one spatial direction in a Cartesian co-ordinate system are considered in this chapter. The concentration, velocity or temperature fields, here denoted field variables, vary along one spatial direction and in time. The 'forcing' for the field variables could be due to internal sources of mass, momentum or energy, or due to the fluxes/stresses at boundaries which are planes perpendicular to the spatial co-ordinate. Though the dependence on one spatial co-ordinate and time appears a gross simplification of practical situations, the solution methods developed here are applicable for problems involving transport in multiple directions as well.

There are two steps in the solution procedure. The first step is a 'shell balance' to derive a differential equation for the field variables. The procedure, discussed in Section 4.1, is easily extended to multiple dimensions and more complex geometries. The second step is the solution of the differential equation subject to boundary and initial conditions. Steady problems are considered in Section 4.2, where the field variable does not depend on time, and the conservation equation is an ordinary differential equation. For unsteady problems, the equation is a partial differential equation involving one spatial dimension and time. There is no general procedure for solving a partial differential equation; the procedure depends on the configuration and the kind of forcing, and physical insight is necessary to solve the problem. The procedures for different geometries and kinds of forcing are explained in Sections 4.4–4.7.

The conservation equations in Sections 4.2 and 4.4–4.7 are linear differential equations in the field variable—that is, the equations contain the field variable to the first power in addition to inhomogeneous terms independent of the field variable.

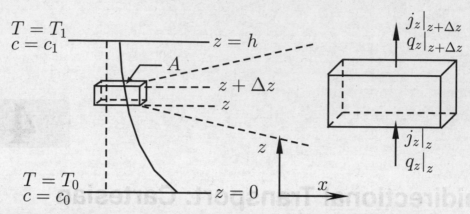

FIGURE 4.1. Configuration and co-ordinate system for analysing unsteady mass and heat transfer in one dimension.

For the special case of multicomponent diffusion in Section 4.3, the equations are non-linear in the field variable. This is because the diffusion of a molecular species generates a flow velocity, which contributes to the flux of the species. The conservation equation for the simple case of diffusion in a binary mixture is derived in Section 4.3, and some simple applications are discussed.

In Section 4.8, correlations for the average fluxes presented in Chapter 2 are used in the spatial or time evolution equations for the field variables. The correlations in Chapter 2 apply to processes at steady state. These can be used for unsteady processes only if the diffusion time is small compared to the time for the variation of the specified temperature/concentration fields at the boundaries, or when the diffusion length scale is small compared to the characteristic length. The conditions for the validity of the correlations, and examples of the use of correlations in balance equations, are discussed in Section 4.8.

4.1 Conservation Equation

Consider a fluid bounded by two plane surfaces of infinite extent separated by distance h, as shown in Fig. 4.1. There is a spatial variation in the value of the field variable, shown schematically in Fig. 4.1, when the values of the field variable at the two boundaries are different, or when there is internal generation of mass, momentum or energy within the domain. The conservation equation is derived using shell balance for a differential volume of thickness Δz in the z direction, and area A in the $x - y$ plane, shown in Fig. 4.1. The equations are derived separately for mass, energy and momentum transport in the following sub-sections, and it is found that a common form emerges for the transport equations.

4.1.1 Mass Transfer

The configuration and co-ordinate system for analysing mass transfer is shown in Fig. 4.1. The total mass within the differential volume is the product of the concentration and the volume, $c(z,t)A\Delta z$. The mass conservation condition for this differential volume is,

$$\begin{pmatrix} \text{Change of mass} \\ \text{in differential volume} \end{pmatrix} = \begin{pmatrix} \text{Mass in} \end{pmatrix} - \begin{pmatrix} \text{Mass out} \end{pmatrix} + \begin{pmatrix} \text{Production of mass} \\ \text{in differential volume} \end{pmatrix}. $$
$$(4.1.1)$$

The change of mass in time Δt is the product of the change in concentration and the volume,

$$\begin{pmatrix} \text{Change of mass} \\ \text{in differential volume} \end{pmatrix} = (c(z, t + \Delta t) - c(z,t))A\Delta z. \qquad (4.1.2)$$

The mass transfer across the surfaces of the differential volume takes place due to molecular diffusion in the z direction through the surfaces at z and $z+\Delta z$. The total mass entering the differential volume through the surface at z in a time interval Δt is the product of the mass flux j_z, the area of transfer A and the time interval Δt,

$$\begin{pmatrix} \text{Mass in} \end{pmatrix} = j_z|_z \, A\Delta t. \qquad (4.1.3)$$

Similarly, the mass leaving the volume at $z + \Delta z$ is,

$$\begin{pmatrix} \text{Mass out} \end{pmatrix} = j_z|_{z+\Delta z} \, A\Delta t. \qquad (4.1.4)$$

There could be production/consumption of mass in the differential volume due to a chemical reaction. The mass produced in the volume $A\Delta z$ within the time Δt is $SA\Delta z\Delta t$, where S is the rate of production of mass per unit volume per unit time. The rate of production is a function of the concentrations of the reacting species and the temperature, which are functions of position and time, and so S is a function on position and time,

$$\begin{pmatrix} \text{Production of mass} \\ \text{in differential volume} \end{pmatrix} = S(z,t)A\Delta z\Delta t. \qquad (4.1.5)$$

Substituting Eqs. 4.1.2–4.1.5 into the conservation condition Eq. 4.1.1, and dividing by $A\Delta z\Delta t$, we obtain

$$\frac{c(z, t + \Delta t) - c(z,t)}{\Delta t} = \frac{(j_z|_z - j_z|_{z+\Delta z})}{\Delta z} + S(z,t). \qquad (4.1.6)$$

The aforementioned 'difference' equation is converted into a differential equation by taking the limit $\Delta t \to 0$ and $\Delta z \to 0$,

$$\boxed{\frac{\partial c}{\partial t} = -\frac{\partial j_z}{\partial z} + \mathcal{S}.} \qquad (4.1.7)$$

Note that there is a negative sign in the above equation, because the partial derivative is defined as $(\partial j_z/\partial z) = \lim_{\Delta z \to 0}(j_z|_{z+\Delta z} - j_z|_z)/\Delta z$ with t fixed.

Using the Fick's law for diffusion, Eq. 1.5.1, expressed in terms of the partial derivative of the concentration with respect to z,

$$\boxed{j_z = -\mathcal{D}\frac{\partial c}{\partial z},} \qquad (4.1.8)$$

the mass conservation Eq. 4.1.7 becomes,

$$\boxed{\frac{\partial c}{\partial t} = \frac{\partial}{\partial z}\left(\mathcal{D}\frac{\partial c}{\partial z}\right) + \mathcal{S},} \qquad (4.1.9)$$

where $\mathcal{D}$ is the mass diffusion coefficient. For the special case where the diffusion coefficient is a constant, Eq. 4.1.9 reduces to

$$\boxed{\frac{\partial c}{\partial t} = \mathcal{D}\frac{\partial^2 c}{\partial z^2} + \mathcal{S}.} \qquad (4.1.10)$$

If the diffusion coefficient is dependent on position or concentration, Eq. 4.1.10 is not valid, and it is necessary to use Eq. 4.1.9.

The combination of the mass conservation condition Eq. 4.1.1 and the Fick's law for diffusion (constitutive relation) Eq. 4.1.8 leads to the diffusion equation, Eq. 4.1.9, which is a second order differential equation in the z co-ordinate and a first order differential equation in time. In order to solve this equation, it is necessary to specify two 'boundary conditions' on the spatial boundaries, and one 'initial condition' for the concentration throughout the domain at the initial time.

4.1.2 Heat Transfer

The equivalent heat transfer problem involves two plates with temperature T_1 at $z = h$, and temperature T_0 at $z = 0$, shown in Fig. 4.1. The energy conservation condition is

$$\left(\begin{array}{c}\text{Change in energy} \\ \text{in differential volume}\end{array}\right) = \left(\text{Energy in}\right) - \left(\text{Energy out}\right)$$

$$+ \left(\begin{array}{c}\text{Production of energy} \\ \text{in differential volume}\end{array}\right). \quad (4.1.11)$$

The change in heat energy in a time Δt is,

$$\left(\begin{array}{c}\text{Change in energy} \\ \text{in differential volume}\end{array}\right) = (\rho C_p T(z, t + \Delta t) - \rho C_p T(z, t)) A \Delta z. \quad (4.1.12)$$

The energy entering at the surface at z in a time interval Δt is the product of the heat flux q_z, the area of transfer A and the time interval Δt,

$$\left(\text{Energy in}\right) = q_z|_z \, A\Delta t. \quad (4.1.13)$$

In a similar manner, the energy leaving the surface at $z + \Delta z$ is,

$$\left(\text{Energy out}\right) = q_z|_{z+\Delta z} \, A\Delta t. \quad (4.1.14)$$

The production/consumption of heat in the differential volume could be due to chemical reaction, phase change, viscous heating or other physical processes. The energy produced in the volume $A\Delta z$ within the time Δt is $\mathcal{S}_e(z,t)A\Delta z\Delta t$, where $\mathcal{S}_e$ is the rate of production per unit volume per unit time,

$$\left(\begin{array}{c}\text{Production of energy} \\ \text{in differential volume}\end{array}\right) = \mathcal{S}_e(z,t)A\Delta z\Delta t. \quad (4.1.15)$$

Substituting Eqs. 4.1.12–4.1.15, into the energy conservation condition Eq. 4.1.11, and dividing by $A\Delta z\Delta t$, we obtain

$$\rho C_p \frac{(T(z, t + \Delta t) - T(z, t))}{\Delta t} = \frac{q_z|_z - q_z|_{z+\Delta z}}{\Delta z} + \mathcal{S}_e. \quad (4.1.16)$$

The above difference equation is converted into a differential equation by taking the limit $\Delta t \to 0$ and $\Delta z \to 0$,

$$\boxed{\rho C_p \frac{\partial T}{\partial t} = -\frac{\partial q_z}{\partial z} + \mathcal{S}_e.} \quad (4.1.17)$$

The flux is expressed using Fourier's law for heat conduction, Eq. 1.5.3, in differential form,

$$q_z = -k \frac{\partial T}{\partial z}, \qquad (4.1.18)$$

to obtain energy conservation equation,

$$\rho C_p \frac{\partial T}{\partial t} = \frac{\partial}{\partial z} \left(k \frac{\partial T}{\partial z} \right) + S_e, \qquad (4.1.19)$$

where k is the thermal conductivity. If the density ρ and specific heat C_p are independent of temperature, position or time, Eq. 4.1.19 can be reduced to the same form as the mass diffusion equation, Eq. 4.1.9,

$$\frac{\partial T}{\partial t} = \frac{\partial}{\partial z} \left(\alpha \frac{\partial T}{\partial z} \right) + \frac{S_e}{\rho C_p}, \qquad (4.1.20)$$

where $\alpha = (k/\rho C_p)$ is the thermal diffusivity. In addition, if the thermal diffusivity is also a constant, the diffusion equation reduces to the same form as Eq. 4.1.10,

$$\frac{\partial T}{\partial t} = \alpha \frac{\partial^2 T}{\partial z^2} + \frac{S_e}{\rho C_p}. \qquad (4.1.21)$$

In writing Eq. 4.1.12, it was assumed that ρC_p is independent of time. While going from Eq. 4.1.19 to Eq. 4.1.20, it was assumed that ρC_p is independent of spatial position. Implicit is the assumption that ρC_p is also independent of T, which is a function of position and time. The density is considered a constant because the analysis here is restricted to 'incompressible' flows. This is a good approximation when the flow velocity is much smaller than the speed of sound. The variation in the specific heat with temperature is also relatively small in many practical situations. For an ideal gas, the molar specific heat is $((d_f + 2)R/2)$, where d_f is the number of degrees of freedom and R is the gas constant; this is independent of temperature. For real gases, the specific heat does vary with temperature, but the variation is numerically small. For example, for air, C_p varies by less than 1 % when the temperature is in the range 0°C to 100°C. The variation in the specific heat for liquids is even less – for water, C_p varies by 0.1 % for temperature in the range 0°C to 100°C. Therefore, the assumptions made in arriving at Eq. 4.1.20 are valid in many practical situations, though they may not be valid in extreme conditions.

Going from Eq. 4.1.20 to Eq. 4.1.21, it is assumed that the thermal diffusivity is independent of z or T. This assumption is less general, and the thermal conductivity

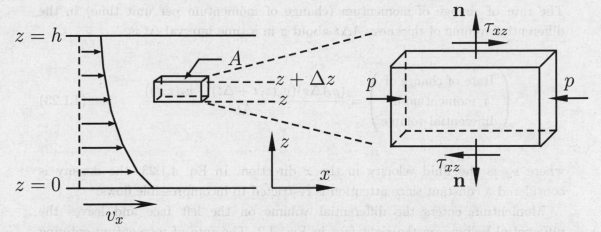

FIGURE 4.2. Configuration and co-ordinate system for analysing unsteady momentum transfer in one dimension.

does vary with temperature in many practical situations. For example, the thermal conductivity of air changes by about 30% over the temperature range 0°C–100°C, while that of water changes by about 20% over the same temperature range.

4.1.3 Momentum Transfer

Though the final equation for the momentum transfer is identical in form to Eqs. 4.1.9 and 4.1.20, the derivation is slightly different. The configuration, shown in Fig. 4.2, consists of a fluid between two surfaces moving with two different velocities in the x direction, which is tangential to the surfaces. The balance equation is written for the x component of the momentum vector. It is important to note that there are two directions in the problem: the direction of the momentum, x, and the direction of momentum variation, z. Since diffusion takes place along the direction where there is a variation of momentum, the momentum flux is in the z direction.

The rate of change of momentum in a differential volume is the difference between the momentum transported in and out plus the sum of the applied forces on the differential volume,

$$
\begin{pmatrix} \text{Rate of change of} \\ x \text{ momentum in} \\ \text{differential volume} \end{pmatrix} = (x \text{ Momentum in}) - (x \text{ Momentum out})
$$
$$
+ \begin{pmatrix} \text{Surface force in} \\ \text{the } x \text{ direction} \end{pmatrix} + \begin{pmatrix} \text{Body force in} \\ \text{the } x \text{ direction} \end{pmatrix}. \quad (4.1.22)
$$

The rate of change of momentum (change of momentum per unit time) in the differential volume of thickness $A\Delta z$ about z in a time interval Δt is,

$$\left(\begin{array}{c} \text{Rate of change of} \\ x \text{ momentum in} \\ \text{differential volume} \end{array}\right) = \frac{(\rho A\Delta z)(v_x(z, t + \Delta t) - v_x(z, t))}{\Delta t}, \qquad (4.1.23)$$

where v_x is the fluid velocity in the x direction. In Eq. 4.1.23, the density is considered a constant since attention is restricted to incompressible flows.

Momentum enters the differential volume on the left face and leaves the differential volume on the right face in Fig. 4.2. The rate of momentum entering per unit area is ρv_x^2, the x momentum density ρv_x times the velocity perpendicular to the surface v_x. Since the velocity is independent of x, the momentum entering left face is equal to that leaving the right face of the differential volume, and therefore there is no change in momentum due to transport across the surfaces perpendicular to the x co-ordinate.

The forces acting are of two types. The first is the 'body force', such as the gravitational or centrifugal force. The body force on a differential volume is

$$\left(\begin{array}{c} \text{Body force} \\ \text{in } x \text{ direction} \end{array}\right) = f_x A\Delta z, \qquad (4.1.24)$$

where f_x is the body force density, or body force per unit volume, in the x direction. If the fluid is subjected to gravitational acceleration, the gravitational force density is the product of the mass density and gravitational acceleration, $f_x = \rho g_x$, where g_x is the component of the acceleration due to gravity in the x direction.

The surface forces in Eq. 4.1.22 act on the bounding surfaces of the differential volume; forces due to the pressure and shear stress are examples of surface forces. The surface forces acting in the x direction on the two surfaces at z and $z + \Delta z$ are the products of the shear stress τ_{xz} and the surface area A. The directions of the forces have to be carefully evaluated in this case, since the force is a vector. The shear stress τ_{xz} is defined as the force per unit area in the x direction acting at a surface whose *outward* unit normal is in the positive z direction. For the surface at $z + \Delta z$, the outward unit normal $\mathbf{n}$ is in the $+z$ direction, as shown in Fig. 4.2. Consequently, the force per unit area at this surface is $+ \tau_{xz}|_{z+\Delta z}$. For the surface at z, the outward unit normal is in the $-z$ direction, and the force per unit area at this surface is $- \tau_{xz}|_z$. Therefore,

$$\left(\begin{array}{c} \text{Surface force} \\ \text{in } x \text{ direction} \end{array}\right) = A(\tau_{xz}|_{z+\Delta z} - \tau_{xz}|_z). \qquad (4.1.25)$$

There is a pressure force acting in the x direction on the left and right faces of the differential volume. The pressure p is compressive, and it acts along the inward perpendicular to the surface, as shown in Fig. 4.2. Here, it is assumed that the flow properties do not change in the x direction. Therefore, the pressure force on the left face is equal in magnitude and opposite in direction to that on the right face, and there is no net force exerted in the x direction due to the pressure.

Substituting Eqs. 4.1.23–4.1.25 into Eq. 4.1.22, the momentum balance equation is,

$$\rho A \Delta z \frac{\Delta v_x}{\Delta t} = A(\tau_{xz}|_{z+\Delta z} - \tau_{xz}|_z) + f_x A \Delta z. \qquad (4.1.26)$$

Dividing throughout by $A \Delta z$, and taking the limit $\Delta t \to 0$ and $\Delta z \to 0$, we obtain the momentum conservation equation,

$$\boxed{\rho \frac{\partial v_x}{\partial t} = \frac{\partial \tau_{xz}}{\partial z} + f_x.} \qquad (4.1.27)$$

Using Newton's law of viscosity, Eq. 1.5.4, in differential form,

$$\boxed{\tau_{xz} = \mu \frac{\partial v_x}{\partial z},} \qquad (4.1.28)$$

the momentum conservation equation becomes,

$$\rho \frac{\partial v_x}{\partial t} = \frac{\partial}{\partial z} \left(\mu \frac{\partial v_x}{\partial z} \right) + f_x. \qquad (4.1.29)$$

For an incompressible flow where the density is a constant, the momentum conservation equation can be expressed in terms of the momentum diffusivity $\nu = (\mu/\rho)$,

$$\boxed{\frac{\partial v_x}{\partial t} = \frac{\partial}{\partial z} \left(\nu \frac{\partial v_x}{\partial z} \right) + \frac{f_x}{\rho}.} \qquad (4.1.30)$$

If the kinematic viscosity is also a constant, the momentum equation reduces to,

$$\boxed{\frac{\partial v_x}{\partial t} = \nu \frac{\partial^2 v_x}{\partial z^2} + \frac{f_x}{\rho}.} \qquad (4.1.31)$$

The assumption of constant kinematic viscosity, used in simplifying from Eq. 4.1.30 to Eq. 4.1.31, is not generally applicable, since the viscosity is a function of temperature in gases and liquids. Eq. 4.1.31 can be used only for Newtonian 'isothermal' flows, where the viscosity is a constant.

The momentum equations, Eqs. 4.1.30–4.1.31, derived above have the same form as the mass and energy equations, Eqs. 4.1.9–4.1.10 and 4.1.20–4.1.21. These provide a unified framework for the study of diffusion of mass, momentum and energy in one dimension.

Summary (4.1)

1. The equations for the rate of change of the field variable Φ_{fv}, Eqs. 4.1.7, 4.1.17 and 4.1.27, have a common form,

$$\frac{\partial \Phi_{fv}}{\partial t} = -\frac{\partial \mathcal{J}_z}{\partial z} + \mathcal{S}, \tag{4.1.32}$$

where the field variable Φ_{fv}, fl   source $\mathcal{S}$ for mass, heat and momentum transfer are summarised in   1.

2. The constitutive relations for the flux, Fick's   (Eq. 4.1.8), Fourier's law (Eq. 4.1.18) and Newton's law (Eq. 4.1.28), have   mmon form

$$\mathcal{J}_z = -\mathcal{D}\frac{\partial \Phi_{fv}}{\partial z}, \tag{4.1.33}$$

where $\mathcal{D}$ is the generalised diffusivity.

3. The constitutive relation Eq. 4.1.33 is substituted into   ation equation, Eq. 4.1.32, to obtain Eqs. 4.1.9, 4.1.20 and 4.1.30

$$\frac{\partial \Phi_{fv}}{\partial t} = \frac{\partial}{\partial z}\left(\mathcal{D}\frac{\partial \Phi_{fv}}{\partial z}\right) + \tag{.34}$$

4. When the diffusivity is also independent of the field variable, position or time, the conservation equations, Eqs. 4.1.10, 4.1.21 and 4.1.31 are,

$$\frac{\partial \Phi_{fv}}{\partial t} = \mathcal{D}\frac{\partial^2 \Phi_{fv}}{\partial z^2} + \mathcal{S}. \tag{4.1.35}$$

TABLE 4.1. The field variables, generalised fluxes, diffusivity and sources for mass, heat and momentum transfer in Eqs. 4.1.32 to 4.1.35, and the constitutive relations.

	Φ_{fv}	$\mathcal{J}_z$	$\mathcal{D}$	$\mathcal{S}$	Constitutive relation
Mass	c	j_z	$\mathcal{D}$	$\mathcal{S}$	Fick's law $j_z = -\mathcal{D}\frac{\partial c}{\partial z}$
Heat	T	q_z	$\alpha = (k/\rho C_p)$	$(\mathcal{S}_e/\rho C_p)$	Fourier's law $q_z = -k\frac{\partial T}{\partial z}$
x Momentum	v_x	$-\tau_{xz}$	$\nu = (\mu/\rho)$	(f_x/ρ)	Newton's law $\tau_{xz} = \mu\frac{\partial v_x}{\partial z}$

4.2 Steady Solutions

The left side of Eq. 4.1.34 is zero at steady state, and the diffusion equation reduces to an ordinary differential equation in z. Two cases can be considered, the first where the diffusivity is a constant, and the second where the diffusivity is a function of z or Φ_{fv}. In the first case, the field variable is determined by integrating two times with respect to z, and applying boundary conditions. In the second, it is necessary to first determine the flux from Eq. 4.1.32 with the left side set equal to zero, and then use the constitutive relation to determine the field variable.

4.2.1 Constant Diffusivity

Conduction in Single/Multiple Slabs

If the diffusivity is a constant and there are no sources in the domain, the solution of Eq. 4.1.35 is a linear function of the z co-ordinate of the form $\Phi_{fv} = C + Dz$, where C and D are constants. These constants are determined from the boundary conditions for Φ_{fv} at the two boundaries. Consider the heat conduction problem in a slab of thickness h, shown in Fig. 4.3(a). If the temperatures on the two boundaries of the slab are T_0 and T_h, the temperature is a linear function of z in the material,

$$T = T_0 + \frac{(T_h - T_0)z}{h}, \tag{4.2.1}$$

and the heat transfer rate Q_z (heat transferred per unit time) is independent of z,

$$Q_z = -\frac{Ak(T_h - T_0)}{h}, \tag{4.2.2}$$

where A is the area of cross section. Note there is a negative sign on the right in Eq. 4.2.2, because heat transfer takes place from higher to lower temperature.

Consider a composite slab consisting of multiple layers, each layer having different thickness and thermal conductivity, as shown in Fig. 4.3(b) and (c). The temperature

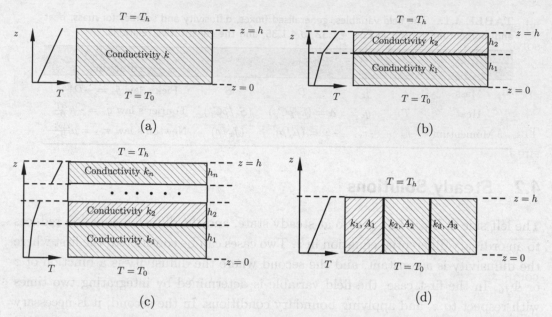

FIGURE 4.3. Heat conduction (a) in a single slab of thickness h and thermal conductivity k, (b) in a composite slab consisting of two layers with thicknesses h_1 and h_2 and conductivities k_1 and k_2, respectively, (c) in a slab consisting of multiple layers in the series configuration, and (d) in a slab consisting of multiple layers in the parallel configuration.

profile is linear within each layer, but the slope of the profile varies because the thermal conductivity of the layers could be different. At the interface between two layers, the temperature and the heat flux in both the materials are equal. If the heat flux leaving one layer was not equal to that entering the adjacent layer, there would be an accumulation of heat at the interface, violating the steady-state condition. Similarly, if the temperatures in the two materials were different when they are in perfect thermal contact, there would be a finite temperature difference across a region of infinitesimal thickness, resulting in an infinite heat flux. Therefore, the condition at the interface between two layers in perfect thermal contact is that the temperature and the heat flux in the two materials at the interface are equal.

EXAMPLE 4.2.1: Consider a composite slab consisting of two layers of thickness h_1 and h_2 and thermal conductivity k_1 and k_2, shown in Fig. 4.3(b). The temperature is T_0 at $z = 0$, and T_h at $z = h = h_1 + h_2$. What is the heat transfer rate Q_z?

Solution: The heat transfer rates through the two materials are equal at steady state, and therefore,

$$Q_z = -\frac{k_1 A(T_f - T_0)}{h_1} = -\frac{k_2 A(T_h - T_f)}{h_2}, \tag{4.2.3}$$

where T_f is the temperature at the interface between the two materials. Eq. 4.2.3 is solved to obtain the unknown interface temperature T_f,

$$T_f = \left(\frac{k_1}{h_1} + \frac{k_2}{h_2} \right)^{-1} \left(\frac{k_1 T_0}{h_1} + \frac{k_2 T_h}{h_2} \right). \qquad (4.2.4)$$

The above expression for T_f is substituted into Eq. 4.2.3 to obtain, after some simplification,

$$Q_z = - \frac{A(T_h - T_0)}{(h_1/k_1) + (h_2/k_2)}. \qquad (4.2.5)$$

□

The heat transfer rate due to an applied temperature difference across a composite slab, shown in Fig. 4.3(c), can be solved by determining the temperature at each interface in a manner similar to the above example. A simpler method is to define the 'heat transfer resistance' across each layer. The heat transfer resistance is the ratio of the negative of the temperature difference across the layer and the heat transfer rate through the layer,

$$\mathcal{R}_i = - \frac{\Delta T_i}{Q_z} = \frac{h_i}{k_i A}, \qquad (4.2.6)$$

where h_i and k_i are the thickness and thermal conductivity of layer i, and ΔT_i is the temperature difference across layer i. Here, the negative of the temperature difference is analogous to the voltage difference, and the heat transfer rate is analogous to the current in an electrical conductor. When there are multiple layers of different materials arranged in series, as shown in Fig. 4.3(c), the total resistance $\mathcal{R}_{total}$ is the sum of the individual resistances,

$$\boxed{\mathcal{R}_{total} = \mathcal{R}_1 + \mathcal{R}_2 + \ldots + \mathcal{R}_n = \frac{h_1}{k_1 A} + \frac{h_2}{k_2 A} + \ldots + \frac{h_n}{k_n A}.} \qquad (4.2.7)$$

The relation between the heat transfer rate and the temperature difference is,

$$Q_z = - \frac{\Delta T}{\mathcal{R}_{total}}, \qquad (4.2.8)$$

where $\Delta T = T_h - T_0$ is the temperature difference across the composite slab shown in Fig. 4.3(c).

In the 'parallel' configuration shown in Fig. 4.3(d), the thickness h is the same for all the layers, but the areas of cross section $A_1, A_2, \ldots$ differ. The temperature difference ΔT is equal for all the layers. The total heat transfer rate is the sum of the heat transfer rates across the individual layers,

$$Q_z = Q_{z1} + Q_{z2} + \ldots + Q_{nz} = -\left(\frac{k_1 A_1}{h} + \frac{k_2 A_2}{h} + \ldots + \frac{k_n A_n}{h}\right)\Delta T. \qquad (4.2.9)$$

This can be expressed in terms of the heat transfer resistances as,

$$Q_z = -\left(\frac{1}{\mathcal{R}_1} + \frac{1}{\mathcal{R}_2} + \ldots + \frac{1}{\mathcal{R}_n}\right)\Delta T. \qquad (4.2.10)$$

Thus, the total resistance follows the inverse sum rule for the parallel arrangement of multiple layers,

$$\boxed{\frac{1}{\mathcal{R}_{\text{total}}} = \frac{1}{\mathcal{R}_1} + \frac{1}{\mathcal{R}_2} + \ldots + \frac{1}{\mathcal{R}_n}.} \qquad (4.2.11)$$

Viscous Heating in a Channel

When there is an internal source, Eq. 4.1.32 is first solved to determine the flux at steady state, and then the field variable is determined from the constitutive relation. For a constant internal source, the flux is a linear function of z from Eq. 4.1.32,

$$\mathcal{J}_z = \mathcal{S}z + C, \qquad (4.2.12)$$

where C is a constant of integration. The field variable is determined from the constitutive relation 4.1.33,

$$\Phi_{fv} = -\frac{\mathcal{S}z^2}{2\mathcal{D}} - \frac{Cz}{\mathcal{D}} + D. \qquad (4.2.13)$$

The constants C and D are determined from the boundary conditions. Thus, the field variable is a quadratic function of z for a constant internal source.

EXAMPLE 4.2.2: There is dissipation of energy due to viscous friction in the shear flow of a viscous liquid, and this increases the temperature of the liquid. Consider a linear shear flow between two plates at $z = 0$ and $z = h$, as shown in Fig. 4.4. The top plate moves with a velocity V, and the bottom plate is stationary. The temperature at both the plates is T_0. The heat energy generation per unit volume of the fluid per unit time is given by,

$$S_e = \tau_{xz}\left(\frac{\mathrm{d}v_x}{\mathrm{d}z}\right), \qquad (4.2.14)$$

where v_x is the fluid velocity, and τ_{xz} is the shear stress. What is the temperature profile in the channel? Determine the correlation for the Nusselt number.

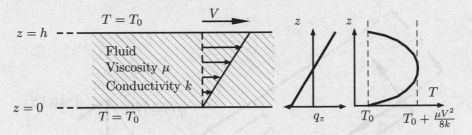

FIGURE 4.4. The viscous heating in a channel due to a linear shear flow.

Solution: For a linear velocity profile, $(dv_x/dz) = (V/h)$, the stress is $\tau_{xz} = \mu(dv_x/dz) = (\mu V/h)$, where μ is the viscosity, and the heat generated per unit volume per unit time is,

$$S_e = \frac{\mu V^2}{h^2}. \qquad (4.2.15)$$

At steady state, the energy balance equation, Eq. 4.1.21 is,

$$k\frac{d^2T}{dz^2} + \frac{\mu V^2}{h^2} = 0. \qquad (4.2.16)$$

This equation is solved, subject to boundary conditions $T = T_0$ at $z = 0$ and $z = h$, to obtain the temperature and flux profiles,

$$T = T_0 + \frac{\mu V^2 z(h-z)}{2h^2 k}, \qquad (4.2.17)$$

$$q_z = -k\frac{dT}{dz} = -\frac{\mu V^2(h-2z)}{2h^2}. \qquad (4.2.18)$$

The solution, Eq. 4.2.17, for the temperature field is a parabolic profile, with maximum temperature $T_0 + (\mu V^2/8k)$ at the centre of the channel, $z = h/2$, as shown in Fig. 4.4. From Eq. 4.2.18, there is an outward heat flux $(\mu V^2/2h)$ across across each wall of the channel, which is in the $+z$ direction at $z = h$ and in the $-z$ direction at $z = 0$.

It should be noted that in addition to the heat flux due to viscous heating, there could be a flux due to the temperature difference ΔT between the fluid and the wall. The Nusselt number for the heat flux due to viscous heating is,

$$\mathrm{Nu} = \frac{q_z}{(k\Delta T/h)} = \frac{\mu V^2}{2k\Delta T} = \frac{\mathrm{Br}}{2}, \qquad (4.2.19)$$

where the Brinkman number, defined in Example 1.6.5, is the ratio of the flux due to viscous heating and that due to the temperature difference. $\qquad \square$

152   *Fundamentals of Transport Processes*

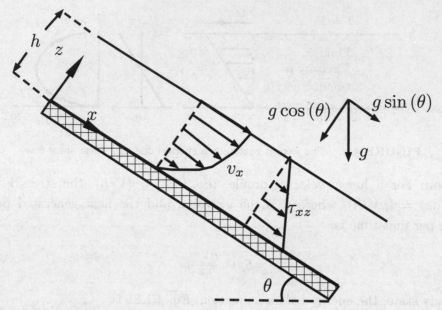

FIGURE 4.5. The velocity and shear stress profiles in the flow down a plane inclined at an angle θ to the horizontal.

Flow down Inclined Plane

The gravitational acceleration on a fluid results in a constant force density which drives the flow down inclined surfaces. When the force density is a constant, the stress is a linear function of height, and the velocity profile is parabolic, as shown in the following example.

EXAMPLE 4.2.3: A fluid layer of thickness h is flowing down a plane inclined at an angle θ to the horizontal, as shown in Fig. 4.5. There is a no-slip (zero velocity) condition at the bottom surface, while the shear stress is zero at the top free interface. Determine the velocity and stress profiles and the Froude number.

Solution: The co-ordinate system shown in Fig. 4.5 is used, where x is in the flow direction along the inclined plane, z is perpendicular to the plane, and the fluid occupies the region $0 \leq z \leq h$. The equation for the stress, Eq. 4.1.32, for this configuration is,

$$\frac{d\tau_{xz}}{dz} + \rho g \sin(\theta) = 0, \quad (4.2.20)$$

where $\rho g \sin(\theta)$ is the component of the gravitational force (momentum source) acting in the x direction, and ρ and g are the density and acceleration due to gravity.

The stress profile is determined by solving Eq. 4.2.20, subject to the condition that the stress is zero at the free interface at $z = h$,

$$\tau_{xz} = \rho g \sin(\theta)(h - z). \tag{4.2.21}$$

Thus, the shear stress varies linearly with z for constant force density, as shown in Fig. 4.5. The velocity field is determined from Newton's law of viscosity,

$$\mu \frac{dv_x}{dz} = \rho g \sin(\theta)(h - z). \tag{4.2.22}$$

This equation is solved subject to the condition $v_x = 0$ at $z = 0$ to obtain the velocity profile,

$$v_x = \frac{\rho g \sin(\theta)(h^2 - (h - z)^2)}{2\mu} = v_{max}\left[1 - \left(1 - \frac{z}{h}\right)^2\right], \tag{4.2.23}$$

where the maximum velocity v_{max} is,

$$v_{max} = v_x|_{z=h} = \frac{\rho g \sin(\theta)h^2}{2\mu}. \tag{4.2.24}$$

The velocity profile is a parabolic profile with zero velocity at the base $z = 0$ and zero slope at the surface $z = h$. The average velocity is,

$$v_{av} = \frac{1}{h}\int_0^h dz\, v_x = \frac{\rho g \sin(\theta)h^2}{3\mu}. \tag{4.2.25}$$

Therefore, the average velocity is two-thirds of the maximum velocity. From the expression for the average velocity, a correlation can be derived for the dimensionless Froude number $Fr = (v_{av}/\sqrt{gh})$,

$$Fr = \frac{\rho g \sin(\theta)h^2}{3\mu\sqrt{gh}} = \frac{Ar^{1/2}\sin(\theta)}{3}, \tag{4.2.26}$$

where the Archimedes number is $Ar = (\rho^2 gh^3/\mu^2)$. $\square$

Electrokinetic Flow

The source of momentum driving the flow is not spatially uniform in electrokinetic flows, where the electrical force on the ions in a solution is used to drive a flow close to a charged surface. When a fluid with ions is confined between charged surfaces, the surface attracts oppositely charged ions. These ions are confined to a 'diffuse

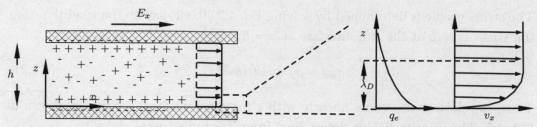

FIGURE 4.6. Flow of a liquid with dissolved ions through a channel with height h with charged walls subjected to a stream-wise electric field E_x. The velocity on the length scale of the channel height h is shown on the left, and the variation of the charge density q_e and the velocity v_x in the Debye layer close to the surface is shown on the right.

layer' adjacent to the surface, within which there is an exponential decay of the ions away from the surface, as shown in Fig. 4.6. In the dilute limit where the ion concentration is small, the charge concentration q_e (charge per unit volume) close to the charged surface is given by,

$$q_e = -\frac{Q_{es}}{\lambda_D}\, e^{(-z/\lambda_D)}, \qquad (4.2.27)$$

where Q_{es} is the surface electrical charge density (charge per unit area) on the charged surface, z is the distance from the surface and λ_D is the Debye screening length defined as,

$$\lambda_D = \sqrt{\frac{\epsilon k_B T}{\sum_i N_i^0 e_i^2}}, \qquad (4.2.28)$$

where i denotes the ion species, N_i^0 is the number density of ion i in the bulk far from the surface, e_i is the electrical charge on ion i, ϵ is the electric permittivity of the medium, and k_B and T are the Boltzmann constant and temperature, respectively. The electric permittivity of the medium is usually expressed as $\epsilon = \epsilon_r \epsilon_0$, where ϵ_0 is the vacuum permittivity (8.85×10^{-12} kg^{-1}m^{-3}s^{4}A^2), and ϵ_r is the (dimensionless) relative permittivity of the material. There is a negative sign in Eq. 4.2.27 because the charges in the solution attracted to the surface have opposite polarity to the surface charge. The charged surface is often characterised by the 'zeta potential' ζ, instead of the surface charge density Q_{es}; the following approximate relation connects the two,

$$\zeta = \frac{Q_{es}\lambda_D}{\epsilon}. \qquad (4.2.29)$$

EXAMPLE 4.2.4: Calculate the Debye length for a 1 millimolar aqueous NaCl solution at 300 K. The relative permittivity of water is 80.

Solution: A 1 millimolar aqueous solution contains 10^{-3} moles each of sodium and chloride ions per litre, or 10^{-3} kilomoles each of sodium and chloride ions per cubic meter. The number density of the sodium and chloride ions is,

$$N_{Na^+} = N_{Cl^-} = 10^{-3} \text{ kilomole/m}^3 \times 6.023 \times 10^{26} \text{ ions/kilomole} = 6.023 \times 10^{23} \text{ ions/m}^3.$$
$$(4.2.30)$$

The charge on an electron is $e = 1.6 \times 10^{-19}$ A s. The term in the denominator on the right of Eq. 4.2.28 is,

$$\sum_i N_i e_i^2 = 2 \times 6.023 \times 10^{23} \text{ ions/m}^3 \times (1.6 \times 10^{-19} \text{ A s})^2 = 3.08 \times 10^{-14} \text{ m}^{-3}\text{s}^2\text{A}^2.$$
$$(4.2.31)$$

The electric permittivity is,

$$\epsilon = \epsilon_r \epsilon_0 = 80 \times 8.85 \times 10^{-12} \text{ kg}^{-1}\text{m}^{-3}\text{s}^4\text{A}^2 = 7.08 \times 10^{-10} \text{ kg}^{-1}\text{m}^{-3}\text{s}^4\text{A}^2. \quad (4.2.32)$$

The product of the Boltzmann constant and temperature is

$$k_B T = (1.38 \times 10^{-23} \text{ J/K}) \times 300 \text{ K} = 4.14 \times 10^{-21} \text{ J}. \quad (4.2.33)$$

The Debye length, Eq. 4.2.28, is

$$\lambda_D = \left(\frac{7.08 \times 10^{-10} \text{ kg}^{-1}\text{m}^{-3}\text{s}^4\text{A}^2 \times 4.14 \times 10^{-21} \text{ J}}{3.08 \times 10^{-14} \text{ m}^{-3}\text{s}^2\text{A}^2} \right)^{1/2} = 9.75 \times 10^{-9} \text{ m}.$$
$$(4.2.34)$$
□

As shown in the preceding example, the Debye length is usually small, approximately 1 nm for a 0.1 molar univalent electrolyte, and about 10 nm for 1 millimolar univalent electrolyte. It should be noted that the electrical double-layer near a charged surface has complex structure, with an adsorbed (Stern) layer of immobile counterions closest to the surface, outside of which is the diffuse layer. The fluid close to the Stern layer is not mobile, and there is a 'slipping plane' outside the Stern layer beyond which the fluid is mobile. The zeta potential is related to the net charge outside the 'slipping plane'. However, a simplified picture is used here, where the relation between the surface charge density and the zeta potential is given by Eq. 4.2.29 outside the Stern layer.

When an electric field E_x is applied in the x direction, as shown in Fig. 4.6, the force density (force per unit volume) exerted on the fluid is $E_x q_e$, where q_e is given in Eq. 4.2.27, and the electric field E_x is the potential drop (decrease in voltage) per unit length along the channel. This electrokinetic force drives fluid flow along the channel. The calculation is simplified by realising that the charge density, Eq. 4.2.27, decreases exponentially within a distance comparable to the Debye length, λ_D, which is in the range of 1–10 nm for normal ion concentrations. This is much smaller than the channel height in practical situations. Therefore, the body force density exerted by the electric field is confined to regions of thickness comparable to λ_D at the walls of the channel. In these regions, the channel can be considered of infinite extent in the direction perpendicular to the surface while solving for the force.

The body force density $E_x q_e$ is substituted into the the momentum conservation equation, Eq. 4.1.31, the time derivative is zero at steady state, and relations, Eqs. 4.2.27 and 4.2.29, are used for the charge density q_e,

$$\mu \frac{d^2 v_x}{dz^2} - \frac{E_x \epsilon \zeta e^{(-z/\lambda_D)}}{\lambda_D^2} = 0. \tag{4.2.35}$$

This equation is solved to obtain the velocity,

$$v_x = \frac{E_x \epsilon \zeta e^{(-z/\lambda_D)}}{\mu} + Cz + D. \tag{4.2.36}$$

The integration constants C and D are chosen to satisfy two boundary conditions. The first is the no-slip condition $v_x = 0$ at the surface $z = 0$. The second condition is the zero shear stress condition far from the surface for $z \gg \lambda_D$. If there is no pressure variation along the flow direction and no relative motion of the walls, then the shear stress in the bulk has to decrease to zero far from the surface. This implies that $(dv_x/dz) = 0$ for $z \gg \lambda_D$. The constants $C = 0$ and $D = -(E_x \epsilon \zeta / \mu \lambda_D)$ are chosen to satisfy the two boundary conditions $v_x = 0$ at $z = 0$ and $(dv_x/dz) = 0$ for $z \to \infty$. The solution for the velocity field is,

$$v_x = v_s(1 - e^{(-z/\lambda_D)}), \tag{4.2.37}$$

where the 'slip' velocity is given by,

$$v_s = -\frac{E_x \epsilon \zeta}{\mu}. \tag{4.2.38}$$

The slip velocity is positive if the zeta potential is negative (positive charges are attracted to the surface) and the electric field is along the positive x direction, as shown in Fig. 4.6.

A similar calculation can be performed in a region of thickness λ_D near the top surface, and the velocity is,

$$v_x = v_s(1 - e^{(-(h-z)/\lambda_D)}),\qquad(4.2.39)$$

where $h - z$ is the distance from the top wall. Eq. 4.2.38 for the velocity implies that the fluid velocity approaches a constant value v_s when the distance from the wall is larger than λ_D. Therefore, over distances comparable to the channel width h, the flow appears as a plug flow with slip velocity v_s at the wall, as shown in Fig. 4.6.

An interesting feature of the expression of the slip velocity, Eq. 4.2.38, is that it does not depend on the channel height, the Debye length or the ion concentration, and depends only on the electric field and the zeta potential. The flow rate increases linearly with the channel height. In practical applications, the electrokinetic slip velocity is rather small even when large electric fields are applied, as shown in the following example.

EXAMPLE 4.2.5: An aqueous solution with viscosity 10^{-3} kg/m/s is confined in a microchannel of height 100 μm in the z direction between two charged surfaces with zeta potential $-$ 30 mV, as shown in Fig. 4.6. The dielectric constant of water is 80. What is the electric field strength required for generating a flow velocity of 1 mm/s?

Solution: The electric permittivity of water is $\epsilon_r \epsilon_0$, where the dielectric constant of water $\epsilon_r = 80$, and the electric permittivity of vacuum $\epsilon_0 = 8.85 \times 10^{-12} \mathrm{kg}^{-1}\mathrm{m}^{-3}\mathrm{s}^4\mathrm{A}^2$. The slip velocity is related to the electric field by Eq. 4.2.38,

$$E_x = -\frac{\mu v_s}{\epsilon \zeta} = -\frac{10^{-3} \text{ kg/m/s} \times 10^{-3} \text{ m/s}}{80 \times 8.85 \times 10^{-12} \text{ kg}^{-1}\mathrm{m}^{-3}\mathrm{s}^4\mathrm{A}^2 \times (-30 \times 10^{-3} \text{ kg m}^2\mathrm{s}^{-3}\mathrm{A}^{-1})}$$

$$= 47 \times 10^3 \text{ kg m s}^{-3}\mathrm{A}^{-1} = 47 \times 10^3 \text{ V/m}.\qquad(4.2.40)$$

$\square$

Diffusion and Reaction

In the examples considered so far, the source could be a function of position, but is independent of the field variable. Mass diffusion coupled with a homogeneous chemical reaction is an example where the source/sink depends on the field variable. The source of mass depends on the concentration, because the rate of reaction is a function of the concentration of the reactant. For a first order reaction, the reaction rate constant $\mathcal{K}$ has dimension $\mathcal{T}^{-1}$. The dimensionless ratio of the reaction and

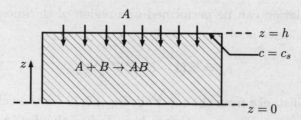

FIGURE 4.7. Diffusion of a gas species into a liquid with homogeneous chemical reaction.

diffusion rates is the Damkohler number, $\mathrm{Da} = (\mathcal{K}l_c^2/\mathcal{D})$, where $\mathcal{D}$ is the diffusion coefficient and l_c is the characteristic length. In this case, the Sherwood number (dimensionless flux) is a function of the Damkohler number. The correlation for the Sherwood number for diffusion and reaction into a liquid film is calculated in the following example.

EXAMPLE 4.2.6: A gaseous reactant A dissolves in a liquid B, and undergoes a first order reaction $A + B \to AB$ with rate constant $\mathcal{K}$ in a tank of height h, as shown in Fig. 4.7. The concentration of reactant on the liquid side at the gas–liquid interface is c_s, and the diffusion coefficient for the diffusion of reactant A into the liquid is $\mathcal{D}$. Determine the correlation for the Sherwood number.

Solution: The governing equation is the mass conservation equation, Eq. 4.1.10, where the rate of production of reactant per unit volume is $\mathcal{S} = -\mathcal{K}c$, and c is the concentration of reactant A. The reaction-diffusion equation at steady state is,

$$\mathcal{D}\frac{\mathrm{d}^2 c}{\mathrm{d}z^2} - \mathcal{K}c = 0. \tag{4.2.41}$$

The concentration of A is c_s at the surface $z = h$, and there is a zero flux condition at the bottom surface at $z = 0$.

The solution of Eq. 4.2.41 is,

$$c = Ce^{(z\sqrt{\mathcal{K}/\mathcal{D}})} + De^{(-z\sqrt{\mathcal{K}/\mathcal{D}})}. \tag{4.2.42}$$

The integration constants C and D are determined from the boundary conditions $c = c_s$ at the top surface $z = h$, and the zero flux condition $(\mathrm{d}c/\mathrm{d}z) = 0$ at the impenetrable bottom surface at $z = 0$,

$$c = c_s \left(\frac{e^{(z\sqrt{\mathcal{K}/\mathcal{D}})} + e^{(-z\sqrt{\mathcal{K}/\mathcal{D}})}}{e^{(h\sqrt{\mathcal{K}/\mathcal{D}})} + e^{(-h\sqrt{\mathcal{K}/\mathcal{D}})}} \right). \tag{4.2.43}$$

The flux at the top surface is,

$$j_z = -\mathcal{D}\frac{dc}{dz}\bigg|_{z=h} = -\mathcal{D}c_s\sqrt{\frac{\mathcal{K}}{\mathcal{D}}}\left(\frac{e^{(h\sqrt{\mathcal{K}/\mathcal{D}})} - e^{(-h\sqrt{\mathcal{K}/\mathcal{D}})}}{e^{(h\sqrt{\mathcal{K}/\mathcal{D}})} + e^{(-h\sqrt{\mathcal{K}/\mathcal{D}})}}\right). \tag{4.2.44}$$

The Sherwood number, $(|j_z|/(\mathcal{D}c_s/h))$, is expressed in terms of the Damkohler number $\mathrm{Da} = (\mathcal{K}h^2/\mathcal{D})$,

$$\mathrm{Sh} = \mathrm{Da}^{1/2}\left[\frac{e^{(\mathrm{Da}^{1/2})} - e^{(-\mathrm{Da}^{1/2})}}{e^{(\mathrm{Da}^{1/2})} + e^{(-\mathrm{Da}^{1/2})}}\right]. \tag{4.2.45}$$

A physical interpretation of the Damkohler number is the square of the ratio of the characteristic length scale h and the penetration depth $\sqrt{\mathcal{D}/\mathcal{K}}$ (with dimension $\mathcal{L}$) to which the reactant diffuses before it is consumed in the reaction. For low Damkohler number $\mathrm{Da} \ll 1$, where the thickness of the liquid film is much smaller than the penetration depth $\sqrt{\mathcal{D}/\mathcal{K}}$, the approximation $e^{(\mathrm{Da}^{1/2})} = 1 + \mathrm{Da}^{1/2}$ is used in Eq. 4.2.45, and the correlation reduces to $\mathrm{Sh} = \mathrm{Da}$. For high Damkohler number, where the thickness of the liquid film is much larger than the penetration depth $\sqrt{\mathcal{D}/\mathcal{K}}$, we can make the approximation $e^{(\mathrm{Da}^{1/2})} \gg e^{(-\mathrm{Da}^{1/2})}$. The term in square brackets in Eq. 4.2.45 tends to 1, and the correlation reduces to $\mathrm{Sh} = \mathrm{Da}^{1/2}$. $\quad\square$

4.2.2 Varying Diffusivity

The constitutive relation for a non-Newtonian fluid is an example where the diffusivity does depend on the field variable. The power-law model is a commonly used non-linear constitutive relation between the stress and strain rate for a non-Newtonian fluid. The following example considers a power-law fluid flowing under gravity down a vertical surface. In this case, the body force density in Eq. 4.1.32 is a constant, and so the stress increases linearly from the free surface in a manner similar to that for a Newtonian fluid (Example 4.2.3). However, the velocity profile is different from a parabolic profile, because the stress is not a linear function of the strain rate.

EXAMPLE 4.2.7: Determine the velocity profile for a power-law fluid of thickness h flowing down a vertical surface at steady state, if the stress in the fluid is given by,

$$\tau_{xz} = \kappa \left(\frac{dv_x}{dz} \right) \left| \frac{dv_x}{dz} \right|^{n-1}, \tag{4.2.46}$$

where n is the power-law index and κ is the flow consistency index. The velocity is zero at the solid–liquid interface $z = 0$, and the shear stress is zero at the liquid–gas interface at $z = h$.

Solution: The configuration and co-ordinate system is the same as that in Fig. 4.5 with $\theta = \pi/2$. The momentum conservation equation for the film is,

$$\frac{d\tau_{xz}}{dz} + \rho g = 0, \tag{4.2.47}$$

where ρ is the density and g is the gravitational acceleration. This is solved, and the zero shear stress condition is applied at $z = h$, to obtain,

$$\tau_{xz} = \rho g(h - z) = \kappa \left(\frac{dv_x}{dz} \right)^n \left| \frac{dv_x}{dz} \right|^{n-1}. \tag{4.2.48}$$

The above equation is integrated, subject to the boundary condition $v_x = 0$ at $z = 0$, to obtain,

$$v_x = \frac{n}{n+1} \left(\frac{\rho g}{\kappa} \right)^{1/n} (h^{((n+1)/n)} - (h - z)^{((n+1)/n)}). \tag{4.2.49}$$

The maximum and average velocity are,

$$v_{max} = v_x|_{z=h} = \frac{n}{n+1} \left(\frac{\rho g h^{n+1}}{\kappa} \right)^{1/n}, \tag{4.2.50}$$

$$v_{av} = \frac{1}{h} \int_0^h dz \, v_x = \frac{n}{2n+1} \left(\frac{\rho g h^{n+1}}{\kappa} \right)^{1/n}. \tag{4.2.51}$$

$\square$

Summary (4.2)

1. For steady transport in the absence of sources/sinks the flux $\mathcal{J}_z$ in Table 4.1 is a constant, and the field variable Φ_{fv} is a linear function of the spatial co-ordinate in the material.

(a) For a composite consisting of materials with different thickness and conductivities in the series configuration shown in Fig. 4.3(c), the total heat transfer resistance $-(\Delta \Phi_{fv}/\mathcal{J}_z)$ is given by the sum rule, Eq. 4.2.7.

(b) For a composite in the parallel configuration shown in Fig. 4.3(d), the total heat transfer resistance $-(\Delta \Phi_{fv}/\mathcal{J}_z)$ is given by the inverse sum rule, Eq. 4.2.11.

2. When there are internal sources/sinks which depend on position but not on the field variable, Eq. 4.1.32 is first solved at steady state for the flux, and then the constitutive relation Eq. 4.1.33 is used to determine the field variable.

3. If the source is a constant independent of the spatial co-ordinate, the flux $\mathcal{J}_z$ is a linear function of the spatial co-ordinate. If the diffusivity is also a constant, the field variable Φ_{fv} is a quadratic function of the spatial co-ordinate.

4.3 Binary Diffusion

In a multicomponent diffusion process, an average fluid velocity could be generated due to the motion of the multiple species, and this has to be incorporated into the expression for the flux. The average velocity could be defined as a mass average velocity or a molar average velocity. The fluid velocity in the momentum conservation equation is the mass average velocity. However, the molar average velocity is commonly used to model multicomponent diffusion processes. The two formulations are described here for the binary diffusion in a two-component system.

Consider the diffusion in the z direction in a binary mixture of species A and B with concentrations (mass densities) c_A and c_B. The total mass density of the fluid ρ is

$$\rho = c_A + c_B. \tag{4.3.1}$$

The mass fractions of the species are $w_A = (c_A/\rho)$ and $w_B = (c_B/\rho)$. The mass fluxes j_{Az} and j_{Bz} are related to the species velocities v_{Az} and v_{Bz}, $j_{Az} = c_A v_{Az}$ and $j_{Bz} = c_B v_{Bz}$. The total mass flux, ρv_z, is the sum of the fluxes of the two species,

$$\rho v_z = c_A v_{Az} + c_B v_{Bz} = j_{Az} + j_{Bz}. \tag{4.3.2}$$

The flux of species A has two components: the first due to the spatial variation of the concentration given by Fick's law, and the second due to the mass average velocity v_z of the fluid,

$$j_{Az} = -\mathcal{D}_{AB}\rho\frac{dw_A}{dz} + c_A v_z = -\mathcal{D}_{AB}\rho\frac{dw_A}{dz} + w_A \rho v_z, \qquad (4.3.3)$$

where $\mathcal{D}_{AB}$ is the mass diffusion coefficient of species A in a mixture containing species A and B, and the last term on the right, $c_A v_z = w_A \rho v_z$, is the flux of species A due to the fluid mass average velocity v_z.

The expression for the diffusion flux, Eq. 4.3.3, is proportional to the total density times the spatial derivative of the mass fraction. This is different from the Fick's law for diffusion, where the flux is proportional to the spatial derivative of the concentration. The thermodynamic driving force for diffusion is the spatial derivative of the chemical potential, and not the concentration. The expression for the flux follows from the model used for the chemical potential in a mixture, and the first term on the right in Eq. 4.3.3 is derived from a model for the chemical potential for ideal mixtures; this is beyond the scope of this text. If the mass density is a constant, the expression for the flux used in Eq. 4.3.3 is the same as Fick's law. When the solute mass fraction is small, the change in the solute mass fraction does not significantly alter the total density, and the expression for the flux in Eq. 4.3.3 reduces to Fick's law. However there is, in general, a difference in the two definitions for processes where the density is not a constant.

Using Eq. 4.3.2 for the mass average velocity, the expression for the flux, Eq. 4.3.3 is,

$$\boxed{j_{Az} = -\mathcal{D}_{AB}\rho\frac{dw_A}{dz} + w_A(j_{Az} + j_{Bz}).} \qquad (4.3.4)$$

The flux of species B is obtained by interchanging the indices A and B in Eq. 4.3.4,

$$j_{Bz} = -\mathcal{D}_{BA}\rho\frac{dw_B}{dz} + w_B(j_{Az} + j_{Bz}), \qquad (4.3.5)$$

where $\mathcal{D}_{BA}$ is the mass diffusion coefficient of species B in a mixture of species A and B. It is easy to show that $\mathcal{D}_{BA} = \mathcal{D}_{AB}$ by adding Eqs. 4.3.4 and 4.3.5, and using the relations $w_A + w_B = 1$ and $(dw_A/dz) + (dw_B/dz) = 0$ for the mass fractions.

The description based on the mole fraction is similar to that based on the mass fraction. The molar concentrations C_A, C_B, the mole fractions W_A, W_B, and the molar fluxes J_{Az}, J_{Bz} are denoted by capital letters. The molar fluxes are the product of the molar concentrations and the species molar average velocities, $J_{Az} = C_A v_{Az}$ and $J_{Bz} = C_B v_{Bz}$, where v_{Az} and v_{Bz} are the species velocities. The total molar concentration is $C = C_A + C_B$. The mole fractions are defined as $W_A = (C_A/C)$ and $W_B = (C_B/C)$. The fluid molar average velocity V_z is defined by the relation,

$$CV_z = C_A v_{Az} + C_B v_{Bz}. \tag{4.3.6}$$

The equivalents of Eqs. 4.3.4 and 4.3.5 for the molar fluxes are,

$$\boxed{J_{Az} = -\mathcal{D}_{AB} C \frac{dW_A}{dz} + W_A(J_{Az} + J_{Bz}),} \tag{4.3.7}$$

$$J_{Bz} = -\mathcal{D}_{BA} C \frac{dW_B}{dz} + W_B(J_{Az} + J_{Bz}). \tag{4.3.8}$$

It is easily verified, by adding Eqs. 4.3.7 and 4.3.8, that $\mathcal{D}_{AB} = \mathcal{D}_{BA}$.

In general, there is a difference between the mass average and mole average velocities. The formulation based on the molar average concentration and velocity is advantageous in situations where the total molar concentration is a constant. In ideal gases, for example, the total molar concentration (moles per unit volume) is a constant at fixed temperature and pressure. When chemical reactions are involved, the molar rates of consumption and production of the different species are related by the stoichiometry of the reaction, and it is more convenient use molar average quantities. In cases where there is a constraint on the total molar flux or the flux of one species, the formulation based on the molar concentration is simpler. Exercise 4.14 illustrates the increased complexity of the mass average flux formulation for a system where the molar density is a constant.

However, the total momentum conservation equation is based on the mass average velocity. The molar average velocity cannot be used in the momentum conservation equation. The mass average velocity is also simpler in formulations where mixture mass density does not change significantly as the mass fractions change.

EXAMPLE 4.3.1: Water evaporates from glass tube into dry air flowing above the tube, as shown in Fig. 4.8. The ambient temperature is 25°C, the height of water in the tube is 0.5 cm, and the length of the tube is 15 cm. What is the rate of decrease of the height of water due to evaporation? The vapour pressure of water at 25°C is 3.15 kPa, and the diffusion coefficient for water vapour in air is 2.2×10^{-5} m^2/s.

Solution: The mole fractions of air and water in the tube are denoted W_a and W_w, and the fluxes are J_{az} and J_{wz}. There is a decrease in the water level due to evaporation, resulting in an inflow of air into the tube. Since the density of liquid water is about 800 times that of air, the decrease in the volume of the liquid water is smaller, by factor of about 8×10^{-2}, than the volume of the water vapour generated. This implies that the flux of air into the tube is much smaller than the flux of water vapour emerging from the tube. Therefore, the flux of air into the tube J_{az} is set equal to zero.

The configuration and co-ordinate system are shown in Fig. 4.8, where the z co-ordinate is directed upwards from the water surface. The mole fraction is the saturation value $W_w = W_s$ at $z = 0$, and the mole fraction decreases to zero at the top of the tube, $z = h$. Eq. 4.3.7 for J_{wz}, with $J_{az} = 0$, is

$$J_{wz} = -C\mathcal{D}_{wa}\frac{\mathrm{d}W_w}{\mathrm{d}z} + W_w J_{wz}, \qquad (4.3.9)$$

where C is the total molar concentration. This equation is rearranged to obtain an explicit expression for the flux,

$$J_{wz} = -\frac{C\mathcal{D}_{wa}}{1 - W_w}\frac{\mathrm{d}W_w}{\mathrm{d}z}. \qquad (4.3.10)$$

Since the flux is a constant at steady state, the above equation is integrated to obtain

$$\ln(1 - W_w) = \frac{J_{wz}z}{\mathcal{D}_{wa}C} + I. \qquad (4.3.11)$$

The constant of integration $I = \ln(1 - W_s)$ is determined from the condition that the water mole fraction is equal to its saturation value W_s at $z = 0$. The expression

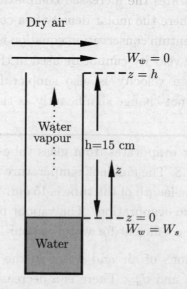

FIGURE 4.8. Evaporation of water from a tube.

for the mole fraction of water is,

$$\ln\left(\frac{1-W_w}{1-W_s}\right) = \frac{J_{wz}z}{\mathcal{D}_{wa}C}. \tag{4.3.12}$$

The flux J_{wz} is determined from the condition that the mole fraction of water $W_w = 0$ at the top of the tube, $z = h$,

$$J_{wz} = -\frac{\mathcal{D}_{wa}C \ln(1-W_s)}{h}. \tag{4.3.13}$$

If air is considered an ideal gas, the molar concentration of air at atmospheric pressure $p = 1.013 \times 10^5$ Pa and absolute temperature $T = 298$ K is

$$C = \frac{p}{RT} = \frac{1.013 \times 10^5 \text{ Pa}}{8.314 \text{ J/mol/K} \times 298 \text{ K}} = 40.9 \text{ moles/m}^3. \tag{4.3.14}$$

The mole fraction of water vapour at the water surface is the ratio of the vapour pressure and the atmospheric pressure, $W_s = 3.1 \times 10^{-2}$. Substituting the $C = 40.9$ moles/m^3, $W_s = 3.1 \times 10^{-2}$, $\mathcal{D}_{wa} = 2.2 \times 10^{-5}$ m^2/s and $h = 0.15$ m into Eq. 4.3.13,

$$J_{wz} = -\frac{2.2 \times 10^{-5} \text{ m}^2/\text{s} \times 40.9 \text{ moles/m}^3 \ln(1 - 0.031)}{0.15 \text{ m}}$$
$$= 1.89 \times 10^{-4} \text{ moles/m}^2/\text{s} = 3.40 \times 10^{-6} \text{ kg/m}^2/\text{s}. \tag{4.3.15}$$

Here, the molecular weight of water is 18×10^{-3} kg/mole.

If the area of cross section of the tube is A m^2, the mass of water evaporated per unit time is 3.40×10^{-6} A kg/s. If the density of water is 10^3 kg/m^3, the volume of water evaporated is 3.40×10^{-9} A m^3/s. The height of water evaporated is the ratio of the volume and the area of cross section, which is 3.40×10^{-9} m/s or 0.29 mm/day.
□

EXAMPLE 4.3.2: The process for producing iron pentacarbonyl gas of molecular weight 196 g/mol involves the flow of carbon monoxide over an iron surface at 1 atm and 150°C, and the chemical reaction

$$\text{Fe} + 5 \text{ CO} \rightarrow \text{Fe (CO)}_5, \tag{4.3.16}$$

takes place at the surface of iron. Consider a reactor of height 1 cm and cross sectional area 100 cm^2, above which carbon monoxide gas is passed as shown in

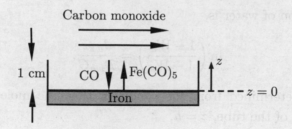

FIGURE 4.9. Production of iron pentacarbonyl.

Fig. 4.9. The carbon monoxide diffuses through a film of thickness 1 cm and reacts with the iron surface, and the iron pentacarbonyl product is carried away with the gas flow above the reactor. The reaction is sufficiently fast that the concentration of carbon monoxide is zero at the iron surface. The velocity of CO gas above the reactor is sufficiently fast that the concentration of iron pentacarbonyl in the flowing gas stream above the reactor is negligible. What is the rate of production of iron pentacarbonyl? The diffusion coefficient of iron pentacarbonyl in the gas stream is 2×10^{-5} m^2/s.

Solution: At steady state, the molar fluxes of carbon monoxide and iron pentacarbonyl are related by the stoichiometry of the chemical reaction Eq. 4.3.16,

$$\frac{J_{CO}}{J_{Fe(CO)_5}} = -5, \qquad (4.3.17)$$

where J_{CO} and $J_{Fe(CO)_5}$ are the molar fluxes of carbon monoxide and iron pentacarbonyl respectively. There is a negative sign on the right in Eq. 4.3.17 because the flux of Fe(CO)$_5$ is upward and that of CO is downward.

Eq. 4.3.7 for the flux of iron pentacarbonyl is,

$$
\begin{aligned}
J_{Fe(CO)_5} &= -\mathcal{D}C\frac{dW_{Fe(CO)_5}}{dz} + W_{Fe(CO)_5}\left(J_{Fe(CO)_5} + J_{CO}\right) \\
&= -\mathcal{D}C\frac{dW_{Fe(CO)_5}}{dz} - 4W_{Fe(CO)_5}J_{Fe(CO)_5}, \qquad (4.3.18)
\end{aligned}
$$

where $W_{Fe(CO)_5}$ is the mole fraction of iron pentacarbonyl, C is the total molar concentration, $\mathcal{D}$ is the diffusion coefficient and Eq. 4.3.17 has been used to express J_{CO} in terms of $J_{Fe(CO)_5}$ in the second step above. Eq. 4.3.18 is simplified as,

$$J_{Fe(CO)_5} = -\frac{\mathcal{D}C}{1 + 4W_{Fe(CO)_5}}\frac{dW_{Fe(CO)_5}}{dz}. \qquad (4.3.19)$$

The above equation is integrated assuming constant molar flux at steady state,

$$\frac{J_{Fe(CO)_5} z}{\mathcal{D}C} = -\frac{\ln\left(1 + 4W_{Fe(CO)_5}\right)}{4} + I, \qquad (4.3.20)$$

where I is the constant of integration. The integration constant is determined from the condition that the mole fraction of $Fe(CO)_5$ is 1 at the surface $z = 0$ because the carbon monoxide is rapidly consumed,

$$I = \frac{\ln(5)}{4}, \qquad (4.3.21)$$

and the expression for the flux is,

$$\frac{J_{Fe(CO)_5} z}{\mathcal{D}C} = \frac{1}{4}\ln\left(\frac{5}{1 + 4W_{Fe(CO)_5}}\right). \qquad (4.3.22)$$

The flux is determined from the condition $W_{Fe(CO)_5} = 0$ in the gas stream at $z = h$,

$$J_{Fe(CO)_5} = \frac{\mathcal{D}C}{h}\frac{\ln(5)}{4}, \qquad (4.3.23)$$

where h is the height of the film through which the $Fe(CO)_5$ diffuses.

The molar density of carbon monoxide at 1 atm and 150° C (423 K) is estimated from the ideal gas law,

$$C = \frac{p}{RT} = \frac{1.013 \times 10^5 \text{ Pa}}{8.314 \text{ J/mole/K} \times 423 \text{ K}} = 28.8 \text{ moles/m}^3. \qquad (4.3.24)$$

Substituting the above value for the molar density, $\mathcal{D} = 2 \times 10^{-5}$ m^2/s, and $h = 0.01$ m in Eq. 4.3.23, the flux is,

$$J_{Fe(CO)_5} = \frac{2 \times 10^{-5}\text{m}^2/\text{s} \times 28.8 \text{ moles/m}^3}{0.01 \text{ m}}\frac{\ln(5)}{4}$$

$$= 2.3 \times 10^{-2} \text{ moles/m}^2/\text{s} = 4.54 \times 10^{-3} \text{ kg/m}^2/\text{s}. \qquad (4.3.25)$$

Here, the molecular mass of iron pentacarbonyl 0.196 kg/mole has been used. The total production rate of iron pentacarbonyl is 4.54×10^{-5} kg/s, the product of the flux and the area of cross section 100 cm^2 = 0.01 m^2. $\qquad \square$

Summary (4.3)

1. For multicomponent diffusion, there is a contribution to the flux of a material due to the average fluid velocity, which is the sum of the species average velocities.

2. The equations for the mass flux and the molar flux of a component in a binary mixture are given by Eqs. 4.3.4 and 4.3.7. The flux due to diffusion is proportional to the spatial derivative of the mass or mole fraction times the mass/molar concentration, instead of the spatial derivative of the concentration.

3. If a relation between J_{Az} and J_{Bz} is known from stoichiometry or other conditions, an explicit relation is obtained for the flux of a species as a function of its mass/mole fraction.

4.4   Transport into an Infinite Medium

The configuration for the unsteady conduction in an infinite medium, shown in Fig. 4.10, consists of a fluid in the region $z > 0$ bounded by a surface at $z = 0$. The initial temperature of the surface and fluid is T_∞. At time $t = 0$, the temperature of the surface is set equal to $T_0 > T_\infty$. The variation of the temperature with z and time is to be determined.

The scaled temperature is defined as,

$$T^* = \frac{T - T_\infty}{T_0 - T_\infty}, \tag{4.4.1}$$

so that $T^* = 1$ at the surface $z = 0$ and $T^* = 0$ for $z \to \infty$. The scaled temperature is determined by solving the energy balance equation (Eq. 4.1.21 where T is replaced by T^*) with constant diffusivity and no source term,

$$\frac{\partial T^*}{\partial t} = \alpha \frac{\partial^2 T^*}{\partial z^2}. \tag{4.4.2}$$

The boundary and initial conditions are,

$$T = T_\infty \Rightarrow T^* = 0 \text{ as } z \to \infty \text{ for all } t, \tag{4.4.3}$$

$$T = T_0 \Rightarrow T^* = 1 \text{ at } z = 0 \text{ for all } t > 0, \tag{4.4.4}$$

$$T = T_\infty \Rightarrow T^* = 0 \text{ at } t = 0 \text{ for all } z > 0. \tag{4.4.5}$$

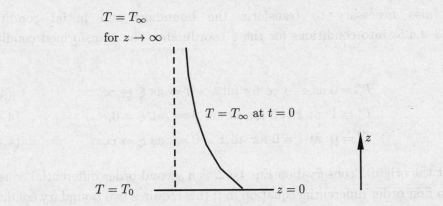

FIGURE 4.10. Configuration for unidirectional transport into an infinite medium.

A 'similarity' solution for Eq. 4.4.2, with the boundary and initial conditions Eqs. 4.4.3–4.4.5, can be obtained by realising that there is no intrinsic length scale in the problem. The plate and fluid are of infinite extent in the plane perpendicular to z, and the boundary conditions are applied at $z = 0$ and $z \to \infty$. Since the temperature T^* is dimensionless, there are only three dimensional variables z, t and α in the problem. These contain two dimensions, $\mathcal{L}$ and $\mathcal{T}$, and it is possible to construct only one dimensionless parameter, $\xi = (z/\sqrt{\alpha t})$. Therefore, just from dimensional analysis, it is expected that the temperature field depends on z and t only through the dimensionless 'similarity' variable,

$$\boxed{\xi = \frac{z}{\sqrt{\alpha t}}.}$$

(4.4.6)

It is postulated that T^* is a function of ξ, and the z and t derivatives are converted into ξ derivatives using differentiation by chain rule,

$$\frac{\partial T^*}{\partial z} = \frac{\partial \xi}{\partial z} \frac{\mathrm{d} T^*}{\mathrm{d}\xi} = \frac{1}{\sqrt{\alpha t}} \frac{\mathrm{d} T^*}{\mathrm{d}\xi},$$

$$\frac{\partial^2 T^*}{\partial z^2} = \frac{\partial \xi}{\partial z} \frac{\mathrm{d}}{\mathrm{d}\xi} \left(\frac{\partial \xi}{\partial z} \frac{\mathrm{d} T^*}{\mathrm{d}\xi} \right) = \frac{1}{\alpha t} \frac{\mathrm{d}^2 T^*}{\mathrm{d}\xi^2},$$

$$\frac{\partial T^*}{\partial t} = \frac{\partial \xi}{\partial t} \frac{\mathrm{d} T^*}{\mathrm{d}\xi} = -\frac{z}{2\sqrt{\alpha} t^{3/2}} \frac{\mathrm{d} T^*}{\mathrm{d}\xi} = -\frac{\xi}{2t} \frac{\mathrm{d} T^*}{\mathrm{d}\xi}.$$

(4.4.7)

The above expressions are substituted into the energy equation, Eq. 4.4.2, and the resulting equation is multiplied by t, to obtain,

$$-\frac{\xi}{2} \frac{\mathrm{d} T^*}{\mathrm{d}\xi} = \frac{\mathrm{d}^2 T^*}{\mathrm{d}\xi^2}.$$

(4.4.8)

Eq. 4.4.8 is consistent with the assumption that the spatial and time dependence of T^* is only through the dependence on the similarity variable ξ.

It is also necessary to transform the boundary and initial conditions, Eqs. 4.4.3–4.4.5, into conditions for the ξ coordinate. The transformed conditions are,

$$T^* = 0 \text{ as } z \to \infty \text{ for all } t > 0 \Rightarrow \text{ as } \xi \to \infty, \qquad (4.4.9)$$

$$T^* = 1 \text{ at } z = 0 \text{ for all } t > 0 \Rightarrow \text{ at } \xi = 0, \qquad (4.4.10)$$

$$T^* = 0 \text{ at } t = 0 \text{ for all } z > 0 \Rightarrow \text{ as } \xi \to \infty. \qquad (4.4.11)$$

Note that the original conservation Eq. 4.4.2, is a second order differential equation in z and a first order differential equation in t; this requires two boundary conditions in the z coordinate and one initial condition at $t = 0$. Eq. 4.4.8 for $T^*(\xi)$ is a second order differential equation, which requires two boundary conditions in the ξ co-ordinate. It is evident that one of the boundary conditions for $z \to \infty$ (Eq. 4.4.9) and the initial condition at $t = 0$ (Eq. 4.4.11) turn out to be identical conditions for $\xi \to \infty$. As required, there are only two independent boundary conditions in the ξ co-ordinate.

Eq. 4.4.8 is a first order ordinary differential equation for $T^{*\prime} = (\mathrm{d}T^*/\mathrm{d}\xi)$,

$$-\frac{\xi T^{*\prime}}{2} = \frac{\mathrm{d}T^{*\prime}}{\mathrm{d}\xi}. \qquad (4.4.12)$$

This is solved to obtain,

$$T^{*\prime} = C_1 \mathrm{e}^{(-\xi^2/4)}, \qquad (4.4.13)$$

where C_1 is a constant of integration. The above equation is integrated with respect to ξ to obtain the temperature,

$$T^* = C_2 + C_1 \int_0^\xi \mathrm{d}\xi' \mathrm{e}^{(-\xi'^2/4)}. \qquad (4.4.14)$$

The integral in the second term on the right cannot be evaluated analytically. However, it can be expressed in terms of a special function called the 'error function'. The error function and its properties are discussed in Appendix 4.A, and the values of the error function are tabulated. The solution for Eq. 4.4.12 is obtained by substituting $A = \frac{1}{2}$ in Eqs. 4.A.12 and 4.A.14 in Appendix 4.A,

$$T^*(\xi) = C_2 + C_1\sqrt{\pi}\mathrm{erf}(\xi/2). \qquad (4.4.15)$$

The constants C_1 and C_2 are determined from the conditions, Eqs. 4.4.9 and 4.4.10. The constant $C_2 = 1$ from the boundary condition Eq. 4.4.10, because $T^* = 1$

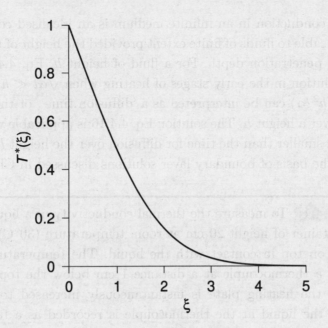

FIGURE 4.11. The solution, Eq. 4.4.16, for $T^*(\xi)$ as a function of ξ.

and erf$(\xi/2) = 0$ at $\xi = 0$. The constant $C_1 = -(1/\sqrt{\pi})$ from the condition Eq. 4.4.9, because $T^* = 0$ and erf$(\xi/2) = 1$ for $\xi \to \infty$. Therefore, the solution for the temperature field is,

$$\boxed{T^*(\xi) = 1 - \text{erf}(\xi/2).} \tag{4.4.16}$$

The heat flux at the surface is,

$$\boxed{q_z = -\,k\frac{\partial T}{\partial z}\Big|_{z=0} = -\,\frac{k(T_0 - T_\infty)}{\sqrt{\alpha t}}\frac{dT^*}{d\xi}\Big|_{\xi=0} = \frac{k(T_0 - T_\infty)}{\sqrt{\pi \alpha t}}.} \tag{4.4.17}$$

Here, Eq. 4.4.13 is substituted for $dT^*/d\xi$, and the constant C_1 is $-(1/\sqrt{\pi})$.

The solution Eq. 4.4.16 for $T^*(\xi)$ is shown as a function of ξ in Fig. 4.11. This is a monotonically decreasing function with value 1 and slope $-(1/\sqrt{\pi})$ at $\xi = 0$. The value of T^* decreases to about 0.48 at $\xi = 1.0$, about 0.16 at $\xi = 2.0$, and further to about 0.034 at $\xi = 3.0$.

The 'penetration depth' $\sqrt{\alpha t}$ is the characteristic length over which there is a temperature variation at time t. The temperature is T_∞ for $z \gg \sqrt{\alpha t}$, and there is a disturbance due to the heat flux from the surface for $z \sim \sqrt{\alpha t}$. The penetration depth increases proportional to $t^{1/2}$, the slope of the temperature profile and the heat flux at the surface decrease proportional to $t^{-1/2}$.

Though the conduction in an infinite medium is an idealised configuration, the solution is applicable to fluids of finite extent provided the height of the fluid is much larger than the penetration depth. For a fluid of height h, Eq. 4.4.16 provides an approximate solution in the early stages of heating when $\sqrt{\alpha t} \ll h$, or $t \ll (h^2/\alpha)$. The timescale (h^2/α) can be interpreted as a 'diffusion time', or the time taken for heat diffusion over a height h. The solution Eq. 4.4.16 is applicable when the time of heating is much smaller than the time for diffusion over the height h. The similarity method forms the basis of boundary layer solutions discussed in Chapter 9.

EXAMPLE 4.4.1: To measure the thermal conductivity of a liquid, the liquid is placed in a container of height 20 cm at room temperature (30°C), and a heating plate is placed on top in contact with the liquid. The temperature of the liquid is measured by a thermocouple at a distance 1 cm below the top. At t = 0, the temperature of the heating plate is instantaneously increased to 70°C, and the temperature of the liquid at the thermocouple is recorded as a function of time. In addition, the total heat transferred from the top surface to the liquid per unit area is also recorded.

The temperature at the thermocouple is 49°C after 10 seconds. The total amount of heat transferred per unit area of the heading surface of the container is 5 kJ/m^2 in 10 seconds. What is the thermal conductivity and thermal diffusivity of the liquid?

Solution: Eq. 4.4.16 is used to determine the thermal diffusivity. At $t = 10$ s,

$$T^* = \frac{T - T_\infty}{T_0 - T_\infty} = \frac{49 - 30}{70 - 30} = 0.475 \Rightarrow \operatorname{erf}(z/2\sqrt{\alpha t}) = (1 - T^*) = 0.525. \quad (4.4.18)$$

From the error function Table 4.4 in Appendix 4.A, we obtain,

$$\frac{z}{2\sqrt{\alpha t}} = 0.5 \Rightarrow \alpha = \frac{z^2}{4 \times 0.5^2 \times t} = \frac{(0.01 \text{ m})^2}{4 \times 0.5^2 \times 10 \text{ s}} = 10^{-5} \text{ m}^2/\text{s}. \quad (4.4.19)$$

The heat flux from the surface is given by Eq. 4.4.17. The heat transferred per unit area (Q_z/A), within time t is,

$$\frac{Q_z}{A} = \int_0^t dt' \, q_z(t') = \frac{2k(T_0 - T_\infty)\sqrt{t}}{\sqrt{\pi \alpha}}. \quad (4.4.20)$$

The above equation is solved to determine the thermal conductivity,

$$k = \frac{(Q_z/A)\sqrt{\pi\alpha}}{2(T_0 - T_\infty)\sqrt{t}} = \frac{5 \times 10^3 \text{ J/m}^2 \times \sqrt{\pi} \times \sqrt{10^{-5} \text{ m}^2/\text{s}}}{2 \times (40^o\text{C}) \times \sqrt{10 \text{ s}}} = 0.11 \text{ W/m/°C}.$$

$$(4.4.21)$$

□

The solution procedure for the mass transfer or momentum transfer problem in an infinite fluid bounded by a surface, with a step change in the field variable at the surface at $t = 0$, is identical to that for the heat transfer problem. For the mass transfer problem, a fluid with initial concentration c_∞ for $t < 0$ is considered, and the concentration at the surface at $z = 0$ is increased to c_0 at $t = 0$. The scaled concentration, defined as $c^* = (c - c_\infty)/(c_0 - c_\infty)$, satisfies the boundary and initial conditions, Eq. 4.4.3 to 4.4.5. The mass conservation equation is identical to the energy conservation Eq. 4.4.2 with T^* replaced by c^*, and thermal diffusivity α replaced by the mass diffusivity $\mathcal{D}$. Therefore, the solution for c^* is the right side of Eq. 4.4.16, with α replaced by $\mathcal{D}$ in the similarity variable.

For the momentum transfer problem, the fluid is stationary for $t < 0$, and the surface at $z = 0$ is instantaneously translated with velocity v_0 in the x direction at $t = 0$. The scaled velocity, defined as $v_x^* = (v_x/v_0)$, satisfies the initial and boundary conditions 4.4.3–4.4.5. The momentum conservation equation is identical to the energy conservation Eq. 4.4.2 with T^* replaced by v_x^*, and the thermal diffusivity α replaced by momentum diffusivity ν. The solution for v_x^* is the right side of 4.4.16, with α replaced by ν.

4.4.1 Steady Diffusion into a Falling Film

A thin liquid film of thickness h and length L flows down a vertical surface with average velocity v_{av} in the x direction, as shown in Fig. 4.12. At the gas–liquid interface, a soluble species in the gas dissolves in the liquid and then diffuses into the film. The concentration of the gas in the liquid at the entrance is c_0, while the concentration of the gas at the liquid–gas interface is the saturation concentration c_s. The concentration difference between the interface and the bulk liquid is the driving force for diffusion. The z coordinate is perpendicular to the gas–liquid interface, which is located at $z = 0$, as shown in Fig. 4.12. As the liquid flows down, the gas is dissolved in the liquid and convected in the stream-wise x direction. Therefore, the concentration varies in the x and z directions. However, the system is at steady state, and the concentration does not vary in time. The scaled concentration, defined as $c^* = (c - c_0)/(c_s - c_0)$, is 1 at the interface, and 0 in the bulk of the liquid.

The diffusion into a falling film described above is not a unidirectional transport problem, since there is diffusion across the film and convection along with the flow. However, the spatially evolving steady state is transformed into a unidirectional unsteady transport problem by shifting to a reference frame moving with the velocity of the liquid at the interface. If the film thickness is much larger than the penetration depth of the solute, the transport can be approximated as diffusion into an infinite medium. The correlation for the Sherwood number is derived here assuming unidirectional transport into an infinite medium. The systematic boundary layer analysis for solving this problem, discussed in Chapter 9, provides the same solution as the approximate method used here.

The velocity profile for the flow down an inclined plane was derived in Example 4.2.3, and the maximum and average velocities are given in Eqs. 4.4.24 and 4.2.25, respectively. The maximum velocity at the interface is $(3v_{av}/2)$, where v_{av} is the average velocity. The spatial derivative of the velocity (slope of the velocity profile) is zero at the interface. If the penetration depth of the solute is much smaller than h, the velocity can be considered a constant equal to $(3v_{av}/2)$ in a thin region close to the interface, as shown in Fig. 4.12. Therefore, the fluid element at the surface translates with the velocity at the surface, and the time of contact is the ratio of the distance travelled and the velocity, $(2x/3v_{av})$, where x is the distance from the start of the contacting section. The solution for concentration is obtained

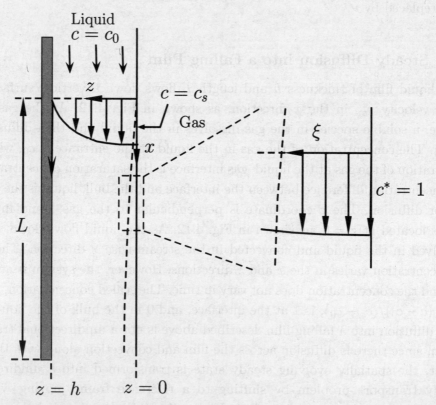

FIGURE 4.12. Diffusion of a solute from the gas phase into a film of liquid of thickness h flowing down a vertical surface under gravity.

by setting $t = (2x/3v_{av})$ in the similarity variable defined in Eq. 4.4.6,

$$\xi = \frac{z}{\sqrt{2\mathcal{D}x/3v_{av}}}. \tag{4.4.22}$$

The solution for the concentration field in terms of this similarity variable is identical to Eq. 4.4.16. The mass flux at the surface is obtained using the same procedure as that for the heat flux, Eq. 4.4.17,

$$j_z|_{z=0} = -\mathcal{D}\left.\frac{\partial c}{\partial z}\right|_{z=0} = -\frac{\mathcal{D}(c_s - c_0)}{\sqrt{2\mathcal{D}x/3v_{av}}}\left.\frac{dc^*}{d\xi}\right|_{\xi=0} = \mathcal{D}(c_s - c_0)\sqrt{\frac{3v_{av}}{2\pi\mathcal{D}x}}. \tag{4.4.23}$$

Here, Eq. 4.4.22 is used to transform the z derivative to the ξ derivative, and $(dc^*/d\xi)|_{\xi=0} = -(1/\sqrt{\pi})$ is the negative of the slope of the error function at $\xi = 0$. The average flux j_{av} over the length L is,

$$j_{av} = \frac{1}{L}\int_0^L dx \, j_z|_{z=0} = \frac{\mathcal{D}(c_s - c_0)}{L}\sqrt{\frac{3v_{av}}{2\pi\mathcal{D}}}\int_0^L dx \, x^{-1/2}$$

$$= \frac{\mathcal{D}(c_s - c_0)}{L}\sqrt{\frac{3v_{av}}{2\pi\mathcal{D}}} \times 2L^{1/2} = \sqrt{\frac{6}{\pi}}\frac{\mathcal{D}(c_s - c_0)}{L}\mathrm{Pe}_L^{1/2}, \tag{4.4.24}$$

where $\mathrm{Pe}_L = (v_{av}L/\mathcal{D})$ is the Peclet number based on the length L. The Sherwood number Sh_L is the non-dimensional average flux based on the length of the film,

$$\boxed{\mathrm{Sh}_L = \frac{j_{av}}{\mathcal{D}(c_s - c_0)/L} = \sqrt{\frac{6}{\pi}}\mathrm{Pe}_L^{1/2} = 1.38 \, \mathrm{Pe}_L^{1/2}.} \tag{4.4.25}$$

In the above expression, the Sherwood number is based on the length L of the film. Alternately, the Sherwood number can be expressed in terms of the thickness h as $\mathrm{Sh}_h = (j_{av}h/\mathcal{D}(c_s - c_0))$,

$$\boxed{\mathrm{Sh}_h = \frac{j_{av}}{\mathcal{D}(c_s - c_0)/h} = 1.38 \, \mathrm{Pe}_h^{1/2}\left(\frac{h}{L}\right)^{1/2},} \tag{4.4.26}$$

where $\mathrm{Pe}_h = (v_{av}h/\mathcal{D})$. Eqs. 4.4.25 and 4.4.26 are the Sherwood number correlations for diffusion into a falling film.

We now examine the 'infinite medium' assumption that the penetration depth $\sqrt{2\mathcal{D}L/3v_{av}}$ (length used to non-dimensionalise z in Eq. 4.4.22) at the outlet of the contactor $x = L$ is small compared to the width of the fluid layer, h,

$$\sqrt{\frac{2\mathcal{D}L}{3v_{av}}} \ll h \Rightarrow \frac{L}{h} \ll \frac{3\text{Pe}_h}{2}. \qquad (4.4.27)$$

The diffusion coefficient $\mathcal{D}$ is about 10^{-9} m^2/s for the diffusion of small molecules in water. For a water film of thickness 1 mm, density 10^3 kg/m^3 and viscosity 10^{-3} kg/m/s flowing down a vertical wall, the average velocity (Eq. 4.2.25) is approximately 3.33 m/s. For these parameter values, the Peclet number is $\text{Pe}_h = 3.33 \times 10^6$. The condition $(L/h) \ll \frac{3}{2}\text{Pe}_h$ is satisfied for $L \ll 5 \times 10^6 h \sim 5 \times 10^3$ m. This calculation shows that the condition Eq. 4.4.27 is satisfied in most practical liquid–gas contactors.

EXAMPLE 4.4.2: In a liquid–gas contactor, a liquid film of thickness 1 mm flowing down a vertical surface with average velocity 0.2 m/s is in contact with a gas dissolving into the liquid. Consider the liquid film to be of sufficiently thick so that the problem can be considered as diffusion into an infinite medium. The diffusion coefficient of the gas in the liquid is 1.6×10^{-9} m^2/s, there is solute in the liquid at the inlet, and the surface concentration of the gas is 40 mg/l. There is no dissolved gas at the inlet of the contactor. It is desired that the average concentration of the gas in the liquid at the exit is 5 mg/l. What is the required length of the contactor?

Solution: Consider a contactor of length L in the vertical direction, width W perpendicular to the plane of flow and thickness h. The total mass leaving the exit per unit time is $c_{av} \times v_{av} \times h \times W$. This is equal to the total mass absorbed per unit time at the liquid–gas interface, which is $j_{av} \times L \times W$. Therefore,

$$j_{av} = \frac{c_{av} v_{av} h}{L}. \qquad (4.4.28)$$

The expression for the length of the contactor is determined by equating the right sides of Eqs. 4.4.28 and 4.4.24,

$$\frac{c_{av} v_{av} h}{L} = \frac{\mathcal{D}(c_s - c_0)}{L} \sqrt{\frac{6}{\pi}} \sqrt{\frac{v_{av} L}{\mathcal{D}}}$$

$$\Rightarrow L = \frac{\pi}{6} \frac{c_{av}^2}{(c_s - c_0)^2} \frac{v_{av} h^2}{\mathcal{D}} = \frac{\pi}{6} \frac{(5 \times 10^{-3} \text{ kg/m}^3)^2}{(4 \times 10^{-2} \text{ kg/m}^3)^2} \frac{(0.2 \text{ m/s})(10^{-3} \text{ m})^2}{1.6 \times 10^{-9} \text{ m}^2/\text{s}}$$

$$= 1.02 \text{ m}. \qquad (4.4.29)$$

□

Summary (4.4)

1. For diffusion from a surface into an infinite fluid, when there is a step change in the field variable Φ_{fv} at the surface at $t = 0$, the unsteady diffusion equation can be reduced to an ordinary differential equation for the similarity variable $\xi = (z/\sqrt{\mathcal{D}t})$, Eq. 4.4.6.

2. If the scaled field variable is 1 at the surface and 0 far from the surface, the solution for Φ_{fv} is $1 - \mathrm{erf}(\xi/2)$, Eq. 4.4.16.

3. The flux at the surface is given by Eq. 4.4.17, where the difference in the field variable between the surface and the medium far from the surface is substituted for $T_0 - T_\infty$ and the appropriate diffusivity is substituted for α.

4. A spatially evolving steady concentration field for diffusion into a falling film, in which diffusion occurs perpendicular to the interface and the fluid is moving with a constant translational velocity v_s parallel to the interface, can be transformed into an unsteady problem using the transformation $t = (x/v_s)$, where x is the stream-wise distance from the location of the step change in the field variable.

5. The Nusselt number correlation, derived using the infinite medium approximation, is Eq. 4.4.25 or Eq. 4.4.26 for the diffusion into a falling film. The infinite medium approximation is valid when the condition Eq. 4.4.27 is satisfied.

4.5   Decay of a Pulse

Discharge from a smokestack into the air or effluent discharges into water involve the injection of concentrated pollutants within a small volume of fluid. The material then spreads into the ambient fluid as time progresses, either due to molecular diffusion or due to dispersion in turbulent flows. A similarity solution can be obtained for the concentration field in the long time limit when the spatial extent of diffusion/dispersion is much larger than that of the initial injection.

Consider an idealised situation in one dimension, where a pulse of solute of mass M per unit area is injected into the fluid at $z = 0$, as shown in Fig. 4.13(a), and this pulse spreads as time progresses due to molecular diffusion. The concentration field is determined by solving the mass diffusion equation, Eq. 4.1.10, in which there are no sources,

$$\frac{\partial c}{\partial t} = \mathcal{D}\frac{\partial^2 c}{\partial z^2}. \tag{4.5.1}$$

Since there are no length or time scales in the problem, the concentration depends the dimensionless combination of z, t and the diffusion coefficient. The solute diffuses to equal extents in the $+z$ and $-z$ directions due to symmetry, and the maximum concentration is at $z = 0$. However, the maximum concentration decreases with time, so that the mass of solute per unit area is M for all time. This is in contrast to the diffusion into an infinite medium in the previous section, where the surface concentration is maintained at a defined value c_s which is used for scaling the fluid concentration. Instead of the boundary condition at a fixed location, we have an integral condition,

$$M = \int_{-\infty}^{\infty} \mathrm{d}z\, c(z,t), \tag{4.5.2}$$

at all times.

Since there are no length or time scales in the problem, the similarity variable is defined as,

$$\xi = \frac{z}{\sqrt{\mathcal{D}t}}. \tag{4.5.3}$$

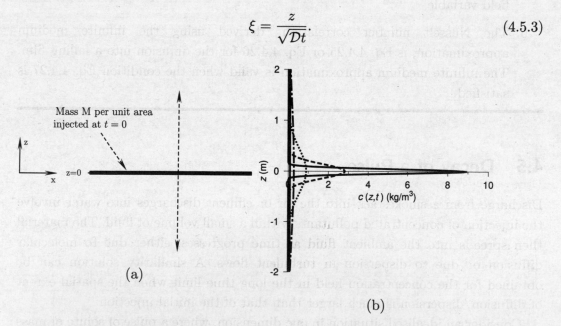

(a)

(b)

FIGURE 4.13. (a) Schematic of a pulse of solute of mass M per unit area injected into the solvent on the plane $z = 0$, and (b) the similarity solution, Eq. 4.5.11, for the concentration profile $c(z,t)$ (kg/m^3) vs. z (m) for $M = 1$ kg, $\mathcal{D} = 1$ m^2/s, and time $t = 0.001$ s (solid line), 0.01 s (dashed line), 0.1 s (dotted line), and 1 s (dot dash line).

Eq. 4.5.2 is expressed in terms of the similarity variable ξ,

$$M = \sqrt{\mathcal{D}t} \int_{-\infty}^{\infty} \mathrm{d}\xi \, c(\xi). \qquad (4.5.4)$$

From Eq. 4.5.4, it is appropriate to scale the concentration by $(M/\sqrt{\mathcal{D}t})$, the ratio of the total mass per unit area and the penetration depth,

$$c^*(\xi) = \frac{c(z,t)}{(M/\sqrt{\mathcal{D}t})}, \qquad (4.5.5)$$

where $c^*(\xi)$ now satisfies the normalisation condition,

$$1 = \int_{-\infty}^{\infty} \mathrm{d}\xi \, c^*(\xi). \qquad (4.5.6)$$

The derivatives of c with respect to z and t in Eq. 4.5.1 are expressed as derivatives of c^* with respect to ξ using differentiation by chain rule,

$$\frac{\partial c}{\partial z} = \frac{M}{\sqrt{\mathcal{D}t}} \frac{\partial \xi}{\partial z} \frac{\mathrm{d}c^*}{\mathrm{d}\xi} = \frac{M}{\mathcal{D}t} \frac{\mathrm{d}c^*}{\mathrm{d}\xi},$$

$$\frac{\partial^2 c}{\partial z^2} = \frac{M}{(\mathcal{D}t)^{3/2}} \frac{\mathrm{d}^2 c^*}{\mathrm{d}\xi^2},$$

$$\frac{\partial c}{\partial t} = -\frac{Mc^*}{2\sqrt{\mathcal{D}}t^{3/2}} + \frac{M}{\sqrt{\mathcal{D}t}} \frac{\partial \xi}{\partial t} \frac{\mathrm{d}c^*}{\mathrm{d}\xi} = -\frac{Mc^*}{2\sqrt{\mathcal{D}}t^{3/2}} - \frac{M}{\sqrt{\mathcal{D}t}} \frac{z}{2\sqrt{\mathcal{D}}t^{3/2}} \frac{\mathrm{d}c^*}{\mathrm{d}\xi}$$

$$= -\frac{M}{2\sqrt{\mathcal{D}}t^{3/2}} \left(c^* + \xi \frac{\mathrm{d}c^*}{\mathrm{d}\xi} \right). \qquad (4.5.7)$$

The above expressions are substituted into Eq. 4.5.1, and the equation is divided by $(M/\sqrt{\mathcal{D}}t^{3/2})$, to obtain

$$\frac{\mathrm{d}^2 c^*}{\mathrm{d}\xi^2} = -\frac{c^*}{2} - \frac{\xi}{2}\frac{\mathrm{d}c^*}{\mathrm{d}\xi} = -\frac{1}{2}\frac{\mathrm{d}(\xi c^*)}{\mathrm{d}\xi} \qquad (4.5.8)$$

The above equation is integrated one time to obtain,

$$\frac{\mathrm{d}c^*}{\mathrm{d}\xi} = -\frac{\xi c^*}{2} + C_1, \qquad (4.5.9)$$

where C_1 is a constant of integration. Since the total mass in the domain is finite, the concentration has to decrease to zero for $\xi \to \pm\infty$. When the concentration is zero, the slope of the concentration $(dc^*/d\xi)$ is also zero. This condition for $\xi \to \pm\infty$ fixes the constant $C_1 = 0$ in Eq. 4.5.9. The solution is then integrated to obtain,

$$c^* = C_2 e^{(-\xi^2/4)}. \qquad (4.5.10)$$

The integral condition, Eq. 4.5.6, is applied to determine the constant of integration $C_2 = 1/(2\sqrt{\pi})$, and the final solution for the concentration field is,

$$\boxed{c^* = \frac{1}{2\sqrt{\pi}} \, e^{(-\xi^2/4)}, \text{ or } c = \frac{M}{2\sqrt{\pi \mathcal{D} t}} \, e^{(-z^2/4\mathcal{D}t)}.} \qquad (4.5.11)$$

The above concentration field is called a Gaussian function, which has maximum $(M/(2\sqrt{\pi \mathcal{D} t}))$ at $z = 0$, decreases to zero at $z \to \pm\infty$, and satisfies the integral condition, Eq. 4.5.4. More details about the Gaussian function, its properties and integrals are provided in Appendix 4.A.

The variance σ_z^2 of the Gaussian function is defined as,

$$\boxed{\sigma_z^2 = \frac{\int_{-\infty}^{\infty} dz \, z^2 \, c(z,t)}{\int_{-\infty}^{\infty} dz \, c(z,t)} = 2\mathcal{D}t.} \qquad (4.5.12)$$

The above result can be obtained using the definite integrals provided in Appendix 4.A. The spatial spread of the concentration field is the standard deviation $\sigma_z = \sqrt{2\mathcal{D}t}$. Thus, the spatial extent of the solute increases proportional to $t^{1/2}$, and the maximum concentration decreases proportional to $t^{-1/2}$, in such a way that the total mass is conserved, as shown in Fig. 4.13(b).

In the limit $t \to 0$, the solution for the concentration field, Eq. 4.5.11, has unusual properties. The maximum value of the concentration goes to infinity proportional to $t^{-1/2}$, while the width goes to zero proportional to $t^{1/2}$ in such a way that the total mass of the solute per unit area is M. Thus, in the idealised concentration field at $t = 0$, a finite amount of mass per unit area is injected into the fluid in a region of infinitesimal thickness along the plane $z = 0$. The concentration is zero for $z \neq 0$, and the total mass per unit area is M. This idealised function, called a 'Dirac delta function' or an impulse, is defined by the relations,

$$\boxed{\begin{aligned} \delta(z) &= 0 \text{ for } z \neq 0, \\ \int_{-\infty}^{\infty} dz \, \delta(z) &= 1. \end{aligned}} \qquad (4.5.13)$$

The initial condition for the diffusion problem can be written as,

$$\boxed{c(z,t) = M\delta(z) \text{ at } t = 0.} \qquad (4.5.14)$$

Eq. 4.5.11 is the solution of the differential Eq. 4.5.1 with the initial condition, Eq. 4.5.14, and boundary condition $c = 0$ for $z \to \pm\infty$.

In real situations, the material is injected in a volume of finite dimension, and so the initial condition is not a Dirac delta function. The delta function approximation for the initial condition is valid only when the spread, $\sqrt{\mathcal{D}t}$, is much larger than the spatial extent l of the initial injection. For $\sqrt{\mathcal{D}t} \gg l$, or $t \gg (l^2/\mathcal{D})$, the concentration field assumes a universal Gaussian form which is independent of the details of the initial injection.

The solutions for the diffusion from an instantaneous point source in two and three dimensions are similar to Eq. 4.5.11. The solutions are presented here without derivation, and it can be verified that these are solutions of the diffusion equation in curvilinear co-ordinates in the exercises in Chapter 5. The two-dimensional configuration is shown in Fig. 4.14(a), where a mass of solute M_2 per unit length perpendicular to the plane is introduced at the origin, and this spreads outward due to diffusion. The concentration field is only a function of time and distance from the origin, $r = \sqrt{x^2 + y^2}$. The initial condition is,

$$c(x, y, t) = M_2 \delta(x, y) \text{ at } t = 0, \tag{4.5.15}$$

where the Dirac delta function in two dimensions is defined as,

$$\delta(x, y) = 0 \text{ for } x \neq 0 \text{ or } y \neq 0,$$
$$\int_{-\infty}^{\infty} \mathrm{d}x \int_{-\infty}^{\infty} \mathrm{d}y \, \delta(x, y) = 1. \tag{4.5.16}$$

Note that the delta function in two dimensions has dimension $\mathcal{L}^{-2}$, so the right side of Eq. 4.5.15 has the dimension of concentration, $\mathcal{ML}^{-3}$. The total mass at any time instant is equal to the initial mass injected, and the constraint due to mass conservation is,

$$\int_{-\infty}^{\infty} \mathrm{d}x \int_{-\infty}^{\infty} \mathrm{d}y \, c(r, t) = M_2. \tag{4.5.17}$$

The solution for the concentration field (Exercise 5.9 in Chapter 5) is a Gaussian function in two dimensions,

$$c = \frac{M_2}{4\pi \mathcal{D}t} \, \mathrm{e}^{(-r^2/4\mathcal{D}t)}. \tag{4.5.18}$$

Using the substitution $r^2 = x^2 + y^2$, Eq. 4.5.18 can be expressed as the product of two Gaussian functions in the x and y directions,

$$c = M_2 \times \frac{1}{2\sqrt{\pi \mathcal{D}t}} \, \mathrm{e}^{(-x^2/4\mathcal{D}t)} \times \frac{1}{2\sqrt{\pi \mathcal{D}t}} \, \mathrm{e}^{(-y^2/4\mathcal{D}t)}. \tag{4.5.19}$$

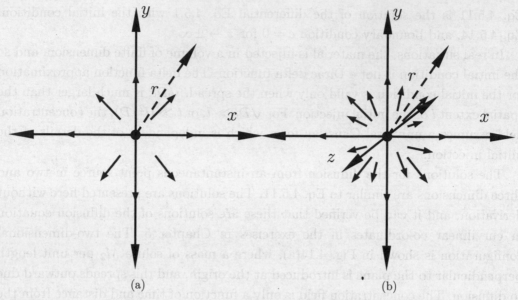

(a)   (b)

FIGURE 4.14. The diffusion of a spot of solute placed at the origin in two dimensions (a) and three dimensions (b). In sub-figure (b), the z axis is perpendicular to the plane.

In Eq. 4.5.19, the integral of each Gaussian function over the respective co-ordinate is 1; therefore, the integral of the concentration field over all space is M_2. The variance of the distribution function in two dimensions is

$$\sigma_r^2 = \frac{\int_{-\infty}^{\infty} \mathrm{d}x \int_{-\infty}^{\infty} \mathrm{d}y \, r^2 \, c(r,t)}{\int_{-\infty}^{\infty} \mathrm{d}x \int_{-\infty}^{\infty} \mathrm{d}y \, c(r,t)} = 4\mathcal{D}t. \qquad (4.5.20)$$

In the three-dimensional configuration shown in Fig. 4.14(b), a spot of solute of mass M_3 is introduced at the origin in three dimensional space. The solute spreads radially outward due to diffusion. The concentration field is a function of time and the radial distance from the origin, $r = \sqrt{x^2 + y^2 + z^2}$. The initial condition is,

$$c(x,y,z,t) = M_3 \delta(x,y,z) \text{ at } t = 0, \qquad (4.5.21)$$

where the Dirac delta function in three dimensions is defined by the relations,

$$\delta(x,y,z) = 0 \text{ for } x \neq 0 \text{ or } y \neq 0 \text{ or } z \neq 0,$$
$$\int_{-\infty}^{\infty} \mathrm{d}x \int_{-\infty}^{\infty} \mathrm{d}y \int_{-\infty}^{\infty} \mathrm{d}z \, \delta(x,y,z) = 1. \qquad (4.5.22)$$

Note that the delta function has dimension $\mathcal{L}^{-3}$. The solution for the concentration field (Exercise 5.14 in Chapter 5) is,

$$c = \frac{M_3}{8(\pi \mathcal{D}t)^{3/2}} \, e^{(-r^2/4\mathcal{D}t)}$$

$$= M_3 \times \frac{1}{2\sqrt{\pi \mathcal{D}t}} \, e^{(-x^2/4\mathcal{D}t)} \times \frac{1}{2\sqrt{\pi \mathcal{D}t}} \, e^{(-y^2/4\mathcal{D}t)}$$

$$\times \frac{1}{2\sqrt{\pi \mathcal{D}t}} \, e^{(-z^2/4\mathcal{D}t)}. \tag{4.5.23}$$

The variance of the distribution function in three dimensions is

$$\sigma_r^2 = \frac{\int_{-\infty}^{\infty} dx \int_{-\infty}^{\infty} dy \int_{-\infty}^{\infty} dz \, r^2 \, c(r,t)}{\int_{-\infty}^{\infty} dx \int_{-\infty}^{\infty} dy \int_{-\infty}^{\infty} dz \, c(r,t)} = 6\mathcal{D}t. \tag{4.5.24}$$

The solution, Eq. 4.5.11, for the diffusion of a pulse is also applicable to situations where the dispersion of a material takes place due to fluid velocity fluctuations, instead of molecular velocity fluctuations. There are two important applications where fluid velocity fluctuations disperse dissolved or suspended materials, the flow through porous media and turbulent flows. In porous media or packed columns, the fluid flows through the interstitial gaps between densely packed particles, and the fluid follows a tortuous path as it passes around the particles. This causes the stretching and distortion of fluid elements, and as a consequence, solutes are dispersed much faster than in a unidirectional flow. A common experiment to characterise the dispersion is to insert a pulse of solute at a location, and then measure the solute concentration as a function of time at a downstream location. If the two locations are separated by a distance L and the fluid velocity is v, the time taken for the pulse to travel between the two locations is (L/v). If the width of the pulse (standard deviation of the concentration distribution) at the downstream location is W, then the relation $W = \sqrt{2\mathcal{D}_{eff}t} = \sqrt{2\mathcal{D}_{eff}L/v}$ can be used to calculate the effective dispersion coefficient $\mathcal{D}_{eff}$.

EXAMPLE 4.5.1: Consider the flow in a packed column with porosity 0.4 and superficial velocity 0.1 m/s. In order to calculate the dispersion coefficient, a pulse of dye is injected at the inlet, and the concentration of the dye is monitored at a distance 1 m downstream of the inlet. The dye concentration is found to have a standard deviation of 1 cm. What is the dispersion coefficient?

Solution: The superficial velocity is the velocity of the fluid in the same cross section as the packed column at the same flow rate when there are no particles present. If the superficial velocity is $v_s = 0.1$ m/s, the average velocity v of the fluid through the interstices between the particles is the ratio of the superficial velocity and the porosity,

$$v = \frac{v_s}{\varepsilon} = \frac{0.1 \text{ m/s}}{0.4} = 0.25 \text{ m/s}. \tag{4.5.25}$$

The time taken for the pulse to travel 1 m is,

$$t = \frac{1 \text{ m}}{v} = \frac{1 \text{ m}}{0.25 \text{ m/s}} = 4 \text{ s}. \tag{4.5.26}$$

The spread of the pulse is $\sigma_z = 0.01$ m after 4 seconds,

$$\sqrt{2 \mathcal{D}_{eff} t} = \sqrt{2 \times \mathcal{D}_{eff} \times 4 \text{ s}} = 0.01 \text{ m}. \tag{4.5.27}$$

Therefore, the dispersion coefficient is $\mathcal{D}_{eff} = 1.25 \times 10^{-5}$ m^2/s. ☐

EXAMPLE 4.5.2: Factories handling toxic chemicals have an exclusion zone around them which is cleared of habitation, so that the concentration of accidentally released chemicals at the edge of the exclusion zone remains below a critical value. What should be the radius of the exclusion zone around a factory handling a gaseous pollutant with dispersion coefficient 10^{-3} m^2/s, such that if 100 kg of pollutant is released, the concentration at the edge of the exclusion zone does not exceed 10^{-5} kg/m^3.

Solution: The concentration in the vicinity of the point of release first increases as the pulse spreads spatially, and then decreases as the maximum concentration decreases. The concentration field due to the release of a pulse of solute in three dimensions is given by Eq. 4.5.23. At a distance r from the point of release, the concentration passes through a maximum when the time derivative of the concentration is zero,

$$\frac{\partial c}{\partial t} = \frac{\partial}{\partial t} \left(\frac{M}{8(\pi \mathcal{D}_{eff} t)^{3/2}} e^{(-r^2/4\mathcal{D}_{eff} t)} \right)$$

$$= \frac{M(r^2 - 6\mathcal{D}_{eff} t)}{32 \, \pi^{3/2} \, \mathcal{D}_{eff}^{5/2} \, t^{7/2}} e^{(-r^2/4\mathcal{D}_{eff} t)} = 0, \tag{4.5.28}$$

where $\mathcal{D}_{eff}$ is the dispersion coefficient. Therefore, the concentration passes through a maximum at

$$t_{max} = \frac{r^2}{6\mathcal{D}_{eff}}, \qquad (4.5.29)$$

and the maximum concentration is obtained by substituting $t = t_{max}$ in Eq. 4.5.23,

$$c_{max} = M \left(\frac{3}{2\pi e r^2}\right)^{3/2}. \qquad (4.5.30)$$

Here, the exponential constant e = 2.718. Eq. 4.5.30 is solved to determine the distance at which the maximum concentration is 10^{-5} kg/m^3,

$$r = \left(\frac{3}{2\pi e}\right)^{1/2} \left(\frac{M}{c_{max}}\right)^{1/3} = \left(\frac{3}{2\pi \times 2.718}\right)^{1/2} \left(\frac{100 \text{ kg}}{10^{-5} \text{ kg/m}^3}\right)^{1/3} = 90.3 \text{ m}$$

$$(4.5.31)$$

Therefore, the extent of the exclusion zone is approximately 90 m.

This calculation assumes that the gaseous pollutant spreads radially from the source. This may not be true in real situations, because pollutants may be released from a certain height above the earth, and the concentration boundary condition at the surface of the earth does affect the concentration field. However, this provides a rough estimate of the distance beyond which the concentration is within the safe limit. □

Summary (4.5)

1. A pulse of mass/energy/momentum M introduced at the origin in one/two/three dimensions spreads radially outward due to diffusion. The initial conditions, Eq. 4.5.14, 4.5.15 or 4.5.21, can be represented as,

$$\Phi_{fv} = M\delta(\mathbf{x}),$$

where M is the mass per unit area, mass per unit length or mass in one, two and three dimensions, respectively, and $\delta(\mathbf{x})$ is the Dirac delta function defined in Eqs. 4.5.13, 4.5.16 and 4.5.22.

2. The concentration field is given by Eq. 4.5.11, 4.5.18 or 4.5.23,

$$\Phi_{fv} = \frac{M}{2^{D_n}(\pi \mathcal{D}t)^{D_n/2}} \, e^{(-r^2/(4\mathcal{D}t))},$$

where D_n is the dimension, $\mathcal{D}$ is the generalised diffusivity and r is the distance from the origin in one/two/three dimensions.

3. The 'spread' or standard deviation σ_r of the concentration, from Eqs. 4.5.12, 4.5.20 and 4.5.24, is proportional to $t^{1/2}$,

$$\sigma_r = \sqrt{2 D_n \mathcal{D} t}.$$

4.6   Time-dependent Length Scale

In Sections 4.4 and 4.5, the unsteady diffusion equation was reduced to an ordinary differential equation because there is no length scale in the problem, and the characteristic length scales as $\sqrt{\mathcal{D}t}$, where $\mathcal{D}$ is the generalised diffusion coefficient. Even when there is a length scale, the unsteady diffusion equation can be reduced to an ordinary differential equation if the characteristic length scale has a specific dependence on time. An example is the expansion of a melt pool, where a solid is heated by a surface with temperature above the melting point, resulting in an expanding melt pool above a solid–liquid interface. Consider the configuration shown in Fig. 4.15, where a solid with melting temperature T_m is brought into contact with a hot surface at a temperature T_0 above the melting temperature. The heat transfer is considered to be unidirectional in the z direction perpendicular to the heated surface. It is assumed that the solid is initially at the melting temperature T_m, and as time progresses, a liquid pool forms whose height $h(t)$ is a function of time, t.

In order to analyse conduction in the melt pool, the z co-ordinate is scaled by $h(t)$, which is a time-dependent length scale. The scaled z co-ordinate and temperature are defined as

$$z^* = \frac{z}{h(t)}, \ \ T^* = \frac{T - T_m}{T_0 - T_m}. \tag{4.6.1}$$

The heated surface is located at $z^* = 0$, and the solid–liquid interface is at $z^* = 1$, as shown in Fig. 4.15. The conduction equation is first expressed in terms of z^* and T^*, and then the necessary condition for a similarity solution is determined. The derivatives with respect to z and t in the unsteady heat conduction equation,

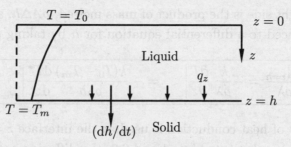

FIGURE 4.15. The melting of a solid at $z = h(t)$ due to the heat flux from a heated surface at $z = 0$ results in the downward motion of the interface with velocity (dh/dt). The temperature T_0 at $z = 0$ is higher than the melting temperature T_m at $z = h$.

Eq. 4.4.2, are expressed as derivatives with respect to z^* using chain rule for differentiation,

$$\frac{\partial T^*}{\partial z} = \frac{\partial z^*}{\partial z}\frac{dT^*}{dz^*} = \frac{1}{h(t)}\frac{dT^*}{dz^*},$$

$$\frac{\partial^2 T}{\partial z^2} = \frac{\partial z^*}{\partial z}\frac{d}{dz^*}\left(\frac{\partial z^*}{\partial z}\frac{dT^*}{dz^*}\right) = \frac{1}{h(t)^2}\frac{d^2 T^*}{dz^{*2}},$$

$$\frac{\partial T^*}{\partial t} = \frac{\partial z^*}{\partial t}\frac{dT^*}{dz^*} = -\frac{z^*}{h(t)}\frac{dh}{dt}\frac{dT^*}{dz^*}. \qquad (4.6.2)$$

The scaled unsteady conduction equation, Eq. 4.4.2, is expressed in terms of T^* and z^*, and divided by (α/h^2),

$$-\frac{h}{\alpha}\frac{dh}{dt}z^*\frac{dT^*}{dz^*} = \frac{d^2 T^*}{dz^{*2}}. \qquad (4.6.3)$$

The boundary conditions are,

$$T = T_0 \Rightarrow T^* = 1 \text{ at } z^* = 0, \qquad (4.6.4)$$

$$T = T_m \Rightarrow T^* = 0 \text{ at } z^* = 1. \qquad (4.6.5)$$

A similarity solution is possible if $(h/\alpha)(dh/dt)$ in Eq. 4.6.3 is independent of time.

The equation for $h(t)$ is determined from the condition that the heat transferred to the solid in time Δt is equal to the product of the latent heat and the mass of solid melted in time Δt.

$$q_z A \, \Delta t = \Delta h \, A \rho \lambda. \qquad (4.6.6)$$

Here, A is the area of cross section, λ is the latent heat (heat required for melting unit mass of the material), ρ is the mass density of the solid, and Δh is the change in the height in time Δt. The left side of Eq. 4.6.6 is the energy transferred to the solid in time Δt. The right side is the product of mass melted, $\rho A \Delta h$, and the latent heat λ. Eq. 4.6.6 is reduced to a differential equation for h by taking the limit $\Delta t \to 0$,

$$
\boxed{\frac{\mathrm{d}h}{\mathrm{d}t} = \frac{q_z|_{z=h}}{\rho\lambda} = -\frac{k}{\rho\lambda}\left.\frac{\partial T}{\partial z}\right|_{z=h} = -\frac{k(T_0 - T_m)}{\rho\lambda h}\left.\frac{\mathrm{d}T^*}{\mathrm{d}z^*}\right|_{z^*=1}.}
$$

(4.6.7)

Here, Fourier's law of heat conduction is used at the interface $z = h$.

When Eq. 4.6.7 is substituted into Eq. 4.6.3, the diffusion equation is,

$$
\boxed{-Pz^*\frac{\mathrm{d}T^*}{\mathrm{d}z^*} = \frac{\mathrm{d}^2T^*}{\mathrm{d}z^{*2}},}
$$

(4.6.8)

where the parameter P is a Peclet number based on the height h and the velocity $(\mathrm{d}h/\mathrm{d}t)$ of the solid–liquid interface,

$$
P = \frac{h}{\alpha}\frac{\mathrm{d}h}{\mathrm{d}t} = -\frac{k(T_0 - T_m)}{\rho\lambda\alpha}\left.\frac{\mathrm{d}T^*}{\mathrm{d}z^*}\right|_{z^*=1} = -\frac{C_p(T_0 - T_m)}{\lambda}\left.\frac{\mathrm{d}T^*}{\mathrm{d}z^*}\right|_{z^*=1} = -\mathrm{Ja}\left.\frac{\mathrm{d}T^*}{\mathrm{d}z^*}\right|_{z^*=1}.
$$

(4.6.9)

Here, Eq. 4.6.7 has been used for $(\mathrm{d}h/\mathrm{d}t)$, and the substitution $\alpha = (k/\rho C_p)$ has been made for the thermal diffusivity. The dimensionless parameter $\mathrm{Ja} = (C_p(T_0 - T_m)/\lambda)$ is the Jakob number, the ratio of the energy required to raise the temperature of unit mass from T_m to T_0 and the latent heat. It should be noted that $(\mathrm{d}T^*/\mathrm{d}z^*)$ is negative, because the temperature decreases with height towards the interface, and therefore, P is positive. The parameter P is a constant if the scaled temperature is only a function of z^*. However, the value of $(\mathrm{d}T^*/\mathrm{d}z^*)$ at $z^* = 1$ is not yet known.

Eq. 4.6.8 is solved using the same steps as in Eq. 4.4.12 to 4.4.15, and the solution is the linear combination of a constant and the error function solution 4.A.14 where $A = P$,

$$
T^* = C_1\mathrm{erf}\left(\sqrt{\frac{Pz^{*2}}{2}}\right) + C_2.
$$

(4.6.10)

The constants C_1 and C_2 are determined from the boundary conditions, Eqs. 4.6.4 and 4.6.5, to obtain the solution for the temperature field,

$$
T^* = 1 - \frac{\mathrm{erf}\left(\sqrt{\frac{Pz^{*2}}{2}}\right)}{\mathrm{erf}\left(\sqrt{\frac{P}{2}}\right)}.
$$

(4.6.11)

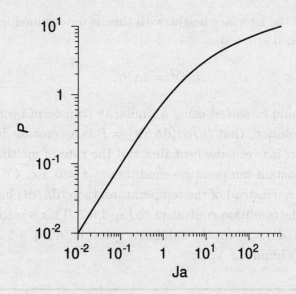

FIGURE 4.16. The variation of P with Ja obtained by solving the implicit Eq. 4.6.13.

The spatial derivative of the temperature is now calculated using the relation 4.A.15,

$$\left.\frac{\mathrm{d}T^*}{\mathrm{d}z^*}\right|_{z^*=1} = -\sqrt{\frac{2P}{\pi}}\frac{\mathrm{e}^{(-P/2)}}{\mathrm{erf}(\sqrt{P/2})}. \qquad (4.6.12)$$

When the above expression for the $(\mathrm{d}T^*/\mathrm{d}z^*)$ is substituted into Eq. 4.6.9, we obtain a relation between P and the Jakob number,

$$P = \mathrm{Ja}\sqrt{\frac{2P}{\pi}}\frac{\mathrm{e}^{(-P/2)}}{\mathrm{erf}(\sqrt{P/2})}. \qquad (4.6.13)$$

The parameter P, obtained by solving the implicit equation Eq. 4.6.13, is shown as a function of Ja in Fig. 4.16. In the limit $\mathrm{Ja} \ll 1$, the Taylor series expansion of Eq. 4.6.13 about $\mathrm{Ja} = 0$ provides the following result correct to $O(\mathrm{Ja}^2)$,

$$P = \mathrm{Ja} - \frac{\mathrm{Ja}^2}{3}. \qquad (4.6.14)$$

For $P \gg 1$, the implicit Eq. 4.6.13 is simplified as,

$$\mathrm{Ja} = \sqrt{\frac{P\pi}{2}}\mathrm{e}^{(P/2)}. \qquad (4.6.15)$$

The variation of the interface height with time is determined from Eq. 4.6.9 with initial condition $h = 0$ at $t = 0$,

$$h(t)^2 = 2\alpha P t. \tag{4.6.16}$$

This problem could be solved using a similarity transform because the condition for the similarity solution, that $(h/\alpha)(dh/dt) = P$ is a constant in Eq. 4.6.3, is the same as the relation between the heat flux and the rate of melting, Eq. 4.6.7. This is a result of the constant temperature condition at $z = 0$, Eq. 4.6.4. If the heat flux were specified at $z = 0$ instead of the temperature, then (dh/dt) has to be a constant in order to satisfy the condition equivalent to Eq. 4.6.7. This is incompatible with the condition $(h/\alpha)(dh/dt) = P$ for the similarity solution in Eq. 4.6.3, and a similarity solution cannot be obtained.

EXAMPLE 4.6.1: An aluminium ingot of height 10 cm has to be melted in 10 minutes by bringing it in contact with a heated surface. What should be the temperature of the surface? The properties of aluminium areas follows: melting temperature $T_m = 660°C$, density $\rho = 2700$ kg/m^3, thermal conductivity $k = 220$ W/m/°C, specific heat $C_p = 900$ J/kg/°C, and latent heat $\lambda = 4 \times 10^5$ J/kg.

Solution: The thermal diffusivity of aluminium is $\alpha = (k/\rho C_p) = 9.05 \times 10^{-5} \text{m}^2/\text{s}$. The parameter P is determined from the relation

$$P = \frac{h^2}{2\alpha t} = 0.092. \tag{4.6.17}$$

Since P is small, Eq. 4.6.14 is used to determine Ja,

$$\text{Ja} = 0.095 = \frac{C_p(T_0 - T_m)}{\lambda}. \tag{4.6.18}$$

The above equation is solved to determine the surface temperature T_0,

$$T_0 = \frac{\lambda \text{Ja}}{C_p} + T_m = \frac{4 \times 10^5 \text{ J/kg} \times 0.095}{900 \text{ J/kg/°C}} + 660°C = 702°C. \tag{4.6.19}$$

□

Summary (4.6)

1. In problems with a time-dependent length scale, the spatial co-ordinate is scaled by this length scale to obtain a similarity variable.

2. A time evolution equation is derived to determine the time dependence of the length scale.

3. If the resulting heat conduction equation, Eq. 4.6.8, depends only position and time only through its dependence on the similarity variable, and this is consistent with the time dependence of the length scale obtained from the macroscopic balance condition such as Eq. 4.6.7, a similarity solution can be obtained for the evolution of the time dependent length scale and the temperature field.

4.7 Domain of Finite Extent

The method of 'separation of variables' is illustrated in the example of the conduction in a channel of finite height. The configuration, shown in Fig. 4.17, consists of a channel of height h in the region $0 < z < h$ bounded by two walls at $z = 0$ and $z = h$, of infinite extent in the plane perpendicular to z, containing a material of thermal diffusivity α. Initially, the temperature of the walls and the material in the channel is T_0. At time $t = 0$, the temperature of the bottom wall at $z = 0$ is instantaneously increased to T_1. The final steady temperature profile is a linear profile, shown by the solid line in Fig. 4.17. At $t = 0$, the temperature in the channel is T_0, but the temperature of the bottom wall has increased to T_1, so the temperature profile is a step function shown by the dotted line in Fig. 4.17. The time evolution of the temperature in the channel is to be determined.

In this case, there is a length scale h in the problem, and the z co-ordinate can be scaled by the channel height. Therefore, it is not possible to use a similarity transform. The scaled temperature is defined as,

$$T^* = \frac{T - T_0}{T_1 - T_0}. \tag{4.7.1}$$

The equation for the temperature is the energy balance Eq. 4.1.21 for a fluid with constant thermal diffusivity. There is one length scale, the height of the channel h, and one dimensional parameter, the thermal diffusivity α in the governing equation.

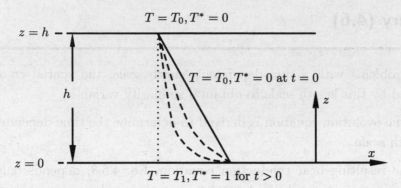

FIGURE 4.17. Configuration and co-ordinate system for analysing heat transfer in a channel of finite height.

Based on dimensional analysis, the scaled z co-ordinate and time are defined as,

$$z^* = \frac{z}{h}, \quad t^* = \frac{t\alpha}{h^2}. \tag{4.7.2}$$

Here, (h^2/α) is the characteristic time for the diffusion of heat over the length h. The governing Eq. 4.1.21, expressed in terms of scaled variables, is

$$\boxed{\frac{\partial T^*}{\partial t^*} = \frac{\partial^2 T^*}{\partial z^{*2}}.} \tag{4.7.3}$$

The boundary and initial conditions are,

$$T^* = 0 \ \text{ at } z^* = 1 \text{ for all } t^*, \tag{4.7.4}$$

$$T^* = 1 \ \text{ at } z^* = 0 \text{ for all } t^* > 0, \tag{4.7.5}$$

$$T^* = 0 \ \text{ at } t^* = 0 \text{ for all } z^* > 0. \tag{4.7.6}$$

When expressed in scaled variables, the governing equation, Eq. 4.7.3, and the boundary and initial conditions, Eqs. 4.7.4–4.7.6, do not contain any parameters.

Equivalent mass and momentum transfer problems can be posed in the same configuration. In the mass transfer problem, the two plates and the fluid are initially at concentration c_0. At $t = 0$, the concentration of the bottom plate is instantaneously increased to c_1. The scaled concentration is $c^* = (c - c_0)/(c_1 - c_0)$, and the mass diffusivity is used to define the scaled time, $t^* = (t\mathcal{D}/h^2)$. In the momentum transfer problem, the fluid and the two plates are initially at rest. At $t = 0$, the bottom plate moves with a constant velocity $v_x = v_0$ in the x direction. The scaled velocity is defined as $v_x^* = (v_x/v_0)$, and the momentum diffusivity is used to define the scaled time, $t^* = (t\nu/h^2)$. The governing equations for c^* and v_x^* are identical to Eq. 4.7.3, and boundary conditions are the same as Eqs. 4.7.4–4.7.6. Therefore, the solution for $c^*(z^*, t^*)$ and $v_x^*(z^*, t^*)$ is the same as that for $T^*(z^*, t^*)$ derived here.

The temperature field is expressed as the sum of a steady (T_s^*) and a transient (T_t^*) part,

$$T^* = T_s^* + T_t^*. \tag{4.7.7}$$

The equation for T_s^* is the steady heat conduction equation,

$$\frac{\mathrm{d}^2 T_s^*}{\mathrm{d} z^{*2}} = 0, \tag{4.7.8}$$

and the solution is a linear profile that satisfies the boundary conditions, Eqs. 4.7.4–4.7.5,

$$T_s^* = (1 - z^*). \tag{4.7.9}$$

The conservation equation for the transient temperature field is obtained by substituting $T^* = T_s^* + T_t^*$ in Eq. 4.7.3. Since the steady temperature field is independent of time and satisfies Eq. 4.7.8, the equation for the transient temperature is identical to Eq. 4.7.3,

$$\frac{\partial T_t^*}{\partial t^*} = \frac{\partial^2 T_t^*}{\partial z^{*2}}. \tag{4.7.10}$$

The boundary and initial conditions for T_t^* are determined by substituting $T^* = T_s^* + T_t^*$ in Eqs. 4.7.4–4.7.6. Comparing Eqs. 4.7.4–4.7.5 and 4.7.9, it is evident that T^* and T_s^* are equal at $z^* = 0$ and $z^* = 1$. Therefore, T_t^* is zero (homogeneous boundary condition) at both spatial boundaries,

$$\boxed{T_t^* = 0 \text{ at } z^* = 1 \text{ for all } t^*,} \tag{4.7.11}$$

$$\boxed{T_t^* = 0 \text{ at } z^* = 0 \text{ for all } t^* > 0.} \tag{4.7.12}$$

In the following analysis, it is shown that homogeneous boundary condition at both the spatial boundaries is an important requirement for the separation of variables procedure for T_t^*.

At $t^* = 0$, $T^* = 0$ throughout the channel. The steady temperature T_s^*, Eq. 4.7.9, is independent of time. Therefore, the transient temperature is,

$$\boxed{T_t^* = T^* - T_s^* = -(1 - z^*) \text{ at } t^* = 0 \text{ for all } z^* > 0.}$$ (4.7.13)

Eq. 4.7.10 is solved by the method of 'separation of variables', where the unsteady temperature field is expressed as the product of two functions, one of which is a function of t^*, and the other is a function of z^*,

$$T_t^* = G(t^*)Z(z^*).$$ (4.7.14)

The above expression is substituted into the conservation equation, Eq. 4.7.10, and the equation is divided by GZ, to obtain

$$\frac{1}{G}\frac{\mathrm{d}G}{\mathrm{d}t^*} = \frac{1}{Z}\frac{\mathrm{d}^2 Z}{\mathrm{d}z^{*2}}.$$ (4.7.15)

In Eq. 4.7.15, the left side is only a function of t^*, while the right side is only a function of z^*. From this, it can be inferred that the left and right sides are both equal to a constant. Consider the contrary, that the left side varies with t^*, and the right side varies with z^*. If we keep z^* a constant and vary t^*, then the left side of Eq. 4.7.15 changes, while the right side remains unchanged, and so the equality is not satisfied. The equality in Eq. 4.7.15 is satisfied for all z^* and t^* only if both sides are equal to the same constant.

The solution for Z is obtained by solving

$$\frac{1}{Z}\frac{\mathrm{d}^2 Z}{\mathrm{d}z^{*2}} = -\beta^2.$$ (4.7.16)

The reason for choosing a negative constant on right side of Eq. 4.7.16 is discussed a little later. The solution of Eq. 4.7.16 is,

$$Z = A \sin(\beta z^*) + B \cos(\beta z^*),$$ (4.7.17)

where A and B are constants to be determined from the boundary conditions. The boundary condition $T_t^* = 0$ ($Z = 0$) at $z^* = 0$ is satisfied for $B = 0$. The boundary condition $T_t^* = 0$ at $z^* = 1$ is satisfied if either $A = 0$ or $\beta_i = (i\pi)$, where i is an integer. The former choice results in the trivial solution $T_t^* = 0$ throughout the domain. Therefore, the only non-trivial solution for Z is,

$$\boxed{\beta_i = i\pi, \quad Z = A\sin(i\pi z^*)}$$ (4.7.18)

where i is an integer and A is a constant.

Note that the solution, Eq. 4.7.18, satisfies the differential Eq. 4.7.16 and the boundary conditions, Eqs. 4.7.11–4.7.12, only when i is an integer, and therefore the constant β_i in Eq. 4.7.16 has discrete values. These discrete values are termed the eigenvalues, and the solutions $\sin(i\pi z^*)$ are called eigenfunctions or basis functions for the solution of the differential Eq. 4.7.16.

The solution for $G(t^*)$ can now be obtained from Eqs. 4.7.15 and 4.7.16,

$$\frac{1}{G}\frac{\mathrm{d}G}{\mathrm{d}t^*} = -\beta_i^2 = -i^2\pi^2. \tag{4.7.19}$$

The solution is,

$$G(t^*) = De^{(-i^2\pi^2 t^*)}, \tag{4.7.20}$$

and the solution for T^* is,

$$T_t^* = G(t^*)Z(z^*) = Ce^{(-i^2\pi^2 t^*)}\sin(i\pi z^*), \tag{4.7.21}$$

where C is a constant. The solution, Eq. 4.7.21, satisfies Eq. 4.7.10 for any value of i. The general solution is a linear combination of the solution, Eq. 4.7.21, for different values of i,

$$\boxed{T_t^* = \sum_{i=1}^{\infty} C_i e^{(-i^2\pi^2 t^*)}\sin(i\pi z^*),} \tag{4.7.22}$$

where C_i are constants.

The reason for choosing a negative constant on the right side of Eq. 4.7.16 is now examined. If we had chosen the right side of Eq. 4.7.16 to be $+\beta_i^2$, the solution for Z consists of an exponentially growing and an exponentially decaying function, $Z = Ae^{(\beta z^*)} + Be^{(-\beta z^*)}$. In this case, it can easily be verified that the boundary conditions $Z = 0$ at $z^* = 0$ and at $z^* = 1$ can be satisfied only for the trivial solution, $A = B = 0$. In addition, the solution for $G(t^*)$ in Eq. 4.7.20 is exponentially increasing in time, and therefore, there is no steady solution. Since we have chosen the constant to be negative, the unsteady solution T_t^* decays exponentially in time, and goes to zero in the long time limit. The solutions for $Z(z^*)$ are sine functions, which satisfy the homogeneous boundary conditions $T_t^* = 0$ at $z^* = 0$ and $z^* = 1$ for specific values of β_i.

The coefficients C_i in Eq. 4.7.22 are determined from the initial condition, Eq. 4.7.13,

$$\sum_{i=1}^{\infty} C_i \sin{(i\pi z^*)} = -(1 - z^*). \tag{4.7.23}$$

It appears infeasible to solve Eq. 4.7.23, because it is one equation for an infinite number of constants. However, each constant can be unambiguously determined using the orthogonality relation for the basis functions $\sin{(i\pi z^*)}$, which is explained in the Appendix 4.B. The 'inner product' or the 'scalar product' of two basis functions, $\sin{(i\pi z^*)}$ and $\sin{(j\pi z^*)}$ is defined as,

$$\langle \sin{(i\pi z^*)}, \sin{(j\pi z^*)} \rangle = \int_0^1 dz^* \, \sin{(i\pi z^*)} \sin{(j\pi z^*)}. \tag{4.7.24}$$

It is easily verified, using the identity,

$$\sin{(i\pi z^*)} \sin{(j\pi z^*)} = \tfrac{1}{2} [\cos{((i-j)\pi z^*)} - \cos{((i+j)\pi z^*)}],$$

that the integral on the right side is zero for $j \neq i$, and it has a value $\tfrac{1}{2}$ for $j = i$. Therefore, the scalar product is,

$$\boxed{\langle \sin{(i\pi z^*)}, \sin{(j\pi z^*)} \rangle = \frac{\delta_{ij}}{2},} \tag{4.7.25}$$

where the Kronecker delta $\delta_{ij} = 0$ for $i \neq j$ and $\delta_{ij} = 1$ for $i = j$.

The orthogonality relation is used to determine the constants in Eq. 4.7.22 as follows. The scalar product of Eq. 4.7.23 and $\sin{(j\pi z^*)}$ is,

$$\sum_{i=1}^{\infty} C_i \int_0^1 dz^* \, \sin{(i\pi z^*)} \sin{(j\pi z^*)} = - \int_0^1 dz^* \, \sin{(j\pi z^*)}(1 - z^*). \tag{4.7.26}$$

The integral on the right is evaluated by integration by parts,

$$-\int_0^1 dz^* \, \sin{(j\pi z^*)}(1 - z^*) = \left. \frac{(1 - z^*)\cos{(j\pi z^*)}}{j\pi} \right|_0^1 + \int_0^1 dz^* \frac{\cos{(j\pi z^*)}}{j\pi}$$

$$= -\frac{1}{j\pi} + \left. \frac{\sin{(j\pi z^*)}}{(j\pi)^2} \right|_0^1 = -\frac{1}{j\pi}. \tag{4.7.27}$$

The scalar product on the left side of Eq. 4.7.26 is zero for $i \neq j$ and is $\tfrac{1}{2}$ for $i = j$ (Eq. 4.7.25). Therefore, the summation on the left side of Eq. 4.7.26 reduces to

$(C_j/2)$, and the result for the coefficient C_j is,

$$C_j = -\frac{2}{j\pi}. \tag{4.7.28}$$

The final solution for the concentration field, which includes the steady part $T_s^* = (1 - z^*)$ (Eq. 4.7.9) and the transient part (Eq. 4.7.22) with the coefficients given in Eq. 4.7.28, is

$$T^* = T_s^* + T_t^* = (1 - z^*) - \sum_{i=1}^{\infty} \frac{2}{i\pi} \sin\left(i\pi z^*\right) e^{\left(-i^2\pi^2 t^*\right)}. \tag{4.7.29}$$

To summarise, the conceptual basis for the separation of variables procedure is as follows. The temperature field is separated into a steady and a transient part, so that the transient part has homogeneous boundary conditions ($T_t^* = 0$) at both the spatial boundaries, $z^* = 0$ and $z^* = 1$. The initial condition ($T_t^* = -T_s^*$ at $t^* = 0$) is inhomogeneous. Therefore, there is no forcing for the transient part of T_t^* at the boundaries, but there is forcing at the initial time $t^* = 0$, due to the difference between the initial and steady temperature profiles.

Homogeneous boundary conditions at the spatial boundaries are necessary for obtaining the solution as the sum of orthogonal basis functions. In Eq. 4.7.18, discrete values for the constant $\beta_i = i\pi$, and the corresponding set of basis functions $\sin\left(i\pi z^*\right)$, were obtained because of the conditions $Z = 0$ at $z^* = 0$ and 1. In Appendix 4.B, it is shown that basis functions are orthogonal when the boundary conditions are homogeneous.

The series in Eq. 4.7.29 contains an infinite number of terms for $1 \leq i \leq \infty$. The terms in the series decrease exponentially proportional to $e^{\left(-i^2\pi^2 t^*\right)}$ as i is increased. An approximate solution is obtained by truncating the series at a sufficiently large value of i, so that the sum of the terms neglected is below the specified maximum error. The results obtained for $T^*(z^*, t^*)$ for different numbers of terms in the expansion, Eq. 4.7.29, are shown as a function of z^* for different values of t^* in Fig. 4.18(a). The solid lines are the 'exact' solutions obtained by including 20 terms in the expansion in Eq. 4.7.29. These smooth curves are indistinguishable from the numerical solutions of the differential Eq. 4.7.3 for $t^* \geq 0.01$. The dot-dash lines are the results when only the first term ($i = 1$) is included in Eq. 4.7.29. These have one extremum, and they exhibit the unphysical feature that T^* is negative for $t \leq 0.03$. These are not in agreement with the exact result for $t \leq 0.03$. However, even the one-term expansion is in quantitative agreement with the exact result for $t^* \geq 0.1$. The dashed lines are the results when three terms are included in the expansion in

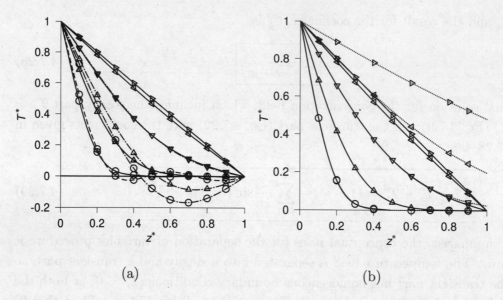

(a)   (b)

FIGURE 4.18. Numerical solutions for $T^*(z^*, t^*)$, Eq. 4.7.29, as a function for z^* for $t^* = 0.01$ (○), $t^* = 0.03$ (△), $t^* = 0.1$ (▽), $t^* = 0.3$ (◁), and $t^* = 1.0$ (▷). In sub-figure (a), the solid, dashed and dot-dash lines are the results when 20, 3 and 1 terms are included in the expansion in Eq. 4.7.29. In sub-figure (b), the solid lines are the results when 20 terms are included in the expansion, and the dotted lines are the results for the conduction into an infinite medium (Eq. 4.4.16) where $\xi = z^*/\sqrt{t^*} = z/\sqrt{\alpha t}$.

Eq. 4.7.29. The three-term approximation is in agreement with the exact solution for $t^* \geq 0.03$, but there is a difference in the results for $t^* = 0.01$, where unphysical oscillations due to the superposition of three sine waves are visible.

In Fig. 4.18(b), the result Eq. 4.7.29 is compared with the result Eq. 4.4.16 for the conduction into an infinite medium, where the similarity variable is $\xi = (z/\sqrt{\alpha t}) = (z^*/\sqrt{t^*})$. For $t^* \ll 1$ ($t \ll (h^2/\alpha)$), it is expected that the 'penetration depth' for the temperature disturbance due to the surface at $z^* = 0$ is small compared to the channel height, and the problem can be treated as conduction into an infinite medium. Fig. 4.18(b) shows that the temperature profile from the similarity solution for conduction into an infinite medium is in quantitative agreement with the exact solution for $t^* \leq 0.03$. At $t^* = 0.1$, there is excellent agreement between the similarity solution and the exact solution throughout the domain, with the exception of the top wall where the former does not satisfy the boundary condition $T^* = 0$. For $t^* \geq 0.3$, the penetration depth is comparable to the channel height or larger, and the similarity solution cannot be used.

The above numerical comparisons show that the one-term approximation for the solution, Eq. 4.7.29, is in quantitative agreement with the exact solution for $t^* \geq 0.3$, and the similarity solution is applicable for $t^* \leq 0.1$. There is only a small range $0.1 < t^* < 0.3$ where a three-term approximation is necessary. Thus, a combination of the similarity solution and the separation of variables solution provides a quantitatively accurate approximation for the temperature field in a finite channel.

EXAMPLE 4.7.1: Two solid slabs made of the same material with thermal diffusivity α and thickness h, but with two different initial temperatures T_1 and T_2, are brought into contact at time $t = 0$, as shown in Fig. 4.19(a). Consider the slabs to be of infinite extent in the plane perpendicular to the z direction, and the surfaces of the slabs at $z = h$ and $z = -h$ are insulated. Determine the evolution of the temperature of the slabs.

Solution: Since the system is insulated, there is no change in the energy of the system between the initial and steady states. The final temperature is the average of the initial temperatures of the two slabs, $T_s = (T_1 + T_2)/2$. The schematic of the evolution of the temperature field is shown in Fig. 4.19(b). Here, $(\mathrm{d}T/\mathrm{d}z)$ is zero at both insulated boundaries to satisfy the no-flux condition, and the temperature and flux are equal at the interface between the two slabs. In Fig. 4.19(b), the dotted line is the initial temperature which changes discontinuously at $z = 0$, and the solid line is the final constant temperature.

The scaled co-ordinate is defined as $z^* = (z/h)$, and the scaled time is defined as $t^* = (t\alpha/h^2)$, where α is the thermal diffusivity. The scaled temperature is defined

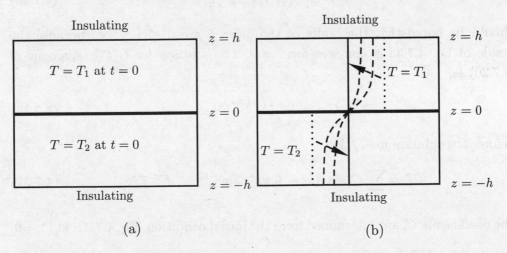

FIGURE 4.19. (a) Schematic of two slabs of the same material of thickness h with temperatures T_1 and T_2 are brought together at $t = 0$, and (b) evolution of the temperature field from the initial state (dotted line) to the final state (solid line) where the temperature is uniform.

as

$$T_t^* = \frac{T - (T_1 + T_2)/2}{(T_1 - T_2)/2}.$$ (4.7.30)

The conservation equation for the transient part of the temperature profile is identical to Eq. 4.7.10. The boundary conditions are,

$$\frac{\partial T_t^*}{\partial z^*} = 0 \text{ at } z^* = \pm 1,$$ (4.7.31)

and the initial condition at $t^* = 0$ is,

$$T_t^* = -1 \text{ for } -1 < z^* < 0$$
$$= 1 \text{ for } 0 < z^* < 1.$$ (4.7.32)

The final steady solution is $T_s = (T_1 + T_2)/2$ or $T_s^* = 0$. Therefore, the transient temperature is the same as T^*.

The transient temperature is expressed as $T^* = Z(z^*)G(t^*)$, where Z is a solution of Eq. 4.7.16. The general solution, Eq. 4.7.17, is the sum of sine and cosine functions. In the present case, the heat flux is zero at the insulated surfaces $z^* = \pm 1$. This implies that the $(\mathrm{d}Z/\mathrm{d}z^*) = 0$ at $z^* = \pm 1$. The zero derivative condition is satisfied if $Z = \sin(\beta_i z^*)$ and $\beta_i = \frac{1}{2}, \frac{3}{2}, \ldots$. The solution for $Z(z^*)$ (analogue of Eq. 4.7.18) is,

$$Z = \sin((2i + 1)\pi z^*/2).$$ (4.7.33)

It should be noted that the limits of the index i are $0 \leq i \leq \infty$, because the right side of Eq. 4.7.33 is non-zero for $i = 0$. The solution for $G(t^*)$ (analogue of Eq. 4.7.20) is,

$$G = \mathrm{e}^{(-(2i+1)^2\pi^2 t^*/4)}.$$ (4.7.34)

Therefore, the solution for T_t^* is,

$$T_t^* = \sum_i C_i \sin((2i+1)\pi z^*/2)\mathrm{e}^{(-(2i+1)^2\pi^2 t^*/4)}.$$ (4.7.35)

The coefficients C_i are determined from the initial condition, Eq. 4.7.32, at $t^* = 0$,

$$\sum_i C_i \sin((2i+1)\pi z^*/2) = -1 \text{ for } -1 < z^* < 0,$$

$$= 1 \text{ for } 0 < z^* < 1.$$ (4.7.36)

Both sides of the above equation are multiplied by $\sin\left((2j+1)\pi z^*/2\right)$ and integrated from $z^* = -1$ to $z^* = 1$,

$$\sum_{i=0}^{\infty} \int_{-1}^{1} dz^* C_i \sin\left((2i+1)\pi z^*/2\right) \sin\left((2j+1)\pi z^*/2\right)$$

$$= -\int_{-1}^{0} dz^* \sin\left((2j+1)\pi z^*/2\right) + \int_{0}^{1} dz^* \sin\left((2j+1)\pi z^*/2\right)$$

$$= \frac{4}{\pi(2j+1)}. \tag{4.7.37}$$

The integral on the left side of the above equation is non-zero only for $i = j$, and is zero otherwise. The orthogonality relation is,

$$\int_{-1}^{1} dz^* \sin\left((2i+1)\pi z^*/2\right) \sin\left((2j+1)\pi z^*/2\right) = \delta_{ij}. \tag{4.7.38}$$

The above integral is calculated using the identity

$$\sin\left((2i+1)\pi z^*/2\right) \sin\left((2j+1)\pi z^*/2\right) = \tfrac{1}{2}[\cos\left((i-j)\pi z^*\right) - \cos\left((i+j+1)\pi z^*\right)].$$

Therefore, the left side of Eq. 4.7.37 is C_j, and the solution for C_j is,

$$C_j = \frac{4}{\pi(2j+1)}. \tag{4.7.39}$$

The solution for the temperature field is,

$$T^* = \sum_{n=0}^{\infty} \frac{4}{\pi(2i+1)} \sin\left((2i+1)\pi z^*/2\right) e^{\left(-(2i+1)^2 \pi^2 t^*/4\right)}. \tag{4.7.40}$$

The one- and three-term approximations of the solution, Eq. 4.7.40, are compared with the 'exact' solutions comprising 20 terms in Fig. 4.20. Since the solutions have odd symmetry about $z^* = 0$, that is, $T^*(-z^*) = -T^*(z^*)$, the results are shown only for $z^* > 0$. The broad conclusions are the same as those for the profiles in Fig. 4.18. The one-term approximation is in reasonable agreement, and the three-term approximation is in quantitative agreement with the exact result for $t^* \geq 0.1$. For $t^* < 0.1$, significant errors are incurred by the one-term approximation. The three-term approximation shows unphysical oscillations for $t^* = 0.01$, but it accurately captures the temperature variation for $t^* = 0.03$. □

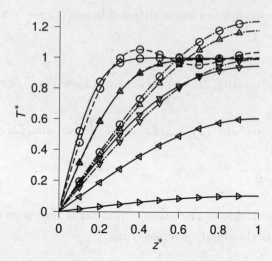

FIGURE 4.20. The solution, Eq. 4.7.40, as a function for z^* for $t^* = 0.01$ (○), $t^* = 0.03$ (△), $t^* = 0.1$ (∇), $t^* = 0.3$ (◁), and $t^* = 1.0$ (▷). The solid, dashed and dot-dash lines are the results when 20, 3 and 1 terms are included in the expansion, Eq. 4.7.40.

Summary (4.7)

1. The governing equation for the diffusion in a channel of finite width h can be reduced to universal form, Eq. 4.7.3, if the scaled co-ordinate is defined as $z^* = (z/h)$, the scaled time is $t^* = (t\mathcal{D}/h^2)$, and the field variable Φ_{fv} is scaled such that it varies between 0 and 1 at the two boundaries.

2. The transient part of the field variable is defined as $\Phi_{fv}^t = \Phi_{fv} - \Phi_{fv}^s$, where Φ_{fv}^s is the steady solution in the long time limit. The transient part has homogeneous boundary conditions at the spatial boundaries, Eqs. 4.7.11–4.7.12, and a non-homogeneous initial condition at initial time, Eq. 4.7.13.

3. The transient part of the field variable Φ_{fv}^t is written as the sum of a series of terms multiplied by constant coefficients. Each term is the product of a function of time and a spatial basis function of the scaled co-ordinate, Eq. 4.7.22.

4. Since the boundary conditions for Φ_{fv}^t are homogeneous, the eigenvalues are discrete, Eq. 4.7.18, and the basis functions satisfy orthogonality relations, Eq. 4.7.25. In a Cartesian co-ordinate system, the basis functions are sine/cosine functions.

5. The coefficients in the expansion for the transient part, Eq. 4.7.28, are determined from the initial condition, Eq. 4.7.13, using the orthogonality relations for the spatial basis functions.

6. The exact solution, 4.7.29, is an infinite series. The series is truncated at a finite number of terms for an approximate solution. The approximate solution exhibits unphysical oscillations, and there is a truncation error in the solution when t^* decreases below a minimum. This minimum t^* for a specified maximum error decreases as the number of terms in the expansion is increased.

4.8   Correlations in Balance Equations

It is often possible to incorporate correlations in differential balances in order to determine the spatial or time evolution of the field variables. The implicit assumption is that the fluxes are determined by the 'local' values of the field variables—that is, the correlation at each time instant or at each value of the spatial co-ordinate depends only on the value of the field variable at that instant/location. For unsteady problems, the necessary condition is that the diffusion time required for the flux to reach its steady value is much smaller than the time for the evolution of the field variable at the boundaries. The condition for steady, spatially varying fields is that the characteristic length scale for the variation in the field variable is much larger than the diffusion length, so that the flux depends only on the local value of the field variable. The diffusion length and time for different convection and diffusion processes are discussed in the following sub-sections.

4.8.1   Transport by Diffusion

For diffusion-dominated transport, when there is a time variation of the field variable at the boundaries, the time required for the concentration/temperature field to reach steady state scales as $(l_c^2/\mathcal{D})$ from dimensional analysis. Here, l_c is the characteristic length scale and $\mathcal{D}$ is the diffusivity. Therefore, correlations derived using the steady-state assumption can be used for unsteady problems if the timescale for variation of the field variable at the boundaries is much larger than $(l_c^2/\mathcal{D})$. The local assumption cannot be used for problems where there is a spatial variation in the field variable, because the length scale for diffusion is the same as the characteristic length of the object/conduit.

For the problem of conduction into a medium with a time-dependent length scale considered in Section 4.6, it is evident from Eq. 4.6.8 that the steady-state approximation can be used if the parameter P is small compared to 1, or alternatively, the Jakob number is small compared to 1 from Eq. 4.6.14. It can be shown (Exercise 4.24) that the temperature is a linear function of z in this limit, as expected for steady diffusion.

The unsteady cooling of a sphere due to conduction is determined using the expression for the heat flux at steady state, and the validity of the steady state approximation is examined in the following example.

EXAMPLE 4.8.1: A solid sphere of diameter d, density ρ_s and specific heat C_{ps} at initial temperature T_i cools in a viscous liquid with density ρ_f, specific heat C_{pf}, thermal conductivity k_f and temperature T_f. What is the characteristic time for cooling if the heat transfer is due to conduction and the steady-state assumption is used for the heat flux? Under what condition is the steady-state assumption valid? Assume that the thermal conductivity of the solid is sufficiently large that the temperature in the solid is uniform.

Solution: The rate of change of thermal energy of the sphere is equal to the heat lost due to conduction,

$$mC_{ps}\frac{dT}{dt} = -q_{av}(\pi d^2),\qquad(4.8.1)$$

where m and C_{ps} are the mass and specific heat of the sphere, T is the sphere temperature, and q_{av} is the average heat flux. The average heat flux is expressed as,

$$q_{av} = \frac{k_f(T-T_f)\mathrm{Nu}}{d} = \frac{2k_f(T-T_f)}{d},\qquad(4.8.2)$$

where the Nusselt number is set equal to 2 for heat conduction (see Section 2.3.3). Substituting Eq. 4.8.2 in Eq. 4.8.1, and dividing the resulting equation by mC_{ps}, we obtain,

$$\frac{dT}{dt} = -\frac{T-T_f}{t_c},\qquad(4.8.3)$$

where the characteristic time t_c is,

$$t_c = \frac{mC_{ps}}{2\pi k_f d} = \frac{d^2\rho_s C_{ps}}{12k_f},\qquad(4.8.4)$$

where the mass is the product of the sphere volume and density, $m = (\pi d^3/6)\rho_s$. The solution of Eq. 4.8.3 is,

$$\frac{T-T_f}{T_i-T_f} = e^{(-t/t_c)}.\qquad(4.8.5)$$

The steady-state approximation used in deriving the Nusselt number correlation is valid when the characteristic time t_c is much larger than (d^2/α_f), where $\alpha_f = (k_f/\rho_f C_{pf})$ is the thermal diffusivity of the fluid,

$$\frac{d^2 \rho_s C_{ps}}{12 k_f} \gg \frac{d^2 \rho_f C_{pf}}{k_f} \Rightarrow \frac{\rho_s C_{ps}}{12 \rho_f C_{pf}} \gg 1. \qquad (4.8.6)$$

Thus, the steady-state approximation is valid when the product of the particle density and specific heat is much larger than that for the surrounding liquid. $\square$

In problems involving evaporation of liquid drops or sublimation of solid particles, the diameter of the particle decreases with time. The time evolution of the diameter is examined using the Sherwood number correlation for the flux in the following example.

EXAMPLE 4.8.2: A spherical camphor particle of diameter 1 mm and mass density 10^3 kg/m^3 sublimes in air at 20°C. The diffusion coefficient of camphor in air in 2.3×10^{-7} m^2/s, the molecular weight of camphor is 152 g/mol, and the vapour pressure is 87 Pa. What is the time required for complete sublimation?

Solution: As the camphor sublimes, the diameter of the particle decreases. The rate of decrease of the particle mass is the product of the flux j_{av} and the surface area πd^2,

$$\frac{dm}{dt} = \frac{d(\rho \pi d^3/6)}{dt} = -\pi d^2 j_{av}, \qquad (4.8.7)$$

where m is the particle mass, d is the diameter, ρ is the camphor density and j_{av} is the average flux. The above equation is simplified to obtain an equation for the time evolution of the diameter,

$$\frac{dd}{dt} = -\frac{2 j_{av}}{\rho} = -\frac{2 \mathcal{D} c_s \mathrm{Sh}}{d \rho}, \qquad (4.8.8)$$

where c_s is the surface concentration, $\mathcal{D}$ is the diffusion coefficient, and Sh is the Sherwood number. The correlation Sh = 2 for transport by diffusion is substituted in the above equation, which is multiplied by d and integrated once,

$$d_0^2 - d^2 = \frac{8\mathcal{D}c_s t}{\rho}, \qquad (4.8.9)$$

where d_0 is the initial diameter at $t = 0$. The time required for the diameter to decrease to zero for complete sublimation is,

$$t_d = \frac{\rho d_0^2}{8\mathcal{D}c_s}. \qquad (4.8.10)$$

The saturation concentration of camphor in air is,

$$c_s = \frac{p_v M}{k_B T} = \frac{87 \text{ Pa} \times (0.152/6.023 \times 10^{23}) \text{ kg}}{1.38 \times 10^{-23} \text{ J/K} \times 293 \text{ K}} = 5.43 \times 10^{-3} \text{ kg/m}^3. \qquad (4.8.11)$$

Here, p_v is the vapour pressure, M is the mass of a molecule, k_B is the Boltzmann constant and T is the absolute temperature. The sublimation time t_d (Eq. 4.8.10) is,

$$t_d = \frac{10^3 \text{ kg/m}^3 \times (10^{-3} \text{ m})^2}{8 \times 2.3 \times 10^{-7} \text{ m}^2/\text{s} \times 5.43 \times 10^{-3} \text{ kg/m}^3} = 10^5 \text{ s} \sim 1.2 \text{ days}. \qquad (4.8.12)$$

The steady-state approximation is valid in this case, because the sublimation time is much larger than $(d_0^2/\mathcal{D}) = ((10^{-3} \text{ m})^2/2.3 \times 10^{-7} \text{ m}^2/\text{s}) \sim 4.35 \text{ s}.$ $\qquad \square$

4.8.2 Forced Convection

For forced convection at high Peclet number, it is shown in Chapter 9 that heat/mass diffusing from a surface is swept downstream due to the flow, and is restricted to a thin 'boundary layer' close to the surface. The diffusion length, or the thickness of the boundary layer, is usually much smaller than the characteristic length. The diffusion time required to reach steady state scales as $(\text{Diffusion length})^2/\mathcal{D}$, where $\mathcal{D}$ is the diffusion coefficient. The Nusselt/Sherwood number correlation based on the steady-state approximation can be used when the diffusion time is much smaller than the timescale for the variation of the temperature at the boundaries or due to internal sources.

The diffusion lengths for different types of laminar flows are summarised in Table 4.2. The diffusion lengths for these flows are derived in Chapters 9 and 10, with the exception of the length scale for flow past a particle or flat plate at high Reynolds number. For high Reynolds number laminar flow past a flat surface shown in Fig. 2.6

TABLE 4.2. The ratio of the diffusion length and the characteristic length l_c for high Peclet number forced convection and high Grashof number natural convection. The dimensional and dimensionless parameters are defined in the caption of Table 2.5. Rigid and mobile interfaces are schematically shown in Fig. 2.10.

	Regime	Interface	$\dfrac{\text{(Diffusion length)}}{l_c}$
Forced convection around a bubble	Laminar	Mobile	$\text{Pe}^{-1/2}$
Forced convection around a particle	$\text{Re} \ll 1$	Rigid	$\text{Pe}^{-1/3}$
Forced convection around a particle, flat plate	$\text{Re} \gg 1$ Laminar	Rigid	$\text{Re}^{-1/6}\text{Pe}^{-1/3}$
Forced convection Pipe flow	$\text{Re} \gg 1, (L/d) \ll \text{Pe}$ Laminar	Rigid	$\text{Pe}^{-1/3}(d/L)^{-1/3}$
Forced convection Falling film	$\text{Re} \gg 1, (L/h) \ll \text{Pe}$ Laminar	Mobile	$\text{Pe}^{-1/2}(h/L)^{-1/2}$
Natural convection	$\text{Gr} \gg 1$ Laminar	Rigid/mobile	$\text{Gr}^{-1/4}$

in Section 2.2.3 (Chapter 2), the boundary layer thickness scales as $\text{Re}^{-1/6}\text{Pe}^{-1/3}$, which is equivalent to $\text{Re}^{-1/2}\text{Pr}^{-1/3}$, since $\text{Pe} = \text{Re Pr}$. This results in correlations for the Nusselt number such as Eq. 2.3.11 for a flat plate. This also explains the first term on the right in Eq. 2.3.15 for the flow past a particle which are proportional to $\text{Re}^{1/2}\text{Pr}^{1/3}$ due to forced convection in the laminar boundary layer in the upstream hemisphere. Equivalent estimates are not available for turbulent flows, due to the complexity of the flow structure.

Simple correlations such as Eq. 2.3.13 can be used to estimate the characteristic time for heat/mass transfer as shown in the following example.

EXAMPLE 4.8.3: A steel sphere of diameter 1 cm with initial temperature 300°C is cooled by a stream of cold air at temperature 0°C with free stream velocity 1 m/s. What is the time taken for the temperature of the sphere to decrease to 100°C? Examine whether the steady-state assumption in deriving Eq. 2.3.13 is valid in this case. The density and specific heat of steel are 8×10^3 kg/m³ and 500 J/kg/°C, respectively, and the density, thermal conductivity and specific heat of air are 1.25 kg/m³, 2.4×10^{-2} W/m/°C and 10^3 J/kg/°C, respectively. Assume the thermal conductivity of steel is sufficiently large that the temperature in the sphere is uniform.

Solution: The thermal diffusivity in air is $\alpha_a = (k_a/\rho_a C_{pa}) = 1.92 \times 10^{-5}$ m^2/s, and the Peclet number is Pe $= (v_{fs}d/\alpha_a) = 520.83$. Here, ρ_a, k_a and C_{pa} are the air density, thermal conductivity and specific heat, $d = 0.01$ m is the sphere diameter and $v_{fs} = 1$ m/s is the free stream velocity of air. Since the Peclet number is high, the Nusselt number correlation Eq. 2.3.13 is used for the heat flux.

The energy balance equation for the sphere is,

$$\rho_s C_{ps} \frac{\pi d^3}{6} \times \frac{\mathrm{d}T_s}{\mathrm{d}t} = -(\pi d^2) q_{av}, \qquad (4.8.13)$$

where ρ_s and C_{ps} are the density and specific heat of steel, T_s is the sphere temperature, and $(\pi d^3/6)$ on the left and πd^2 on the right are the volume and surface area of the sphere. The left side of Eq. 4.8.13 is the rate of change of the thermal energy of the sphere, and the right side is the energy leaving the sphere per unit time. Eq. 4.8.13 is divided by $\rho_s C_{ps} \pi d^3/6$, and the correlation Eq. 2.3.13 is used for the average heat flux q_{av},

$$\frac{\mathrm{d}T_s}{\mathrm{d}t} = -\frac{6}{\rho_s C_{ps} d} \frac{k_a(T_s - T_f) \mathrm{Nu}}{d} = -\frac{6k_a(T_s - T_f) \times 0.992\, \mathrm{Pe}^{1/3}}{\rho_s C_{ps} d^2}, \qquad (4.8.14)$$

where T_f is the fluid temperature. Eq. 4.8.14 is expressed as,

$$\frac{\mathrm{d}T_s}{\mathrm{d}t} = -\frac{T_s - T_f}{t_c}, \qquad (4.8.15)$$

where the characteristic time t_c is,

$$t_c = \frac{\rho_s C_{ps} d^2}{6k_a(0.992\, \mathrm{Pe}^{1/3})} = \frac{8 \times 10^3 \text{ kg/m}^3 \times 500 \text{ J/kg/}^\circ\text{C} \times (0.01 \text{ m})^2}{6 \times 2.4 \times 10^{-2} \text{ W/m/}^\circ\text{C} \times 0.992 \times (520.83)^{1/3}} = 348 \text{ s.} \qquad (4.8.16)$$

Eq. 4.8.15 is integrated,

$$\frac{T_{sf} - T_f}{T_{si} - T_f} = e^{(-t/t_c)}, \qquad (4.8.17)$$

where $T_{si} = 300^\circ$C and $T_{sf} = 100^\circ$C are the initial and final temperatures, and $T_f = 0^\circ$C is the air temperature. From Eq. 4.8.17, the time taken is

$$t = 1.1\, t_c = 382 \text{ s} \sim 6.4 \text{ minutes.} \qquad (4.8.18)$$

The validity of the steady-state assumption is examined. From Table 4.2, the diffusion length for the flow around a solid particle is Pe$^{-1/3}d$, and the diffusion time is $\mathrm{Pe}^{-2/3}(d^2/\alpha_a)$,

$$\frac{\mathrm{Pe}^{-2/3}d^2}{\alpha_a} = \frac{(520.83)^{-2/3} \times (0.01 \text{ m})^2}{1.9 \times 10^{-5} \text{ m}^2/\text{s}} = 0.081 \text{ s}. \tag{4.8.19}$$

The error due to the steady-state approximation is small because the characteristic time is three orders of magnitude larger than the diffusion time. □

In the case of the dissolution of a sphere due to forced convection, the variation in diameter can be incorporated by equating the rate of dissolution of mass with the product of the flux and the surface area. In the following example, it is shown that an analytical solution for the time evolution of the diameter can be derived using a correlation such as Eq. 2.3.13.

EXAMPLE 4.8.4: A pill of initial diameter d_0 and density ρ_p made of a soluble material dissolves in the surrounding fluid with diffusivity $\mathcal{D}$. If the pill is moving with velocity v_c, what is the time taken for the pill to dissolve? Assume the Reynolds number is small, the Peclet number is large, the concentration of the solute at the pill surface is c_s, and the solute concentration far from the pill is zero.

Solution: The rate of decrease of the mass of the pill is equal to the mass dissolved per unit time due to diffusion,

$$\frac{\mathrm{d}}{\mathrm{d}t}\left(\frac{\rho_p \pi d^3}{6}\right) = -\pi d^2 j_{av} = -\pi d^2 \left(\frac{\mathcal{D}c_s \mathrm{Sh}}{d}\right), \tag{4.8.20}$$

where d is the diameter of the pill at time t and j_{av} is the average flux at the surface. The left side of the above equation is the rate of change of mass (density times volume), and the right side is the surface area times the average flux, the latter expressed in terms of the Sherwood number. The above equation is simplified, and the correlation Eq. 2.3.13 is used for the Sherwood number, to obtain,

$$\frac{\mathrm{d}d}{\mathrm{d}t} = -\frac{2\mathcal{D}c_s \mathrm{Sh}}{\rho_p d} = -\frac{2\mathcal{D}c_s}{\rho_p d} \times 0.992 \left(\frac{v_c d}{\mathcal{D}}\right)^{1/3} = -\frac{1.984\, v_c^{1/3}\mathcal{D}^{2/3}c_s}{\rho_p d^{2/3}} \tag{4.8.21}$$

The above equation is multiplied by $d^{2/3}$ and integrated with respect to time to obtain,

$$\tfrac{3}{5}(d^{5/3} - d_0^{5/3}) = -\frac{1.984\, v_c^{1/3}\mathcal{D}^{2/3}c_s t}{\rho_p}, \tag{4.8.22}$$

where d_0 is the initial diameter of the pill at $t = 0$. The time for dissolution t_d is the time at which $d = 0$,

$$t_d = \frac{0.3\rho_p d_0^{5/3}}{c_s v_c^{1/3} \mathcal{D}^{2/3}}. \tag{4.8.23}$$

The steady-state approximation for the concentration field is examined in Exercise 4.29. $\qquad\qquad\qquad\qquad\qquad\qquad\qquad\qquad\qquad\qquad\qquad\square$

4.8.3   Natural Convection

The Nusselt number correlations, Eqs. 2.4.12, 2.4.13 and 2.4.14, for natural convection at high Grashof number are derived in Section 10.2 (Chapter 10), where the Grashof and Prandtl numbers are defined in Eqs. 2.4.2 and 1.6.36, respectively. The boundary layer thickness for natural convection is proportional to $\mathrm{Gr}^{-1/4} l_c$ in the limit $\mathrm{Gr} \gg 1$. Therefore, the timescale for the development of the boundary layer is $\mathrm{Gr}^{-1/2}(l_c^2/\alpha)$, and the correlation derived assuming steady state can be applied for $t_c \gg \mathrm{Gr}^{-1/2}(l_c^2/\alpha)$, where t_c is the characteristic time. The decrease in the temperature of a heated object due to natural convection is calculated in the following example.

EXAMPLE 4.8.5: A heated sphere of mass m, diameter d and specific heat C_{ps} at initial temperature T_0 cools in a fluid of density ρ_f, viscosity μ_f, thermal conductivity k_f, specific heat C_{pf} and thermal expansion coefficient β at ambient temperature T_f due to natural convection. Derive an expression for the time dependence of the temperature. Assume that the conductivity of the sphere is sufficiently large that the temperature is uniform within the sphere.

Solution: The energy balance equation for the sphere is,

$$mC_{ps}\frac{\mathrm{d}T_s}{\mathrm{d}t} = -\pi d^2 q_{av} = -\pi d k_f (T_s - T_f) \mathrm{Nu}$$

$$= -\pi d k_f (T_s - T_f) C(\mathrm{Pr}) \left(\frac{\rho_f^2 \beta g (T_s - T_f) d^3}{\mu_f^2} \right)^{1/4}, \tag{4.8.24}$$

where T_s is the sphere temperature at time t, the left side of Eq. 4.8.24 is the rate of change of the thermal energy of the sphere, and the right side is the rate of energy transfer from the sphere to the fluid per unit time which is the product of the average flux and the surface area. The correlation, Eq. 2.4.12, is used for the Nusselt number which is based on the difference in the instantaneous sphere temperature T_s and the fluid temperature T_f. The energy balance equation is expressed in terms of the scaled temperature $T^* = (T_s - T_f)/(T_0 - T_f)$, where T_0 is the initial temperature of the sphere at $t = 0$,

$$\frac{\mathrm{d}T^*}{\mathrm{d}t} = -\frac{\pi C(\mathrm{Pr})k_f\rho_f^{1/2}(\beta(T_0 - T_f)g)^{1/4}d^{7/4}}{\mu_f^{1/2}mC_{ps}}(T^*)^{5/4} = -\frac{(T^*)^{5/4}}{t_c}, \quad (4.8.25)$$

where the time constant t_c is,

$$t_c = \frac{\mu_f^{1/2}mC_{ps}}{\pi C(\mathrm{Pr})k_f\rho_f^{1/2}(\beta g(T_0 - T_f))^{1/4}\, d^{7/4}}. \quad (4.8.26)$$

Eq. 4.8.25 is integrated subject to the initial condition, $T^* = 1$ $(T_s = T_0)$ at $t = 0$,

$$4((T^*)^{-1/4} - 1) = \frac{t}{t_c} \Rightarrow T^* = \left(1 + \frac{t}{4t_c}\right)^{-4}. \quad (4.8.27)$$

$\square$

The Nusselt number correlation for natural convection can also be used to examine the spatial variation of the temperature for elongated objects in the limit of high Grashof number, because the diffusion length is $\mathrm{Gr}^{-1/4}$ times the characteristic cross-sectional dimension of the object. The equation for the spatial evolution of the temperature for a heat transfer from a fin is derived in the following example.

EXAMPLE 4.8.6: A fin of length L and cross-sectional area A_{cs} is attached to the outer wall of a pipe in order to enhance the heat transfer rate by natural convection, as shown in Fig. 4.21. The wall of the pipe is at temperature T_0, the insulated end of the fin at $x = L$ is at temperature T_L, and the ambient temperature far from the fin is T_f. Determine the rate of heat transfer from the fin.

Solution: Consider the energy balance at steady state for a cuboidal differential volume of thickness Δx at the location x. There is heat entering at the location x and leaving at the location $x + \Delta x$ due to conduction in the solid, and there is heat transfer due to natural convection on the surface in contact with the fluid with area $l_w\Delta x$, where l_w is the wetted perimeter of the cross section. The energy balance equation is,

$$(q_x|_x - q_x|_{x+\Delta x})A_{cs} - q_{av}l_w\Delta x = 0, \quad (4.8.28)$$

where q_{av} is the average heat flux due to natural convection. The above equation is divided by $l_w\Delta x$, and converted into a differential equation by taking the limit

$\Delta x \to 0$,

$$-\frac{\mathrm{d}q_x}{\mathrm{d}x} - \frac{4q_{av}}{d_e} = 0. \tag{4.8.29}$$

Here, the definition $d_e = (4A_{cs}/l_w)$ for the equivalent diameter has been used (see Section 2.1.3). The Fourier's law of heat conduction is used for the flux q_x, and the flux due to natural convection is expressed in terms of the Nusselt number $q_{av} = (\mathrm{Nu}k_f(T - T_f)/d_e)$,

$$k_s\frac{\mathrm{d}^2T}{\mathrm{d}x^2} - \frac{4k_f(T - T_f)\mathrm{Nu}}{d_e^2} = 0. \tag{4.8.30}$$

Here, k_s and k_f are, respectively, the thermal conductivities of the solid and fluid, and T is the temperature of the fin at the location x. Substituting the Nusselt number correlation Eq. 2.4.12, $\mathrm{Nu} = C(\mathrm{Pr})\mathrm{Gr}^{1/4}$, the energy balance equation is,

$$\frac{\mathrm{d}^2T}{\mathrm{d}x^2} = \frac{4k_f(T - T_f)C(\mathrm{Pr})}{d_e^2 k_s}\left(\frac{\rho_f^2 g\beta(T - T_f)d_e^3}{\mu_f^2}\right)^{1/4}, \tag{4.8.31}$$

where ρ_f and μ_f are the fluid density and viscosity, and the equivalent diameter is used as the length scale in the Grashof number. The scaled temperature is defined

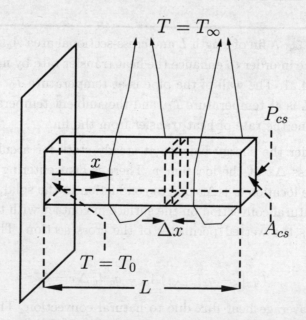

FIGURE 4.21. Natural convection from a fin attached to the wall of a pipe.

as $T^* = (T - T_f)/(T_0 - T_f)$, to obtain the equation for the temperature evolution,

$$\frac{d^2 T^*}{dx^2} = \frac{(T^*)^{5/4}}{l_c^2}, \tag{4.8.32}$$

where the characteristic length l_c is,

$$l_c = \sqrt{\frac{k_s}{4 k_f C(\text{Pr})}} \frac{d_e^{5/8} \mu_f^{1/4}}{(\rho_f^2 g \beta (T_0 - T_f))^{1/8}}. \tag{4.8.33}$$

Eq. 4.8.32 can be integrated one time with respect to x by first multiplying the equation by (dT^*/dx). The left and right sides of the equation are,

$$\frac{dT^*}{dx} \frac{d^2 T^*}{dx^2} = \frac{1}{2} \frac{d}{dx} \left(\frac{dT^*}{dx} \right)^2, \tag{4.8.34}$$

$$\frac{dT^*}{dx} \frac{(T^*)^{5/4}}{l_c^2} = \frac{4}{9 l_c^2} \frac{d(T^*)^{9/4}}{dx}. \tag{4.8.35}$$

Substituting Eqs. 4.8.34 and 4.8.35 into Eq. 4.8.32, and integrating once with respect to x, we obtain,

$$\left(\frac{dT^*}{dx} \right)^2 = \frac{8 (T^*)^{9/4}}{9 l_c^2} + C. \tag{4.8.36}$$

The constant of integration is determined from the temperature condition $T^* = T_L^*$ and the zero flux condition $(dT^*/dx) = 0$ at the end of the fin, $x = L$. Here, the scaled temperature at the end of the fin is defined as $T_L^* = (T_L - T_f)/(T_0 - T_f)$. The relation between the temperature gradient and the temperature is,

$$\frac{dT^*}{dx} = -\sqrt{\frac{8((T^*)^{9/4} - (T_L^*)^{9/4})}{9 l_c^2}}. \tag{4.8.37}$$

The negative square root on the right side of Eq. 4.8.36 is selected above so that the temperature decreases as x increases. The above equation cannot be integrated analytically to determine the temperature field. However, the total heat transfer rate from the fin can be determined, because it is equal to the product of the flux $- k_s (dT/dx)$ and the area of cross section A_{cs} at $x = 0$,

$$Q = - k_s A_{cs} (T_0 - T_f) \left. \frac{dT^*}{dx} \right|_{x=0} = \frac{2\sqrt{2} k_s A_{cs} (T_0 - T_f) \sqrt{1 - (T_L^*)^{9/4}}}{3 l_c}. \tag{4.8.38}$$

Here, we have used the condition $T^* = 1$ at $x = 0$. $\qquad\square$

4.8.4 Packed Column

The definition of the dimensionless groups and the correlations for a packed column were discussed in Section 2.3.4. For deriving a one-dimensional balance equation, it is necessary to assume that the concentration/temperature field varies only in the flow direction—that is, cross-stream variations are small compared to stream-wise variations in the field variable. Since the correlations are applicable over a length scale large compared to the particle diameter, a one-dimensional balance equation is valid only if the length scale for the concentration/temperature variation is large compared to the particle diameter. The following example illustrates the use of the heat transfer correlation Eq. 2.3.34 for calculating the length scale for the variation of the temperature and the heat transfer rate in the column.

EXAMPLE 4.8.7: In a heat recovery process, water flows upwards through a column of particles of diameter 1 cm made of paraffin wax, a phase change material. The void fraction in the packed column is 0.4, and the superficial velocity is 1 cm/s. The surface temperature of the particles is the melting temperature of the paraffin wax, 80°C. If the inlet temperature of water is 20°C, estimate the length of column required to heat the water at the outlet to 75°C. The density, viscosity, thermal conductivity and specific heat of water are $\rho = 10^3$ kg/m^3, $\mu = 10^{-3}$ kg/m/s, $k = 0.6$ W/m/°C and $C_p = 4.2 \times 10^3$ J/kg/°C, respectively.

Solution: The terminology and the empirical correlation for a packed column used here were discussed in Section 2.3.4. The schematic of the packed column is shown in Fig. 4.22. Consider a differential volume of area of cross section A and thickness dz in the packed column. The rate of change of the thermal energy of the water between the locations z and $z + \Delta z$ is the product of the volumetric flow rate and the change in the energy density,

$$\text{Rate of change of energy} = (v_s/\varepsilon)(A\varepsilon)\rho C_p(T(z + \Delta z) - T(z))$$
$$= v_s A\rho C_p(T(z + \Delta z) - T(z)). \qquad (4.8.39)$$

Here, the fluid velocity through the packed bed is (v_s/ε), where v_s is the superficial velocity and ε is the void fraction, and the area of cross section for fluid flow is $A\varepsilon$, the area of cross section of the packed column times the void fraction. The heat transfer rate from the particles to the fluid in the differential volume is given by Eq. 2.3.32, where $V = A\Delta z$ is the differential volume of the column,

$$\text{Heat transfer rate} = q_{av} \times \frac{6A\Delta z(1 - \varepsilon)}{d}, \qquad (4.8.40)$$

where d is the particle diameter. Equating the change in energy with the heat transfer rate, dividing by $A\Delta z$ and taking the limit $\Delta z \to 0$, the expression for the fluid temperature is,

$$v_s \rho C_p \frac{\mathrm{d}T}{\mathrm{d}z} = \frac{6(1-\varepsilon)q_{av}}{d}. \tag{4.8.41}$$

The above equation is divided by $v_s \rho C_p$, and Eq. 2.3.33 is used to express the average flux in terms of the Nusselt number,

$$\frac{\mathrm{d}T}{\mathrm{d}z} = \frac{6(1-\varepsilon)}{\rho C_p v_s d} \frac{\mathrm{Nu}\, k(T_p - T)(1-\varepsilon)}{d\varepsilon} = \frac{T_p - T}{l_c}, \tag{4.8.42}$$

where T_p is the temperature of the particles, $\Delta T = T_p - T$ is the temperature difference between the particles and the fluid, and the characteristic length l_c is,

$$l_c = \frac{\rho C_p v_s d^2 \varepsilon}{6 \mathrm{Nu} k (1-\varepsilon)^2}. \tag{4.8.43}$$

Eq. 4.8.42 is solved to obtain the fluid temperature as a function of z,

$$T_p - T = (T_p - T_i)\, \mathrm{e}^{(-z/l_c)}, \tag{4.8.44}$$

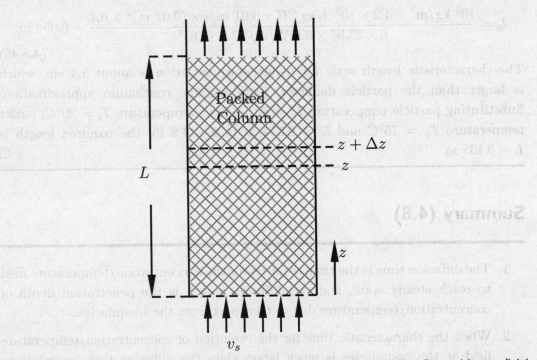

FIGURE 4.22. Schematic of a packed column of height L through which fluid flows with superficial velocity v_s. The balance equation, Eq. 4.8.41, is written for the differential volume of area of cross section A and thickness Δz.

where T_i is the fluid temperature at the inlet. The outlet temperature T_o at $z = L$ is,

$$T_o = T_p - (T_p - T_i) \, e^{(-L/l_c)}. \qquad (4.8.45)$$

The characteristic length l_c is calculated as follows. The Reynolds (Eq. 2.2.36) and Prandtl numbers for the flow of water through the packed column are,

$$\text{Re} = \frac{\rho v_s d}{(1 - \varepsilon)\mu} = \frac{10^3 \text{ kg/m}^3 \times 0.01 \text{ m/s} \times 10^{-2} \text{ m}}{0.6 \times 10^{-3} \text{ kg/m/s}} = 1.67 \times 10^2,$$

$$\text{Pr} = \frac{4.2 \times 10^3 \text{ J/kg/}^\circ\text{C} \times 10^{-3} \text{ kg/m/s}}{0.6 \text{ W/m/}^\circ\text{C}} = 7.0. \qquad (4.8.46)$$

The Nusselt number from correlation, Eq. 2.3.34, is

$$\text{Nu} = (0.5 \times (1.67 \times 10^2)^{1/2} + 0.2 \times (1.67 \times 10^2)^{2/3}) \times (7.0)^{1/3} = 23.96. \quad (4.8.47)$$

The characteristic length, Eq. 4.8.43, is

$$l_c = \frac{10^3 \text{ kg/m}^3 \times 4.2 \times 10^3 \text{ J/kg/}^\circ\text{C} \times 0.01 \text{ m/s} \times (0.01 \text{ m})^2 \times 0.4}{6 \times 23.96 \times 0.6 \text{ W/m/}^\circ\text{C} \times (0.6)^2} = 0.054 \text{ m}.$$
$$(4.8.48)$$

The characteristic length scale for temperature variation is about 5.4 cm, which is larger than the particle diameter, justifying the continuum approximation. Substituting particle temperature $T_p = 80^\circ\text{C}$, inlet temperature $T_i = 20^\circ\text{C}$, outlet temperature $T_o = 75^\circ\text{C}$ and $l_c = 0.054$ m in Eq. 4.8.45, the required length is $L = 0.135$ m. $\qquad\qquad\qquad\qquad\qquad\qquad\qquad\qquad\qquad\qquad\qquad\qquad\square$

Summary (4.8)

1. The diffusion time is the time required for the concentration/temperature field to reach steady state, and the diffusion length is the penetration depth of concentration/temperature due to transport from the boundaries.

2. When the characteristic time for the evolution of concentration/temperature field at the boundaries is much larger than the diffusion time, correlations derived using the steady-state approximation can be used for the instantaneous flux in balance equations.

3. When the length scale for the variation in the concentration/temperature field is much larger than the diffusion length, correlations based on the local value of the concentration/temperature can be used in balance equations.

Exercises

EXERCISE 4.1 Consider a composite slab consisting of two layers of different thicknesses and thermal conductivities shown in Fig. 4.3(b). The thermal contact between the two layers is not perfect, and the temperatures in the two layers at the interface are different. The 'contact resistance' is defined as $\mathcal{R}_c = -(\Delta T/q_z)$, where ΔT is the difference in the temperatures of the two layers at the interface and q_z is the heat flux. The thermal conductivities and thicknesses of the two slabs are $k_1 = 0.6$ W/m/$°$C, $k_2 = 2$ W/m/$°$C, $h_1 = 1$ cm, $h_2 = 1.5$ cm. When the temperature difference $T_h - T_0 = 80°$C, the heat flux is $q_z = 3.2 \times 10^3$ W/m^2. What is the contact resistance?

EXERCISE 4.2 Instead of a multi-layer configuration shown in Fig. 4.3(c), consider a continuous variation of the thermal conductivity $k(z)$ with z. The slab can be divided layers of infinitesimal thickness Δz, and Eq. 4.2.6 relates the temperature difference ΔT and Δz for each of these layers. A differential equation is obtained by taking the limit $\Delta z \to 0$, and the temperature difference $T_h - T_0$ is determined by integrating this equation. Determine the effective thermal conductivity k_{eff} defined as

$$k_{eff} = -\frac{Q_z}{(A(T_h - T_0)/h)}.$$

What is k_{eff} if $k(z) = k_0 + k_1 z$?

EXERCISE 4.3 The gap of thickness h between two conducting plates of area A and temperatures T_0 and T_h can be filled by slabs of two materials, of thermal conductivities k and rk, in two possible configurations shown in Fig. 4.23. Here, r is the ratio of the conductivities of the two materials. In the parallel configuration, shown in 4.23 (a), the thickness of both slabs is h, and the areas of cross section of the two slabs are Af and $A(1 - f)$, respectively, where $f < 1$. In the series configuration, shown in 4.23 (b), the slabs have equal area, and thickness fh and $(1 - f)h$ respectively. Which configuration has a higher heat transfer rate? *Hint:* Write down the expression for the difference in the heat transfer rates in the two configurations, and examine whether this is positive or negative.

EXERCISE 4.4 A solute diffuses through a stationary gas film of thickness h_g, and a liquid layer of thickness h_l, and reacts at a catalyst surface as shown in Fig. 4.24. The diffusion coefficients of the solute in the gas and liquid layer are $\mathcal{D}_g$ and $\mathcal{D}_l$, respectively. At the interface, the ratio of the concentrations in the liquid and gas is $(c_l/c_g) = P$, where P is the partition coefficient. At the catalyst surface $z = 0$, the flux is equal to the reaction rate at the surface (mass reacted per unit area per unit time), $j_{lz}|_{z=0} = -\mathcal{K}c_l$, where j_{lz} is the

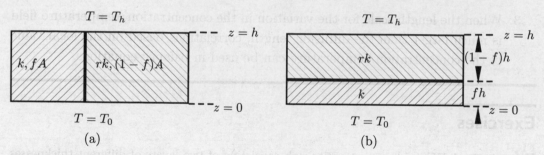

FIGURE 4.23. The parallel (a) and series (b) configurations of two slabs with thermal conductivities k and rk between two conducting plates with area A separated by a distance h.

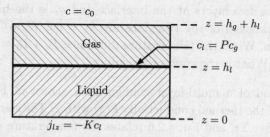

FIGURE 4.24. A solute diffuses though a stationary gas layer of height h_g, a liquid layer of height h_l and then reacts at a catalyst surface at $z = 0$.

mass flux in the liquid and $\mathcal{K}$ is the reaction rate. At the top of the gas film, $z = h_g + h_l$, the concentration of the solute is c_0. Determine the rate of reaction of the solute per unit area of the catalyst surface at steady state.

EXERCISE 4.5 Two fluids with viscosities μ_1 and μ_2 and heights h_1 and h_2 are confined in a channel, as shown in Fig. 4.25. The boundary condition at the bottom of the channel $z = 0$ is,

$$\frac{dv_x}{dz} = \frac{v_s}{l_s},$$

where l_s is the slip length, as shown in Fig. 4.25. If a shear stress τ_0 is applied on the top plate, determine the velocity of the top plate.

EXERCISE 4.6 A layer of fluid of thickness 1 cm in the z direction and of infinite extent in the other two directions, with thermal conductivity 0.6 W/m/K, has one insulating surface at z=1 cm, and a surface at constant temperature 80°C, at z=0, as shown in Fig. 4.26. The fluid is irradiated by microwave, with the volumetric heat source (heat input per unit volume per unit time) of 0.12 W/cm³. What is the temperature at the insulating surface at $z = 1$ cm in °C at steady state?

EXERCISE 4.7 A lubricating oil with density 900 kg/m³, viscosity 2 kg/m/s, specific heat 1.8×10^3 J/kg/°C and thermal conductivity 0.15 W/m/°C is used to lubricate the gap

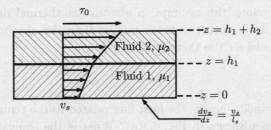

FIGURE 4.25. The flow of two fluid layers of height h_1 and h_2 and viscosities μ_1 and μ_2 in a channel with a constant stress τ_0 at the top plate, and a slip condition at the bottom surface.

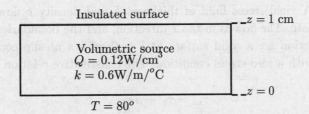

FIGURE 4.26. Microwave heating of a fluid layer.

between two surfaces separated by a distance of 1 mm moving with velocity V relative to each other. The temperature of the surfaces is 25°C. If the maximum temperature within the oil cannot exceed 75°C, what is the maximum velocity?

EXERCISE 4.8 For the flow down an inclined plane, the velocity profile $v_x(z)$ is given by Eq. 4.2.23. The rate of dissipation of energy per unit volume due to fluid friction is given by Eq. 4.2.14. Determine the temperature in the fluid from the energy balance equation, with boundary conditions $T = T_0$ at the bottom surface $z = 0$ and the no-flux condition $q_z = 0$ at the top surface $z = h$. Determine the relation between the Nusselt number and Brinkman number.

EXERCISE 4.9 For the reaction-diffusion system considered in Example 4.2.6, determine the concentration field c_A for a second order reaction, where the reaction rate is $r_A = -\mathcal{K}c_A^2$. Assume that the penetration depth of the reactant is much smaller than the height h, so that the boundary conditions are $c_A = c_s$ at $z = 0$, and $c_A \to 0$ and $(\mathrm{d}c_A/\mathrm{d}z) \to 0$ as $z \to \infty$, where z is the distance from the interface. What is the expression for the Sherwood number?

EXERCISE 4.10 A fluid film of thickness 2 mm is flowing down an inclined plane under gravity at steady state with a flow rate of 25 ml/s. The film is bounded by a solid surface with a no-slip condition at the bottom, and there is a free surface with zero shear stress on top. When the film thickness is increased to 4 mm, the flow rate is 141.42 ml/s. If the fluid is described by a power-law constitutive relation Eq. 4.2.46, what is the power-law index n?

EXERCISE 4.11 In Chapter 3, it was shown that the thermal diffusivity does increase with absolute temperature $\propto T^{1/2}$ for hard sphere molecules, and as a power of T for other molecular models. Therefore, the assumption of constant thermal diffusivity is valid only when the difference in temperature is much smaller than the absolute temperature. If we assume a power-law model for the thermal conductivity,

$$k = k_T T^n,$$

where n is a power-law index, how is the heat flux related to the temperature difference for a single slab configuration shown in Fig. 4.3(a)? What is the equivalent of Eqs. 4.2.8 and 4.2.10 for a multi-layer configuration shown in Fig. 4.3(c), if the values of k_T are different for each layer, but n is the same for all the layers.

EXERCISE 4.12 A yield stress fluid of thickness h and density ρ flows down a vertical surface at steady state. The flow is in the x direction, and the boundaries of the film in the cross-stream z direction are a solid surface at $z = 0$ with a no-slip condition, and a free interface at $z = h$ with a zero stress condition. The constitutive relation for the stress is,

$$\tau_{xz} = \tau_Y + \mu \frac{dv_x}{dz},$$

when the stress is greater than the yield stress τ_Y, and the strain rate is zero when the stress is less than the yield stress. Here, μ is the viscosity.

a) Determine the stress profile across the fluid layer.

b) What is the minimum film thickness for flow?

c) Determine the velocity profile and the maximum and average velocity.

EXERCISE 4.13 A container of volume 1 l, filled with pure oxygen at 1 atm and 300 K, has an outlet in the form of a tube of length 20 cm and diameter 3 mm. When the outlet is opened, there is diffusion of oxygen from the container into the air with molar composition 20% oxygen and 80% nitrogen. How long does it take for the mole fraction of oxygen in the container to decrease to 50%? The diffusion coefficient for oxygen in air is 2×10^{-5} m^2/s.

EXERCISE 4.14 Consider a binary mixture of A and B with molecular weights m_A and m_B, in which the total molar density C is a constant. How is the total mass density ρ related to the mass fraction w_A? If j_{Az} and j_{Bz} are related, $j_{Bz} = S_c j_{Az}$, where S_c is a stoichiometric coefficient, derive the equivalent of Eq. 4.3.4 for the mass flux so that it depends only on the mass fraction of A, total molar density, the diffusion coefficient and the ratio of molecular weights.

EXERCISE 4.15 In order to determine the viscosity of the liquid of density 10^3 kg/m^3, the liquid is placed on a surface which is initially stationary. At $t = 0$, the surface is moved with a velocity 1 cm/s, and the stress on the surface is measured as a function of time. If the stress is 0.01 N/m^2 at time $t = 10$ s, what is the viscosity? Consider the liquid to be of infinite extent in the cross-stream direction.

EXERCISE 4.16 A fluid with thermal conductivity 0.6 W/m/°C, density 10^3 kg/m^3 and specific heat 4.2×10^3 J/kg/°C is placed in a container at room temperature 20°C and the top of the container is in contact with a heated surface. The temperature of the surface is instantaneously increased to 40°C at $t = 0$. How much time does it take to transfer 100 kJ/m^2 heat from the surface to the fluid? Assume that conduction takes place into an infinite medium.

EXERCISE 4.17 Consider the heat conduction in an infinite medium, shown in Fig. 4.10, where the temperature is a constant far from the surface, and the surface temperature at $z = 0$ has the form $T_0 = f(t)$, where $f(t)$ is a function of time. For what class of functions $f(t)$ can the similarity transform be used? What is the similarity equation, the equivalent of Eq. 4.4.8, in this case?

EXERCISE 4.18 In a liquid–gas contactor, a pollutant from the gas diffuses into a downward flowing vertical liquid film of length L and thickness h. The mass of gas absorbed per unit time is m when the volumetric flow rate in the liquid film is $\dot{V}$. If it is desired to increase the mass of gas absorbed per unit time to $2m$ by increasing the film thickness while keeping the length unchanged, what should be the flow rate of the liquid? Assume that the flow in the film is laminar.

EXERCISE 4.19 A liquid stream of length L and diameter d (which decreases as the stream accelerates due to gravity) falls downwards in a gas containing a soluble species, as shown in Fig. 4.27. The velocity of the liquid is zero and the diameter is d_0 at the start of the contacting section $x = 0$, and the velocity is $\sqrt{2gx}$ at the location x, where g is the acceleration due to gravity. The soluble species dissolves in the liquid stream and diffuses in with diffusion coefficient $\mathcal{D}$. The concentration of the solute at the gas–liquid surface is c_s. If the penetration depth of the solute is much smaller than the diameter of the stream, determine the expression for the Sherwood number. **Hint:** Use the analogy $v_x\,dt = dx$ similar to that used in Section 4.4.1, assume that the similarity variable is $(z/(Ax^\beta))$, and determine the values of A and β for a similarity solution.

EXERCISE 4.20 A pulse of solute of mass 1 g and dispersion coefficient 10^{-5} m^2/s of infinitesimal width is injected uniformly across a packed bed reactor of cross sectional area 100 cm^2. The pulse disperses axially as it is carried downstream at a velocity of 1 cm/s. After 20 s, what is the solute concentration in kg/m^3 at a point 18 cm from the point of injection? Assume that the dispersion is one dimensional.

EXERCISE 4.21 A pulse of pollutant of mass 100 kg and dispersion coefficient 10^{-2} m^2/s is injected from a smoke stack at a height of 10 m. After 1 minute, what is the concentration in kg/m^3 of the pollutant at a distance 1 m from the smokestack? Assume the dispersion is three-dimensional.

EXERCISE 4.22 Consider the decay of a pulse of solute, discussed in Section 4.5, where the solute reacts with the medium and disappears with first order rate constant $\mathcal{K}$. The conservation Eq. 4.5.1 is modified as,

$$\frac{\partial c}{\partial t} = \mathcal{D}\frac{\partial^2 c}{\partial z^2} - \mathcal{K}c,$$

with initial condition Eq. 4.5.14.

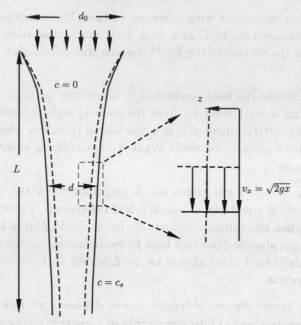

FIGURE 4.27. Absorption of a gas into a liquid stream freely falling under gravity with velocity $v_x = \sqrt{2gx}$ and diameter $d(x)$, where x is the downstream distance from the outlet, and d_0 is the diameter at the outlet.

a) The total mass of solute is no longer a constant. What is the variation of the total mass with respect to time?

b) Modify Eq. 4.5.5 for the scaled concentration to incorporate the effect of the reaction.

c) What is the solution for the concentration field?

EXERCISE 4.23 How long does it take to melt a copper ingot of height 20 cm, if it is brought into contact with a surface of temperature 1200°C? The properties of copper are as follows: melting point $T_m = 1085$°C, latent heat $\lambda = 2.08 \times 10^5$ J/kg, thermal conductivity $k = 350$ W/m/°C, density $\rho = 9000$ kg/m^3 and specific heat $C_p = 385$ J/kg/°C.

EXERCISE 4.24 The solution for the temperature field in Section 4.6, Eq. 4.6.11, depends on the parameter $P = (h/\alpha)(dh/dt)$, which is the Peclet number based on the height and the velocity of the interface.

a) When the Peclet number is small ($P \ll 1$), the motion of the interface can be neglected in the energy balance equation, and energy conduction can be considered a steady-state process. Verify that Eq. 4.6.11 for the scaled temperature profile reduces to the steady profile between two flat surfaces at different temperatures, Eq. 4.7.9, in this limit.

b) When the Peclet number is large ($P \gg 1$), the velocity of the interface is large compared to the ratio of the diffusion coefficient and the height. The heat transfer at

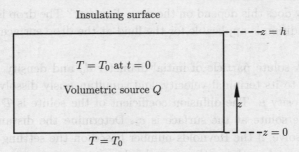

FIGURE 4.28. Microwave heating of a fluid bounded by a constant temperature wall at $z = 0$ and insulating wall at $z = h$. The fluid is initially at temperature T_0, and the volumetric heat source per unit volume per unit time is Q.

the interface resembles the transport into an infinite medium. Verify that Eq. 4.6.11 reduces to Eq. 4.4.16 for the conduction into an infinite medium.

EXERCISE 4.25 A fluid of density ρ, specific heat C_p, thermal conductivity k, and thermal diffusivity α is in a rectangular box of height h, where the bottom wall at $z = 0$ is at a fixed temperature T_0, the top wall at $z = h$ is insulating. The fluid temperature is T_0 at $t = 0$. The fluid is placed in a microwave, and heated with a volumetric heat source Q per unit volume per unit time, as shown in Fig. 4.28. The box is sufficiently wide that the temperature can be considered uniform along planes of constant z. Determine the steady and transient temperature profiles.

EXERCISE 4.26 The concentration profile at steady state for the diffusion and reaction of a solute in a liquid was calculated in Example 4.2.6. Determine the concentration field during the development of the concentration profile from the initial state of zero concentration, $c = 0$ for $0 < z < h$ and $c = c_s$ at $z = h$.

EXERCISE 4.27 The velocity profile for the flow down an inclined plane at steady state was calculated in Example 4.2.3. For this configuration, calculate the velocity profile during flow development starting from rest till steady state.

EXERCISE 4.28 A naphthalene ball of diameter 0.5 cm and mass density 1000 kg/m^3 sublimes in air at 20°C in 40 days. What is the diffusion coefficient of naphthalene in air in cm^2/s? The molecular weight of naphthalene is 128 g/mol, and the vapour pressure is 8 Pa.

EXERCISE 4.29 In Example 4.8.4, what is the condition for the timescale for the decrease of the particle diameter to be large compared to the timescale for the development of the concentration field in the boundary layer?

EXERCISE 4.30 Molten metal of drops of diameter d, density ρ_m, viscosity μ_m, melting temperature T_m and latent heat λ_m are dropped into a cold oil with density ρ_f, viscosity μ_f, temperature T_f, thermal conductivity k_f and specific heat C_{pf}. Assume the drops fall at their terminal velocity, the Reynolds number is small and the Peclet number based on this terminal velocity is large. What is the distance travelled by the drop before it freezes completely, and how does this depend on the drop diameter? The drop is sufficiently viscous that the no-slip condition is applicable for the fluid at the drop surface.

EXERCISE 4.31 A solute particle of initial diameter d_0 and density ρ_p settles in a fluid with velocity equal to its terminal velocity, and simultaneously dissolves in the fluid with density ρ_f and viscosity μ. The diffusion coefficient of the solute is $\mathcal{D}$ and the saturation concentration of the solute at the surface is c_s. Determine the distance required for the particle to fully dissolve, if the Reynolds number based on the settling speed is small, and the Peclet number based on the settling speed is large.

EXERCISE 4.32 For the configuration shown in Fig. 4.21, if the heat transfer is due to forced convection by fluid flow with free stream velocity v_{fs}, determine the characteristic length l_c in the conduction Eq. 4.8.32.

EXERCISE 4.33 A spherical particle of diameter 0.5 cm at initial temperature 200°C is cooled in air due to natural convection at temperature 25°C. It takes 5 minutes for the temperature to decrease to 75°C. If a spherical particle of the same material of diameter 1 cm with the same material properties and initial temperature is cooled under the same ambient conditions, how long does it take to reach a final temperature of 75°C?

EXERCISE 4.34 A steel sphere of diameter 1 cm at initial temperature 100°C is cooled due to natural convection in air at temperature 20°C. Estimate the time taken for the sphere temperature to decrease to 50°C. If the cooling were due to heat conduction instead of natural convection, what is the time taken for cooling? The density and specific heat of steel are $\rho_s = 8 \times 10^3$ kg/m^3 and $C_{ps} = 500$ J/kg/°C, respectively, the density, viscosity, thermal conductivity, specific heat and thermal expansion coefficient of air are $\rho_a = 1.2$ kg/m^3, $\mu_a = 2 \times 10^{-5}$ kg/m/s, $k_a = 0.025$ W/m/°C, $C_{pa} = 10^3$ J/kg/°C and $\beta = 2 \times 10^{-3}$(°C^{-1}), respectively. Assume that $C(\text{Pr}) = 1$ in Eq. 2.4.12.

EXERCISE 4.35 Consider the heat transfer from a fin analysed in Example 4.8.6. The area of cross section is increased to $2A_{cs}$ while the shape of the cross section is unchanged. What should be the length so that the temperatures at the pipe surface and the insulated end of the fin are unchanged? What is the heat flux in this case?

EXERCISE 4.36 A catalytic converter is used to oxidise carbon monoxide in the exhaust of an automobile. The converter consists of catalyst particles of diameter 2 mm packed in a cylindrical column of diameter 4 cm with void fraction 0.45. The exhaust, which is generated at the rate 0.01 kg/s, temperature 400°C and atmospheric pressure, contains about 70% nitrogen, 14% water, 14% carbon dioxide and 2% carbon monoxide. What should be the length of the catalytic converter so that the carbon monoxide concentration decreases to 0.1%? Assume that the reaction at the catalyst surface is fast, and the diffusion of carbon monoxide to the catalyst surface is the rate-limiting step. The viscosity of the exhaust gases is 3.25×10^{-5} kg/m/s, and the diffusivity of carbon monoxide is 8.75×10^{-5} m^2/s.

Appendix

4.A   Gaussian and Error Functions

The normalised Gaussian function with unit variance is defined as,

$$\mathcal{G}_0(x) = \frac{1}{\sqrt{2\pi}}\, e^{(-x^2/2)}. \tag{4.A.1}$$

This satisfies the normalisation condition,

$$\int_{-\infty}^{\infty} dx\, \mathcal{G}_0(x) = 1, \tag{4.A.2}$$

has zero mean,

$$\int_{-\infty}^{\infty} dx\, x\, \mathcal{G}_0(x) = 0, \tag{4.A.3}$$

and it has variance and standard deviation equal to 1,

$$\int_{-\infty}^{\infty} dx\, x^2 \mathcal{G}_0(x) = 1. \tag{4.A.4}$$

The Gaussian function is a bell-shaped curve, as shown in Fig. 4.29(a). It is an even function of x, which decreases to zero for $x \to \pm\infty$, with a maximum value $(1/\sqrt{2\pi}) \sim 0.399$ at the $x = 0$.

The general form of the Gaussian function is,

$$\mathcal{G}(x) = \frac{1}{\sqrt{2\pi}\sigma}\, e^{(-(x-\mu)^2/2\sigma^2)}. \tag{4.A.5}$$

where μ is the mean and σ is the standard deviation. The general form can be reduced to a Gaussian function with zero mean and unit variance using the transformation $x' = (x - \mu)/\sigma$. It is easy to show that the Gaussian function 4.A.5 is normalised with mean and standard deviation,

$$\int_{-\infty}^{\infty} dx\, \mathcal{G}(x) = 1, \tag{4.A.6}$$

$$\int_{-\infty}^{\infty} dx\, x\, \mathcal{G}(x) = \mu, \tag{4.A.7}$$

$$\int_{-\infty}^{\infty} dx\, x^2 \mathcal{G}(x) = \mu^2 + \sigma^2. \tag{4.A.8}$$

The integrals of the Gaussian function from $x = 0$ to ∞ are required for some calculations, such as those for the moments in curvilinear co-ordinates. The general

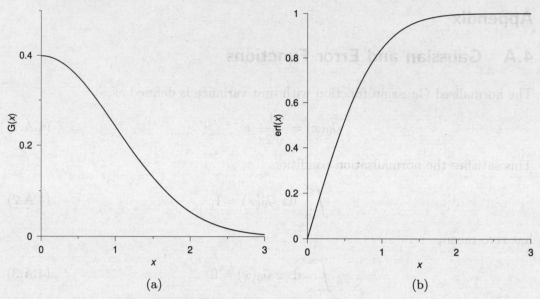

FIGURE 4.29. The Gaussian function $\mathcal{G}(x)$ (a) and the error function erf(x) (b) as a function of x.

TABLE 4.3. The values of the integral, Eq. 4.A.9, for different values of n where the expression for $\mathcal{G}_0$ is given in Eq. 4.A.1.

n	$\int_0^\infty \mathrm{d}x\, x^n\, \mathcal{G}_0(x)$	n	$\int_0^\infty \mathrm{d}x\, x^n\, \mathcal{G}_0(x)$
0	$\frac{1}{2}$	1	$\frac{1}{\sqrt{2\pi}}$
2	$\frac{1}{2}$	3	$\sqrt{\frac{2}{\pi}}$
4	$\frac{3}{2}$	5	$4\sqrt{\frac{2}{\pi}}$

form of these definite integrals is,

$$\int_0^\infty \mathrm{d}x\, x^n\, \mathcal{G}_0(x) = \frac{2^{(n-1)/2}}{\sqrt{2\pi}} \left(\frac{n-1}{2}\right)! \text{ for odd} \, n,$$

$$= \frac{(n-1)!!}{2} \text{ for even} \, n, \qquad (4.A.9)$$

where $m!! = m \times (m-2) \times (m-4) \times \ldots \times 3 \times 1$. The values of the right side of 4.A.9 for $n = 0$ to 5 are listed in Table 4.3.

The error function is defined as,

$$\mathrm{erf}(x) = \frac{2}{\sqrt{\pi}} \int_0^x \mathrm{d}x' \mathrm{e}^{(-x'^2)}, \qquad (4.A.10)$$

where, x' is the variable of integration. The prefactor of the integral in 4.A.10 is chosen so that the erf(x) = 1 for $x \to \infty$. This error function is one of the solutions of the differential equation

$$\frac{\mathrm{d}^2 f}{\mathrm{d}x^2} = -2x\frac{\mathrm{d}f}{\mathrm{d}x}, \qquad (4.A.11)$$

for the function $f(x)$; the other solution is a constant. The solution for the equation

$$\frac{\mathrm{d}^2 f}{\mathrm{d}x^2} = -Ax\frac{\mathrm{d}f}{\mathrm{d}x}, \qquad (4.A.12)$$

where A is a constant, can be obtained by transforming the independent variable as $x' = (x\sqrt{A/2})$; the differential equation for x' is,

$$\frac{\mathrm{d}^2 f}{\mathrm{d}x'^2} = -2x'\frac{\mathrm{d}f}{\mathrm{d}x'}. \qquad (4.A.13)$$

The solution of the above equation is

$$f = \mathrm{erf}(x') = \mathrm{erf}(x\sqrt{A/2}). \qquad (4.A.14)$$

The derivative of the error function with respect to x is the value of the integrand on the right side of 4.A.10,

$$\frac{\mathrm{derf}(x)}{\mathrm{d}x} = \frac{2e^{(-x^2)}}{\sqrt{\pi}}. \qquad (4.A.15)$$

The higher derivatives can be determined from Eq. 4.A.10 and its derivatives. The error function has value 0 and slope $2/\sqrt{\pi}$ at $x = 0$, and it monotonically increases to 1 for $x \to \infty$, as shown in Fig. 4.29 (b).

The numerical values of the Gaussian and error functions are provided in Table 4.4.

4.B   Orthogonality Relation

The reason for the 'orthogonality' relation, Eq. 4.7.25, is as follows. The basis functions $\Psi(\beta_i z^*) = A\sin(\beta_i z^*) + B\cos(\beta_i z^*)$ are solutions of the equation

$$\frac{\mathrm{d}^2\Psi(\beta_i z^*)}{\mathrm{d}z^{*2}} = -\beta_i^2\Psi(\beta_i z^*), \qquad (4.B.1)$$

with homogeneous boundary conditions

$$\Psi(\beta_i z^*) = 0 \;\text{ or }\; \frac{\mathrm{d}\Psi(\beta_i z^*)}{\mathrm{d}z^*} = 0 \;\text{ or }\; \Psi(\beta_i z^*) = C\frac{\mathrm{d}\Psi(\beta_i z^*)}{\mathrm{d}z^*}, \qquad (4.B.2)$$

at the two boundaries, which are considered $z^* = 0$ and $z^* = 1$, where z^* is the scaled co-ordinate and C is a constant. The 'scalar product' of two functions $\Psi(\beta_i z^*)$ and

TABLE 4.4. The Gaussian and error functions.

x	$\frac{1}{\sqrt{2\pi}}e^{(-x^2/2)}$	$\mathrm{erf}(x)$
0.	0.3989	0.00000
0.025	0.3988	0.02820
0.05	0.3984	0.05637
0.075	0.3978	0.08447
0.1	0.3970	0.1125
0.125	0.3958	0.1403
0.15	0.3945	0.1680
0.175	0.3929	0.1955
0.2	0.3910	0.2227
0.225	0.3890	0.2497
0.25	0.3867	0.2763
0.275	0.3841	0.3027
0.3	0.3814	0.3286
0.325	0.3784	0.3542
0.35	0.3752	0.3794
0.375	0.3719	0.4041
0.4	0.3683	0.4284
0.425	0.3645	0.4522
0.45	0.3605	0.4755
0.475	0.3564	0.4983
0.5	0.3521	0.5205
0.525	0.3476	0.5422
0.55	0.3429	0.5633
0.575	0.3382	0.5839
0.6	0.3332	0.6039
0.625	0.3282	0.6232
0.65	0.3230	0.6420
0.675	0.3177	0.6602
0.7	0.3123	0.6778
0.725	0.3067	0.6948
0.75	0.3011	0.7112
0.775	0.2955	0.7269
0.8	0.2898	0.7421
0.825	0.2839	0.7567
0.85	0.2780	0.7707
0.875	0.2721	0.7841

x	$\frac{1}{\sqrt{2\pi}}e^{(-x^2/2)}$	$\mathrm{erf}(x)$
0.9	0.2661	0.7969
0.925	0.2601	0.8092
0.95	0.2541	0.8209
0.975	0.2480	0.8321
1.0	0.2420	0.8427
1.1	0.2179	0.8802
1.2	0.1942	0.9103
1.3	0.1714	0.9340
1.4	0.1497	0.9523
1.5	0.1295	0.9661
1.6	0.1109	0.9764
1.7	0.09405	0.9838
1.8	0.07895	0.9891
1.9	0.06562	0.9928
2.0	0.05399	0.9953
2.1	0.04398	0.9970
2.2	0.03548	0.9981
2.3	0.02833	0.9989
2.4	0.02240	0.9993
2.5	0.01753	0.9996
2.6	0.01358	0.9998
2.7	0.01042	0.9999
2.8	0.007916	0.9999
2.9	0.005953	1.0000
3.0	0.004432	1.0000
3.1	0.003267	1.0000
3.2	0.002384	1.0000
3.3	0.001723	1.0000
3.4	0.001232	1.0000
3.5	0.0008727	1.0000
3.6	0.0006119	1.0000
3.7	0.0004248	1.0000
3.8	0.0002920	1.0000
3.9	0.0001987	1.0000
4.0	0.0001338	1.0000

$\Psi(\beta_j z^*)$ is defined as,

$$\langle \Psi(\beta_i z^*), \Psi(\beta_j z^*) \rangle = \int_0^1 \mathrm{d}z^* \Psi(\beta_i z^*) \Psi(\beta_j z^*). \tag{4.B.3}$$

The scalar product of $\Psi(\beta_j z^*)$ times Eq. 4.B.1 is,

$$\int_0^1 \mathrm{d}z^* \Psi(\beta_j z^*) \frac{\mathrm{d}^2 \Psi(\beta_i z^*)}{\mathrm{d}z^2} = -\beta_i^2 \int_0^1 \mathrm{d}z^* \Psi(\beta_i z^*) \Psi(\beta_j z^*). \tag{4.B.4}$$

The left side of the above equation is expressed in an alternate form by integration by parts,

$$
\int_0^1 dz^* \Psi(\beta_j z^*) \frac{d^2 \Psi(\beta_i z^*)}{dz^{*2}} = \Psi(\beta_j z^*) \frac{d\Psi(\beta_i z^*)}{dz^*}\bigg|_0^1 - \Psi(\beta_i z^*) \frac{d\Psi(\beta_j z^*)}{dz^*}\bigg|_0^1
$$

$$
+ \int_0^1 dz^* \Psi(\beta_i z^*) \frac{d^2 \Psi(\beta_j z^*)}{dz^2}
$$

$$
= \Psi(\beta_j z^*) \frac{d\Psi(\beta_i z^*)}{dz^*}\bigg|_0^1 - \Psi(\beta_i z^*) \frac{d\Psi(\beta_j z^*)}{dz^*}\bigg|_0^1
$$

$$
- \beta_j^2 \int_0^1 dz^* \Psi(\beta_j z^*) \Psi(\beta_i z^*). \tag{4.B.5}
$$

Equating the right sides of 4.B.4 and 4.B.5, and dividing the resulting equation by $\beta_j^2 - \beta_i^2$, we obtain,

$$
\int_0^1 dz^* \Psi(\beta_j z^*) \Psi(\beta_i z^*) = \frac{\Psi(\beta_j z^*) \frac{d\Psi(\beta_i z^*)}{dz^*}\big|_0^1 - \Psi(\beta_i z^*) \frac{d\Psi(\beta_j z^*)}{dz^*}\big|_0^1}{\beta_j^2 - \beta_i^2}. \tag{4.B.6}
$$

The numerator of the right side of the above equation is zero for homogeneous boundary conditions specified in 4.B.2. Therefore, the integral on the left in 4.B.6 is zero for $i \neq j$, and the scalar product 4.B.3 is,

$$
\langle \Psi(\beta_i z^*), \Psi(\beta_j z^*) \rangle = N_i \delta_{ij}, \tag{4.B.7}
$$

where Kronecker delta $\delta_{ij} = 1$ for $i = j$, and $\delta_{ij} = 0$ for $i \neq j$. The normalisation factor N_i depends on the specific form of the basis functions.

5

Unidirectional Transport: Curvilinear Co-ordinates

In the previous chapter, a Cartesian co-ordinate system was used to analyse the transport between surfaces of constant co-ordinate, and the boundary conditions were specified at a fixed value of the co-ordinate z. For configurations with curved boundaries, such as a cylindrical pipe or a spherical particle, the boundaries are not surfaces of constant co-ordinate in a Cartesian system. It is necessary to apply boundary conditions at, for example, $x^2 + y^2 + z^2 = R^2$ for the diffusion around a spherical particle of radius R. It is simpler to use a co-ordinate system where one of the co-ordinates is a constant on the boundary, so that the boundary condition can be applied at a fixed value of the co-ordinate. Such co-ordinate systems, where one or more of the co-ordinates is a constant on a curved surface, are called curvilinear co-ordinate systems.

The procedure for deriving balance laws for a Cartesian co-ordinate system can be easily extended to a curvilinear co-ordinate system. First, we identify the differential volume or 'shell' between surfaces of constant co-ordinate separated by an infinitesimal distance along the co-ordinate. The balance equation is written for the change in mass/momentum/energy in this differential volume in a small time interval Δt. The balance equation is divided by the volume and Δt to derive the differential equation for the field variable. The balance equations for the cylindrical and spherical co-ordinate system are derived in this chapter, and the solution procedures discussed in Chapter 4 are applied to curvilinear co-ordinate systems.

5.1 Cylindrical Co-ordinates

5.1.1 Conservation Equation

A cylindrical surface is characterised by a constant distance from an axis, which is the x axis in Fig. 5.1. It is natural to define one of the co-ordinates r as the distance from the axis, and a second co-ordinate x as the distance along the axis. The third co-ordinate ϕ, which is the angle around the x axis, is considered later in Chapter 7. For unidirectional transport, we consider a variation of concentration, temperature, or velocity only in the r direction and in time, and there is no dependence on ϕ and x.

The energy balance equation is derived for the temperature variation in a cylindrical shell of height Δx and thickness Δr between the cylindrical surfaces at r and $r + \Delta r$, shown in Fig. 5.1. The balance equation, Eq. 4.1.11, is applied to the annular differential volume. The surface areas of the inner and outer surfaces of the shell are $2\pi r \Delta x$ and $2\pi(r + \Delta r)\Delta x$, respectively, and the volume is $2\pi r \Delta r \Delta x$. The change in energy between the time instants t and $t + \Delta t$ is the product of the difference in the energy density $\rho C_p T$ and the volume of the cylindrical shell $2\pi r \Delta r \Delta x$,

$$\begin{pmatrix} \text{Change in energy} \\ \text{in differential volume} \end{pmatrix} = \rho C_p (T(r, t + \Delta t) - T(r, t)) 2\pi r \Delta r \Delta x. \qquad (5.1.1)$$

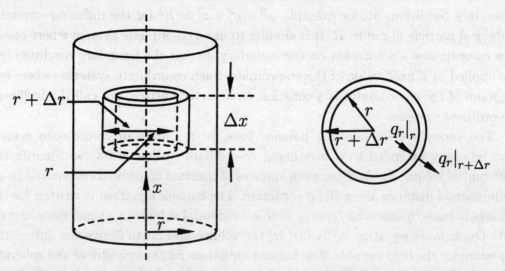

FIGURE 5.1. The co-ordinates and the differential volume for deriving the conservation equation in a cylindrical co-ordinate system.

The energy entering the differential volume at the surface at r is the product of the heat flux, the surface area $2\pi r \Delta x$, and the time interval Δt,

$$\left(\text{Energy in}\right) = q_r|_r \left(2\pi r \Delta x \Delta t\right), \qquad (5.1.2)$$

where $q_r|_r$ is the heat flux entering the differential volume at the surface at r. The energy leaving the differential volume at the surface at $r + \Delta r$ is the product of the flux, the surface area $2\pi(r + \Delta r)\Delta x$ and the time interval Δt,

$$\left(\text{Energy out}\right) = q_r|_{r+\Delta r} \left(2\pi(r + \Delta r)\Delta x \Delta t\right), \qquad (5.1.3)$$

where $q_r|_{r+\Delta r}$ is the heat flux leaving the differential volume at the surface at $r + \Delta r$. The production of energy in the differential volume is,

$$\left(\begin{array}{l}\text{Production of energy} \\ \text{in differential volume}\end{array}\right) = \mathcal{S}_e(2\pi r \Delta r \Delta x \Delta t), \qquad (5.1.4)$$

where $\mathcal{S}_e$ is the amount of energy generated per unit volume per unit time. When these are substituted into the conservation Eq. 4.1.11, and divided by $2\pi r \Delta r \Delta x \Delta t$, the difference equation is,

$$\rho C_p \frac{(T(z, t + \Delta t) - T(z, t))}{\Delta t} = \frac{1}{r \Delta r}\left(r\, q_r|_r - (r + \Delta r)\, q_r|_{r+\Delta r}\right) + \mathcal{S}_e(r, t). \quad (5.1.5)$$

Taking the limit $\Delta r \to 0$ and $\Delta t \to 0$, the partial differential equation for the temperature field is,

$$\boxed{\rho C_p \frac{\partial T}{\partial t} = -\frac{1}{r}\frac{\partial(r q_r)}{\partial r} + \mathcal{S}_e.} \qquad (5.1.6)$$

It is important to note that the first term on the right side, which is the change in energy due to the heat flux in and out of the volume, has a more complicated form than the spatial derivative of the heat flux in Eq. 4.1.17 for a Cartesian co-ordinate system. This is because the surface area of the cylindrical shell increases proportional to r as the radius increases; in contrast, in a Cartesian co-ordinate system, the surface area is independent of the z co-ordinate. The rate of energy entering or leaving is the product of the flux and the surface area, and the change in surface area is accounted for by the r within the derivative in the first term on the right in Eq. 5.1.6.

The heat flux q_r is related to the spatial derivative of the temperature in the radial direction by the Fourier's law for heat conduction,

$$q_r = -k\frac{\partial T}{\partial r},\qquad(5.1.7)$$

where k is the thermal conductivity. With this, the energy balance equation becomes,

$$\rho C_p\frac{\partial T}{\partial t} = \frac{1}{r}\frac{\partial}{\partial r}\left(rk\frac{\partial T}{\partial r}\right) + \mathcal{S}_e.\qquad(5.1.8)$$

When the density and specific heat are constant, the energy balance equation reduces to,

$$\frac{\partial T}{\partial t} = \frac{1}{r}\frac{\partial}{\partial r}\left(r\alpha\frac{\partial T}{\partial r}\right) + \frac{\mathcal{S}_e}{\rho C_p},\qquad(5.1.9)$$

where $\alpha = (k/\rho C_p)$ is the thermal diffusivity.

The mass conservation equation for the concentration field is identical to Eq. 5.1.9, with the temperature T replaced by the concentration c, the thermal diffusivity α replaced by the mass diffusivity $\mathcal{D}$ and the energy source $(\mathcal{S}_e/\rho C_p)$ replaced by the mass source $\mathcal{S}$,

$$\frac{\partial c}{\partial t} = \frac{1}{r}\frac{\partial}{\partial r}\left(r\mathcal{D}\frac{\partial c}{\partial r}\right) + \mathcal{S}.\qquad(5.1.10)$$

The radial mass flux is given by Fick's law,

$$j_r = -\mathcal{D}\frac{\partial c}{\partial r}.\qquad(5.1.11)$$

In a Cartesian co-ordinate system, the momentum conservation equation was written for the velocity v_x, which is a constant in the x direction, but is a function of the cross-stream direction z. In a cylindrical co-ordinate system, there are two velocity components perpendicular to r. The first is the axial velocity v_x shown in Fig. 5.2(a). This is driven, for example, by a body force in the x direction. The momentum balance equation for v_x is similar to Eqs. 5.1.9 and 5.1.10,

$$\rho\frac{\partial v_x}{\partial t} = \frac{1}{r}\frac{\partial}{\partial r}\left(r\mu\frac{\partial v_x}{\partial r}\right) + f_x,\qquad(5.1.12)$$

where ρ and μ are the density and viscosity, and f_x is the body force density in the x direction. Newton's law for the non-zero component of the shear stress in an axial

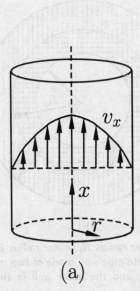

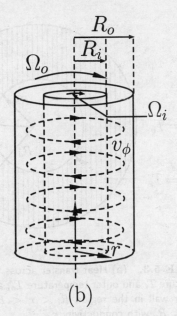

(a)   (b)

FIGURE 5.2. The velocity v_x along the axial x direction in a pipe (a) and v_ϕ along the polar direction between coaxial cylinders (b).

flow is,

$$\tau_{xr} = \mu \frac{\partial v_x}{\partial r}. \tag{5.1.13}$$

The axial velocity for the flow in a pipe is examined in the context of pressure-driven flows in the next chapter.

The second velocity component perpendicular to the r direction is v_ϕ along the polar angle ϕ, which is the velocity along circular streamlines around the axis, as shown in Fig. 5.2(b). This velocity is driven, for example, by coaxial inner and outer cylinders rotating at different velocities. The momentum conservation equation for v_ϕ is different from the energy and mass balance equations, Eqs. 5.1.9 and 5.1.10. For a fluid with constant viscosity, the ϕ momentum conservation equation is,

$$\rho \frac{\partial v_\phi}{\partial t} = \mu \frac{\partial}{\partial r} \left(\frac{1}{r} \frac{\partial (r v_\phi)}{\partial r} \right) + f_\phi, \tag{5.1.14}$$

where f_ϕ is the force per unit volume acting in the ϕ direction. Note that the differential operator on the right side of Eq. 5.1.14 is different from that in Eqs. 5.1.9, 5.1.10 or 5.1.12. Newton's law for the stress $\tau_{\phi r}$, which is the transport of ϕ

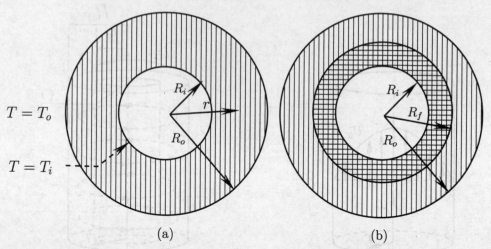

FIGURE 5.3. (a) Heat transfer across the wall of a pipe of inner radius R_i, outer radius R_o, inner temperature T_i and outer temperature T_o, and (b) across a composite pipe wall made of two materials, the inner wall in the region $R_i < r < R_f$ with conductivity k_i, and the outer wall in the region $R_f < r < R_o$ with conductivity k_o.

momentum in the r direction is,

$$\tau_{\phi r} = \mu \left(\frac{\partial v_\phi}{\partial r} - \frac{v_\phi}{r} \right). \tag{5.1.15}$$

The derivation of Eqs. 5.1.14 and 5.1.15 is beyond the scope of this text.

5.1.2   Steady Solution

Consider a cylindrical pipe with inner radius R_i, outer radius R_o and thermal conductivity k, as shown in Fig. 5.3(a). The inner surface is at temperature T_i, while the outer surface is at temperature T_o. It is necessary to determine the heat transfer rate across the wall of the pipe at steady state.

The scaled radius and temperature are,

$$r^* = \frac{r}{R_i}, \ \ T^* = \frac{T - T_i}{T_o - T_i}. \tag{5.1.16}$$

The energy balance equation, Eq. 5.1.9, at steady state expressed in scaled variables is,

$$\frac{1}{r^*} \frac{\mathrm{d}}{\mathrm{d}r^*} \left(r^* \frac{\mathrm{d}T^*}{\mathrm{d}r^*} \right) = 0 \tag{5.1.17}$$

and the boundary conditions are,

$$T^* = 0 \quad \text{at} \quad r^* = 1, \tag{5.1.18}$$

$$T^* = 1 \quad \text{at} \quad r^* = (R_o/R_i). \tag{5.1.19}$$

The solution of Eq. 5.1.17 is

$$T^* = A \ln(r^*) + B, \tag{5.1.20}$$

where A and B are constants of integration. The constant $B = 0$ from the condition, Eq. 5.1.18, and $A = (1/\ln(R_o/R_i))$ from Eq. 5.1.19. The solution for the temperature is,

$$T^* = \frac{\ln(r^*)}{\ln(R_o/R_i)}. \tag{5.1.21}$$

The heat flux q_r is,

$$q_r = -k\frac{\partial T}{\partial r} = -\frac{k(T_o - T_i)}{R_i}\frac{\partial T^*}{\partial r^*}$$

$$= -\frac{k(T_o - T_i)}{r^* R_i \ln(R_o/R_i)} = -\frac{k(T_o - T_i)}{r\ln(R_o/R_i)}. \tag{5.1.22}$$

The total heat transfer rate across the pipe wall is the product of the heat flux and surface area $2\pi r L$, where L is the length of the pipe,

$$\boxed{Q_r = (2\pi r L) \times \left(-\frac{k(T_o - T_i)}{r\ln(R_o/R_i)}\right) = -\frac{2\pi k(T_o - T_i)L}{\ln(R_o/R_i)}.} \tag{5.1.23}$$

Note that the heat transfer rate, Eq. 5.1.23, is independent of the radius. This is because, at steady state and in the absence of internal sources, the total heat entering and leaving an annular volume are equal. The heat flux, Eq. 5.1.22, does depend on r.

In analogy with the heat transfer across a flat plate, the total heat transfer rate across a pipe wall is sometimes written as,

$$Q_r = -\frac{kA_L(T_o - T_i)}{(R_o - R_i)}, \tag{5.1.24}$$

where the area A_L is,

$$A_L = \frac{2\pi L(R_o - R_i)}{\ln(R_o/R_i)} = 2\pi L R_L, \tag{5.1.25}$$

and the logarithmic mean radius R_L is

$$R_L = \frac{R_o - R_i}{\ln(R_o/R_i)}. \tag{5.1.26}$$

The logarithmic mean radius, R_L, is the value of the radius to be substituted into the expression for the area for heat transfer in Eq. 5.1.25 to obtain the correct expression for the heat transfer rate.

EXAMPLE 5.1.1: Consider a composite tube, with the inner tube and outer tube walls occupying the regions $R_i < r < R_f$ and $R_f < r < R_o$, made with materials of thermal conductivity k_i and k_o, respectively, as shown in Fig. 5.3(b). The temperature at $r = R_i$ is T_i, and the temperature at $r = R_o$ is T_o. What is the heat transfer rate across the tube?

Solution: The inner and outer tube walls provide resistances in series for heat transfer across the tube. The heat transfer rate at any radial location in the inner and outer walls of the tube is the same at steady state in the absence of internal sources,

$$Q_r = -\frac{2\pi k_i (T_f - T_i)L}{\ln(R_f/R_i)} = -\frac{2\pi k_o (T_o - T_f)L}{\ln(R_o/R_f)}, \tag{5.1.27}$$

where T_f is the temperature at $r = R_f$. From Eq. 5.1.27, the total temperature difference can be written as,

$$T_o - T_i = (T_o - T_f) + (T_f - T_i) = -Q_r \left(\frac{\ln(R_o/R_f)}{2\pi k_o L} + \frac{\ln(R_f/R_i)}{2\pi k_i L} \right). \tag{5.1.28}$$

Thus, the heat relation between the heat transfer rate and the temperature difference is,

$$Q_r = -\frac{T_o - T_i}{\frac{\ln(R_o/R_f)}{2\pi k_o L} + \frac{\ln(R_f/R_i)}{2\pi k_i L}}. \tag{5.1.29}$$

$\square$

The heat transfer resistance, which is the ratio of the temperature difference and the heat transfer rate, is

$$\boxed{\mathcal{R} = -\frac{\Delta T}{Q_r} = \frac{\ln(R_o/R_i)}{2\pi k L},} \tag{5.1.30}$$

for an annulus with thermal conductivity k and inner and outer radii R_i and R_o. The heat transfer resistance for a composite tube consisting of n layers with resistances $\mathcal{R}_1, \mathcal{R}_2, \ldots \mathcal{R}_n$ is given by the sum rule, Eq. 4.2.7.

EXAMPLE 5.1.2: A vacuum flask consists of concentric inner and outer cylinders of diameter 4 cm and 4.2 cm, respectively, and height 10 cm. Hot water is stored in the inner cylinder, and the region between the inner and outer cylinders is evacuated. Conduction heat transfer takes place across the low pressure air between the inner and outer cylinders with thermal conductivity $k = 2.4 \times 10^{-3}$ W/m/°C. Water, with density $\rho = 10^3$ kg/m^3 and specific heat $C_p = 4.2 \times 10^3$ J/kg/°C, initially at 100°C, is stored in the flask, while the temperature of the outer wall is 25°C. How long does it take for the temperature inside the flask to decrease to 50°C? Assume the temperature of water inside the flask is uniform, conduction takes place at steady state, and there is no heat transfer from the end caps.

Solution: The mass of water in the flask is

$$m = \pi R_i^2 h \rho = \pi (2 \times 10^{-2} \text{ m})^2 \times 0.1 \text{ m} \times 10^3 \text{ kg/m}^3 = 0.126 \text{ kg}, \qquad (5.1.31)$$

where $R_i = 2$ cm $= 0.02$ m is the diameter of the inner cylinder and $h = 10$ cm $= 0.1$ m is the height of the flask. The rate of change of the temperature of water T_w in the flask, which is the same as the temperature at $R = R_i$, is

$$mC_p \frac{dT_w}{dt} = -Q_r = \frac{2\pi k h (T_o - T_w)}{\ln(R_o/R_i)}, \qquad (5.1.32)$$

where T_o is the outside temperature, the total heat transfer rate Q_r is given by Eq. 5.1.23, and there is a negative sign in the first equality because the temperature decreases when Q_r is positive. Eq. 5.1.32 is simplified as,

$$\frac{dT_w}{dt} = -\frac{T_w - T_o}{t_c}, \qquad (5.1.33)$$

where the characteristic time t_c is,

$$t_c = \frac{mC_p \ln(R_o/R_i)}{2\pi k h} = \frac{0.126 \text{ kg} \times 4.2 \times 10^3 \text{ J/kg/°C} \times \ln(0.021 \text{ m}/0.02 \text{ m})}{2\pi \times 2.4 \times 10^{-3} \text{ W/m/°C} \times 0.1 \text{ m}}$$

$$= 1.7 \times 10^4 \text{ s}. \qquad (5.1.34)$$

Eq. 5.1.33 is solved to obtain,

$$t = -t_c \ln \left(\frac{T_w(t) - T_o}{T_w(t=0) - T_o} \right) = 1.88 \times 10^4 \text{ s} \approx 5.2 \text{ hours}, \qquad (5.1.35)$$

where $T_w(t=0) = 100$°C is the initial temperature, $T_w(t) = 50$°C is the final temperature and $T_o = 25$°C is the outside temperature. $\qquad \square$

EXAMPLE 5.1.3: Determine the velocity and the torque for the viscous flow between two concentric cylinders of length L and radii R_i and R_o rotating at angular velocities Ω_i and Ω_o as shown in Fig. 5.2(b).

Solution: Eq. 5.1.14 is solved at steady state with $f_\phi = 0$ for the tangential velocity v_ϕ,

$$v_\phi = Ar + \frac{B}{r}, \tag{5.1.36}$$

where A and B are constants. Since the velocity v_ϕ is the product of the angular velocity and the distance from the axis, the boundary conditions are,

$$v_\phi = R_i\Omega_i \text{ at } r = R_i, \tag{5.1.37}$$

$$= R_o\Omega_o \text{ at } r = R_o. \tag{5.1.38}$$

The constants in Eq. 5.1.36 are determined from the boundary conditions,

$$A = \frac{\Omega_o R_o^2 - \Omega_i R_i^2}{R_o^2 - R_i^2}, \tag{5.1.39}$$

$$B = -\frac{(\Omega_o - \Omega_i)R_o^2 R_i^2}{R_o^2 - R_i^2}, \tag{5.1.40}$$

and the solution for the velocity field is,

$$v_\phi = \frac{(\Omega_o R_o^2 - \Omega_i R_i^2)r}{R_o^2 - R_i^2} - \frac{R_i^2 R_o^2 (\Omega_o - \Omega_i)}{r(R_o^2 - R_i^2)}. \tag{5.1.41}$$

On the right side of the solution, Eq. 5.1.36, the first term is a solid body rotation, where v_ϕ is proportional to r and the angular velocity is a constant.

The stress $\tau_{\phi r}$ is given by Eq. 5.1.15,

$$\tau_{\phi r} = \mu\left(\frac{dv_\phi}{dr} - \frac{v_\phi}{r}\right) = -\frac{2\mu B}{r^2} = \frac{2\mu R_i^2 R_o^2(\Omega_o - \Omega_i)}{r^2(R_o^2 - R_i^2)}. \tag{5.1.42}$$

The torque per unit area is the product of the stress $\tau_{\phi r}$ (force per unit area) and the distance r from the axis,

$$\frac{\text{Torque}}{\text{Area}} = r \times \tau_{\phi r} = \frac{2\mu R_i^2 R_o^2(\Omega_o - \Omega_i)}{r(R_o^2 - R_i^2)}. \tag{5.1.43}$$

The torque is the product of the right side of Eq. 5.1.43 and the surface area $2\pi r L$,

$$\boxed{\text{Torque} = 2\pi r L \times r \times \tau_{\phi r} = \frac{4\pi L \mu R_i^2 R_o^2 (\Omega_o - \Omega_i)}{(R_o^2 - R_i^2)}.} \tag{5.1.44}$$

The torque, Eq. 5.1.44, is independent of r, and the torque on any cylindrical surface, including the inner and outer cylinders, is the same. $\qquad\square$

5.1.3 Heat Conduction from a Wire

In the analysis on conduction into an infinite medium in Section 4.4, a similarity variable could be identified because there were no length or time scales in the problem. A similar procedure cannot be used for the conduction from a cylinder into an infinite medium, because the radius of the cylinder can be used to non-dimensionalise length. A similarity variable can be identified for the heat conduction from a thin wire of infinitesimal radius. This problem is of relevance in real applications, because resistance heating is a common method for heating fluids. In order to design the apparatus, it is necessary to determine the temperature in the fluid as a function of the heat transfer rate from the wire.

The configuration, shown in Fig. 5.4, consists of a wire of infinitesimal thickness and of infinite length perpendicular to the $x - y$ plane in an infinite medium with thermal conductivity k. The wire and the fluid are initially at temperature T_∞. At time $t = 0$, the heating is switched on and the heat generated per unit length of the wire is Q.

There is an important difference between the present problem and the problem of conduction into an infinite medium in Cartesian co-ordinates. In the latter, the temperature at the surface was instantaneously increased to from T_∞ to T_0 at time $t = 0$. Here, the heat produced per unit length of the wire, Q, is specified. Since the wire is considered to be of infinitesimal thickness, the surface area of the wire is also infinitesimal. A finite heat transfer rate is required to increase the fluid temperature, even though the wire is of infinitesimal thickness.

For analysing the conduction from a flat surface to an infinite medium, the temperature difference between the surface and the medium far from the surface was used for scaling the temperature. In the present example, the temperature is not specified at the wire. Therefore, the temperature is not scaled at present; the scaling will become evident after the temperature profile is calculated. The reduced temperature is defined as $T^* = T - T_\infty$, so that the reduced temperature is zero in the limit $r \to \infty$. The heat conduction equation, Eq. 5.1.9, for constant thermal

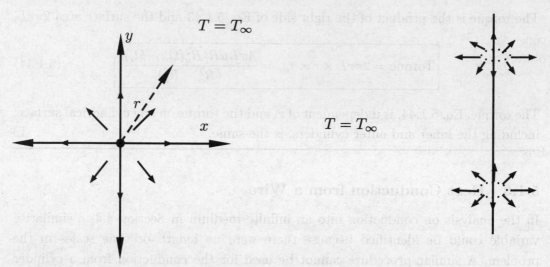

FIGURE 5.4. Top (left) and side (right) view of the heat conduction from a wire of infinitesimal thickness.

diffusivity expressed in terms of the T^* is

$$\frac{\partial T^*}{\partial t} = \frac{\alpha}{r}\frac{\partial}{\partial r}\left(r\frac{\partial T^*}{\partial r}\right). \tag{5.1.45}$$

Of the two boundary conditions in the r co-ordinate, one is the condition that the temperature is T_∞ far from the wire,

$$T = T_\infty \Rightarrow T^* = 0 \text{ for } r \to \infty. \tag{5.1.46}$$

The second condition is that the total heat generated at the wire per unit length is Q. Since the wire has infinitesimal thickness, this boundary condition is to be applied for $r \to 0$,

$$-2\pi r k \frac{\partial T^*}{\partial r} = Q \text{ for } r \to 0. \tag{5.1.47}$$

The left side of the above equation is the product of the radial flux of energy from the wire, $-k(\partial T^*/\partial r)$, and the surface area per unit length of the wire, $2\pi r$. Eq. 5.1.47 indicates that the temperature gradient goes to $-\infty$ for $r \to 0$, so that Q is finite. The initial condition is,

$$T = T_\infty \Rightarrow T^* = 0 \text{ for } r > 0 \text{ at } t = 0. \tag{5.1.48}$$

Based on dimensional analysis, the similarity variable is defined as $\xi = (r/\sqrt{\alpha t})$. The expressions in Eq. 4.4.7, with z replaced by r, are used to express the r and

t derivatives in terms of the similarity variable ξ. These are substituted in the conservation equation, Eq. 5.1.45, and the resulting equation is multiplied by t, to obtain,

$$\frac{d^2 T^*}{d\xi^2} + \left(\frac{1}{\xi} + \frac{\xi}{2}\right)\frac{dT^*}{d\xi} = 0. \tag{5.1.49}$$

Eqs. 5.1.46 and 5.1.48 reduce to the same condition when expressed in terms of ξ,

$$T^* = 0 \text{ for } \xi \to \infty. \tag{5.1.50}$$

The boundary condition, Eq. 5.1.47, expressed in terms of ξ, is

$$\xi\frac{dT^*}{d\xi} = -\frac{Q}{2\pi k} \text{ for } \xi \to 0. \tag{5.1.51}$$

Since Eq. 5.1.49 and the boundary conditions Eqs. 5.1.50–5.1.51 depend on ξ alone, a similarity solution can be obtained for the temperature field.

Eq. 5.1.49 is integrated one time to obtain the ξ derivative of the temperature,

$$\frac{dT^*}{d\xi} = \frac{C}{\xi}e^{(-\xi^2/4)}. \tag{5.1.52}$$

The constant C is determined from the boundary condition, Eq. 5.1.51, at the wire,

$$C = -\frac{Q}{2\pi k}. \tag{5.1.53}$$

Eq. 5.1.52 is integrated with respect to the ξ,

$$T^* = -\frac{Q}{2\pi k}\int_\infty^\xi d\xi' \frac{1}{\xi'}e^{(-\xi'^2/4)}. \tag{5.1.54}$$

Here, we have take the lower limit of integration as $\xi = \infty$, so that the result satisfies the condition $T^* = 0$ for $\xi \to \infty$. Eq. 5.1.54 shows that the characteristic temperature scale is $Q/(2\pi k)$, where Q is the heat generation rate per unit length.

The solution for the scaled temperature $T^*/(Q/2\pi k)$ is shown as a function of ξ in Fig. 5.5. For $r \to 0$, it is evident from Eq. 5.1.54 that T^* increases proportional to $-(Q\ln(\xi)/2\pi k)$. The heat flux, $q_r = -k(\partial T/\partial r) = (Q/2\pi r)$, and the total heat generation rate per unit length, $2\pi r q_r = Q$, is finite. The temperature decreases rapidly as ξ is increased, and the scaled temperature decreases below 0.01 for $\xi > 3.26$ or $r > 3.26\sqrt{\alpha t}$.

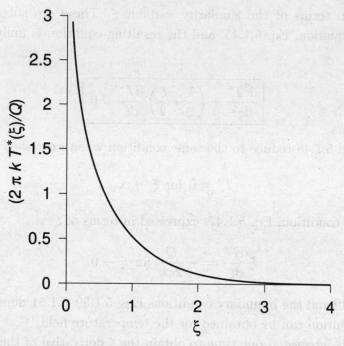

FIGURE 5.5. The solution, Eq. 5.1.54, for $(2\pi k T^*/Q)$ as a function of ξ for the heat conduction from a wire of infinitesimal thickness.

5.1.4   Unsteady Heat Conduction into a Cylinder

A cylindrical container of radius R containing fluid with temperature T_1 is dipped into a bath at temperature T_0, as shown in Fig. 5.6(a). The temperature is to be determined as a function of time and position as the material in the container cools.

The conservation Eq. 5.1.45 is scaled as follows. The scaled radius and time are defined as $r^* = (r/R)$ and $t^* = (tR^2/\alpha)$, since R is the length scale in the radial direction. The scaled temperature is defined as, $T^* = (T - T_0)/(T_1 - T_0)$. Expressed in these coordinates, the conservation equation is

$$\frac{\partial T^*}{\partial t^*} = \frac{1}{r^*}\frac{\partial}{\partial r^*}\left(r^* \frac{\partial T^*}{\partial r^*}\right). \qquad (5.1.55)$$

The boundary condition at the surface of the cylinder, $r^* = 1$, is

$$T^* = 0 \text{ at } r^* = 1 \text{ for all } t^*. \qquad (5.1.56)$$

Along the axis of the cylinder $r^* = 0$, we specify a 'symmetry condition',

$$\frac{\partial T^*}{\partial r^*} = 0 \text{ at } r^* = 0 \text{ for all } t^*. \qquad (5.1.57)$$

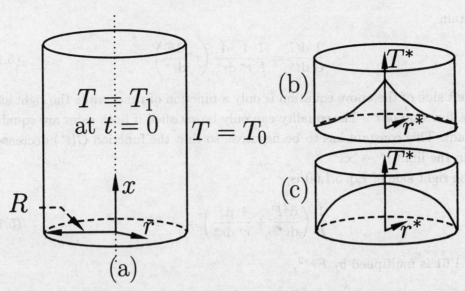

FIGURE 5.6. Configuration and co-ordinate system for heat conduction into a cylinder (a). A non-zero slope of the temperature profile at the axis results in an unphysical cusp in the temperature profile at $r^* = 0$ (b), whereas the slope of the temperature profile is uniquely defined at $r^* = 0$ only if it is zero (c).

If Eq. 5.1.57 is not satisfied, as shown in Fig. 5.6(b), the slope of the temperature profile is finite at the centre. For an axisymmetric temperature profile, there is a cusp in the temperature profile at the $r^* = 0$, and the derivative of the temperature is not unique along the axis. The derivative is unique only if Eq. 5.1.57 is satisfied, as shown in Fig. 5.6(c).

At time $t^* = 0$, the temperature is T_1 throughout the cylinder except at the cylinder wall, and so the initial condition is,

$$T^* = 1 \quad \text{at } t^* = 0 \text{ for all } r^* < 1. \tag{5.1.58}$$

The spatial boundary conditions at $r^* = 1$ and $r^* = 0$, Eqs. 5.1.56 and 5.1.57, are both homogeneous, and the initial condition, Eq. 5.1.58, is inhomogeneous.

Eq. 5.1.55 is solved using separation of variables. The final steady solution is a uniform temperature, $T_s^* = 0$, throughout the cylinder. Therefore, the transient temperature is the same as T^*. The temperature is expressed as,

$$T^* = F(r^*)G(t^*), \tag{5.1.59}$$

where F is a function of the radial coordinate, and G is a function of time. Eq. 5.1.59 is substituted into Eq. 5.1.55, and the resulting equation is divided by $F(r^*)G(t^*)$, to obtain

$$\frac{1}{G}\frac{\mathrm{d}G}{\mathrm{d}t^*} = \frac{1}{F}\frac{1}{r^*}\frac{\mathrm{d}}{\mathrm{d}r^*}\left(r^*\frac{\mathrm{d}F}{\mathrm{d}r^*}\right). \tag{5.1.60}$$

The left side of the above equation is only a function of time, while the right side is only a function of r^*. The equality can only be satisfied if both sides are equal to a constant. This constant has to be negative, so that the function $G(t^*)$ decreases to zero in the limit $t^* \to \infty$.

The right side of Eq. 5.1.60 is,

$$\frac{1}{F}\left(\frac{\mathrm{d}^2F}{\mathrm{d}r^{*2}} + \frac{1}{r^*}\frac{\mathrm{d}F}{\mathrm{d}r^*}\right) = -\beta^2. \tag{5.1.61}$$

Eq. 5.1.61 is multiplied by Fr^{*2},

$$r^{*2}\frac{\mathrm{d}^2F}{\mathrm{d}r^{*2}} + r^*\frac{\mathrm{d}F}{\mathrm{d}r^*} + \beta^2 r^{*2}F = 0. \tag{5.1.62}$$

The above equation is different from Eq. 4.7.16 for Cartesian co-ordinates in Section 4.7. However, the requirement that the solution satisfies homogeneous boundary conditions at the boundaries leads to a discrete set of values for the constants $\beta = \beta_i$, and a set of orthogonal basis functions for $F = J_0(\beta_i r^*)$. The calculation procedure is described below, and the orthogonality relation for a cylindrical co-ordinate system is derived in Appendix 5.A.

Eq. 5.1.62 is reduced to a form independent of parameters by the substitution $r^\dagger = \beta r^*$,

$$\boxed{r^{\dagger 2}\frac{\mathrm{d}^2F}{\mathrm{d}r^{\dagger 2}} + r^\dagger\frac{\mathrm{d}F}{\mathrm{d}r^\dagger} + r^{\dagger 2}F = 0.} \tag{5.1.63}$$

Eq. 5.1.63 is a special case of the Bessel equation,

$$r^{\dagger 2}\frac{\mathrm{d}^2F}{\mathrm{d}r^{\dagger 2}} + r^\dagger\frac{\mathrm{d}F}{\mathrm{d}r^\dagger} + (r^{\dagger 2} - n^2)F = 0, \tag{5.1.64}$$

with $n = 0$. The solutions for Eq. 5.1.63 are 'zeroeth order' Bessel functions,

$$F = CJ_0(r^\dagger) + DY_0(r^\dagger) = CJ_0(\beta r^*) + DY_0(\beta r^*). \tag{5.1.65}$$

The Bessel functions $J_0(r^\dagger)$ and $Y_0(r^\dagger)$ are shown as a function of $r^\dagger$ in Fig. 5.7. The Bessel functions have an oscillatory dependence on $r^\dagger$, the maximum amplitude of the oscillations decreases and the period increases as $r^\dagger$ increases. The Bessel

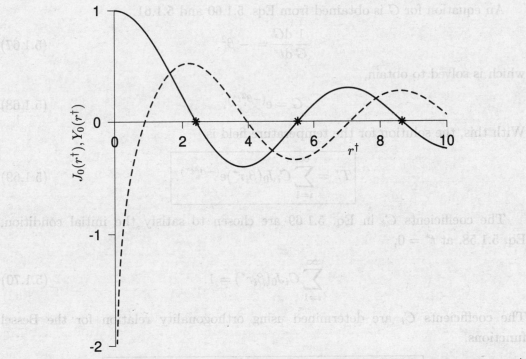

FIGURE 5.7. The Bessel functions, $J_0(r^\dagger)$ (solid line) and $Y_0(r^\dagger)$ (dashed line) as a function of $r^\dagger$.

function $J_0(r^\dagger)$ has a maximum value of 1 at $r^\dagger = 0$. The Bessel function $Y_0(r^\dagger)$ tends to $-\infty$ for $r^\dagger \to 0$.

The values of the constants C and D are determined from the boundary conditions at $r^* = 0$ and $r^* = 1$. Eq. 5.1.57 requires that $(\mathrm{d}F/\mathrm{d}r^*) = 0$ at $r^* = 0$. From Fig. 5.7, it is observed that the derivative of $J_0(r^\dagger)$ is zero, whereas the radial derivative of $Y_0(r^\dagger)$ diverges for $r^\dagger \to 0$. Therefore, the boundary condition, Eq. 5.1.57, is satisfied only if $D = 0$ in Eq. 5.1.65.

The boundary condition, $T^* = 0$ at $r^\dagger = 1$ (Eq. 5.1.56), is used to determine the value of β in Eq. 5.1.61,

$$J_0(\beta) = 0. \qquad (5.1.66)$$

Since $J_0(\beta)$ is an oscillatory function of β, there are multiple solutions for Eq. 5.1.66, which are the points where $J_0(\beta)$ crosses zero, shown by the $*$ symbols in Fig. 5.7. The first few solutions are, $\beta_1 = 2.40483$, $\beta_2 = 5.52008$, $\beta_3 = 8.65373$, $\beta_4 = 11.79150$ and $\beta_5 = 14.93090$. Thus, the requirement that $F(r^*) = 0$ at $r^* = 1$ results in a discrete set of eigenvalues β_i. This is analogous to the discrete values $\beta_i = (i\pi)$ in the unsteady heat conduction between two parallel plates in Cartesian co-ordinates in Section 4.7.

An equation for G is obtained from Eqs. 5.1.60 and 5.1.61,

$$\frac{1}{G}\frac{dG}{dt^*} = -\beta_i^2,$$ (5.1.67)

which is solved to obtain,

$$G = e^{(-\beta_i^2 t^*)}.$$ (5.1.68)

With this, the solution for the temperature field is

$$T^* = \sum_{i=1}^{\infty} C_i J_0(\beta_i r^*) e^{(-\beta_i^2 t^*)}.$$ (5.1.69)

The coefficients C_i in Eq. 5.1.69 are chosen to satisfy the initial condition, Eq. 5.1.58, at $t^* = 0$,

$$\sum_{i=1}^{\infty} C_i J_0(\beta_i r^*) = 1.$$ (5.1.70)

The coefficients C_i are determined using orthogonality relation for the Bessel functions,

$$\int_0^1 r^* dr^* J_0(\beta_i r^*) J_0(\beta_j r^*) = \frac{1}{2}(J_1(\beta_j))^2 \text{ for } i = j,$$
$$= 0 \text{ for } i \neq j.$$ (5.1.71)

The orthogonality relation for the Bessel functions is derived in Appendix 5.A. In order to determine the coefficients C_i, the right and left sides of Eq. 5.1.70 are multiplied by $r^* J_0(\beta_j r^*)$, and integrated from $r^* = 0$ to $r^* = 1$, to obtain

$$\frac{1}{2}(J_1(\beta_j))^2 C_j = \int_0^1 r^* dr^* J_0(\beta_j r^*)$$
$$= \frac{J_1(\beta_j)}{\beta_j}.$$ (5.1.72)

The above integral is available in standard tables of Bessel functions, and it can also be evaluated numerically. This provides the solution for C_j,

$$C_j = \frac{2}{\beta_j J_1(\beta_j)},$$ (5.1.73)

and the final solution for the temperature field is,

$$T^* = \sum_{i=1}^{\infty} \frac{2}{\beta_i J_1(\beta_i)} J_0(\beta_i r^*) e^{(-\beta_i^2 t^*)}.$$ (5.1.74)

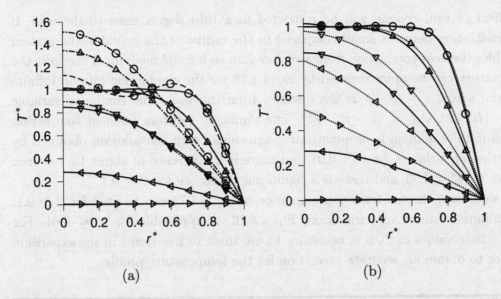

(a)   (b)

FIGURE 5.8. In sub-figure (a), the solution, Eq. 5.1.74, for the heat conduction in a cylinder for $t^* = 0.01$ ($\circ$), $t^* = 0.03$ ($\triangle$), $t^* = 0.1$ (∇), $t^* = 0.3$ ($\triangleleft$) and $t^* = 1$ ($\triangleright$). The solid lines — are the results when five terms are included in the expansion, the dashed lines – – – are the results when three terms are included in the expansion, and the dash-dot lines – · — · – are the results when one term is included in the expansion. In sub-figure (b), the symbols refer to the same t^* as (a), the solid lines are the solution, Eq. 5.1.74, and the dotted lines are the solutions for the temperature field for conduction into an infinite medium, Eq. 4.4.16, where the similarity variable is defined as $\xi = (R - r)/\sqrt{\alpha t} = (1 - r^*)/\sqrt{t^*}$.

The solution, Eq. 5.1.74, is shown as a function of r^* for different values of t^* in the range 0.01–1 in Fig. 5.8(a). The solid lines are the results when the first five terms are included in the expansion in Eq. 5.1.74, the dashed lines are the results when the first three terms are included, and the dash-dot lines are the results when only the first term is included. The one-term approximation is in perfect agreement with the three- and five-term approximations for $t^* \geq 0.3$; at $t^* = 0.1$, the one-term approximation incurs a maximum error of about 5.5% for the temperature at the centre of the cylinder. The one-term approximation is in significant error for $t^* \leq 0.03$; the temperature at the centre is greater than the initial temperature by more than 50%. The three-term approximation is quantitatively accurate for $t^* = 0.03$, but there is significant error and unphysical oscillations at $t^* = 0.01$. The five-term approximation captures the smooth decrease in temperature close to the wall even at $t^* = 0.01$.

The dotted lines in Fig. 5.8(b) are the results for the conduction into an infinite medium, Eq. 4.4.16, where $\xi = ((1 - r^*)/\sqrt{t^*})$ is the similarity variable based on the distance from the wall. At $t^* = 0$, the temperature is $T^* = 1$ throughout the cylinder, and the wall temperature is $T^* = 0$. For $t^* \ll 1$, it is expected that the effect of wall cooling will be restricted to a thin region close to the wall. If the penetration depth is small compared to the radius of the cylinder, the system resembles the heat conduction from a surface into an infinite medium. Therefore, the temperature can be approximated by Eq. 4.4.16 for the conduction into an infinite medium, where $z = R - r$ is the distance from the wall. The similarity variable is $\xi = (R - r)/\sqrt{\alpha t} = (1 - r^*)/\sqrt{t^*}$. The similarity solution for heat conduction into an infinite domain is in quantitative agreement with the solution obtained by separation of variables for $t^* = 0.01$, but there is a difference of about 4% between the two for $t^* = 0.03$, and there is a significant difference for $t^* \geq 0.1$.

In summary, the one-term approximation of Eq. 5.1.74 can be used for $t^* \geq 0.1$. The infinite-medium approximation, Eq. 4.4.16, is applicable for $t^* \leq 0.01$. For intermediate values of t^*, it is necessary to use three to five terms in the expansion in order to obtain an accurate prediction for the temperature profiles.

EXAMPLE 5.1.4: Consider the start-up flow of a fluid of density ρ and viscosity μ in a cylinder of radius R. The cylinder and fluid are stationary for $t < 0$. At $t = 0$, the cylinder is rotated with angular velocity Ω. Determine the velocity v_ϕ as a function of the radial co-ordinate r and time.

Solution: In Eq. 5.1.14 for the velocity v_ϕ, the scaled radius is defined as $r^* = (r/R)$, the scaled time is $t^* = (t\nu/R^2)$ where $\nu = (\mu/\rho)$ is the kinematic viscosity, and the scaled velocity is defined as $v_\phi^* = (v_\phi/(R\Omega))$. The scaled equation is,

$$\frac{\partial v_\phi^*}{\partial t^*} = \frac{\partial}{\partial r^*} \left(\frac{1}{r^*} \frac{\partial (r^* v_\phi^*)}{\partial r^*} \right). \tag{5.1.75}$$

The boundary conditions are the no-slip condition at the cylinder wall,

$$v_\phi^* = 1 \text{ at } r^* = 1, \tag{5.1.76}$$

and zero velocity at the centre,

$$v_\phi^* = 0 \text{ at } r^* = 0. \tag{5.1.77}$$

If the symmetry condition, Eq. 5.1.77, is not satisfied—that is, if the velocity is non-zero at $r^* = 0$—the scaled angular velocity (v_ϕ^*/r^*) is infinite.

The velocity is expressed as the sum of the steady and transient parts, $v_\phi^* = v_{\phi s}^* + v_{\phi t}^*$. The final steady solution in the long time limit $t^* \to \infty$ is a solid body rotation, $v_\phi = r\Omega$,

$$v_{\phi s}^* = r^*.$$ (5.1.78)

The boundary conditions for the steady solution $v_{\phi s}$ at $r^* = 0$ and $r^* = 1$ are identical to those for v_ϕ^*, Eqs. 5.1.76 and 5.1.77. Therefore, the boundary conditions for the transient part of the velocity field are homogeneous,

$$v_{\phi t}^* = 0 \text{ at } r^* = 0,$$ (5.1.79)

$$v_{\phi t}^* = 0 \text{ at } r^* = 1.$$ (5.1.80)

The initial condition is $v_\phi^* = 0$ at $t^* = 0$ throughout the fluid, and therefore the initial condition for the transient velocity is

$$v_{\phi t}^* = -v_{\phi s}^* = -r^* \text{ at } t^* = 0.$$ (5.1.81)

The separation of variables procedure is used to express $v_{\phi t}^* = F(r^*)G(t^*)$ in Eq. 5.1.75, and the equation is divided by $F(r^*)G(t^*)$, to obtain,

$$\frac{1}{G}\frac{dG}{dt^*} = \frac{1}{F}\frac{d}{dr^*}\left(\frac{1}{r^*}\frac{d(r^*F)}{dr^*}\right).$$ (5.1.82)

Since the left side is only a function of t^* and the right side is only a function of r^*, both sides are equal to a constant. The boundary conditions, Eqs. 5.1.79 and 5.1.80, are homogeneous in the r^* direction. The basis functions in the r^* direction are determined by equating the right side of Eq. 5.1.82 to a constant $-\beta^2$,

$$\frac{d^2F}{dr^{*2}} + \frac{1}{r^*}\frac{dF}{dr^*} - \frac{F}{r^{*2}} + \beta^2 F = 0.$$ (5.1.83)

Multiplying Eq. 5.1.83 by r^{*2}, and using the substitution $r^\dagger = \beta r^*$, we obtain,

$$\boxed{r^{\dagger 2}\frac{d^2F}{dr^{\dagger 2}} + r^\dagger\frac{dF}{dr^\dagger} + (r^{\dagger 2} - 1)F = 0,}$$ (5.1.84)

with boundary conditions

$$F(r^\dagger) = 0 \text{ at } r^\dagger = 0,$$ (5.1.85)

$$F(r^\dagger) = 0 \text{ at } r^\dagger = \beta \ (r^* = 1).$$ (5.1.86)

Eq. 5.1.84 is the Bessel equation, Eq. 5.1.64, with $n = 1$, and the solutions are,

$$F(r^\dagger) = CJ_1(r^\dagger) + DY_1(r^\dagger) = CJ_1(\beta r^*) + DY_1(\beta r^*).$$ (5.1.87)

The Bessel functions $J_1(r^\dagger)$ and $Y_1(r^\dagger)$, shown in Fig. 5.9, are oscillating functions in which the amplitude decreases as $r^\dagger$ increases. The function $J_1(r^\dagger)$ is zero at $r^\dagger = 0$, while the function $Y_1(r^\dagger)$ diverges proportional to $(r^\dagger)^{-1}$ for $r^\dagger \to 0$.

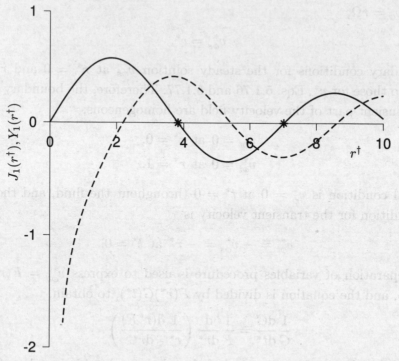

FIGURE 5.9. The Bessel functions $J_1(r^\dagger)$ (solid line) and $Y_1(r^\dagger)$ (dashed line) as a function of $r^\dagger$.

The constant D in the solution, Eq. 5.1.87, is zero from the condition, Eq. 5.1.85, that the rotational velocity is zero at the centre, since the function $Y_1(\beta r^*)$ diverges at $r^* = 0$. The discrete eigenvalues β_i are determined from the boundary condition, Eq. 5.1.86,

$$J_1(\beta_i) = 0. \tag{5.1.88}$$

These solutions are shown by the $*$ symbols in Fig. 5.9, and the first five solutions for Eq. 5.1.88 are $\beta_i = 3.83171, 7.01559, 10.17346, 13.32668$ and 16.47063. The functions $J_1(\beta_i r^*)$ satisfy the orthogonality relation,

$$\boxed{\int_0^1 r^* \, dr^* \, J_1(\beta_i r^*) J_1(\beta_j r^*) = \frac{(J_2(\beta_j))^2 \delta_{ij}}{2}.} \tag{5.1.89}$$

The solution for the $G(t^*)$ is determined from Eq. 5.1.82,

$$G(t^*) = e^{(-\beta_i^2 t^*)}. \tag{5.1.90}$$

Thus, the solution for the transient part of the velocity field is,

$$\boxed{v_{\phi t}^* = \sum_i C_i J_1(\beta_i r^*) e^{(-\beta_i^2 t^*)}.} \tag{5.1.91}$$

The coefficients C_i are determined from the initial condition, Eq. 5.1.81, at $t^* = 0$,

$$\sum_i C_i J_1(\beta_i r^*) = -r^*. \qquad (5.1.92)$$

Multiplying the left and right sides of the above equation by $J_1(\beta_j r^*) r^*$, integrating over the interval $0 \leq r^* \leq 1$, and using the orthogonality relation, Eq. 5.1.89, we obtain,

$$\frac{C_i (J_2(\beta_i))^2 \delta_{ij}}{2} = -\frac{J_2(\beta_j)}{\beta_j}. \qquad (5.1.93)$$

The integral on the right side is available in standard tables for the Bessel function; it can also be evaluated numerically. Thus, the coefficients C_j are,

$$\boxed{C_j = -\frac{2}{\beta_j J_2(\beta_j)}.} \qquad (5.1.94)$$

The final solution for the time-dependent velocity is the sum of the steady part, Eq. 5.1.78, and the transient part, Eq. 5.1.91

$$\boxed{v_\phi^* = r^* - \sum_i \frac{2 J_1(\beta_i r^*) e^{(-\beta_i^2 t^*)}}{\beta_i J_2(\beta_i)}.} \qquad (5.1.95)$$

The velocity profiles at different values of t^* are shown in Fig. 5.10. The velocity profile is close to the steady profile, Eq. 5.1.78, for $t^* \geq 0.3$. The one-term approximation of Eq. 5.1.95 is in good agreement with the three- and five-term approximations for $t^* \geq 0.1$, indicating that it is sufficient to retain just one term in the series for $t \geq (0.1 R^2 / \nu)$. The one-term approximation results in spurious oscillations and a negative velocity for $t^* \leq 0.03$. The three-term approximation gives smooth results for $t^* \geq 0.03$, while the five-term approximation is accurate even for $t^* = 0.01$.

The dotted lines in Fig. 5.10 are the similarity solutions for diffusion into an infinite medium, Eq. 4.4.16 with T^* replaced by v_ϕ^* and the similarity variable based on distance from the wall and the momentum diffusivity, $\xi = (R - r)/\sqrt{\nu t}$. This solution is in quantitative agreement with the five-term approximation for $t^* = 0.01$, though there is a difference between the two for $t^* = 0.03$. Thus, the linear velocity profile is quantitatively accurate for $t^* \geq 0.3$, the one-term approximation of Eq. 5.1.95 for $t^* \geq 0.1$, the five-term approximation of Eq. 5.1.95 for $t^* \geq 0.01$ and the similarity solution for diffusion into an infinite medium, Eq. 4.4.16, for $t^* \leq 0.01$.

□

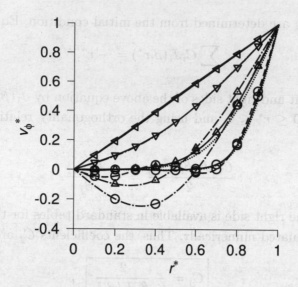

FIGURE 5.10. The solution, Eq. 5.1.95, for the time evolution of the velocity v_ϕ^* in a rotating cylinder cylinder for $t^* = 0.01$ (o), $t^* = 0.03$ ($\triangle$), $t^* = 0.1$ (∇), and $t^* = 0.3$ ($\triangleleft$). The results of the five-term approximation are shown by the solid lines —, the three-term approximation by the the dashed lines $- - -$ and the one-term approximation by the dot-dash lines $- \cdot - \cdot -$. The dotted lines are the results for $t^* = 0.01$ and 0.03 for the momentum diffusion into an infinite medium, Eq. 4.4.16, with T^* replaced by v_ϕ^*, and the similarity variable defined as $\xi = (R - r)/\sqrt{\nu t} = (1 - r^*)/\sqrt{t^*}$.

Summary (5.1)

1. The important results from the derivation of the conservation equation in cylindrical co-ordinates are,

   (a) The rate of change of the field variable Φ_{fv} for unsteady transport in cylindrical co-ordinates, Eq. 5.1.6 and its analogues for mass and axial momentum transfer, has the common form,

   $$\frac{\partial \Phi_{fv}}{\partial t} = -\frac{1}{r}\frac{\partial(r\mathcal{J}_z)}{\partial r} + \mathcal{S}, \tag{5.1.96}$$

   where the field variable Φ_{fv}, flux $\mathcal{J}_z$ and source $\mathcal{S}$ for mass, heat and momentum transfer are summarised in Table 4.1.

   (b) When the flux is given by the constitutive relations, Eqs. 5.1.7, 5.1.11 or 5.1.13, the conservation Eqs. 5.1.9, 5.1.10 and 5.1.12 reduce to,

   $$\frac{\partial \Phi_{fv}}{\partial t} = \frac{1}{r}\frac{\partial}{\partial r}\left(\mathcal{D}r\frac{\partial \Phi_{fv}}{\partial r}\right) + \mathcal{S}, \tag{5.1.97}$$

   where $\mathcal{D}$ is the generalised diffusivity.

(c) When the diffusivity is also independent of the field variable, position or time, the conservation equation is further simplified as,

$$\frac{\partial \Phi_{fv}}{\partial t} = \frac{D}{r} \frac{\partial}{\partial r} \left(r \frac{\partial \Phi_{fv}}{\partial r} \right) + \mathcal{S}. \tag{5.1.98}$$

(d) The momentum conservation Eq. 5.1.14, and the shear stress, Eq. 5.1.15, for the polar velocity in a cylinder have a form that is different from those for heat, mass and axial momentum.

2. At steady state,

   (a) the solution for the steady diffusion equation in the absence of sources is of the form $\Phi_{fv} = A \ln(r) + B$, where A and B are constants determined from the boundary conditions.

   (b) the heat transfer rate across the wall of a pipe of length L, inner radius R_i, outer radius R_o and thermal conductivity k is Eq. 5.1.23. For a composite pipe wall, the heat transfer resistance is given by Eq. 5.1.30.

   (c) the solution for the steady polar velocity between rotating cylinders is Eq. 5.1.36, where the first term proportional to A results in solid body rotation, and does not contribute to the stress.

   (d) the total torque, Eq. 5.1.44, is independent of r for the flow between concentric cylinders.

3. For conduction from an infinitesimal wire, the similarity variable is defined as $\xi = r/\sqrt{\alpha t}$ for the heat conduction from a wire of infinitesimal thickness. The equation for the temperature is Eq. 5.1.49, and the solution is Eq. 5.1.54, where Q is the total heat produced per unit length of the wire per unit time.

4. For unsteady transport into a cylinder of finite size,

   (a) In the separation of variables procedure in cylindrical co-ordinates, the equations for the r co-ordinate are Bessel Equations, Eqs. 5.1.63 and 5.1.84. The discrete eigenvalues β_i are the locations where the appropriate Bessel function, Figs. 5.7 and 5.9, pass through zero.

   (b) For homogeneous boundary conditions, the orthogonality relations for the Bessel functions are Eqs. 5.1.71 and 5.1.89.

   (c) The solutions are expressed as the product of Bessel functions in r and exponentially decreasing functions in time, Eqs. 5.1.69 and 5.1.91, and the coefficients in these expansions, Eqs. 5.1.73 and 5.1.94, are determined from the orthogonality relations.

(d) Only 1–3 terms need to be included in the series solutions, Eqs. 5.1.74 and 5.1.95, for an accurate solution for $t^* \geq 0.1$. The similarity solution, Eq. 4.4.16, can be used for $t^* \leq 0.01$ when the penetration depth is small compared to the radius.

5.2   Spherical Co-ordinates

The spherical co-ordinate system is used to analyse transport in or around objects with spherical symmetry, such as spherical particles, bubbles or drops immersed in a fluid. For a spherical particle with radius R, shown in Fig. 5.11, the origin is fixed at the centre of the particle, and the co-ordinate r is the distance from the centre. Surfaces of constant r are spherical surfaces, and the surface of the particle is a surface of constant co-ordinate, $r = R$.

5.2.1   Conservation Equation

The conservation condition is applied for a spherical shell bounded by two surfaces at r and $r + \Delta r$ shown in Fig. 5.11. The volume of the shell is $4\pi r^2 \Delta r$, and the areas of the inner and outer surfaces are $4\pi r^2$ and $4\pi (r + \Delta r)^2$, respectively. The terms

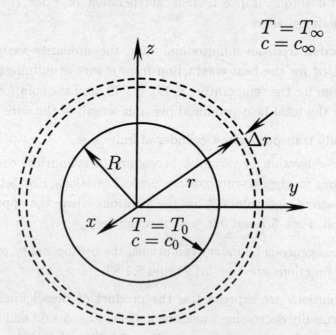

FIGURE 5.11.   The co-ordinate system and the differential volume for analysing the transport around a spherical particle.

in the mass balance equation, Eq. 4.1.1, for this differential volume are as follows. The change in mass in the time Δt is,

$$\left(\begin{array}{c}\text{Change in mass}\\\text{in differential volume}\end{array}\right) = (c(r, t + \Delta t) - c(r, t))4\pi r^2 \Delta r. \qquad (5.2.1)$$

The mass entering the spherical surface at r is the product of the mass flux, the surface area, and the time interval Δt,

$$\left(\text{Mass in}\right) = j_r|_r \left(4\pi r^2 \Delta t\right), \qquad (5.2.2)$$

where j_r is the mass flux in the radial direction. The mass leaving at the surface at $r + \Delta r$ is

$$\left(\text{Mass out}\right) = j_r|_{r+\Delta r} \left(4\pi (r + \Delta r)^2 \Delta t\right). \qquad (5.2.3)$$

The production of mass in the differential volume is,

$$\left(\begin{array}{c}\text{Production of mass}\\\text{in differential volume}\end{array}\right) = \mathcal{S}(4\pi r^2 \Delta r \Delta t), \qquad (5.2.4)$$

where $\mathcal{S}$ is the mass generated per unit volume per unit time.

The conservation equation is obtained by substituting Eqs. 5.2.1–5.2.4 into the balance condition Eq. 4.1.1, and dividing by the volume $4\pi r^2 \Delta r \Delta t$,

$$\frac{(c(r, t + \Delta t) - c(r, t))}{\Delta t} = \frac{\left(r^2 j_r|_r - (r + \Delta r)^2 j_r|_{r+\Delta r}\right)}{r^2 \Delta r} + \mathcal{S}. \qquad (5.2.5)$$

Note that, as in the case of the cylindrical co-ordinate system, the surface area is different for the surfaces at r and $r + \Delta r$. Taking the limit $\Delta r \to 0$ and $\Delta t \to 0$, the partial differential equation for the concentration field is,

$$\boxed{\frac{\partial c}{\partial t} = -\frac{1}{r^2}\frac{\partial}{\partial r}\left(r^2 j_r\right) + \mathcal{S}.} \qquad (5.2.6)$$

The mass flux j_r is related to $(\partial c/\partial r)$ by Fick's law,

$$\boxed{j_r = -\mathcal{D}\frac{\partial c}{\partial r}.} \qquad (5.2.7)$$

With this substitution, the mass balance equation becomes,

$$\boxed{\frac{\partial c}{\partial t} = \frac{1}{r^2}\frac{\partial}{\partial r}\left(r^2 \mathcal{D}\frac{\partial c}{\partial r}\right) + \mathcal{S}.} \qquad (5.2.8)$$

When the diffusion coefficient is a constant, the mass balance equation reduces to,

$$\frac{\partial c}{\partial t} = \mathcal{D}\frac{1}{r^2}\frac{\partial}{\partial r}\left(r^2\frac{\partial c}{\partial r}\right) + \mathcal{S}. \tag{5.2.9}$$

The energy balance equation for unidirectional transport in a spherical co-ordinate system, derived in a similar manner, is

$$\frac{\partial T}{\partial t} = \frac{1}{r^2}\frac{\partial}{\partial r}\left(\alpha r^2\frac{\partial T}{\partial r}\right) + \frac{S_e}{\rho C_p}. \tag{5.2.10}$$

where $\alpha = (k/\rho C_p)$ is the thermal diffusivity. Here, it is assumed that ρC_p is a constant. The momentum balance equation in a spherical co-ordinate system is beyond the scope of this text.

5.2.2 Steady Solution

A spherical particle of radius R, with temperature T_0 at the surface, is immersed in a fluid in which the temperature is T_∞ far from the particle, as shown in Fig. 5.11. The system is at steady state, and the temperature field does not vary with time. It is necessary to determine the Nusselt number correlation for the average heat flux at the surface.

The scaled radius and temperature are defined as, $r^* = (r/R)$ and $T^* = (T - T_\infty)/(T_0 - T_\infty)$. In terms of these scaled co-ordinates, the balance equation at steady state is,

$$\frac{1}{r^{*2}}\frac{d}{dr^*}\left(r^{*2}\frac{dT^*}{dr^*}\right) = 0, \tag{5.2.11}$$

with boundary conditions,

$$T^* = 0 \quad \text{for} \quad r^* \to \infty, \tag{5.2.12}$$

$$T^* = 1 \quad \text{at} \quad r^* = 1. \tag{5.2.13}$$

Eq. 5.2.11 is solved to obtain,

$$T^* = \frac{C}{r^*} + D. \tag{5.2.14}$$

where the constants, C and D are determined from the boundary conditions, Eqs. 5.2.12 and 5.2.13. The solution for the temperature that satisfies the boundary conditions is,

$$T^* = \frac{1}{r^*}.$$ (5.2.15)

The dimensional temperature T is,

$$T = T_\infty + \frac{(T_0 - T_\infty)R}{r}.$$ (5.2.16)

The heat flux from the surface of the sphere is,

$$q_r = -k\frac{\partial T}{\partial r} = -\frac{k(T_0 - T_\infty)}{R}\frac{\partial T^*}{\partial r^*}$$

$$= \frac{k(T_0 - T_\infty)}{Rr^{*2}}.$$ (5.2.17)

An important result is obtained if the temperature field is expressed in terms of the total heat generated by the spherical surface, Q_r, instead of the temperature at the surface. The total heat passing through a spherical shell of radius r is the product of the heat flux and the surface area of the shell,

$$Q_r = 4\pi r^2 q_r = 4\pi Rk(T_0 - T_\infty).$$ (5.2.18)

This is independent of the radius of the shell, since there is no generation or absorption of energy within the fluid. Using Eq. 5.2.18 to express $(T_0 - T_\infty)$ in terms of Q_r in Eq. 5.2.16, we obtain,

$$\boxed{T - T_\infty = \frac{Q_r}{4\pi Kr}.}$$ (5.2.19)

When expressed in this manner, the temperature field does not depend on the radius of the sphere R, but only on the total heat emitted by the sphere per unit time Q_r. Therefore, this solution is valid outside a sphere of any radius, provided the total heat emitted per unit time is Q_r. Specifically, it is valid in the limit $R \to 0$, the 'point source'. The temperature field due to a point source will be an important concept in our discussion of spherical harmonic expansions in Chapter 8.

The Nusselt number is,

$$\text{Nu} = \frac{q_r|_{r=R}}{k(T_0 - T_\infty)/d},$$ (5.2.20)

where $d = 2R$ is the diameter of the sphere. Substituting Eq. 5.2.17 in Eq. 5.2.20, the Nusselt number is a constant equal to 2,

$$\boxed{\text{Nu} = 2.}$$ (5.2.21)

This result applies in the absence of convective effects when the energy flux is entirely due to diffusion.

5.2.3   Conduction in a Spherical Shell

Consider a spherical shell with inner radius R_i and outer radius R_o, temperature T_i at the inner surface and T_o at the outer surface. The cross section is identical to Fig. 5.3(a), but the surfaces at $r = R_i$ and $r = R_o$ are spherical surfaces instead of cylindrical surfaces. The scaled temperature and radius are defined as $T^* = (T - T_i)/(T_o - T_i)$ and $r^* = (r/R_i)$. The steady solution is Eq. 5.2.14. The boundary conditions,

$$T^* = 0 \;\; (T = T_i) \;\; \text{at} \;\; (r = R_i) \;\; \text{or} \;\; r^* = 1, \qquad (5.2.22)$$

$$T^* = 1 \;\; (T = T_o) \;\; \text{at} \;\; (r = R_o) \;\; \text{or} \;\; r^* = (R_o/R_i), \qquad (5.2.23)$$

are used to determine the constants in Eq. 5.2.14, and the scaled temperature field is,

$$T^* = \frac{1 - (1/r^*)}{1 - (R_i/R_o)} \;\; \text{or} \;\; \frac{T - T_i}{T_o - T_i} = \frac{1 - (R_i/r)}{1 - (R_i/R_o)}. \qquad (5.2.24)$$

It is easily verified that Eq. 5.2.24 satisfies both boundary conditions, Eqs. 5.2.22 and 5.2.23.

The heat flux is,

$$q_r = -k \frac{\partial T}{\partial r} = -\frac{k(T_o - T_i)}{R_i} \frac{\partial T^*}{\partial r^*} = -\frac{k(T_o - T_i)}{r^2((1/R_i) - (1/R_o))}. \qquad (5.2.25)$$

The total heat transfer rate is the product of the heat flux and the surface area,

$$\boxed{Q_r = 4\pi r^2 q_r = -\frac{4\pi k(T_o - T_i)}{((1/R_i) - (1/R_o))}.} \qquad (5.2.26)$$

The heat transfer resistance, equivalent of Eq. 4.2.6 for a flat plate and Eq. 5.1.30 for a cylinder, is

$$\boxed{\mathcal{R} = -\frac{\Delta T}{Q_r} = \frac{((1/R_i) - (1/R_o))}{4\pi k}.} \qquad (5.2.27)$$

The heat transfer resistance due to concentric spherical shells is given by Eq. 4.2.7.

5.2.4   Unsteady Conduction in a Sphere

This problem is the analogue, in spherical co-ordinates, of the unsteady conduction into a cylinder in Section 5.1.4. A spherical particle of radius R and initial temperature T_1 is immersed in a fluid with temperature T_0. The volume of the surrounding fluid is large enough that the heat generated from the particle does not increase the temperature of the fluid. We would like to find out the variation of the temperature in the particle with time.

In the long time limit, the temperature of the particle is equal to the fluid temperature T_0. The scaled temperature is defined as $T^* = (T - T_0)/(T_1 - T_0)$, so that the scaled temperature is zero everywhere in the long time limit. The scaled radius is $r^* = (r/R)$, and the scaled time is $t^* = (t\alpha/R^2)$. With this, the unsteady energy balance equation is,

$$\frac{\partial T^*}{\partial t^*} = \frac{1}{r^{*2}} \frac{\partial}{\partial r^*} \left(r^{*2} \frac{\partial T^*}{\partial r^*} \right),$$

(5.2.28)

with the boundary condition $T = T_0$ at $r = R$,

$$T^* = 0 \quad \text{at} \quad r^* = 1.$$

(5.2.29)

The second boundary condition is the 'symmetry' condition at $r^* = 0$, similar to Eq. 5.1.57, at the axis of the cylindrical co-ordinate system. Since the temperature field is spherically symmetric, the derivative of the temperature at the centre will be independent of the direction of approach only if it is zero,

$$\frac{\partial T^*}{\partial r^*} = 0 \quad \text{at} \quad r^* = 0.$$

(5.2.30)

The initial condition is $T = T_1$ everywhere within the sphere at $t = 0$,

$$T^* = 1 \quad \text{at} \quad t^* = 0 \quad \text{for} \quad r^* < 1.$$

(5.2.31)

Eq. 5.2.28 is solved using the separation of variables procedure, where the temperature field is written as,

$$T^*(r^*, t^*) = F(r^*)G(t^*).$$

(5.2.32)

The above expression is substituted into Eq. 5.2.28, and the equation is divided by $F(r^*)G(t^*)$,

$$\frac{1}{F} \frac{1}{r^{*2}} \frac{\mathrm{d}}{\mathrm{d}r^*} \left(r^{*2} \frac{\mathrm{d}F}{\mathrm{d}r^*} \right) = \frac{1}{G} \frac{\mathrm{d}G}{\mathrm{d}t^*} = -\beta^2.$$

(5.2.33)

where β is the constant to be determined from the boundary condition at the surface of the sphere.

Eq. 5.2.33 for F is multiplied by r^{*2} and simplified,

$$r^{*2} \frac{\partial^2 F}{\partial r^{*2}} + 2r^* \frac{\partial F}{\partial r^*} = -\beta^2 r^{*2} F. \qquad (5.2.34)$$

The solution for this equation is,

$$\boxed{F(r^*) = \frac{C \sin(\beta r^*)}{r^*} + \frac{D \cos(\beta r^*)}{r^*},} \qquad (5.2.35)$$

where C and D are constants to be determined from the boundary conditions. The symmetry condition, Eq. 5.2.30, is satisfied only if D is zero. The discrete eigenvalues β are determined from the boundary condition, Eq. 5.2.29, which is equivalent to $F = 0$ at $r^* = 1$,

$$\boxed{\beta_i = i\pi,} \qquad (5.2.36)$$

where i is an integer. The solution of Eq. 5.2.33 for $G(t^*)$ is,

$$G(t^*) = e^{(-\beta_i^2 t^*)} = e^{(-i^2 \pi^2 t^*)}. \qquad (5.2.37)$$

The temperature field is,

$$T^* = \sum_{i=1}^{\infty} C_i \left(\frac{\sin(i\pi r^*)}{r^*} \right) e^{(-i^2 \pi^2 t^*)}, \qquad (5.2.38)$$

where $(\sin(i\pi r^*)/r^*)$ are the basis functions.

The constants C_i are determined from the initial condition,

$$T^* = 1 \quad \text{at} \quad t^* = 0. \qquad (5.2.39)$$

which is equivalent to,

$$\sum_{i=1}^{\infty} \frac{C_i \sin(i\pi r^*)}{r^*} = 1. \qquad (5.2.40)$$

The orthogonality relation for the basis functions $(\sin(i\pi r^*)/r^*)$ in a spherical co-ordinate system is,

$$\boxed{\langle (\sin(i\pi r^*)/r^*), (\sin(j\pi r^*)/r^*) \rangle = \int_0^1 r^{*2} \, dr^* \, \frac{\sin(i\pi r^*)}{r^*} \frac{\sin(j\pi r^*)}{r^*} = \frac{\delta_{ij}}{2}.}$$

$$(5.2.41)$$

There is a factor r^{*2} in the above integral because the surface area in a spherical co-ordinate system is proportional to r^{*2}. Also, note that the orthogonality relation, Eq. 5.2.41, is identical to Eq. 4.7.25 for a Cartesian co-ordinate system.

The left and right sides of Eq. 5.2.40 are multiplied by $r^{*2} \times (\sin{(j\pi r^*)}/r^*)$, and integrated over the domain $0 \leq r^* \leq 1$,

$$\sum_{i=1}^{\infty} C_i \int_0^1 r^{*2} \, dr^* \, \frac{\sin{(i\pi r^*)}}{r^*} \frac{\sin{(j\pi r^*)}}{r^*} = \int_0^1 r^{*2} \, dr^* \frac{\sin{(j\pi r^*)}}{r^*}. \qquad (5.2.42)$$

From the orthogonality relation, Eq. 5.2.41, the scalar product on the left is $(C_j/2)$, while the right side is evaluated by integration by parts,

$$\int_0^1 r^* dr^* \sin{(j\pi r^*)} = -\frac{r^* \cos{(j\pi r^*)}}{j\pi} \bigg|_{r^*=0}^{1} + \int_0^1 dr^* \frac{\cos{(j\pi r^*)}}{j\pi}$$

$$= -\frac{\cos{(j\pi)}}{j\pi} + \frac{\sin{(j\pi r^*)}}{(j\pi)^2} \bigg|_{r^*=0}^{1} = \frac{(-1)^{j-1}}{j\pi}. \qquad (5.2.43)$$

The coefficients C_j are,

$$C_j = \frac{2(-1)^{j-1}}{j\pi}. \qquad (5.2.44)$$

The final result for the temperature field is,

$$\boxed{T^* = \sum_{i=1}^{\infty} \frac{2(-1)^{i-1}}{i\pi} \left(\frac{\sin{(i\pi r^*)}}{r^*} \right) e^{(-i^2\pi^2 t^*)}.} \qquad (5.2.45)$$

The approximations for the solution, Eq. 5.2.45, truncated after one, three and five terms, are compared in Fig. 5.12(a). Converged solutions are obtained for the one-term approximation for $t^* \geq 0.3$, and the numerical error in the one-term approximation is less than 5% at $t^* = 0.1$. The one-term approximation predicts the wrong temperature at the centre for $t^* \leq 0.03$, but the three-term approximation is in good agreement with the five-term approximation. For $t^* = 0.01$, unphysical oscillations in the temperature profile are observed in the three-term approximation, but a smoother temperature profile is predicted by the five-term approximation.

In Fig. 5.12(b), the five-term approximation is compared with the solution for the diffusion into an infinite medium. For $t^* \ll 1$, it is expected that the temperature disturbance due to the surface is restricted to a thin region, and the temperature profile resembles that for diffusion into an infinite medium, Eq. 4.4.16. Here, $\xi = (R - r)/\sqrt{\alpha t} = (1 - r^*)/\sqrt{t^*}$ is the similarity variable based on the distance from the surface. The similarity solution, Eq. 4.4.16, is in quantitative agreement with the five-term approximation for $t^* = 0.01$, though the agreement is not as good at $t^* = 0.03$.

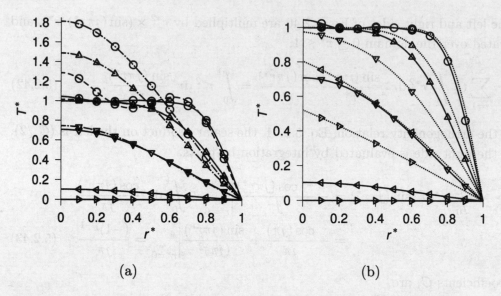

(a)   (b)

FIGURE 5.12. (a) The solution, Eq. 5.2.45Dissolution of a pillequation.5.2.45, for the heat conduction in a sphere for $t^* = 0.01$ ($\circ$), $t^* = 0.03$ ($\triangle$), $t^* = 0.1$ (∇), $t^* = 0.3$ ($\triangleleft$) and $t^* = 1$ ($\triangleright$). The solid lines — are the results when five terms are included in the expansion, the dashed lines $---$ are the results when three terms are included in the expansion, and the dot-dash lines $-\cdot-\cdot-$ are the results when one term is included in the expansion. (b) The five-term approximation, solid lines are compared with the temperature field for conduction into an infinite medium, Eq. 4.4.16, shown by the dotted lines, where the similarity variable is defined as $\xi = (R - r)/\sqrt{\alpha t} = (1 - r^*)/\sqrt{t^*}$.

Summary (5.2)

1. The important results in the derivation of the conservation equation in spherical co-ordinates are:

   (a) The rate of change of the field variable Φ_{fv} for an unsteady spherically symmetric configuration, Eq. 5.2.6, has the common form,

   $$\frac{\partial \Phi_{fv}}{\partial t} = -\frac{1}{r^2}\frac{\partial(r^2 \mathcal{J}_z)}{\partial r} + \mathcal{S}, \qquad (5.2.46)$$

   where the field variable Φ_{fv}, flux $\mathcal{J}_z$ and source $\mathcal{S}$ for mass and heat transfer are summarised in Table 4.1.

   (b) The constitutive relation Eq. 5.2.7 is substituted into Eq. 5.2.6 to obtain Eq. 5.2.8,

   $$\frac{\partial \Phi_{fv}}{\partial t} = \frac{1}{r^2}\frac{\partial}{\partial r}\left(\mathcal{D}r^2\frac{\partial \Phi_{fv}}{\partial r}\right) + \mathcal{S}, \qquad (5.2.47)$$

   where $\mathcal{D}$ is the generalised diffusivity.

(c) When the diffusivity is independent of the field variable, position or time, the simplified conservation equation, Eq. 5.2.9 is,

$$\frac{\partial \Phi_{fv}}{\partial t} = \frac{D}{r^2} \frac{\partial}{\partial r} \left(r^2 \frac{\partial \Phi_{fv}}{\partial r} \right) + S. \tag{5.2.48}$$

2. At steady state,

(a) the solution for the diffusion equation, Eq. 5.2.9 in the absence of sources, is Eq. 5.2.14, $\Phi_{fv} = (C/r) + D$, where C and D are constants determined from the boundary conditions. When expressed in terms of the total heat transfer rate from the particle, the steady solution, Eq. 5.2.19, is independent of the particle radius.

(b) the Nusselt number correlation for conduction heat transfer from a particle is Eq. 5.2.21.

(c) the heat transfer rate across the wall of a spherical shell of inner radius R_i, outer radius R_o and thermal conductivity k is Eq. 5.2.26. The resistance to heat transfer is given in Eq. 5.2.27.

3. For the unsteady heat conduction in a sphere:

(a) The separation of variables solution for the radial co-ordinate has the general form given in Eq. 5.2.35, and homogeneous boundary conditions are satisfied for discrete eigenvalues provided in Eq. 5.2.36. The basis functions are orthogonal when the definition, Eq. 5.2.41, is used for the scalar product.

(a) The series solution, Eq. 5.2.45Dissolution of a pillequation.5.2.45, satisfies the balance equation and the boundary and initial conditions.

(b) The series solution with 1–3 terms provides quantitatively accurate results for $t^* \geq 0.1$. The similarity solution for the diffusion into an infinite medium, Eq. 4.4.16, is valid for $t^* \leq 0.01$, when the penetration depth of the temperature disturbance into the sphere is much smaller than the radius.

Exercises

EXERCISE 5.1 Heat transfer takes place across the wall of a pipe with length 1 m, inner diameter 1 cm and outer diameter 1.1 cm. If the heat transfer rate is 2.3 kW/s when the

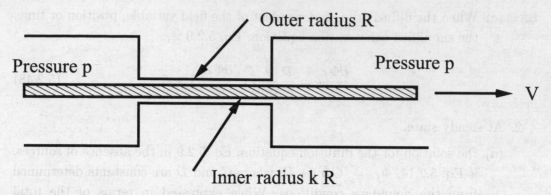

FIGURE 5.13. Wire coating in a die.

temperature difference across the wall is 50°C, what is the thermal conductivity of the pipe wall in W/m/°C?

EXERCISE 5.2 The wall of a pipe of inner diameter 1 cm and outer diameter 1.8 cm consists of an inner layer of thickness 2 mm made of steel with thermal conductivity 10 W/m/°C, and an outer layer of thickness 2 mm made of ceramic with thermal conductivity 40 W/m/°C. If the inner wall of the pipe is at a temperature of 100°C, and the outer wall of the pipe is at a temperature 20°C, what is the temperature at the interface between the ceramic and stainless steel in °C at steady state?

EXERCISE 5.3 A cylindrical pressure tank of diameter 10 cm and length 1 m is covered with a layer of insulation of thickness 1 cm with thermal conductivity 0.05 W/m/°C. Pressurised water initially at a temperature of 200°C is stored in the tank, and the ambient temperature is 20°C. The density and specific heat of water are 10^3 kg/m^3 and 4.2×10^3 J/kg/°C, respectively. How long will it take for the water temperature in the tank to decrease to 100°C?

EXERCISE 5.4 In wire-coating of dies, there is molten polymer in a thin annular region between a wire, which is an inner cylinder of radius kR, and a die which is an outer cylinder of radius R, as shown in Fig. 5.13. The wire is pulled with a constant velocity V, and the pressure is equal on both ends. Assuming the flow is steady, determine the fluid velocity and the flow rate if the molten polymer is modelled as a Newtonian fluid and a power-law fluid.

EXERCISE 5.5 For the flow between concentric cylinders discussed in Example 5.1.3, the flow is steady with concentric circular streamlines for certain ratios of the angular velocity of the inner and outer cylinders. However, for certain other ratios, the flow breaks up into 'convection rolls'. The Rayleigh criterion for break-up into convection rolls is,

$$\frac{\mathrm{d}(rv_\phi)^2}{\mathrm{d}r} < 0,$$

somewhere in the flow. For the special case where both cylinders are moving in the same direction—that is, $\Omega_i > 0$ and $\Omega_o > 0$—what is the condition on the ratio of the angular velocities of the inner and outer cylinders for the break-up into convection rolls?

EXERCISE 5.6 Atmospheric phenomena such as cyclones and tornadoes contain a 'vortex core', a long and thin region of rapid rotation which generates flow in the surrounding fluid. A vortex core can be approximated as a region of infinitesimal thickness where the angular velocity goes to infinity, and the 'circulation' $\Gamma = 2\pi r v_\phi$ is a constant in the limit $r \to 0$. If a vortex core with circulation Γ has formed at $r = 0$ in a fluid at $t = 0$, while the surrounding fluid for $r > 0$ is stationary, determine the velocity profile for the flow around the vortex core as a function of r and t using a similarity transform. **Hint:** Instead of the velocity v_ϕ, use the circulation $\Gamma(r, t) = 2\pi r v_\phi(r, t)$ as the dependent variable.

EXERCISE 5.7 In a pasteurisation process, milk is passed through a heated tube of diameter 2 mm. The tube wall is at 85°C, and the initial temperature of the milk is 25°C. What is the time taken for the milk at the centre of the tube to increase to the pasteurisation temperature of 72°C? Assume the thermal diffusivity of milk is 1.42×10^{-7} m^2/s. Compare the results for one and three terms in the expansion, Eq. 5.1.74. What is the maximum time in seconds for $t^* = 0.01$ where similarity solution, Eq. 4.4.16, can be used? What is the minimum time in seconds for $t^* = 0.1$ where the one-term approximation of Eq. 5.1.74 is quantitatively accurate?

EXERCISE 5.8 Consider the unsteady heat transfer in an annulus of inner radius R_i and outer radius R_o shown in Fig. 5.3(a). The temperatures at the inner and outer surfaces are both equal to T_i for $t < 0$. At $t = 0$, the temperature of the outer surface is increased to T_o. The steady solution is given by Eq. 5.1.21. What are the boundary and initial conditions for the transient solution? What are the eigenvalues and the orthogonal basis functions for the transient solution?

EXERCISE 5.9 Verify that Eq. 4.5.18 is a solution for the diffusion equation in two dimensions, and the variance of the distribution is given by Eq. 4.5.20.

EXERCISE 5.10 Heat transfer takes place across the wall of a spherical shell with inner radius 1 cm and outer radius 1.2 cm. If the heat transfer rate is 151 W when the temperature difference across the wall is 10°C, what is the thermal conductivity of the shell wall in W/m/°C?

EXERCISE 5.11 A 'bomb calorimeter', used for measuring the calorific value of fuels, consists of a spherical shell of inner diameter 2 cm and outer diameter 2.4 cm, made of stainless steel of thermal conductivity $k_s = 50$ W/m/°C. The fuel is charged into the cavity within the shell, the calorimeter is immersed into a bath of fluid, and the fuel is ignited. The bath is cooled and well stirred, and the temperature of the bath is maintained at 20°C. Immediately after ignition of the fuel, the temperature of the fuel in the calorimeter increases to 200°C, and after 1 minute, the temperature decreases to 100°C. What is the total heat generated in the calorimeter? Assume that conduction across the shell is a steady state process.

EXERCISE 5.12 A spherical particle of diameter 1 mm, density 8×10^3 kg/m^3, thermal conductivity 4 W/m/°C, and specific heat 285 J/kg/°C is placed in a water with density 10^3 kg/m^3, thermal conductivity 0.6 W/m/°C, and specific heat 4.2×10^3 J/kg/°C. The particle is heated with a microwave with uniform power density 2.8×10^8 W/m^3. At steady state, what is the temperature at the surface of the particle, and the maximum temperature in the particle, if the ambient temperature is 25°C?

EXERCISE 5.13 Consider the equivalent, in spherical co-ordinates, of the heat conduction from a wire discussed in Section 5.1.3. A point source of heat is placed at the origin in an infinite domain where the temperature is a constant, T_∞, for $t < 0$. At $t = 0$, the point source starts emitting energy Q per unit time. Is it possible to obtain a similarity solution for the temperature field? What should be the time variation of Q to obtain a similarity solution?

EXERCISE 5.14 Verify that Eq. 4.5.23 is a solution of the diffusion equation in three dimensions, and that the variance is given by Eq. 4.5.24.

EXERCISE 5.15 Peas of diameter 0.5 cm are heated in water at 60°C. The initial temperature of the pea is 25°C and the surface temperature is equal to the water temperature. What is the time taken for the minimum temperature in the pea to increase to 50°C? What is the maximum time in seconds for $t^* = 0.01$ where similarity solution Eq. 4.4.16, can be used? What is the minimum time in seconds for $t^* = 0.1$ where the one-term approximation of Eq. 5.2.45 is valid? The density, specific heat and thermal conductivity of the pea are 500 kg/m^3, 7.6×10^3 J/kg/°C and 0.175 W/m/°C, respectively.

Appendix

5.A   Orthogonality Relation in Cylindrical Co-ordinates

The derivation of the orthogonality relation for Bessel functions is as follows. The basis functions, defined as $\Psi(\beta_i r^*) = A J_n(\beta_i r^*) + B Y_n(\beta_i r^*)$, satisfy the Bessel Eq. 5.1.64, and homogeneous boundary conditions at the two boundaries $r^* = r_1^*$ and $r^* = r_2^*$. This corresponds to the heat conduction in an annulus of inner and outer radius r_1^* and r_2^* considered in Exercise 5.8. The heat conduction in a cylinder in Section 5.1.4 is a special case for which $r_1^* = 0$.

Eq. 5.1.64 is rewritten using the substitution $r^\dagger = \beta_i r^*$, and the equation is divided by r^{*2},

$$\frac{1}{r^*}\frac{\mathrm{d}}{\mathrm{d}r^*}\left(r^*\frac{\mathrm{d}\Psi(\beta_i r^*)}{\mathrm{d}r^*}\right) - \frac{n^2\Psi(\beta_i r^*)}{r^{*2}} + \beta_i^2\Psi(\beta_i r^*) = 0. \qquad (5.A.1)$$

The basis functions satisfy the homogeneous boundary conditions

$$\Psi(\beta_i r^*) = 0 \text{ or } \frac{\mathrm{d}\Psi(\beta_i r^*)}{\mathrm{d}r^*} = 0 \text{ or } \Psi(\beta_i r^*) = C\frac{\mathrm{d}\Psi(\beta_i r^*)}{\mathrm{d}r^*}, \qquad (5.A.2)$$

at both boundaries $r^* = r_1^*$ and $r^* = r_2^*$ for all i, where C is a constant. The scalar product is defined as,

$$\langle \Psi(\beta_i r^*), \Psi(\beta_j r^*) \rangle = \int_{r_1^*}^{r_2^*} r^* \, dr^* \, \Psi(\beta_i r^*) \Psi(\beta_j r^*). \tag{5.A.3}$$

The scalar product of 5.A.1 with $\Psi(\beta_j r^*)$ is,

$$\left\langle \left[\frac{1}{r^*} \frac{d}{dr^*} \left(r^* \frac{d\Psi(\beta_i r^*)}{dr^*} \right) - \frac{n^2 \Psi(\beta_i r^*)}{r^{*2}} \right], \Psi(\beta_j r^*) \right\rangle = -\beta_i^2 \langle \Psi(\beta_i r^*), \Psi(\beta_j r^*) \rangle. \tag{5.A.4}$$

The left side of Eq. 5.A.4 can also be evaluated using integration by parts,

$$\left\langle \left[\frac{1}{r^*} \frac{d}{dr^*} \left(r^* \frac{d\Psi(\beta_i r^*)}{dr^*} \right) - \frac{n^2 \Psi(\beta_i r^*)}{r^{*2}} \right], \Psi(\beta_j r^*) \right\rangle$$

$$= \int_{r_1^*}^{r_2^*} r^* \, dr^* \left[\frac{1}{r^*} \frac{d}{dr^*} \left(r^* \frac{d\Psi(\beta_i r^*)}{dr^*} \right) - \frac{n^2 \Psi(\beta_i r^*)}{r^{*2}} \right] \Psi(\beta_j r^*)$$

$$= \left[r^* \Psi(\beta_j r^*) \frac{d\Psi(\beta_i r^*)}{dr^*} \right] \Big|_{r_1^*}^{r_2^*} - \left[r^* \Psi(\beta_i r^*) \frac{d\Psi(\beta_j r^*)}{dr^*} \right] \Big|_{r_1^*}^{r_2^*}$$

$$+ \int_{r_1^*}^{r_2^*} r^* \, dr^* \Psi(\beta_i r^*) \left[\frac{1}{r^*} \frac{d}{dr^*} \left(r^* \frac{d\Psi(\beta_j r^*)}{dr^*} \right) - \frac{n^2 \Psi(\beta_j r^*)}{r^{*2}} \right]$$

$$= \left[r^* \Psi(\beta_j r^*) \frac{d\Psi(\beta_i r^*)}{dr^*} \right] \Big|_{r_1^*}^{r_2^*} - \left[r^* \Psi(\beta_i r^*) \frac{d\Psi(\beta_j r^*)}{dr^*} \right] \Big|_{r_1^*}^{r_2^*}$$

$$- \beta_j^2 \int_{r_1^*}^{r_2^*} r^* \, dr^* \, \Psi(\beta_i r^*) \Psi(\beta_j r^*). \tag{5.A.5}$$

Here, the final step follows from Eq. 5.A.1. The first two terms on the right in Eq. 5.A.5 are zero if the homogeneous boundary conditions 5.A.2 are satisfied. Equating 5.A.4 and 5.A.5, we obtain,

$$(\beta_j^2 - \beta_i^2) \int_{r_1^*}^{r_2^*} r^* \, dr^* \, \Psi(\beta_i r^*) \Psi(\beta_j r^*) = 0. \tag{5.A.6}$$

Therefore, the scalar product in Eq. 5.A.6 is zero if i and j are not equal, and the solutions of the Bessel equation, Eq. 5.1.64, satisfy the orthogonality relations, Eqs. 5.1.71 and 5.1.89.

Pressure-driven Flow

The momentum flux or the force per unit area on a surface within a fluid can be separated into two components: the pressure and the shear stress. The latter is due to variations in the flow velocity, while the former is present even when there is no flow. Pressure has no analogue in mass and heat transfer, where the fluxes are entirely due to the variations in the concentration/temperature fields. The fluid pressure is the compressive force per unit area exerted on a surface within the fluid in the direction perpendicular to the surface. At a point within the fluid, the pressure is a scalar which is independent of the orientation of the surface; the direction of the force exerted due to the pressure is along the perpendicular to the surface.

There is a distinction between the thermodynamic pressure and the dynamical pressure that drives fluid flow. The thermodynamic pressure is an absolute pressure which is calculated, for example, using the ideal gas equation of state. In contrast, flow is driven by the pressure difference between two locations in an incompressible flow. The velocity field depends on the variations in the dynamical pressure, and the flow field is unchanged if a constant pressure is added everywhere in the domain for an incompressible flow.

A potential flow is a limiting case of a pressure-driven flow where viscous effects are neglected. Some applications of potential flows are first reviewed in Section 6.1. The velocity profile and the friction factor for the laminar flow in a pipe is derived in Section 6.2. As discussed in Chapter 2, there is a transition from a laminar to a turbulent flow when the Reynolds number exceeds a critical value. The salient features of a turbulent flow are discussed in Section 6.3. The oscillatory flow in a pipe due to a sinusoidal pressure variation across the ends is considered in Section 6.4. This flow is used to illustrate the use of complex variables for oscillatory flows, and the approximations and analytical techniques used in the convection-dominated and diffusion-dominated regimes.

6.1 Potential Flow: The Bernoulli Equation

At high Reynolds number, viscous effects are neglected in the bulk of the flow, and there is a balance between the pressure, inertial and body forces. The Bernoulli equation relates the pressure and the velocity in potential flow. Since dissipation of energy occurs due to fluid viscosity, energy is conserved in a potential flow when viscosity is neglected. There is no shear stress tangential to a surface when the viscosity or momentum diffusion is neglected. Due to this, it is not possible to apply the no-slip condition for the tangential velocity at the surface, and only the normal velocity (no-penetration) condition is applied in potential flow.

Formally, the Bernoulli equation is valid for an incompressible (constant density), inviscid (zero viscosity), and irrotational flow, where the vorticity or the curl of the velocity (defined in Chapter 7) is zero. At steady state, the Bernoulli equation is,

$$p + \tfrac{1}{2}\rho v^2 + \rho g z = p_0, \qquad (6.1.1)$$

where p and ρ are the pressure and density, v is the fluid velocity, g is the acceleration due to gravity, z is the height which increases in the direction opposite to gravity, and p_0 is a constant. Eq. 6.1.1 is often interpreted as an energy conservation equation, where the second term on the left $e_k = \tfrac{1}{2}\rho v^2$ is the kinetic energy per unit volume. The third term on the left $e_p = \rho g z$ is the potential energy per unit volume in a gravitational field, and the force exerted on the fluid in the vertical direction is $f_z = -(\mathrm{d}e_p/\mathrm{d}z) = -\rho g$. The first term on the left is interpreted as the pressure energy per unit volume. It is important to note that Eq. 6.1.1 is valid for steady flows; there is an additional acceleration term in the equation for an unsteady flow which is derived in Section 7.3.5 (Chapter 7).

EXAMPLE 6.1.1: A tank of area of cross section 2 m^2 is filled with water (density 10^3 kg/m^3 and viscosity 10^{-3} kg/m/s) up to a height of 2 m. The tank contains an outlet of diameter 2 cm with a valve at a height of 0.5 m, as shown in Fig. 6.1. How long does it take for the level in the tank to decrease to 0.5 m if the valve is opened? What is the maximum horizontal distance L travelled by the stream? Assume the flow is inviscid, and the fluid velocity is in the horizontal direction at the outlet.

If the velocity of the fluid exiting the outlet is v_o, the volumetric outflow rate is $\dot{V} = A_o V_o$, where $A_o = \pi(0.02 \text{ m})^2/4 = 3.14 \times 10^{-4}$ m^2 is the area of cross section

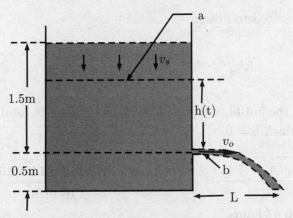

FIGURE 6.1. Flow of water from a tank of area of cross section 2 m^2 through an outlet at a height 0.5 m above the bottom.

of the outlet. The outflow results in a decrease in the height of the tank, and the velocity of the surface is $v_s = (\dot{V}/A_t)$, where $A_t = 2$ m^2 is the tank area. The relation between the outlet velocity and surface velocity is,

$$v_s = \frac{\mathrm{d}h}{\mathrm{d}t} = -\frac{A_o v_o}{A_t}. \tag{6.1.2}$$

There is a negative sign on the right in Eq. 6.1.2 because the surface height decreases for positive v_o as the fluid discharges from the outlet.

The pressure at the location a on the free surface of water in Fig. 6.1 is atmospheric pressure p_{atm}. When the outlet is opened, the pressure at the outlet b is also p_{atm}. The Bernoulli equation is applied between the location a at the surface of water and b at the outlet outside the tank to obtain the following relation between the velocity and the height,

$$p_{atm} + \tfrac{1}{2}\rho v_s^2 + \rho g h = p_{atm} + \tfrac{1}{2}\rho v_o^2. \tag{6.1.3}$$

The relation between the outlet and surface velocities, Eq. 6.1.2, is substituted in Eq. 6.1.3, and the latter is solved to obtain the outlet velocity,

$$v_o = \sqrt{\frac{2gh}{1 - (A_o/A_t)^2}}. \tag{6.1.4}$$

The velocity of the surface of the tank is determined using Eq. 6.1.2,

$$\frac{\mathrm{d}h}{\mathrm{d}t} = -\sqrt{\frac{2gh}{(A_t/A_o)^2 - 1}}. \tag{6.1.5}$$

Eq. 6.1.5 is integrated to obtain

$$2(h_0^{1/2} - h^{1/2}) = t\sqrt{\frac{2g}{(A_t/A_o)^2 - 1}}. \tag{6.1.6}$$

Here, we have used the initial condition $h = h_0$ at $t = 0$. The total time of discharge, t_d, is the time at which $h = 0$.

$$t_d = \sqrt{\frac{2h_0((A_t/A_o)^2 - 1)}{g}} = \sqrt{\frac{2 \times 1.5 \text{ m} \times ((2 \text{ m}^2/3.14 \times 10^{-4} \text{ m}^2)^2 - 1)}{10 \text{ m/s}^2}}$$

$$= 3489 \text{ s} \sim 0.97 \text{ hours}. \tag{6.1.7}$$

In the present example, the area of the outlet is much smaller than the area of the tank. Therefore, little error is incurred if the term $(A_o/A_t)^2$ is neglected in comparison to 1 in the denominator of Eq. 6.1.4; this approximation simplifies the calculation. This term has to be included in the calculation when the tank and outlet areas are comparable.

The maximum discharge length L is determined from the horizontal velocity of discharge at the outlet v_o and the time taken for the stream to descend the distance 0.5 m due to gravitational acceleration. The maximum discharge velocity v_{om} at the outlet at $t = 0$ is determined from Eq. 6.1.4,

$$v_{om} = \sqrt{\frac{2 \times 10 \text{ m/s}^2 \times 1.5 \text{ m}}{(1 - (3.14 \times 10^{-4} \text{ m}^2/2 \text{ m}^2)^2)}} = 5.48 \text{ m/s}. \tag{6.1.8}$$

The velocity, Eq. 6.1.8, is in the horizontal direction at the outlet, and the initial vertical velocity is zero. The outlet stream travels downwards due to gravity, and the time taken t_s to travel a vertical distance of $\Delta z = 0.5$ m is determined from the relation $\Delta z = \frac{1}{2}gt_s^2$,

$$t_s = \sqrt{\frac{2 \times 0.5 \text{ m}}{10 \text{ m/s}^2}} = 0.316 \text{ s}. \tag{6.1.9}$$

The maximum horizontal distance travelled is $v_{om}t_s = 1.732$ m. ☐

In Example 6.1.1, it is assumed that the diameter of the exit stream is the same as that of the outlet. This is not strictly true; the diameter of the exit stream depends on the detailed shape of the outlet[5]. If we examine the exit stream from a hole in a flat wall shown in Fig. 6.2(a), the exit stream converges. The diameter of the

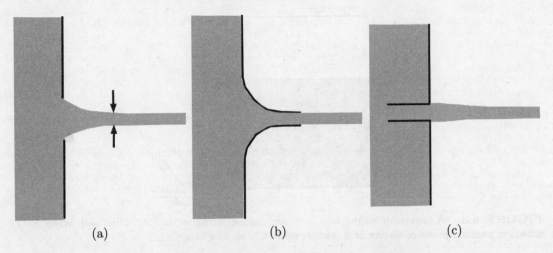

(a) (b) (c)

FIGURE 6.2. Exit streams from different outlet shapes.

stream is smaller than the hole in the wall at the location shown by the arrows where the streamlines are parallel. Therefore, the actual flow rate is about 0.6 times that predicted if the diameter of the exit stream is equal to that of the outlet. In contrast, for a slowly converging outlet shown in Fig. 6.2(b), the streamlines are parallel at the exit, and the diameter of the exit stream is equal to that predicted based on the area of cross section of the outlet. For a recessed exit where the outlet is through a pipe within the container as shown in Fig. 6.2(c), it can be shown that the actual flow rate is one half of that predicted assuming that the area of the exit stream is the same as the area of the exit. Therefore, it is necessary to incorporate a correction factor which depends on the outlet shape in order to accurately predict the flow rate of the exit stream using Bernoulli's equation.

EXAMPLE 6.1.2: A tank with area of cross section 1 m^2 and height 2 m is initially filled with air at atmospheric pressure. Water is filled in from the bottom using a constant pressure pump which operates at a pressure of 2 atm, and air is not allowed to escape, as shown in Fig. 6.3. Up to what height will water fill into the tank? The density of water is 10^3 kg/m^3, the hydrostatic pressure due to the air can be neglected because the density of air is much smaller than that of water, the pressure-volume relationship for air is given by the ideal gas law and the temperature is a constant.

Solution: If the initial height of the tank is h and water fills up to a height of h_w, the height $h - h_w$ above the water is filled with air. Since the air is not allowed to escape, the air is pressurised at constant temperature. The initial pressure and

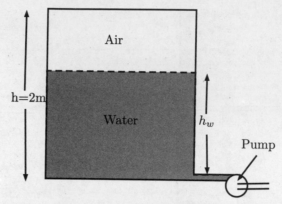

FIGURE 6.3. A tank with height $h = 2$ m and area of cross section A_t, filled with water by a constant pressure pump operating at 2 atm pressure, fills up to a height h_w.

volume of air are p_{atm} and hA_t, respectively, where A_t is the area of cross section of the tank. After the water is filled to a height h_w, the volume of air is $A_t(h - h_w)$. From the ideal gas law, the product of pressure and volume is a constant under isothermal conditions, and so the final pressure of air is $(hp_{atm}/(h - h_w))$.

The hydrostatic pressure of water filled up to height h_w is $\rho g h_w$. Therefore, the total pressure $(hp_{atm}/(h - h_w)) + \rho g h_w$, is equal to the pressure generated by the pump, $2p_{atm}$.

$$\frac{hp_{atm}}{h - h_w} + \rho g h_w = 2p_{atm}. \tag{6.1.10}$$

The above equation is multiplied by $h - h_w$ to obtain a quadratic equation for h_w,

$$hp_{atm} + \rho g (h - h_w)h_w - 2p_{atm}(h - h_w) = 0. \tag{6.1.11}$$

The quadratic equation for h_w, Eq. 6.1.11, is solved to obtain,

$$h_w = \frac{\rho g h + 2p_{atm} - \sqrt{4p_{atm}^2 + (\rho g h)^2}}{2\rho g}. \tag{6.1.12}$$

Substituting $p_{atm} = 1.013 \times 10^5$ Pa, $h = 2$ m, $g = 10$ m/s^2, and $\rho = 10^3$ kg/m^3, the result $h_w = 0.95$ m is obtained. □

6.1.1   Venturi Meter

The Bernoulli equation is the basis for the flow rate measurement in a Venturi meter, which consists of a constriction or 'throat' in a pipe within which the flow is to be measured, as shown in Fig. 6.4. There are pressure tappings in the pipe

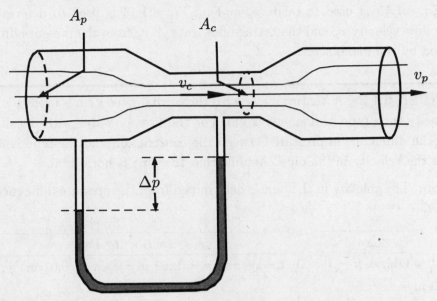

FIGURE 6.4. A Venturi meter consists of a constriction in a pipe. The flow rate is measured from the difference in pressure between the pipe and the constriction.

and in the constriction connected to a manometer, which is used to measure the pressure difference Δp between the constriction and the pipe. The area cross section of the pipe and constriction are A_p and A_c, respectively. The flow velocity in the constriction and in the pipe are related by the condition that the volumetric flow rate across both cross sections is the same at steady state,

$$A_p v_p = A_c v_c, \tag{6.1.13}$$

where v_p and v_c are the flow velocities in the pipe and constriction, as shown in Fig. 6.4. The Bernoulli equation applied between two locations in the pipe and the constriction is,

$$p_p + \tfrac{1}{2}\rho v_p^2 = p_c + \tfrac{1}{2}\rho v_c^2, \tag{6.1.14}$$

where p_p and p_c are the pressures in the pipe and constriction, respectively. Since the velocity in the constriction is higher than that in the pipe, the pressure in the constriction is lower. The difference in pressure between the pipe and constriction is,

$$\Delta p = p_p - p_c = \tfrac{1}{2}\rho(v_c^2 - v_p^2) = \tfrac{1}{2}\rho v_p^2 \left(\frac{A_p^2}{A_c^2} - 1 \right). \tag{6.1.15}$$

Here, Eq. 6.1.13 is used to relate v_p and v_c. Eq. 6.1.15 is used to determine the average flow velocity v_p and the average flow rate $A_p v_p$ from the pressure difference measured by the manometer.

EXAMPLE 6.1.3: A Venturi meter with throat diameter 1 cm is used to measure the velocity in a pipe of diameter 2 cm. For the flow of water with density 1000 kg/m^3, the difference in pressure between the constriction and the pipe is 30 kPa. What is the velocity in the pipe? Assume that the pipe is horizontal.

Solution: The velocity in the pipe is determined from the pressure difference using Eq. 6.1.15,

$$v_p = \sqrt{\frac{2\Delta p}{\rho[(A_p/A_c)^2 - 1]}} = \sqrt{\frac{2 \times 3 \times 10^4 \text{ Pa}}{10^3 \text{ kg/m}^2[(\pi \times (0.02 \text{ m})^2)^2/(\pi \times (0.01 \text{ m})^2)^2 - 1]}}$$
$$= 2 \text{ m/s.} \qquad (6.1.16)$$

$\square$

6.1.2   Flow over a Weir

The Bernoulli equation is used to determine the flow rate in an open channel using a weir, which is a flow obstruction placed in the path of the fluid. The flow rate can be calculated for a 'broad-crested' weir which has a gentle slope, as shown in Fig. 6.5(a). The height of the obstruction h is a function of the stream-wise co-ordinate x. Due to the obstruction, there is an increase in the depth of the water upstream of the weir, denoted d_u. The depth of the water above the bottom of channel, denoted d, decreases as the water passes over the weir. If we assume that the water is stationary far upstream of the weir, the velocity of water at the free surface above the weir is,

$$v = \sqrt{2g(d_u - h - d)}, \qquad (6.1.17)$$

where $d_u - h - d$ is the decrease in height as the water passes over the weir. The flow rate $\dot{V}$ per unit width perpendicular to the weir is related to the velocity,

$$\dot{V} = dv. \qquad (6.1.18)$$

The difference between the upstream depth and the height of the weir is expressed as,

$$d_u - h = (d_u - h - d) + d = \frac{v^2}{2g} + \frac{\dot{V}}{v}. \qquad (6.1.19)$$

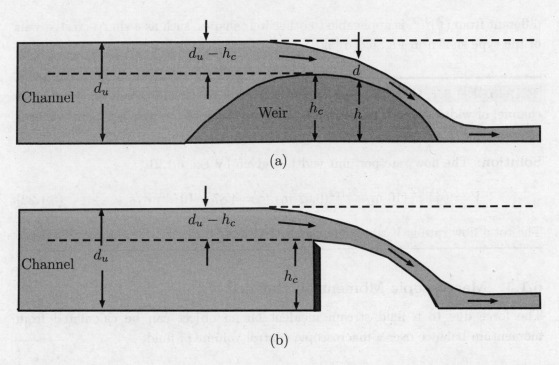

FIGURE 6.5. Flow over a channel smooth-crested weir (a) and a sharp-crested weir (b). The height of the crest of the weir is h_c, and the depth of water in the channel upstream of the weir is d_u.

Here, Eqs. 6.1.17 and 6.1.18 are used for $(d_u - h - d)$ and d, respectively. The difference $d_u - h$ (left side of Eq. 6.1.19) has a minimum with respect to the stream-wise co-ordinate for

$$\frac{v}{g} - \frac{\dot{V}}{v^2} = 0 \;\Rightarrow\; v = (\dot{V}g)^{1/3}. \tag{6.1.20}$$

This minimum occurs at the crest of the weir, where the height is $h = h_c$. Substituting the solution for the velocity, Eq. 6.1.20, into Eq. 6.1.19 at the crest of the weir, $h = h_c$,

$$d_u - h_c = \frac{3}{2}\left(\frac{\dot{V}^2}{g}\right)^{1/3} \;\Rightarrow\; \dot{V} = \left(\frac{2}{3}\right)^{3/2} g^{1/2}(d_u - h_c)^{3/2}. \tag{6.1.21}$$

Thus, the flow rate is determined from the difference between the upstream depth and the height of the crest of the weir. It should be noted that $\dot{V}$, which is the flow rate per unit width, has to be multiplied by the total width to determine the volumetric flow rate.

The calculation was carried out for a smooth-crested weir above which there are smooth fluid streamlines. Eq. 6.1.21, with a modified numerical prefactor slightly different from $\left(\frac{2}{3}\right)^{3/2}$, is applicable to other weir shapes, such as a sharp-crested weir of the type shown in Fig. 6.5(b)[5].

EXAMPLE 6.1.4: A broad-crested weir of height 20 cm is placed in a water channel of width 0.5 m. The depth of water upstream of the weir is 22 cm. What is the flow rate?

Solution: The flow rate per unit width is given by Eq. 6.1.21,

$$\dot{V} = \left(\tfrac{2}{3}\right)^{3/2} (10 \text{ m/s}^2)^{1/2}(0.02 \text{ m})^{3/2} = 4.87 \times 10^{-3} \text{ m}^2/\text{s}. \qquad (6.1.22)$$

The total flow rate is $\dot{V}$ times the width, $2.43 \times 10^{-3} \text{ m}^3/\text{s}$. $\qquad\qquad$ $\square$

6.1.3 Macroscopic Momentum Balance

The force due to a fluid stream incident on an object can be calculated from momentum balance over a macroscopic control volume of fluid,

Rate of momentum in − Rate of momentum out + Force on fluid = 0. $\quad$ (6.1.23)

Since the force on the object is the negative of the force on the fluid,

$$\boxed{\text{Force on object} = \text{Rate of momentum in} - \text{Rate of momentum out.}} \quad (6.1.24)$$

It is important to note that force is a vector, and Eq. 6.1.24 applies for each component of the force.

EXAMPLE 6.1.5: A horizontal jet of liquid of density ρ, velocity v and diameter d is incident on a vertical wall, as shown in Fig. 6.6. Determine the force perpendicular to the wall. Neglect the gravitational force.

Solution: Consider the control volume shown by the dotted line in Fig. 6.6. In the direction perpendicular to the wall, the momentum flux entering the control volume per unit time is $\rho v^2(\pi d^2/4)$, the product of the volumetric flux (product of the velocity perpendicular to the surface v and the area of cross section $(\pi d^2/4)$) and the momentum density perpendicular to the surface ρv. The momentum leaving the volume in the direction perpendicular to the wall is zero, because the momentum of the stream leaving the control volume is parallel to the wall. Therefore, the force on the wall is $\rho v^2(\pi d^2/4)$. $\qquad\qquad$ $\square$

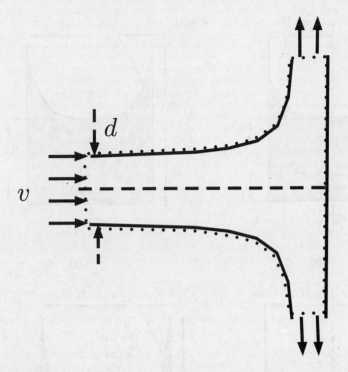

FIGURE 6.6. A jet of liquid of diameter d incident on a vertical wall.

6.1.4 Rotating Liquids

When a tank containing a liquid is rotated, there is an outward centrifugal force due to which the height decreases at the centre and increases at the periphery. The liquid is at rest in the rotating reference frame, but there is an outward centrifugal force exerted per unit volume, $f_r = \rho\Omega^2 r$, due to the rotation, where Ω is the angular velocity and r is the distance from the axis of rotation. If the centrifugal force density is expressed as $f_r = -(de_r/dr)$, then the potential energy per unit volume due to rotation is $e_r = -\frac{1}{2}\rho\Omega^2 r^2$. The Bernoulli Eq. 6.1.1 has to be modified to incorporate the potential energy per unit volume,

$$\boxed{p + \rho g z - \tfrac{1}{2}\rho\Omega^2 r^2 = p_0.} \tag{6.1.25}$$

Eq. 6.1.25 is used to determine the equation of the free surface in a rotating tank of liquid shown in Fig. 6.7(a). Since the pressure is equal to the atmospheric pressure at all points on the surface,

$$\rho g z - \tfrac{1}{2}\rho\Omega^2 r^2 = p_0 - p_{atm}. \tag{6.1.26}$$

The difference $p_0 - p_{atm} = \rho g\, h|_{r=0}$ in the above equation is determined from the condition that the height is $h|_{r=0}$ on the axis of rotation of the tank, $r = 0$.

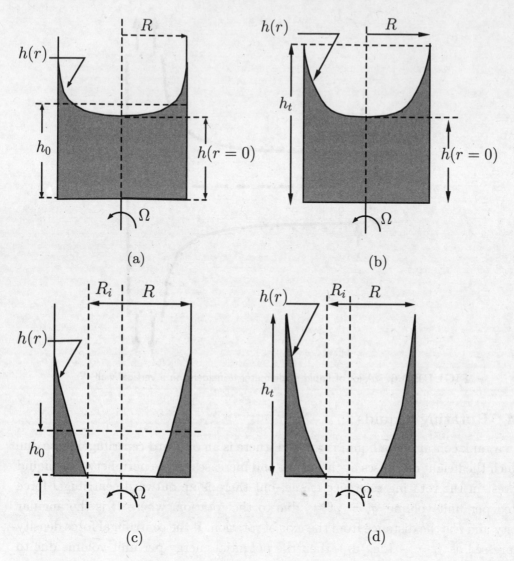

FIGURE 6.7. The shapes of the surface of a liquid in a rotating tank. Here, h_t and R are the height and radius of the tank, h_0 is the initial height of the liquid when there is no rotation, and Ω is the angular velocity.

Therefore, the equation of the surface of the liquid is,

$$h(r) = h|_{r=0} + \frac{\Omega^2 r^2}{2g}, \qquad (6.1.27)$$

where $h(r)$ is the height at the distance r from the axis of rotation. The total volume of liquid in the rotating tank is,

$$V = \int_0^R 2\pi r \, dr \, h(r) = \pi R^2 \left(h|_{r=0} + \frac{\Omega^2 R^2}{4g} \right). \qquad (6.1.28)$$

If h_0 is the height of the liquid in the tank when it is stationary, the volume is $V = \pi R^2 h_0$. Equating the the volumes of the stationary and rotating fluids,

$$\pi R^2 h_0 = \pi R^2 \left(h|_{r=0} + \frac{\Omega^2 R^2}{4g} \right) \Rightarrow h|_{r=0} = h_0 - \frac{\Omega^2 R^2}{4g}. \tag{6.1.29}$$

Substituting Eq. 6.1.29 in Eq. 6.1.27, the equation for the surface of the liquid is,

$$h(r) = h_0 + \frac{\Omega^2}{2g} \left(r^2 - \frac{R^2}{2} \right). \tag{6.1.30}$$

Eq. 6.1.30 applies for $h_0 > (\Omega^2 R^2 / 4g)$, so that the height at $r = 0$ is positive, and when the height of the tank is greater than $h|_{r=R} = h_0 + (\Omega^2 R^2 / 4g)$, as shown in Fig. 6.7(a). The cases shown in Figs. 6.7(b), (c) and (d), where there is either spillover from the top of the tank or where the bottom is exposed, are considered in Exercise 6.8.

Summary (6.1)

---

1. The Bernoulli equation, Eq. 6.1.1, relates the pressure and the fluid velocity for a potential flow (incompressible, inviscid and irrotational flow).

2. For a rotating fluid, the modified Bernoulli equation, Eq. 6.1.25, incorporates the centrifugal energy.

3. The force on an object due to an incident fluid stream is determined from the macroscopic balance, Eq. 6.1.24.

6.2   Laminar Flow in a Pipe

The configuration and co-ordinate system for the flow in a cylindrical pipe of radius R driven by a pressure difference between the two ends is shown in Fig. 6.8. The axis of the pipe is along the x direction, and the radial co-ordinate r is the distance from the axis. The flow is unidirectional; the only non-zero component of the velocity is v_x in the axial direction. The zero velocity condition is applied at the pipe wall, $r = R$. It is assumed that the flow is axisymmetric and fully developed (no velocity variation in the axial direction).

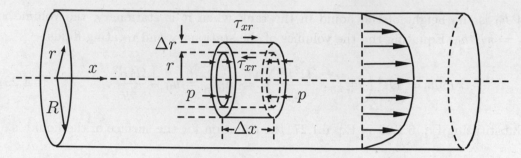

FIGURE 6.8. Flow in a cylindrical pipe due to a pressure difference between the inlet and outlet.

The axial momentum conservation equation is derived for an annular shell of thickness Δr, length Δx and volume $2\pi r\,\Delta r\,\Delta x$ shown in Fig. 6.8. The equation for the x component of the momentum is

$$\begin{bmatrix} \text{Rate of change} \\ \text{of } x \text{ momentum} \end{bmatrix} = \begin{bmatrix} \text{Rate of } x \\ \text{momentum in} \end{bmatrix} - \begin{bmatrix} \text{Rate of } x \\ \text{momentum out} \end{bmatrix}$$

$$+ \begin{bmatrix} \text{Sum of } x \\ \text{forces} \end{bmatrix}. \tag{6.2.1}$$

The rate of change of momentum within the differential volume in a time interval Δt is,

$$\begin{matrix} \text{Rate of change} \\ \text{of } x \text{ momentum} \end{matrix} = \frac{\Delta(\rho v_x)}{\Delta t}(2\pi r\Delta r\Delta x), \tag{6.2.2}$$

where $\Delta(\rho v_x)$ is the change in the momentum per unit volume.

Of the four surfaces for the differential volume, two are perpendicular to the x axis and are located at x and $x+\Delta x$, and two perpendicular to the radial co-ordinate and are located at r and $r + \Delta r$. The force in the x direction acting on the surfaces at x and $x + \Delta x$ is due to the fluid pressure,

$$\begin{matrix} \text{Force due to pressure} \\ \text{on surface at } x \end{matrix} = p|_x\,(2\pi r\Delta r),$$

$$\begin{matrix} \text{Force due to pressure} \\ \text{on surface at } x + \Delta x \end{matrix} = -\,p|_{x+\Delta x}\,(2\pi r\Delta r). \tag{6.2.3}$$

Note that there is a negative sign for the force at $x + \Delta x$, because the pressure acts along the inward unit normal at the surface, and the inward unit normal at $x + \Delta x$

is in the negative x direction. The forces exerted on the curved surfaces at r and $r + \Delta r$ are the product of the shear stress, τ_{xr}, and the surface area,

$$
\begin{array}{l}
\text{Force due to shear stress} \\
\text{on surface at } r
\end{array} = - \tau_{xr}|_r \, (2\pi r \Delta x),
$$

$$
\begin{array}{l}
\text{Force due to shear stress} \\
\text{on surface at } r + \Delta r
\end{array} = \tau_{xr}|_{r+\Delta r} \, (2\pi (r + \Delta r) \Delta x). \qquad (6.2.4)
$$

The force at the surface at $r + \Delta r$ is the product of the stress τ_{xr} and the surface area $(2\pi(r + \Delta r)\Delta x)$. There is a negative sign in the expression for the force at r, because the outward unit normal at this surface is in the $-r$ direction. It should be noted that τ_{xr} is the force per unit area in the x direction at a surface with outward unit normal in the positive r direction; if the outward unit normal is in the $-r$ direction, the force per unit area is $-\tau_{xr}$.

The momentum flux is the product of the momentum density ρv_x (momentum per unit volume), the normal velocity to the surface v_x and the surface area. The rate of momentum in and out of the differential volume is,

$$
\begin{array}{l}
\text{Rate of momentum in} \\
\text{at the surface at } x
\end{array} = (\rho v_x) v_x (2\pi r \Delta r)|_{r,x} \, ,
$$

$$
\begin{array}{l}
\text{Rate of momentum out} \\
\text{at surface at } x + \Delta x
\end{array} = (\rho v_x) v_x (2\pi r \Delta r)|_{r,x+\Delta x} \, . \qquad (6.2.5)
$$

Eqs. 6.2.2, 6.2.3, 6.2.4, and 6.2.5 are substituted into the momentum balance Eq. 6.2.1,

$$
\frac{\Delta(\rho v_x)}{\Delta t}(2\pi r \Delta r \Delta x) = -2\pi r \Delta x \tau_{xr}|_r + 2\pi (r + \Delta r) \Delta x \tau_{xr}|_{r+\Delta r} + 2\pi r \Delta r \rho v_x^2|_x
$$
$$
- 2\pi r \Delta r \rho v_x^2|_{x+\Delta x} + 2\pi r \Delta r \, p|_x - 2\pi r \Delta r \, p|_{x+\Delta x} \, .
$$
$$
(6.2.6)
$$

Eq. 6.2.6 is divided by the volume $(2\pi r \Delta r \Delta x)$, and the momentum conservation equation is obtained by taking the limit $\Delta t, \Delta r, \Delta x \to 0$,

$$
\rho \frac{\partial v_x}{\partial t} = -\frac{\partial p}{\partial x} + \frac{1}{r}\frac{\partial (r\tau_{xr})}{\partial r}. \qquad (6.2.7)
$$

Here, the density is considered a constant for an incompressible flow. The third and fourth terms on the right of Eq. 6.2.6 cancel because v_x is independent of x for a fully developed flow. Using Newton's law of viscosity, $\tau_{xr} = \mu(\partial v_x/\partial r)$, we obtain the momentum conservation equation,

$$\rho\frac{\partial v_x}{\partial t} = -\frac{\partial p}{\partial x} + \frac{1}{r}\frac{\partial}{\partial r}\left(\mu r \frac{\partial v_x}{\partial r}\right). \qquad (6.2.8)$$

Eq. 6.2.8 is the conservation equation for momentum in the x direction. In the r direction, the momentum conservation equation contains the inertial, viscous and pressure terms. When the velocity in the radial direction is zero, $v_r = 0$, the inertial and viscous terms in the radial direction are identically zero. The only term in the radial momentum conservation equation is $(\partial p/\partial r)$ which has to to be zero for momentum conservation. Therefore, the pressure is independent of r and is only a function of x.

It can be shown, from Eq. 6.2.8, that $(\partial p/\partial x)$ is independent of x. If we take the x derivative of Eq. 6.2.8, the left side and the viscous term on the right are both zero, because v_x is independent of x.[1] Therefore, the x derivative of Eq. 6.2.8 is $(\partial^2 p/\partial x^2) = 0$, and the pressure gradient $(\partial p/\partial x)$ is independent of x. However, $(\partial p/\partial x)$ could be a function of time for an unsteady flow. An example of a flow driven by a time-dependent pressure gradient is considered in Section 6.4.

The boundary condition for the velocity is the no-slip condition,

$$v_x = 0 \text{ at } r = R. \qquad (6.2.9)$$

There is no physical boundary at the axis $r = 0$. However, the axisymmetry of the flow requires that the slope of the velocity profile is zero at the axis,

$$\frac{\partial v_x}{\partial r} = 0 \text{ at } r = 0. \qquad (6.2.10)$$

At steady state, the velocity is only a function of r, the pressure is only a function of x and the pressure gradient $(\mathrm{d}p/\mathrm{d}x)$ is a constant. The momentum conservation Eq. 6.2.8 reduces to,

$$-\frac{\mathrm{d}p}{\mathrm{d}x} + \mu\frac{1}{r}\frac{\mathrm{d}}{\mathrm{d}r}\left(r\frac{\mathrm{d}v_x}{\mathrm{d}r}\right) = 0. \qquad (6.2.11)$$

Eq. 6.2.11 is integrated to obtain,

$$v_x = \frac{r^2}{4\mu}\frac{\mathrm{d}p}{\mathrm{d}x} + C_1\ln(r) + C_2. \qquad (6.2.12)$$

---

[1]Since x, r and t are independent co-ordinates, the order of differentiation with respect to any two co-ordinates can be interchanged.

The boundary conditions, Eqs. 6.2.9 and 6.2.10, are satisfied for,

$$C_1 = 0, \;\; C_2 = -\frac{R^2}{4\mu}\frac{\mathrm{d}p}{\mathrm{d}x}. \tag{6.2.13}$$

The final expression for the velocity field is,

$$\boxed{v_x = -\frac{1}{4\mu}\frac{\mathrm{d}p}{\mathrm{d}x}(R^2 - r^2) = v_{max}\left(1 - \left(\frac{r}{R}\right)^2\right).} \tag{6.2.14}$$

This is the parabolic velocity profile for the flow in a pipe, called the Hagen–Poiseuille profile. Note that the velocity is positive when the pressure difference is negative, that is, when the pressure decreases with increasing x. The maximum velocity at the centre of the pipe is,

$$v_{max} = -\frac{R^2}{4\mu}\frac{\mathrm{d}p}{\mathrm{d}x}. \tag{6.2.15}$$

The shear stress is,

$$\tau_{xr} = \mu\frac{\mathrm{d}v_x}{\mathrm{d}r} = -\frac{2\mu v_{max} r}{R^2} = \frac{\mathrm{d}p}{\mathrm{d}x}\frac{r}{2}. \tag{6.2.16}$$

The shear stress is zero at the centre of the pipe, and it magnitude increases linearly with radius. At the wall, the wall shear stress is given by,

$$\tau_w = \tau_{xr}|_{r=R} = -\frac{2\mu v_{max}}{R} \tag{6.2.17}$$

The above shear stress is negative because it is the force per unit area at a surface whose outward unit normal is in the radial direction. At the wall, this represents the force per unit area exerted on the fluid by the wall, which is in the $-x$ direction. There is a force of equal magnitude exerted on the wall by the fluid, which is in the $+x$ direction in accordance with Newton's third law.

The volumetric flow rate $\dot{V}$ is the integral of the velocity $v_x(r)$ times the differential area of an annular strip perpendicular to the flow, $(2\pi r dr)$ (Fig. 6.8) from $r = 0$ to $r = R$,

$$\dot{V} = \int_0^R 2\pi r\,\mathrm{d}r\,v_x = -\frac{\pi R^4}{8\mu}\frac{\mathrm{d}p}{\mathrm{d}x}. \tag{6.2.18}$$

The average velocity v_{av} is defined as the ratio of the flow rate $\dot{V}$ and the area of cross section (πR^2),

$$v_{av} = \frac{\dot{V}}{\pi R^2} = -\frac{R^2}{8\mu}\frac{\mathrm{d}p}{\mathrm{d}x} = \frac{v_{max}}{2}. \tag{6.2.19}$$

The maximum velocity is two times the average velocity for the flow in a pipe.

The friction factor[2] for the flow through a pipe is defined as the ratio of the wall shear stress and $(\rho v_{av}^2/2)$,

$$f = \frac{|\tau_w|}{(\rho v_{av}^2/2)}. \qquad (6.2.20)$$

Eq. 6.2.17 for the wall shear stress is substituted into the above equation,

$$\boxed{f = \frac{2\mu v_{max}}{R(\rho v_{av}^2/2)} = \frac{16\mu}{\rho v_{max} R} = \frac{16}{\text{Re}}.} \qquad (6.2.21)$$

Here, we have used Eq. 6.2.19 to relate v_{max} and v_{av}, and $\text{Re} = (\rho v_{max} R/\mu)$, is the Reynolds number based on the maximum fluid velocity and the radius of the pipe. This is the same as $\text{Re} = (\rho v_{av} d/\mu)$, the Reynolds number based on the average velocity and the pipe diameter, $d = 2R$.

The velocity profile and friction factor calculation have been carried out assuming there is no body force. A gravitational force is easily included in the calculation by defining a modified pressure, $p' = p - \rho \boldsymbol{g} \cdot \boldsymbol{x}$ where $\boldsymbol{g}$ is the gravitational acceleration vector and $\boldsymbol{x}$ is the position vector. The spatial derivative of the pressure in the axial direction is modified as $-(\partial p'/\partial x) = -(\partial p/\partial x) + \rho g_x$, where g_x is the component of the gravitational acceleration along the axis of the pipe. Thus, a body force along the pipe axis is equivalent to the negative of a pressure gradient for an incompressible flow. Since the fluid velocity is zero in the r direction, the radial derivative of the modified pressure $(\partial p'/\partial r)$ is zero. Therefore, p' is a function of x and time, and $(\partial p'/\partial x)$ is a function of time for a pipe flow, and neglect the flow resistance due to the valve.

EXAMPLE 6.2.1: Two tanks, a and b, with areas of cross section $A_a = 1$ m^2 and $A_b = 2$ m^2, respectively, are connected by a pipe of diameter $d = 1$ cm and length $L = 5$ m fitted with a valve, as shown in Fig. 6.9. Fluid with density 10^3 kg/m^3 and viscosity 10^{-2} kg/m/s is filled in tank a up to a height of $h_{a0} = 3$ m with the valve closed. The valve is opened at $t = 0$. Determine the evolution of the heights h_a and h_b in the two tanks. Assume the volume of the pipe is much smaller than that in the tanks.

---

[2]Eq. 6.2.20 is the Fanning friction factor. The Darcy friction factor is four times the Fanning friction factor.

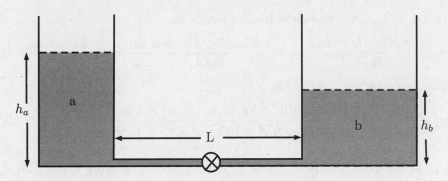

FIGURE 6.9. Two tanks a and b connected by a pipe.

Solution: The total volume of liquid is conserved when the liquid is redistributed in the two tanks. If h_a and h_b are the heights of liquid in the two tanks, then

$$A_a h_a + A_b h_b = A_a h_{a0} \quad \Rightarrow \quad h_b = (3 \text{ m} - h_a)/2, \qquad (6.2.22)$$

where $h_{a0} = 3$ m is the initial height in tank a. In the final equilibrium state, the heights in the two tanks are equal, $h_a = h_b = 1$ m. The transient flow in the pipe is determined from the pressure difference across the two ends of the pipe. The pressures at the bottom of the tanks are $p_a = \rho g h_a$ and $p_b = \rho g h_b$. The pressure difference between the two ends of the pipe is,

$$p_a - p_b = \rho g(h_a - h_b) = \frac{3\rho g(h_a - 1 \text{ m})}{2}. \qquad (6.2.23)$$

If the flow is laminar, the difference in pressure is related to the flow rate by Eq. 6.2.18,

$$\frac{p_a - p_b}{L} = \frac{8\mu \dot{V}}{\pi R^4}, \qquad (6.2.24)$$

where $\dot{V}$ is the flow rate from tank a to tank b. The assumption of laminar flow is examined at the end of this example.

The flow rate $\dot{V}$ is related to the change in height of tank a,

$$\dot{V} = -A_a \frac{dh_a}{dt}. \qquad (6.2.25)$$

There is a negative sign on the right in Eq. 6.2.25 because flow from tank a to tank b results in a decrease in the height in tank a. Substituting Eqs. 6.2.22, 6.2.23 and

6.2.24 in Eq. 6.2.25, the equation for the height is

$$\frac{dh_a}{dt} = -\frac{\pi R^4 \rho g (h_a - h_b)}{8\mu A_a L} = -\frac{\pi R^4 \rho g (3/2)(h_a - 1\text{ m})}{8\mu A_a L} = -\frac{h_a - 1\text{ m}}{t_c}, \quad (6.2.26)$$

where the characteristic time t_c is,

$$t_c = \frac{16\mu A_a L}{3\pi \rho g R^4} = \frac{16 \times 10^{-2}\text{ kg/m/s} \times 1\text{ m}^2 \times 5\text{ m}}{3\pi \times 10^3\text{ kg/m}^3 \times 10\text{ m/s}^2 \times (5 \times 10^{-3}\text{ m})^4} = 1.36 \times 10^4\text{s}.$$
$$(6.2.27)$$

The time variation of the height h_a is determined by solving Eq. 6.2.26,

$$h_a - 1\text{ m} = (h_{a0} - 1\text{ m})e^{(-t/t_c)}. \quad (6.2.28)$$

The maximum Reynolds number is estimated as follows. The flow rate is maximum at $t = 0$, when the difference in heights of the two tanks is maximum,

$$\dot{V}\Big|_{t=0} = -A_a \frac{dh_a}{dt}\Big|_{t=0} = \frac{A_a(h_{a0}-1)}{t_c} = \frac{1\text{ m}^2 \times 2\text{ m}}{1.36 \times 10^4\text{ s}} = 1.47 \times 10^{-4}\text{ m}^3/\text{s}. \quad (6.2.29)$$

The Reynolds number for the flow at $t = 0$ is,

$$\text{Re}_0 = \frac{\rho d v_{av}}{\mu} = \frac{\rho d \dot{V}\big|_{t=0}}{\mu \pi R^2} = \frac{10^3\text{ kg/m}^3 \times 10^{-2}\text{ m} \times 1.47 \times 10^{-4}\text{ m}^3/\text{s}}{10^{-2}\text{ kg/m/s} \times \pi \times (5 \times 10^{-3}\text{ m})^2} = 1875,$$
$$(6.2.30)$$

where d is the pipe diameter and $v_{av} = (\dot{V}/\pi R^2)$ is the average velocity. Since the maximum Reynolds number is less than 2100, the flow is laminar. This justifies the use of the use of Eq. 6.2.24 relating the pressure difference and flow rate for a laminar flow. □

EXAMPLE 6.2.2: For the pressure-driven laminar flow of a power-law fluid of density ρ in a pipe of diameter d, the constitutive relation is Eq. 4.2.46. Determine the correlation for the friction factor. How is the Reynolds number defined?

Solution: The equation for the shear stress in the fluid is,

$$-\frac{dp}{dx} + \frac{1}{r}\frac{d(r\tau_{xr})}{dr} = 0. \quad (6.2.31)$$

This is solved subject to the condition that $\tau_{xr} = 0$ at $r = 0$ to obtain a solution for the stress,

$$\tau_{xr} = \left(-\frac{dp}{dx}\right)\left(-\frac{r}{2}\right). \quad (6.2.32)$$

Here, it is important to note that (dp/dx) is negative if the velocity is in the $+x$ direction, and so we have considered $-(dp/dx)$ as a positive quantity. The wall shear stress is,

$$\tau_w = \frac{\mathrm{d}p}{\mathrm{d}x} \frac{R}{2}. \tag{6.2.33}$$

The velocity is determined from the equation,

$$\kappa \left(\frac{\mathrm{d}v_x}{\mathrm{d}r} \right) \left| \frac{\mathrm{d}v_x}{\mathrm{d}r} \right|^{n-1} = -r \left(-\frac{1}{2} \frac{\mathrm{d}p}{\mathrm{d}x} \right). \tag{6.2.34}$$

This is solved, subject to the no-slip condition $v_x = 0$ at $r = R$, to obtain,

$$v_x = \left(-\frac{1}{2\kappa} \frac{\mathrm{d}p}{\mathrm{d}x} \right)^{1/n} \frac{n}{n+1} (R^{(n+1)/n} - r^{(n+1)/n}), \tag{6.2.35}$$

where R is the pipe radius. The average velocity is,

$$v_{av} = \frac{1}{\pi R^2} \int_0^R 2\pi r \, \mathrm{d}r \, v_x = \frac{2}{R^2} \left(-\frac{1}{2\kappa} \frac{\mathrm{d}p}{\mathrm{d}x} \right)^{1/n} \frac{n}{n+1} \left(\frac{R^{(n+1)/n} r^2}{2} - \frac{n r^{(3n+1)/n}}{3n+1} \right) \Bigg|_{r=0}^R$$

$$= \left(-\frac{1}{2\kappa} \frac{\mathrm{d}p}{\mathrm{d}x} \right)^{1/n} \frac{n R^{((n+1)/n)}}{(3n+1)}. \tag{6.2.36}$$

From dimensional analysis, the Reynolds number based on the average velocity v_{av} and the diameter d is,

$$\mathrm{Re} = \frac{\rho v_{av}^{2-n} d^n}{\kappa}. \tag{6.2.37}$$

The friction factor is,

$$f = \frac{|\tau_w|}{(\rho v_{av}^2/2)} = \frac{\kappa}{\rho v_{av}^{2-n} d^n} \times \frac{|\tau_w|}{(\kappa v_{av}^n d^{-n}/2)} = \frac{1}{\mathrm{Re}} \frac{|\tau_w|}{(\kappa v_{av}^n d^{-n}/2)}. \tag{6.2.38}$$

Since the friction factor is expected to be inversely proportional to Re for a laminar flow, the friction factor has been expressed as the product of Re^{-1} and a dimensionless factor in the above equation. The above expression can be simplified using Eqs. 6.2.33 for τ_w and 6.2.36 for v_{av},

$$f = \frac{1}{\mathrm{Re}} \frac{|\mathrm{d}p/\mathrm{d}x|(R/2)}{(\kappa(n/(3n+1))^n((1/2\kappa)|\mathrm{d}p/\mathrm{d}x|)R^{n+1}d^{-n})/2} = \frac{2}{\mathrm{Re}} \left(\frac{2(3n+1)}{n} \right)^n. \tag{6.2.39}$$

Summary (6.2)

1. The momentum conservation equation for a pipe, Eq. 6.2.8, is determined by a balance over a cylindrical shell.

2. For a fully developed unidirectional laminar flow in a cylindrical pipe, the velocity is independent of the axial co-ordinate, the pressure is independent of the radial co-ordinate, and $(\partial p/\partial x)$, the axial derivative of the pressure, is only a function of time.

3. The velocity profile for a steady flow, Eq. 6.2.14, is a parabolic profile in which the maximum velocity is two times the average velocity.

4. The friction factor, Eq. 6.2.21, for a laminar flow is inversely proportional to the Reynolds number, $f = 16/\text{Re}$.

6.3 Turbulent Flow in a Pipe

The laminar velocity profile, Eq. 6.2.14, is applicable when the Reynolds number is lower than 2100. When the Reynolds number exceeds 2100, there is a spontaneous transition to a more complicated 'turbulent' velocity profile. The characteristics of the laminar and turbulent flows are very different. In a laminar flow, the streamlines are smooth and steady, and momentum transport in the cross-stream direction is due to molecular diffusion. This mechanism was discussed in detail in Chapter 3. In contrast, in a turbulent flow, the velocity field is time-dependent. There are significant velocity fluctuations in both the stream-wise and the cross-stream directions, though the cross-stream velocity is zero on average. The momentum transport is caused by the correlated motion of parcels of fluid called 'eddies'. This mechanism of momentum transport is more efficient than the molecular mechanism, and so this results in a higher friction factor than that in a laminar flow. Due to this efficient cross-stream momentum transport, the velocity profile in a turbulent flow is flatter and more plug-like at the centre in comparison to the parabolic velocity profile for a laminar flow.

In some flows, the transition from a laminar to a turbulent flow occurs due to an instability of the laminar flow. A simplistic analogy for the distinction between stable and unstable flows is shown in Fig. 6.10. An object at the bottom of a well, shown in Fig. 6.10(a), is in mechanical equilibrium. This is a stable equilibrium; if

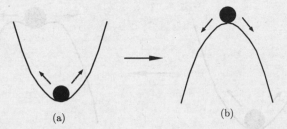

(a)   (b)

FIGURE 6.10. Analogy for the instability of the laminar flow of a fluid when the flow parameter exceeds a critical value.

the object is displaced to the left or the right, the restoring forces return the object to the equilibrium state. An object at the top of a hill, shown in Fig. 6.10(b), is in mechanical equilibrium, because the forces are balanced. However, if the object is displaced to the left or right, it does not return to its original state, but transitions to some other state. Therefore, the state shown in Fig. 6.10(b) is an 'unstable' equilibrium state. The instability occurs when the relevant dimensionless parameter (the Reynolds number or the scaled characteristic flow velocity) exceeds a critical value. It should be noted that the analogy shown in Fig. 6.10 is simplistic; a fluid flow is specified by the velocity field which is a continuous function of space, in contrast to the single degree of freedom shown in Fig. 6.10. However, the instability of a laminar flow does occur through a mechanism of the type shown in Fig. 6.10 for flows such as the rotating flow between concentric cylinders (Taylor–Couette instability) or a fluid heated from below (Rayleigh–Benard instability).

The laminar-turbulent transition in a pipe flow is different from that shown in Fig. 6.10. The laminar velocity profile derived in Section 6.2 is a solution of the equations of motion at all Reynolds numbers. However, when the Reynolds number exceeds the transition value of about 2100, the flow becomes unstable due to disturbances that exceed a small but finite amplitude, as shown in Fig. 6.11. This is different from the transition due to infinitesimal disturbances shown in Fig. 6.10. Since disturbances always exist in nature, the transition in almost all practical situations is repeatably observed at a transition Reynolds number of about 2100. However, if care is taken to reduce disturbances at the entrance of the pipe and in the environment, the flow could be maintained in the laminar state even at Reynolds numbers as high as 10^5.

The separation of the fluid velocity in a turbulent flow into the mean and the fluctuating velocities was discussed in Section 3.2.1 (Chapter 3). The mean and fluctuating velocities at a location were shown schematically in Fig. 3.5 and defined in Eqs. 3.2.1–3.2.2. The Reynolds stress for a turbulent flow was defined in Eq. 3.2.15. Here, the structure of turbulence is discussed in further detail. It should be noted

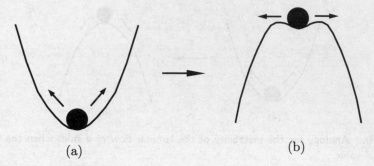

(a)   (b)

FIGURE 6.11. Analogy for the laminar-turbulent transition in a pipe when the Reynolds number exceeds the transition value.

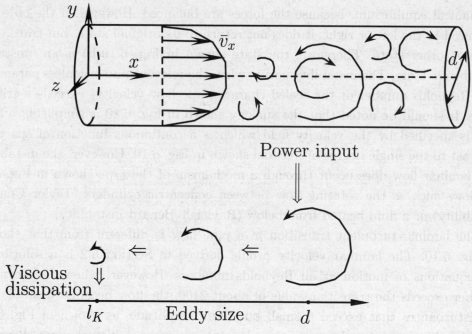

FIGURE 6.12. The 'turbulence cascade' which consists of the input of energy at the flow scale, the transfer of energy through eddies of progressively smaller sizes and the dissipation of energy at the Kolmogorov scale.

that there is a lot of active research in the field of turbulence, and much is yet to be understood. The latest advances in the statistical theories of turbulence are outside the scope of this text, and therefore the discussion is qualitative and based on order-of-magnitude estimates.

A turbulent flow consists of transient 'eddies' or parcels of fluid in correlated motion of different sizes and 'integral' times (time for which the motion is correlated). This is shown schematically in the top panel in Fig. 6.12. The size of the largest eddies is comparable to the pipe diameter d, and the timescale is comparable (d/v_{av}), where v_{av} is the average velocity. The large-scale velocity fluctuations are anisotropic—the velocity fluctuations in the flow direction are typically larger than those in the cross-stream direction. Since the Reynolds number based on the flow scales is high, viscous effects are not important for the large-scale motion. In contrast, the small eddies are isotropic and homogeneous, and the viscous stress does affect the dynamics of the smallest eddies. The velocity variations are smoothed out by momentum diffusion, and there is viscous dissipation of energy at the smallest scales.

The power required for pumping the fluid is the product of the pressure drop and the volumetric flow rate, $-\Delta p \times \dot{V}$, where Δp, the difference between the outlet and inlet pressure, is negative. The total mass of fluid in the pipe is the product of the density and volume, $\rho L(\pi d^2/4)$. The power per unit mass can be estimated as,

$$\frac{\text{Power input}}{\text{Mass}} = -\frac{\Delta p \times \dot{V}}{\rho L(\pi d^2/4)} = -\frac{\Delta p}{L} \times \frac{v_{av}}{\rho} = \frac{2v_{av}^3 f}{d}. \tag{6.3.1}$$

Here, f is the friction factor, and Eq. 2.1.21 is used to relate the pressure drop to the friction factor. For a turbulent flow, the Moody diagram (Fig. 2.5 in Chapter 2) shows that the friction factor decreases very slowly as the Reynolds number increases, in contrast to the Re^{-1} decrease for a laminar flow. If we consider the friction factor to be approximately constant, the power input per unit mass can be estimated as (v_{av}^3/d), as shown in Table 6.1. At steady state, the power input per unit mass, also called 'turbulent energy production rate', is equal to the rate of dissipation of energy per unit mass, ϵ. Therefore,

$$\epsilon \sim \frac{v_{av}^3}{d}. \tag{6.3.2}$$

The length and time scales of the smallest eddies are calculated based on the assumption that they do not depend on the average velocity and pipe diameter, since the size of the eddies is much smaller than the large-scale flow, and the dynamics of the eddy is influenced only by the local environment. These depend only on the energy dissipation rate per unit mass ϵ with dimension $\mathcal{L}^2 \mathcal{T}^{-3}$, since energy has to be dissipated at the small scales, and the kinematic viscosity ν with dimension $\mathcal{L}^2 \mathcal{T}^{-1}$, which causes dissipation. The length l_K, time t_K and velocity v_K of the smallest eddies are determined using dimensional analysis,

$$l_K = \left(\frac{\nu^3}{\epsilon}\right)^{1/4}, \quad t_K = \left(\frac{\nu}{\epsilon}\right)^{1/2}, \quad v_K = (\nu\epsilon)^{1/4}. \tag{6.3.3}$$

The subscript K here refers to the 'Kolmogorov scales', the smallest scales in a turbulent flow, since these were first proposed by Kolmogorov[22].

The rate of energy dissipation, Eq. 6.3.2, is substituted in the expressions for the Kolmogorov scales, Eq. 6.3.3, and these are divided by the flow length, time and velocity scales, to obtain

$$\frac{l_K}{d} \sim \left(\frac{\nu}{v_{av}d}\right)^{3/4} \sim \mathrm{Re}^{-3/4},$$

$$\frac{t_K}{(d/v_{av})} \sim \left(\frac{\nu}{v_{av}d}\right)^{1/2} \sim \mathrm{Re}^{-1/2}, \qquad (6.3.4)$$

$$\frac{v_K}{v_{av}} \sim \left(\frac{\nu}{v_{av}d}\right)^{1/4} \sim \mathrm{Re}^{-1/4}.$$

The Kolmogorov length, time and velocity scales are $\mathrm{Re}^{-3/4}$, $\mathrm{Re}^{-1/2}$ and $\mathrm{Re}^{-1/4}$ smaller than those of the largest eddies. The Reynolds number based on the Kolmogorov scales is 1,

$$\mathrm{Re}_K \sim \frac{l_K v_K}{\nu} \sim \frac{\mathrm{Re}^{-3/4}d \times \mathrm{Re}^{-1/4}v_{av}}{\nu} = 1. \qquad (6.3.5)$$

The Kolmogorov and flow scales, and their ratio, are summarised in Table 6.1.

TABLE 6.1. The characteristic flow scales, the Kolmogorov scales and their ratios for different dynamical quantities for the turbulent flow in a pipe with average velocity v_{av} and diameter d.

Quantity	Dimension	Flow scale	Kolmogorov scale	Kolmogorov scale / Flow scale
Length	$\mathcal{L}$	d	$l_K = (\nu^3/\epsilon)^{1/4}$	$\mathrm{Re}^{-3/4}$
Velocity	$\mathcal{L}\mathcal{T}^{-1}$	v_{av}	$v_K = (\nu\epsilon)^{1/4}$	$\mathrm{Re}^{-1/4}$
Time	$\mathcal{T}$	(d/v_{av})	$t_K = (\nu/\epsilon)^{1/2}$	$\mathrm{Re}^{-1/2}$
Strain rate	$\mathcal{T}^{-1}$	(v_{av}/d)	$(v_K/l_K) = (\epsilon/\nu)^{1/2}$	$\mathrm{Re}^{1/2}$
Reynolds number	Dimensionless	$(v_{av}d/\nu) = \mathrm{Re}$	$(v_K l_K/\nu) = 1$	Re^{-1}
Kinetic energy / Mass	$\mathcal{L}^2\mathcal{T}^{-2}$	v_{av}^2	$v_K^2 = (\nu\epsilon)^{1/2}$	$\mathrm{Re}^{-1/2}$
Power input / Mass	$\mathcal{L}^2\mathcal{T}^{-3}$	$(v_{av}^3/d) = \epsilon$		
Dissipation rate / Mass	$\mathcal{L}^2\mathcal{T}^{-3}$	$\nu v_{av}^2/d^2$	$(\nu v_K^2/l_K^2) = \epsilon$	Re
Dissipation rate / Power input	Dimensionless	Re^{-1}	1	Re

From Table 6.1, it evident that the kinetic energy per unit mass in the Kolmogorov scale eddies is $\mathrm{Re}^{-1/2}$ smaller than than those in the large eddies. However, the strain rate in the Kolmogorov scale eddies is $O(\mathrm{Re}^{1/2})$ larger than that in the large-scale flow. The energy dissipation rate per unit mass for the large-scale eddies is estimated as $(\nu v_{av}^2/d^2)$, since the characteristic velocity and length are v_{av} and d, respectively. This can be expressed as $\mathrm{Re}^{-1}\epsilon$, where ϵ is given in Eq. 6.3.2. The energy dissipation rate per unit mass in the Kolmogorov scales is $(\nu v_K^2/l_K^2) \sim \epsilon$. Therefore, the energy dissipation rate per unit mass in the Kolmogorov scale eddies is $O(\mathrm{Re})$ larger than that in the large-scale flow and is comparable to the power input. Thus, most of the kinetic energy is in the large-scale flow, but most of the energy dissipation is due to the motion at the Kolmogorov scale.

The ratio of the power input and dissipation rate in Table 6.1 suggests the following physical picture. At high Reynolds number, the energy dissipation rate in the largest eddies is smaller, by a factor of Re^{-1}, in comparison to the power input. Eddies of progressively smaller sizes and velocities are created in the flow, so that the power input is dissipated by the smallest (Kolmogorov) scale eddies. The size and velocity of the smallest eddies are independent of the flow scales, and they depend only on the power input per unit mass and the kinematic viscosity. For the Kolmogorov scale motion, the Reynolds number is ~ 1. The energy input at the large scales is transmitted through progressively smaller eddies and then dissipated at the smallest Kolmogorov scale eddies at steady state. This transfer of energy through eddies of progressively smaller size, called the 'turbulence cascade', is shown schematically in Fig. 6.12.

Though the Kolmogorov length is much smaller than the pipe diameter, it is still large compared to the mean free path for a gas for a high Reynolds number incompressible flow, as illustrated by the following example.

EXAMPLE 6.3.1: How do the ratio of the Kolmogorov length and the mean free path, and the ratio of the Kolmogorov time and the characteristic molecular time scale, depend on the Reynolds and Mach numbers for a gas.

Solution: The ratio of the Kolmogorov length and the mean free path is,

$$\frac{l_K}{\lambda} = \frac{(\nu^3/\epsilon)^{1/4}}{\lambda} \sim \frac{(\nu^3 d/v_{av}^3)^{1/4}}{\lambda}. \tag{6.3.6}$$

Here, Eq. 6.3.2 is substituted for the energy dissipation rate per unit mass. The above ratio can be expressed as,

$$\frac{l_K}{\lambda} \sim \left(\frac{v_{av}d}{\nu}\right)^{1/4} \frac{\nu}{v_{av}\lambda} \sim \left(\frac{v_{av}d}{\nu}\right)^{1/4} \frac{v_{rms}}{v_{av}} \sim \frac{\mathrm{Re}^{1/4}}{\mathrm{Ma}}. \tag{6.3.7}$$

Here, the kinematic viscosity is approximated as $\nu \sim v_{rms}\lambda$, the product of the molecular root-mean-square velocity and the mean free path. The ratio $(v_{av}/v_{rms}) \sim$ Ma, the Mach number, because the molecular fluctuating velocity is comparable to the speed of sound.

The ratio of the Kolmogorov timescale and the molecular timescale (λ/v_{rms}) is,

$$\frac{t_K}{(\lambda/v_{rms})} = \frac{(\nu/\epsilon)^{1/2}}{(\lambda/v_{rms})} \sim \left(\frac{\nu d}{v_{av}^3}\right)^{1/2} \frac{v_{rms}}{\lambda} \sim \left(\frac{v_{av}d}{\nu}\right)^{1/2} \frac{\nu v_{rms}}{v_{av}^2\lambda}. \tag{6.3.8}$$

Here, Eq. 6.3.2 is used for ϵ. Substituting $\nu \sim \lambda v_{rms}$, the ratio of the Kolmogorov and molecular timescales is,

$$\frac{t_K}{(\lambda/v_{rms})} \sim \left(\frac{v_{av}d}{\nu}\right)^{1/2} \frac{v_{rms}^2}{v_{av}^2} \sim \frac{\mathrm{Re}^{1/2}}{\mathrm{Ma}^2}. \tag{6.3.9}$$

For a high Reynolds number and low Mach number flow, the Kolmogorov length is much larger than the mean free path, and the Kolmogorov timescale is much larger than the characteristic molecular timescale. □

Near the wall of the pipe, the fluid velocity fluctuations necessarily decrease to zero due to the no-slip condition at the wall, and the Reynolds stress also decreases to zero. In a 'viscous sub-layer' adjacent to the wall of thickness much smaller than the pipe radius, the shear stress is viscous (due to momentum diffusion), and convective transport of momentum is not important because the velocity fluctuations are small. In this region, the pipe radius and the average flow velocity are not relevant for determining the velocity gradient. The magnitude of the shear stress in the fluid, Eq. 6.2.16, increases proportional to r from the centre to the wall of the pipe. The shear stress is approximately equal to the wall shear stress in the viscous sub-layer, where the thickness is much smaller than the pipe radius. From dimensional analysis, the velocity scale within the viscous sub-layer that depends on the wall shear stress and the fluid properties is the 'friction velocity' $v_* = (\tau_w/\rho)^{1/2}$, where τ_w is the wall shear stress. The thickness of the viscous sub-layer, which is much smaller than the pipe diameter, depends only on the friction velocity and the kinematic viscosity, since viscous effects are important here. Based on dimensional analysis, the relevant

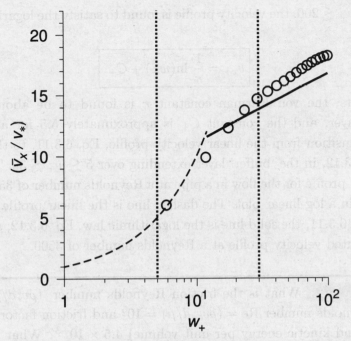

FIGURE 6.13. The von Karman plot for the scaled velocity $\bar{v}_{x+} = (\bar{v}_x/v_*)$ as a function of the scaled distance from the wall w_+ (Eq. 6.3.10) on a log-linear plot. The linear profile $\bar{v}_{x+} = w_+$ (Eq. 6.3.11). is shown by the dashed line on the left, and the log profile $\bar{v}_{x+} = 2.44 \ln(w_+) + 5.5$ is shown by the solid line. The vertical dotted lines are $w_+ = 5$ and $w_+ = 30$. The points are the velocity profile for a flow with Reynolds number 3500.

length scale is (ν/v_*), where ν is the kinematic viscosity. This length scale is called the 'wall unit'. The scaled distance from the wall w_+ and the scaled velocity $\bar{v}_{x+}$ are defined as,

$$w_+ = \frac{R-r}{\nu/v_*}, \quad \bar{v}_{x+} = \frac{\bar{v}_x}{v_*}. \tag{6.3.10}$$

Within the viscous sub-layer where the shear stress is τ_w, the slope of the velocity profile close to the wall is $(d\bar{v}_x/dr) = -(\tau_w/\mu)$, or $(d\bar{v}_{x+}/dw_+) = 1$. Therefore, the linear velocity profile within the viscous sub-layer is

$$\boxed{\bar{v}_{x+} = w_+.} \tag{6.3.11}$$

The velocity profile, Eq. 6.3.11, is observed in experiments for $w_+ \lesssim 5$ close to the wall.

For $30 \lesssim w_+ \lesssim 200$, the velocity profile is found to satisfy the logarithmic law[23],

$$\bar{v}_{x+} = \frac{1}{\kappa} \ln(w_+) + C_+.$$ (6.3.12)

In experiments, the von Karman constant κ is found to be about 0.41 in the logarithmic layer, and the constant C_+ is approximately 5.5 for a smooth wall. There is a transition from the linear velocity profile, Eq. 6.3.11, to the logarithmic profile, Eq. 6.3.12, in the 'buffer' layer extending over $5 \lesssim w_+ \lesssim 30$. The schematic of the velocity profile for the flow in a pipe at a Reynolds number of 3500 is shown in the Fig. 6.13 in a log-linear plot. The dashed line is the linear profile in the viscous sub-layer, Eq. 6.3.11, the solid line is the logarithmic law, Eq. 6.3.12, and the points are the computed velocity profile at a Reynolds number of 3500.

EXAMPLE 6.3.2: What is the friction Reynolds number, $(\rho v_* d/\mu)$, for a pipe flow with Reynolds number $\mathrm{Re} = (\rho v_{av} d/\mu) = 10^5$ and friction factor (ratio of wall shear stress and kinetic energy per unit volume) 4.5×10^{-3}? What is the ratio of the wall unit and the pipe diameter?

Solution: The wall shear stress is related to the Reynolds number by,

$$\tau_w = f \times \frac{\rho v_{av}^2}{2} = \rho v_*^2.$$ (6.3.13)

Therefore, the friction velocity is,

$$v_* = v_{av}\sqrt{f/2}.$$ (6.3.14)

The friction Reynolds number is,

$$(\rho v_* d/\mu) = (\rho v_{av} d/\mu)\sqrt{f/2} = \mathrm{Re}\sqrt{f/2} = 10^5 \times \sqrt{4.5 \times 10^{-3}/2} = 4743.$$ (6.3.15)

The ratio of the wall unit and the pipe diameter is $(\nu/v_* d) = \mathrm{Re}_*^{-1} = 2.1 \times 10^{-4}$. $\qquad\qquad\square$

From the above example, it is evident that the friction factor can be calculated if the average flow velocity and the friction velocity are known. However, there is no analytical relation for the friction velocity or the friction factor as a function of Reynolds number. The friction factor depends on the pipe roughness in addition to the Reynolds number. Several correlations have been proposed, including the Colebrook correlation, Eq. 2.2.2, in Chapter 2. The friction factor is plotted as a function of the Reynolds number for different pipe roughness in the Moody diagram, Fig. 2.5 in Chapter 2.

Summary (6.3)

1. The flow in a pipe is turbulent when the Reynolds number exceeds about 2100. Turbulence is characterised by fluctuations in the instantaneous velocity in all three directions, comparable in magnitude to the average flow velocity.

2. Cross-stream momentum transport occurs due to the correlated motion of parcels of fluid called eddies.

3. The cross-stream momentum transport due to fluid velocity fluctuations is much more efficient than that due to molecular diffusion. Consequently, the velocity profile is flatter with a smaller curvature at the centre in comparison to a laminar flow. For equal flow rate, the friction factor for a turbulent flow is significantly higher than that for a laminar flow.

4. The power required for pumping the fluid per unit mass, is equal to the rate of viscous dissipation of energy per unit mass, which is estimated in Eq. 6.3.2.

5. The length and velocity scales of the smallest eddies, termed the Kolmogorov scales (Eq. 6.3.3), depend only on the rate of dissipation of energy and the kinematic viscosity. The ratios of different quantities in the Kolmogorov and large scales are given in Eq. 6.3.4, and the Reynolds number based on the Kolmogorov length and velocity is 1 (Eq. 6.3.5).

6. The kinetic energy in the Kolmogorov scale eddies is $Re^{-1/2}$ smaller than that in the largest eddies (Table 6.1). However, the rate of dissipation of energy in the Kolmogorov scale eddies is larger, by a factor Re, in comparison to that in the largest eddies. The energy dissipation rate in the Kolmogorov scales is comparable to the power input.

7. The power input at the large scales is transferred through eddies of successively smaller size until it is dissipated at the Kolmogorov scale. This transfer mechanism is termed the 'turbulence cascade'.

8. There is a viscous sub-layer close to the wall where the stress is approximately a constant. Here, the relevant velocity scale is the 'friction velocity', $v_* = \sqrt{\tau_w/\rho}$, where τ_w is the wall shear stress and ρ is the fluid density. The relevant length scale is the wall unit (ν/v_*), where ν is the kinematic viscosity.

9. The viscous sub-layer extends for $0 \lesssim w_+ \lesssim 5$, where w_+ is the distance from the wall scaled by (ν/v_*). The velocity increases proportional to the distance from the wall, $(v_x/v_*) = w_+$ (Eq. 6.3.11), in the viscous sub-layer.

10. There is a logarithmic layer for $30 \lesssim w_+ \lesssim 200$ where the logarithmic law, Eq. 6.3.12, applies for the velocity profile.

6.4 Oscillatory Flow in a Pipe

Consider an unsteady, time-periodic and fully developed flow driven by an oscillatory pressure gradient $(\partial p/\partial x) = K \cos(\omega t)$ in a pipe of radius R. The velocity is dependent on time but is independent of the x co-ordinate. The momentum conservation equation, Eq. 6.2.8 is,

$$\rho \frac{\partial v_x}{\partial t} = \frac{\mu}{r} \frac{\partial}{\partial r} \left(r \frac{\partial v_x}{\partial r} \right) - K \cos(\omega t). \tag{6.4.1}$$

The boundary conditions at the centre and the wall of the pipe are the same as those for a steady flow, Eqs. 6.2.9 and 6.2.10. It is natural to define a scaled radial co-ordinate as $r^* = (r/R)$. The scaled time can be defined as $t^* = (\omega t)$, since ω is the frequency of the pressure variation. To determine the scaling for the velocity, Eq. 6.4.1 is expressed in terms of r^* and t^*, and divided by K, to obtain,

$$\frac{\rho \omega}{K} \frac{\partial v_x}{\partial t^*} = \frac{\mu}{KR^2} \frac{1}{r^*} \frac{\partial}{\partial r^*} \left(r^* \frac{\partial v_x}{\partial r^*} \right) - \cos(t^*). \tag{6.4.2}$$

It is clear that all terms in the above equation are dimensionless, and therefore, we can define a scaled velocity either as $v_x^* = (\mu v_x/KR^2)$, or as $v_x^* = (\rho \omega v_x/K)$. The former is obtained by balancing the viscous and pressure forces, and is appropriate for low Reynolds numbers where inertial forces are negligible. The latter is obtained by balancing inertial and pressure forces, and is appropriate when the Reynolds number is large so that viscous forces are negligible. We will proceed with the viscous scaling, and return later to the high Reynolds number limit. Using the scaled velocity $v_x^* = (\mu v_x/KR^2)$, Eq. 6.4.2 becomes,

$$\mathrm{Re}_\omega \frac{\partial v_x^*}{\partial t^*} = \frac{1}{r^*} \frac{\partial}{\partial r^*} \left(r^* \frac{\partial v_x^*}{\partial r^*} \right) - \cos(t^*), \tag{6.4.3}$$

where $\mathrm{Re}_\omega = (\rho \omega R^2/\mu)$ is the Reynolds number based on the frequency of pressure oscillations and the pipe radius.

Eq. 6.4.3 is a linear inhomogeneous partial differential equation, in which the forcing term is proportional to $\cos(t^*)$. The cosine function $\cos(t^*)$ is the real part of $e^{(\imath t^*)}$, where $\imath = \sqrt{-1}$. It is convenient to solve Eq. 6.4.3 with $e^{(\imath t^*)}$ as the inhomogeneous term, and then take the real part of the solution. A complex velocity field $v_x^\dagger$ is defined as the solution of the equation,

$$\mathrm{Re}_\omega \frac{\partial v_x^\dagger}{\partial t^*} = \frac{1}{r^*}\frac{\partial}{\partial r^*}\left(r^*\frac{\partial v_x^\dagger}{\partial r^*}\right) - \mathrm{e}^{(\imath t^*)}, \qquad (6.4.4)$$

with boundary conditions

$$v_x^\dagger = 0 \quad \text{at} \quad r^* = 1, \qquad (6.4.5)$$

$$\frac{\partial v_x^\dagger}{\partial r^*} = 0 \quad \text{at} \quad r^* = 0. \qquad (6.4.6)$$

Clearly, Eq. 6.4.3 is the real part of Eq. 6.4.4, and the boundary conditions, Eqs. 6.2.9 and 6.2.10, are the real parts of the boundary conditions, Eqs. 6.4.5 and 6.4.6. Therefore, the solution v_x^* is the real part of the complex velocity $v_x^\dagger$.

Since Eq. 6.4.4 is a linear inhomogeneous equation for $v_x^\dagger$ with a time-periodic forcing proportional to $\mathrm{e}^{(\imath t^*)}$, the solution $v_x^\dagger$ is also periodic in time with the same frequency,

$$v_x^\dagger = \tilde{v}_x(r^*)\mathrm{e}^{(\imath t^*)}, \qquad (6.4.7)$$

where $\tilde{v}_x(r^*)$ is only a function of the radial co-ordinate. Substituting Eq. 6.4.7 into Eq. 6.4.4, and dividing the resulting equation by $\mathrm{e}^{(\imath t^*)}$, the equation for $\tilde{v}_x$ is,

$$\frac{1}{r^*}\frac{\mathrm{d}}{\mathrm{d}r^*}\left(r^*\frac{\mathrm{d}\tilde{v}_x}{\mathrm{d}r^*}\right) - \imath\mathrm{Re}_\omega\tilde{v}_x = 1. \qquad (6.4.8)$$

The solution $\tilde{v}_x$ for the inhomogeneous ordinary differential equation, Eq. 6.4.8, is expressed as the sum of a general and a particular solution, $\tilde{v}_x = \tilde{v}_x^g + \tilde{v}_x^p$. The general solution is the solution of the homogeneous equation,

$$\frac{1}{r^*}\frac{\mathrm{d}}{\mathrm{d}r^*}\left(r^*\frac{\mathrm{d}\tilde{v}_x^g}{\mathrm{d}r^*}\right) - \imath\mathrm{Re}_\omega\tilde{v}_x^g = 0, \qquad (6.4.9)$$

while the particular solution is any one solution of the inhomogeneous Eq. 6.4.8.

The simplest particular solution, which satisfies Eq. 6.4.8, is a constant, $\tilde{v}_x^p = \imath\mathrm{Re}_\omega^{-1}$. Eq. 6.4.9 is multiplied by r^{*2}, and the substitution $r^\dagger = \sqrt{-\imath\mathrm{Re}_\omega}\,r^*$ is made, to obtain,

$$r^{\dagger 2}\frac{\mathrm{d}^2\tilde{v}_x^g}{\mathrm{d}r^{\dagger 2}} + r^{\dagger}\frac{\mathrm{d}\tilde{v}_x^g}{\mathrm{d}r^{\dagger}} + r^{\dagger 2}\tilde{v}_x^g = 0. \tag{6.4.10}$$

The above equation is the Bessel equation of zeroth order, Eq. 5.1.63, discussed in Section 5.1.4, and the solution of this equation is,

$$\tilde{v}_x^g = C_1 J_0(\sqrt{-\imath \mathrm{Re}_\omega}\, r^*) + C_2 Y_0(\sqrt{-\imath \mathrm{Re}_\omega}\, r^*), \tag{6.4.11}$$

where J_0 and Y_0 are the Bessel functions of zeroth order, and C_1 and C_2 are constants. The solution of Eq. 6.4.8 is the sum of the general solution, Eq. 6.4.11, and the particular solution $\imath \mathrm{Re}_\omega^{-1}$,

$$\tilde{v}_x = C_1 J_0(\sqrt{-\imath \mathrm{Re}_\omega}\, r^*) + C_2 Y_0(\sqrt{-\imath \mathrm{Re}_\omega}\, r^*) + \imath \mathrm{Re}_\omega^{-1}. \tag{6.4.12}$$

The constants C_1 and C_2 in Eq. 6.4.12 are evaluated from the boundary conditions, Eqs. 6.4.5 and 6.4.6. The constant C_2 is zero to satisfy Eq. 6.4.6, the boundary condition at the axis, because $Y_0 \to -\infty$ for $r^* \to 0$. The constant C_1, evaluated from the boundary condition at the wall, Eq. 6.4.5, is $C_1 = -\imath(\mathrm{Re}_\omega J_0(\sqrt{-\imath \mathrm{Re}_\omega}))^{-1}$. The final solution for $\tilde{v}_x$ is,

$$\tilde{v}_x = \imath \mathrm{Re}_\omega^{-1}\left(1 - \frac{J_0(\sqrt{-\imath \mathrm{Re}_\omega}\, r^*)}{J_0(\sqrt{-\imath \mathrm{Re}_\omega})}\right). \tag{6.4.13}$$

The solution for the velocity field v_x^* is the real part of $\tilde{v}_x \mathrm{e}^{(\imath t^*)}$,

$$v_x^* = \mathrm{Real}\left[\imath \mathrm{Re}_\omega^{-1}\left(1 - \frac{J_0(\sqrt{-\imath \mathrm{Re}_\omega}\, r^*)}{J_0(\sqrt{-\imath \mathrm{Re}_\omega})}\right)\mathrm{e}^{(\imath t^*)}\right]. \tag{6.4.14}$$

Eq. 6.4.14 for the velocity field has to be evaluated numerically, and the numerical solutions are shown as a function of time in Fig. 6.14 for four different values of Re_ω. The qualitative nature of the velocity profile is very different for low and high Reynolds number. Consider the case $\mathrm{Re}_\omega = 0.1$. The velocity profiles are close to a parabolic profile throughout the pressure gradient cycle. The magnitude of the velocity profile has a maximum in the $-x$ direction for $(\mathrm{d}p/\mathrm{d}x) = +K$, and a maximum in the $+x$ direction for $(\mathrm{d}p/\mathrm{d}x) = -K$, as expected for a steady flow. The velocity is close to zero when the pressure gradient is zero. In contrast, at $\mathrm{Re}_\omega = 100$, the velocity profile is independent of r^* near the centre of the channel, but there is a large variation close to the wall for for $r^* \gtrsim 0.6$. The magnitude of the velocity at the centre is close to zero for $(\mathrm{d}p/\mathrm{d}x) = \pm K$, and it has a maximum when $(\mathrm{d}p/\mathrm{d}x)$ passes through zero.

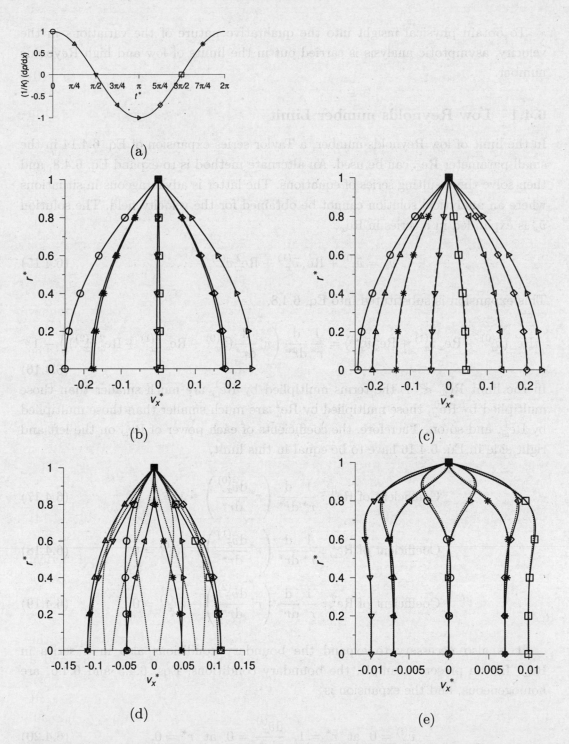

FIGURE 6.14. The velocity v_x^*, Eq. 6.4.14, as a function of r^* for $\mathrm{Re}_\omega = 0.1$ (b), $\mathrm{Re}_\omega = 1.0$ (c), $\mathrm{Re}_\omega = 10.0$ (d) and $\mathrm{Re}_\omega = 100.0$ (e), at different instants in the sinusoidal modulation of $(\mathrm{d}p/\mathrm{d}x)$ shown in (a). The low Reynolds number solution, Eq. 6.4.22, is shown by the dashed lines in (b) and (c), and the high Reynolds number solution, Eq. 6.4.39, is shown by the dotted lines in (d) and (e).

To obtain physical insight into the qualitative nature of the variations in the velocity, asymptotic analysis is carried out in the limits of low and high Reynolds number.

6.4.1 Low Reynolds number Limit

In the limit of low Reynolds number, a Taylor series expansion of Eq. 6.4.14 in the small parameter Re_ω can be used. An alternate method is to expand Eq. 6.4.8, and then solve the resulting series of equations. The latter is advantageous in situations where an analytical solution cannot be obtained for the velocity field. The solution $\tilde{v}_x$ is expanded in a series in Re_ω,

$$\tilde{v}_x = \tilde{v}_x^{(0)} + \mathrm{Re}_\omega \tilde{v}_x^{(1)} + \mathrm{Re}_\omega^2 \tilde{v}_x^{(2)} + \dots . \tag{6.4.15}$$

This expansion is substituted into Eq. 6.4.8,

$$\imath \mathrm{Re}_\omega (\tilde{v}_x^{(0)} + \mathrm{Re}_\omega \tilde{v}_x^{(1)} + \mathrm{Re}_\omega^2 \tilde{v}_x^{(2)}) = \frac{1}{r^*} \frac{\mathrm{d}}{\mathrm{d}r^*} \left(r^* \frac{\mathrm{d}}{\mathrm{d}r^*} (\tilde{v}_x^{(0)} + \mathrm{Re}_\omega \tilde{v}_x^{(1)} + \mathrm{Re}_\omega^2 \tilde{v}_x^{(2)}) \right) - 1. \tag{6.4.16}$$

In the limit $\mathrm{Re}_\omega \ll 1$, the terms multiplied by Re_ω^1 are much smaller than those multiplied by Re_ω^0, those multiplied by Re_ω^2 are much smaller than those multiplied by Re_ω^1, and so on. Therefore, the coefficients of each power of Re_ω on the left and right side in Eq. 6.4.16 have to be equal in this limit,

$$\text{Coefficient of } \mathrm{Re}_\omega^0 : \frac{1}{r^*} \frac{\mathrm{d}}{\mathrm{d}r^*} \left(r^* \frac{\mathrm{d}\tilde{v}_x^{(0)}}{\mathrm{d}r^*} \right) - 1 = 0, \tag{6.4.17}$$

$$\text{Coefficient of } \mathrm{Re}_\omega^1 : \frac{1}{r^*} \frac{\mathrm{d}}{\mathrm{d}r^*} \left(r^* \frac{\mathrm{d}\tilde{v}_x^{(1)}}{\mathrm{d}r^*} \right) - \imath \tilde{v}_x^{(0)} = 0, \tag{6.4.18}$$

$$\text{Coefficient of } \mathrm{Re}_\omega^2 : \frac{1}{r^*} \frac{\mathrm{d}}{\mathrm{d}r^*} \left(r^* \frac{\mathrm{d}\tilde{v}_x^{(2)}}{\mathrm{d}r^*} \right) - \imath \tilde{v}_x^{(1)} = 0. \tag{6.4.19}$$

It is also necessary to expand the boundary conditions also in a series in Re_ω. In the present example, the boundary conditions, Eqs. 6.4.5 and 6.4.6, are homogeneous, and the expansion is

$$\tilde{v}_x^{(i)} = 0 \text{ at } r^* = 1, \quad \frac{\mathrm{d}\tilde{v}_x^{(i)}}{\mathrm{d}r^*} = 0 \text{ at } r^* = 0, \tag{6.4.20}$$

for $i = 0, 1, 2, \dots$

Eqs. 6.4.17, 6.4.18, and 6.4.19 are solved sequentially, subject to the boundary conditions, Eq. 6.4.20, to obtain solutions for $\tilde{v}_x^{(0)}$, $\tilde{v}_x^{(1)}$, $\tilde{v}_x^{(2)}$,

$$\tilde{v}_x^{(0)} = -\frac{(1 - r^{*2})}{4}, \quad \tilde{v}_x^{(1)} = \frac{\imath(3 - 4r^{*2} + r^{*4})}{64}, \quad \tilde{v}_x^{(2)} = \frac{(19 - 27r^{*2} + 9r^{*4} - r^{*6})}{2304}.$$

(6.4.21)

The final solution for v_x^* up to $O(\mathrm{Re}_\omega^2)$ is obtained by multiplying $\tilde{v}_x$ (Eq. 6.4.21) by $\mathrm{e}^{(\imath t^*)}$ and taking the real part,

$$v_x^* = \mathrm{Real}(\tilde{v}_x \mathrm{e}^{(\imath t^*)}) = -\frac{(1 - r^{*2}) \cos{(t^*)}}{4} - \frac{\mathrm{Re}_\omega (3 - 4r^{*2} + r^{*4}) \sin{(t^*)}}{64}$$
$$+ \frac{\mathrm{Re}_\omega^2 (19 - 27r^{*2} + 9r^{*4} - r^{*6}) \cos{(t^*)}}{2304}.$$

(6.4.22)

This procedure, denoted a 'regular perturbation' expansion, provides a solution for the velocity field as a series in the small parameter Re_ω. In a regular perturbation expansion, the approximate solution of an equation containing a small parameter can be obtained by setting the parameter equal to zero in the equation. The approximation can be systematically improved by including the first and higher powers of the small parameter.

For zero Reynolds number, the scaled velocity field is,

$$v_x^* = -\frac{(1 - r^{*2})}{4} \cos{(t^*)},$$

(6.4.23)

or the dimensional velocity field v_x is,

$$v_x = -\frac{K \cos{(\omega t)} R^2}{4\mu} \left(1 - \frac{r^2}{R^2}\right).$$

(6.4.24)

This solution is identical to the steady solution, Eq. 6.2.14, with the $(\mathrm{d}p/\mathrm{d}x)$ replaced by $(K \cos{(\omega t)})$, which is the instantaneous value of $(\partial p/\partial x)$ at time t. The steady solution is recovered because the inertial term on the left side of Eq. 6.2.8 has been neglected in the limit of small Reynolds number. Physically, the Reynolds number $\mathrm{Re}_\omega = (\rho \omega R^2/\mu) = (\omega R^2/\nu)$ can be considered as the ratio of two timescales, the timescale (R^2/ν) for momentum diffusion over a distance R, and the time period of oscillation of the pressure gradient $2\pi\omega^{-1}$. When the Reynolds number is small, the timescale for momentum diffusion is small compared to the period of oscillation of the pressure gradient. Therefore, the velocity field responds instantaneously to the change in the pressure, and we obtain a solution that is identical to the steady solution for the instantaneous value of $(\partial p/\partial x)$.

The low Reynolds number asymptotic solution, Eq. 6.4.22, is compared with the exact solution, Eq. 6.4.14, in Fig. 6.14(b) and (c). At $\text{Re}_\omega = 0.1$, Fig. 6.14(b) shows that there is no visible difference between the asymptotic and the analytical solution. Even at $\text{Re}_\omega = 1$, Fig. 6.14(c) shows that the asymptotic solution (shown by the dashed lines) is in quantitative agreement with the analytical solution.

6.4.2 High Reynolds number Limit

In the limit $\text{Re}_\omega \gg 1$, the velocity is scaled by the inertial scale $(K/\rho\omega)$ in Eq. 6.4.2, since we expect a balance between the inertial and pressure forces when the viscous forces are small compared to the inertial forces. The non-dimensional velocity is defined as $v_x^+ = (\rho v_x \omega / K)$, and Eq. 6.4.2 becomes,

$$\frac{\partial v_x^+}{\partial t^*} = \frac{1}{\text{Re}_\omega} \frac{1}{r^*} \frac{\partial}{\partial r^*} \left(r^* \frac{\partial v_x^+}{\partial r^*} \right) - \cos\left(t^*\right). \tag{6.4.25}$$

The substitution,

$$v_x^+ = \text{Real}(\tilde{v}_x^\ddagger(r^*) e^{(\iota t^*)}), \tag{6.4.26}$$

is used to obtain an ordinary differential equation for $\tilde{v}^\ddagger$,

$$\iota \tilde{v}_x^\ddagger = \frac{1}{\text{Re}_\omega} \frac{1}{r^*} \frac{\mathrm{d}}{\mathrm{d}r^*} \left(r^* \frac{\mathrm{d}\tilde{v}_x^\ddagger}{\mathrm{d}r^*} \right) - 1. \tag{6.4.27}$$

A naive approximation would be to neglect the viscous term on the right in comparison to the inertial term on the left in Eq. 6.4.27 in the limit $\text{Re}_\omega \gg 1$,

$$\iota \tilde{v}_x^\ddagger = -1. \tag{6.4.28}$$

The solution for the velocity field is

$$\tilde{v}_x^\ddagger = \iota. \tag{6.4.29}$$

This solution satisfies Eq. 6.4.6, the boundary condition at the axis. However, it is evident that the solution, Eq. 6.4.29, does not satisfy the boundary condition at the wall, Eq. 6.4.5.

The mathematical reason for the inability to satisfy the boundary condition at the wall is as follows. Eq. 6.4.27 is a second order differential equation in r^*, and so the solution of this equation contains two constants of integration. In going from Eq. 6.4.27 to Eq. 6.4.28, the highest derivative was neglected, because it was multiplied by the small parameter Re_ω^{-1}. In doing so, the differential equation is transformed into an algebraic equation, and there are no constants in the solution.

The physical reason for the inability to satisfy the no-slip is as follows. The velocity at the wall decreases to zero due to the stress exerted by the wall on the fluid, that is, the momentum diffusion from the wall to the fluid. Momentum diffusion is neglected in going from Eq. 6.4.27 to Eq. 6.4.28; consequently there is no shear stress exerted by the pipe wall on the fluid. Due to this, it is not possible to satisfy the no-slip boundary condition. This apparent inconsistency is resolved as follows.

The Reynolds number, $\mathrm{Re}_\omega = (\omega R^2/\nu)$, is the ratio of the time required for diffusion over a length scale R, (R^2/ν), and the period of oscillation $2\pi\omega^{-1}$. When this number is large, the spatial extent momentum diffusion is smaller than R. Over a time period $\sim \omega^{-1}$, the spatial extent of momentum diffusion is $(\nu/\omega)^{1/2} \sim (R\mathrm{Re}_\omega^{-1/2})$. Therefore, if we scale the distance from the wall by this diffusion length scale, there will be a balance between the inertial and the viscous terms in the momentum balance equation, Eq. 6.4.27.

Mathematically, this is accomplished by postulating a 'boundary layer' at the wall with thickness $\delta^* R$, where $\delta^* \ll 1$. The condition for finding the value of δ^* is that in the limit $\mathrm{Re}_\omega \gg 1$, there continues to be a balance between the inertial and viscous terms in Eq. 6.4.25, within the region of thickness $\delta^* R$. In order to apply this condition, we focus on a thin region near the wall and define the distance from the wall, $(R - r) = \delta^* R w$, where w is the scaled distance from the wall,

$$w = \frac{(1 - r^*)}{\delta^*}. \tag{6.4.30}$$

The scaled co-ordinate is defined so that w is $O(1)$ within the region where there is a balance between inertial and viscous forces in the limit $\mathrm{Re}_\omega \gg 1$. The radius r^* is expressed in terms of the scaled co-ordinate w in Eq. 6.4.27,

$$\imath \tilde{v}_x^\ddagger = \frac{1}{\mathrm{Re}_\omega} \frac{1}{(1 - \delta^* w)} \frac{1}{\delta^*} \frac{\mathrm{d}}{\mathrm{d}w} \left((1 - \delta^* w) \frac{1}{\delta^*} \frac{\mathrm{d}\tilde{v}_x^\ddagger}{\mathrm{d}w} \right) - 1. \tag{6.4.31}$$

Terms multiplied by powers of δ^* are neglected in comparison to 1 in the limit $\delta^* \ll 1$, and the equation reduces to,

$$\imath \tilde{v}_x^\ddagger = \frac{1}{\mathrm{Re}_\omega \delta^{*2}} \frac{\mathrm{d}^2 \tilde{v}^\ddagger}{\mathrm{d}w^2} - 1. \tag{6.4.32}$$

It is clear that $\delta^* \sim \mathrm{Re}_\omega^{-1/2}$ for the inertial and viscous terms to be of the same magnitude in the limit $\mathrm{Re}_\omega \gg 1$. The thickness of the 'boundary layer' near the wall, where viscous and inertial effects are of the same magnitude, is $(\delta^* R) \sim (\nu/\omega)^{1/2}$, as anticipated in the previous paragraph on the basis of physical arguments.

One could set δ^* equal to some constant multiplied by $\mathrm{Re}_\omega^{-1/2}$ and proceed to solve the problem; with this choice, the inertial and viscous terms are comparable so long as the constant C is $O(1)$ in the limit $\mathrm{Re}_\omega \gg 1$. While the solution of Eq. 6.4.32 in terms of the scaled co-ordinate w will depend on C, the solution in terms of the original co-ordinate r^* does not depend on this constant. In order to illustrate this, we set $\delta^* = C\mathrm{Re}_\omega^{-1/2}$ without specifying C in the present problem.

With the substitution $\delta^* = C\mathrm{Re}_\omega^{-1/2}$, Eq. 6.4.32 becomes

$$\imath \tilde{v}_x^{\ddagger} = \frac{1}{C^2} \frac{\mathrm{d}^2 \tilde{v}_x^{\ddagger}}{\mathrm{d}w^2} - 1. \tag{6.4.33}$$

This is solved to obtain,

$$\tilde{v}_x^{\ddagger} = \imath + C_1 e^{(\sqrt{\imath}Cw)} + C_2 e^{(-\sqrt{\imath}Cw)}, \tag{6.4.34}$$

where C_1 and C_2 are constants of integration.

The constants in Eq. 6.4.34 are evaluated from the boundary conditions, which are to be re-expressed in terms of the scaled co-ordinate w. The boundary condition at the wall of the pipe, Eq. 6.4.5, expressed in terms of w, is

$$\tilde{v}_x^{\ddagger} = 0 \quad \text{at} \quad w = 0. \tag{6.4.35}$$

The second boundary condition, Eq. 6.4.6, is applied at $r^* = 0$ which is equivalent to $w = \delta^{*-1}$. In the limit $\delta^* \ll 1$ ($\mathrm{Re}_\omega \gg 1$), this is equivalent to

$$\frac{\mathrm{d}\tilde{v}_x^{\ddagger}}{\mathrm{d}w} = 0 \quad \text{for} \quad w \to \infty. \tag{6.4.36}$$

The above two boundary conditions are satisfied for $C_1 = 0$ and $C_2 = -\imath$ in Eq. 6.4.34, and the final solution for $\tilde{v}^{\ddagger}$ is

$$\tilde{v}_x^{\ddagger} = \imath[1 - e^{(-\sqrt{\imath}Cw)}]. \tag{6.4.37}$$

Substituting $w = (1 - r^*)/\delta^*$ and $\delta^* = C(\mathrm{Re}_\omega)^{-1/2}$, the solution for $\tilde{v}^{\ddagger}$ in terms of r^* is,

$$\tilde{v}_x^{\ddagger} = \imath[1 - e^{(-\sqrt{\imath \mathrm{Re}_\omega}(1-r^*))}]. \tag{6.4.38}$$

As anticipated earlier, the solution for $\tilde{v}_x^{\ddagger}$ in terms of r^* is independent of the constant C used in the definition of δ^*. Therefore, without loss of generality, we can set this constant to any value. In practice, the constant is usually set equal to 1.

The solution for v_x^+ is,

$$
\begin{aligned}
v_x^+ &= \mathrm{Real}(\tilde{v}^{\ddagger}e^{(\imath t^*)}) \\
&= \mathrm{Real}\left[(\cos(t^*) + \imath \sin(t^*)) \times \imath \left[1 - e^{\left(-\sqrt{\mathrm{Re}_\omega/2}(1-r^*)\right)} \right.\right. \\
&\qquad \left.\left. \left(\cos\left(\frac{\sqrt{\mathrm{Re}_\omega}(1-r^*)}{\sqrt{2}}\right) - \imath \sin\left(\frac{\sqrt{\mathrm{Re}_\omega}(1-r^*)}{\sqrt{2}}\right)\right)\right]\right] \\
&= -\sin(t^*)\left[1 - e^{\left(-\sqrt{\mathrm{Re}_\omega/2}(1-r^*)\right)}\cos\left(\frac{\sqrt{\mathrm{Re}_\omega}(1-r^*)}{\sqrt{2}}\right)\right] \\
&\quad - \cos(t^*)e^{\left(-\sqrt{\mathrm{Re}_\omega/2}(1-r^*)\right)}\sin\left(\frac{\sqrt{\mathrm{Re}_\omega}(1-r^*)}{\sqrt{2}}\right). \qquad (6.4.39)
\end{aligned}
$$

Here, the identity $\sqrt{\imath} = \frac{1+\imath}{\sqrt{2}}$ has been used.

The asymptotic solution, Eq. 6.4.39, is compared with the analytical solution, Eq. 6.4.14, in Figs. 6.14(d) and (e). Fig. 6.14(d) shows that the asymptotic solution (dotted lines) is not in good agreement with the analytical solution for $\mathrm{Re}_\omega = 10$. However, for $\mathrm{Re}_\omega = 100$, the asymptotic solution is in quantitative agreement with the analytical solution, as shown in Fig. 6.14(e). The asymptotic solution does capture the salient features of the velocity profile, including the constant velocity in the central region of the pipe and the sharp variation in the velocity close to the wall.

The procedure used here, called a 'singular perturbation' analysis, is applicable to problems containing a small parameter where the approximate solution cannot be obtained by setting the small parameter equal to zero. Since the small parameter multiplies the highest derivative, it is not possible to satisfy all the boundary conditions when the small parameter is set equal to zero. There are thin regions within the flow where the contribution of the term containing the small parameter is comparable to the other terms even when the small parameter tends to zero. In convection–diffusion problems, this occurs in the limit of high Reynolds or Peclet number, where the coefficient of the diffusion term containing the highest derivative is small. Diffusion can be neglected in the bulk of the flow, but it is comparable to convection in thin boundary layers at the boundaries.

Summary (6.4)

1. For an unsteady and fully developed unidirectional oscillatory flow with pressure gradient $(\partial p/\partial x)$ proportional to $\cos(\omega t)$, the complex velocity is

expressed as $e^{(\iota\omega t)}$ times a complex function of the spatial co-ordinates. The physical velocity is the real part of the complex velocity.

2. There are two timescales, the time period of the pressure variation $2\pi\omega^{-1}$ and the timescale for diffusion across the channel $(\rho R^2/\mu)$, where R is the pipe radius. The Reynolds number $\mathrm{Re}_\omega = (\rho\omega R^2/\mu)$ can be interpreted as the ratio of the timescale of diffusion and that of the pressure variation.

3. For $\mathrm{Re}_\omega \ll 1$ when the time period of the pressure variation is much larger than the diffusion time, the velocity profile responds quickly to variations in the pressure. The instantaneous velocity profile is close to the steady profile for the instantaneous value of $(\partial p/\partial x)$, and a regular perturbation expansion is used to determine the corrections for small Re_ω.

4. For $\mathrm{Re}_\omega \gg 1$, the time period of the pressure variation is much smaller than that for momentum diffusion. Momentum diffusion is restricted to a layer of thickness $\mathrm{Re}_\omega^{-1/2} R$ at the wall, and there is a plug flow in the bulk of the pipe. Since the diffusion term contains the highest derivative, it is not possible to satisfy all boundary conditions if the diffusion term is neglected. To obtain a solution, it is necessary to rescale the cross-stream distance from the wall so that convection and diffusion are comparable in the momentum 'boundary layer'.

Exercises

EXERCISE 6.1 A dam stores water up to a height h, as shown in Fig. 6.15. What is the force on the dam wall, and what is the torque exerted by the water about the base of the dam.

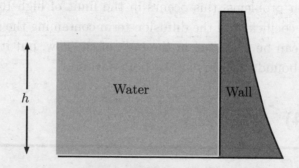

FIGURE 6.15.   Water stored up to height h in a dam.

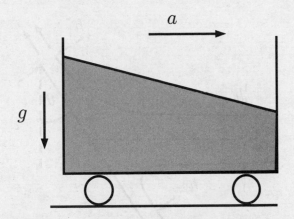

FIGURE 6.16. A container of water pushed with acceleration a.

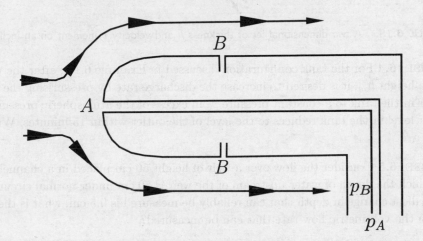

FIGURE 6.17. Schematic of a pitot tube.

EXERCISE 6.2 A container of water is pushed horizontally with an acceleration of magnitude a, as shown in Fig. 6.16. What is the slope of the interface? *Hint:* Determine the direction of the resultant acceleration due to gravity and due to the horizontal acceleration.

EXERCISE 6.3 A pitot tube, shown in Fig. 6.17, is used to measure the flow velocity from the difference in pressure between the location A, where the fluid is incident on the tube, and the location B, where the fluid is flowing parallel to the surface of the tube. There is a stagnation point at location A where the fluid velocity decreases to zero, whereas the fluid velocity is approximately equal to the free stream velocity at location B. What is the relation between the pressure difference $p_A - p_B$ and the flow velocity?

Consider a pitot tube in an incident air flow of density 1.25 kg/m^3. If the difference in pressure is 140 Pa, what is the air velocity?

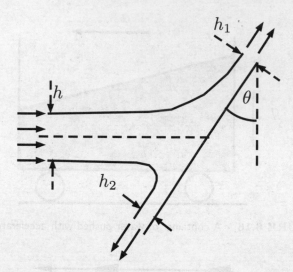

FIGURE 6.18. A two-dimensional jet of thickness h and velocity v incident on an inclined wall.

EXERCISE 6.4 For the tank configuration discussed in Example 6.1.1, after the water has filled to a height h_w, it is desired to increase the discharge rate by pressurising the air above the water in the tank to a constant pressure p_0 in excess of the atmospheric pressure so that the water level in the tank reduces to the level of the outlet within 15 minutes. What is the pressure p_0?

EXERCISE 6.5 Consider the flow over a weir of height 50 cm placed in a channel of width 2 m in which the depth of water upstream of the weir is 70 cm under normal circumstances. If the smallest change in depth that can reliably be measured is 0.5 cm, what is the smallest change in the volumetric flow rate that can be measured?

EXERCISE 6.6 A horizontal two-dimensional jet of thickness h and velocity v is incident on a wall that is inclined at an angle θ to the vertical, as shown in Fig. 6.18. Determine the force perpendicular to the wall. Determine the thicknesses h_1 and h_2 of the fluids streams parallel to the walls. Neglect gravitational effects. *Hint:* Use the condition that, in the absence of viscous stresses, there is no force exerted in the direction parallel to the wall, and consequently there is no change in the component of the fluid momentum parallel to the wall.

EXERCISE 6.7 The exhaust pipe of a carbon dioxide fire extinguisher is a right-angled tube of diameter 2 cm connected to a diverging hose of outer diameter 4 cm, as shown in Fig. 6.19. The mass of carbon dioxide stored is 5 kg, and this mass is discharged in 2 minutes when the valve is opened. Determine the force on the discharge hose if the flow is considered to be inviscid and steady. Assume the temperature is 25° C, and use the ideal gas law for calculating the density of carbon dioxide.

EXERCISE 6.8 Consider a rotating cylinder of radius R and height h_t in which the fluid is filled up to height h_0, and the cylinder is rotated with angular velocity Ω. What is the

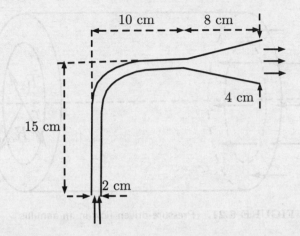

FIGURE 6.19. Flow from the exhaust of a fire extinguisher.

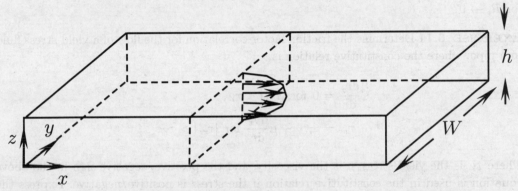

FIGURE 6.20. Pressure-driven flow in a channel of height h in the z direction, width $W \gg h$ in the y direction and of infinite extent in the stream-wise x direction.

threshold angular velocity at which the fluid spills over the top of the cylinder, as shown in Fig. 6.7(b)? What is the equation of the surface when this threshold is exceeded? What is the threshold angular velocity at which the bottom of the cylinder is exposed, as shown in Fig. 6.7(c)? What is the equation of the surface in this case? What is the equation of the surface when the fluid spills over the top and the bottom is exposed, as shown in Fig. 6.7(d)?

EXERCISE 6.9 Determine the relation between the friction factor and the Reynolds number for the pressure-driven flow of a fluid with density ρ and viscosity μ in a channel of height h and width $W \gg h$ shown in Fig. 6.20. Here, the Reynolds number is $(\rho v_{av} h / \mu)$, and the friction factor is $(\tau_w / (\rho v_{av}^2 / 2))$, where v_{av} is the average flow velocity and τ_w is the wall shear stress.

EXERCISE 6.10 Consider the pressure-driven flow in an annulus of inner radius R_i and outer radius R_o shown in Fig. 6.21. The flow is in the stream-wise x direction, and the stream-wise velocity is only a function of the radial co-ordinate r. What is the velocity

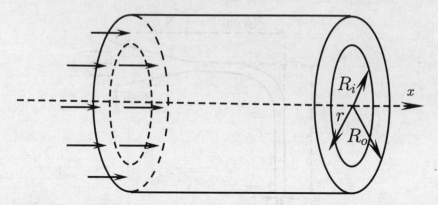

FIGURE 6.21. Pressure-driven flow in an annulus.

profile that satisfies the governing equation and the boundary conditions? How is the average velocity related to the pressure drop? Will the velocity profile reduce to a parabolic profile for $R_i \to 0$?

EXERCISE 6.11 Determine the friction factor correlation for the flow of a yield-stress fluid in a pipe, where the constitutive relation is,

$$\frac{dv_x}{dr} = 0 \text{ for } |\tau_{xr}| < \tau_Y,$$

$$\tau_{xr} = \pm\tau_Y + \mu\frac{dv_x}{dr} \text{ for } |\tau_{xr}| > \tau_Y,$$

where τ_Y is the yield stress, μ is the viscosity, and the positive/negative sign in the above equation is used in the constitutive relation if the stress is positive/negative. Express the friction factor in terms of the Reynolds number, $\mathrm{Re} = (\rho v_{av} d/\mu)$, and the Bingham number $\mathrm{Bi} = (\tau_Y/\tau_w)$, the ratio of the yield stress and the wall shear stress.

EXERCISE 6.12 For the turbulent flow in a pipe discussed in Example 6.3.2, what is (Q_t/Q_l), the ratio of the flow rate for the turbulent flow Q_t and the laminar flow Q_l for the same pressure drop.

EXERCISE 6.13 Consider the flow of water of density 10^3 kg/m^3 and viscosity 10^{-3} kg/m/s in a pipe of length 2 m and diameter 1 cm. The average flow velocity is 1 m/s, and the applied pressure difference between the ends is 3.1 kPa. What is the Reynolds number, the friction factor, the friction velocity and the wall unit ν/v_*?

EXERCISE 6.14 What are the ratios of the Kolmogorov and the macroscopic scales, (l_K/d) and (v_K/v_{av}), for a pipe flow with Reynolds number 10^4 and friction factor 7.7×10^{-3}? Use Eq. 6.3.1 for the rate of dissipation of energy per unit mass.

EXERCISE 6.15 Consider eddies of intermediate size l_e in the turbulence cascade shown in Fig. 6.12, which are much smaller than the pipe diameter and much larger than the Kolmogorov scale, $l_K \ll l_e \ll d$. The characteristic velocity and time depend only on ϵ, the

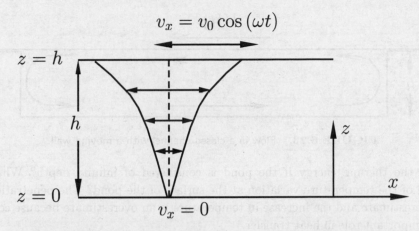

FIGURE 6.22. Flow due to the oscillation of the wall of a channel.

rate of energy transfer down the cascade, and the eddy size l_e. Estimate the velocity and time scales for these eddies, and the kinetic energy and the rate of energy dissipation per unit mass.

EXERCISE 6.16 Express the Colebrook correlation, Eq. 2.2.2, in terms of the flow Reynolds number and the friction Reynolds number.

EXERCISE 6.17 An irrigation tank of area of cross section 1 m^2 and height 2 m is connected to irrigation channels through a horizontal pipe of length 5 m and diameter 2 cm. The tank is filled in with a specified but time-dependent flow rate $\dot{V}_{in}$ which can be approximated as,

$$\dot{V}_{in} = \dot{V}_0 + \dot{V}' \cos(\omega t),$$

where $\dot{V}_0 = 1$ litre/min is a steady flow rate, $\dot{V}' = 0.5$ litres/min is the amplitude of the fluctuations in the flow rate and ω is the frequency of the fluctuations. What is the characteristic time t_c for the height variations in the tank? What is the variation of the exit flow rate and the height variation? What are the approximations for the exit flow rate for $\omega t_c \ll 1$ and $\omega t_c \gg 1$? The density and viscosity of water are 10^3 kg/m^3 and 10^{-3} kg/m/s, respectively.

EXERCISE 6.18 For the oscillatory flow in a pipe discussed in Section 6.4.2, what is the maximum wall shear stress?

EXERCISE 6.19 Determine the velocity profile for the flow between two flat plates at $z = 0$ and $z = h$, shown in Fig. 6.22, where the top plate has an oscillatory velocity $v_x = v_0 \cos(\omega t)$.

EXERCISE 6.20 The temperature at the surface of a pond varies in an oscillatory manner due to solar radiation. The incident solar radiation heats the surface during the day, and the radiation is emitted from the surface to the atmosphere at night. The maximum amplitude of the solar radiation is 350 W/m^2, the period of oscillation is the length of one day, 8.64×10^4 s, and the thermal conductivity and the thermal diffusivity of water are 0.6 W/m/$^\circ$C and 1.5×10^{-7} m^2/s, respectively. If heat transfer is due to conduction, what is the penetration

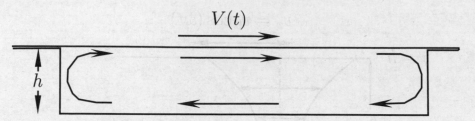

FIGURE 6.23. Flow in a closed channel with a moving wall.

depth for the thermal energy if the pond is considered of infinite depth? What is the amplitude of the temperature variation at the surface of the pond? The penetration depth is an underestimate and the increase in temperature is an overestimate because convection plays an important role in heat transfer.

EXERCISE 6.21 A spherical particle of radius R in a battery is subjected to periodic charge/discharge cycles where the concentration of ions on the surface is of the form $c = c_s \cos(\omega t)$. Determine the concentration of ions as a function of position and time in the particle if transport takes place due to diffusion with diffusion coefficient $\mathcal{D}$. Discuss the limits $(\mathcal{D}/\omega R^2) \ll 1$ and $(\mathcal{D}/\omega R^2) \gg 1$.

EXERCISE 6.22 Consider a long and narrow channel two-dimensional of length L and height H, where $H \ll L$, shown in Fig. 6.23. The ends of the channel are closed so that no fluid can enter or leave the channel. The bottom and side walls of the channel are stationary, while the top wall moves with a velocity $V(t)$. Since the length of the channel is large compared to the height, the flow in the central region away from the side walls can be considered as one dimensional. Near the side walls there will be some circulation, but this can be neglected far from the sides. For the flow far from the side walls of the channel,

(a) Write the equations for the unidirectional flow. What are the boundary conditions? What restriction is placed on the velocity profile because the ends are closed and fluid cannot enter or leave the channel?

(b) If the top wall has a steady velocity V which is independent of time, solve the equations (neglecting the time derivative term). Calculate the pressure gradient.

(c) If the top wall has an oscillatory velocity $V \cos(\omega t)$, determine the velocity profile and the pressure gradient.

7

Conservation Equations

The mass/energy conservation laws are derived for two commonly used co-ordinate systems—the Cartesian co-ordinate system in Section 7.1 and the spherical co-ordinate system in Section 7.2. For unidirectional transport, we have seen that the conservation equation has different forms in different co-ordinate systems. Here, conservation equations are first derived using shell balance in three dimensions for the Cartesian and spherical co-ordinate systems. The conservation equations have a common form when expressed in terms of vector differential operators, the gradient, divergence, and Laplacian operators; the expressions for these operators are different in different co-ordinate systems. The conservation equation derived using shell balance is used to identify the differential operators in the the Cartesian and spherical co-ordinate system, and the procedure for deriving these in a general orthogonal co-ordinate system is explained.

Since the conservation equation is universal when expressed using vector differential operators, it is not necessary to go through the shell balance procedure for each individual problem; it is sufficient to substitute the appropriate vector differential operators in the conservation equation expressed in vector form. It is important to note that the derivation here is restricted to orthogonal co-ordinate systems, where the three co-ordinate directions are perpendicular to each other at all locations.

The discussion in Section 7.1 and 7.2 is restricted to mass/energy transfer. The constitutive relation (Newton's law) for momentum transfer for general three-dimensional flows is more complicated than that for mass/heat transfer. Mass and heat are scalars, and the flux of mass/heat is a vector along the direction of decreasing concentration/temperature. Since momentum is a vector, the flux of momentum has two directions associated with it: the direction of the momentum

319

vector and the direction in which the momentum is transported. Due to this, the stress or momentum flux is a 'second order tensor' with two physical directions—the direction of momentum and the orientation of the perpendicular to the surface across which momentum is transported.

The velocity difference between nearby locations in the flow, which results in deformation of the fluid elements, is related to a second order tensor called the 'velocity gradient' or the 'rate of deformation' tensor. This tensor also contains two directions, the direction of the velocity and the direction of variation. This is in contrast to the gradients of concentration and temperature, which are vectors. Therefore, the constitutive relation for the stress is a relation between two second order tensors. There are constraints on the form of the constitutive relation for the stress tensor, because it has to be invariant under co-ordinate transformations.

In Section 7.3.2, the force on a surface within the fluid is related to the stress. Angular momentum conservation requires that the stress tensor is symmetric. The stress is separated into the pressure (which is non-zero even in the absence of flow) and the shear stress due to the velocity gradient. In Section 7.3.3, the deformation of a volume element of fluid is expressed in terms of the velocity gradient or rate of deformation tensor. This is separated into invariant components, the radial expansion/compression, rotation and extensional strain.

In Section 7.3.4, Newton's law is expressed as a relation between the stress tensor and the velocity gradient tensor. Newton's law is then substituted into the momentum conservation equation to explain the form of the Navier–Stokes equation for an incompressible fluid. The special case of a potential flow is discussed in Section 7.3.5, where the Bernoulli equation for the pressure is derived.

7.1   Cartesian Co-ordinate System

7.1.1   Mass Conservation Equation

In the Cartesian co-ordinate system, the three axes, x, y and z are perpendicular to each other, as shown in Fig. 7.1(a). The unit vectors e_x, e_y and e_z are directed along the x, y and z axes. These form an orthogonal set—that is, each unit vector is perpendicular to the other two at each point in space, as shown in Fig. 7.1(b). In order to derive the mass conservation equation, we consider a differential volume ΔV centred at the location (x, y, z), of extent Δx, Δy and Δz in three co-ordinate directions. This volume is bounded by six surfaces of constant co-ordinate in three dimensions. The location and the surface areas of these six surfaces shown are given in Table 7.1.

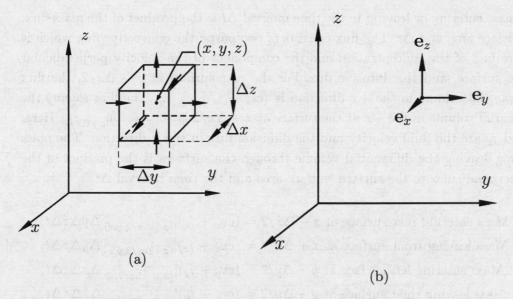

FIGURE 7.1. Differential volume used for deriving the mass conservation equation in a Cartesian co-ordinate system (a), and the unit vectors e_x, e_y and e_z at the location (x, y, z) (b).

TABLE 7.1. The surfaces and corresponding surface areas of the differential volume in the Cartesian co-ordinate system shown in Fig. 7.1.

Surface	Location	Perpendicular to	Surface area ΔS
Rear	$x - \Delta x/2$	e_x	$\Delta y \Delta z$
Front	$x + \Delta x/2$	e_x	$\Delta y \Delta z$
Left	$y - \Delta y/2$	e_y	$\Delta z \Delta x$
Right	$y + \Delta y/2$	e_y	$\Delta z \Delta x$
Bottom	$z - \Delta z/2$	e_z	$\Delta x \Delta y$
Top	$z + \Delta z/2$	e_z	$\Delta x \Delta y$

The mass conservation condition is derived using shell balance for the change in the mass in a time interval Δt between the times t and $t + \Delta t$,

$$\begin{pmatrix} \text{Change of mass} \\ \text{in time } \Delta t \end{pmatrix} = \begin{pmatrix} \text{Mass} \\ \text{in} \end{pmatrix} - \begin{pmatrix} \text{Mass} \\ \text{out} \end{pmatrix} + \begin{pmatrix} \text{Production of mass} \\ \text{in differential volume} \end{pmatrix}. \quad (7.1.1)$$

The change in mass in the time interval Δt is the difference in the concentration multiplied by the volume,

$$\begin{pmatrix} \text{Change of mass} \\ \text{in time } \Delta t \end{pmatrix} = \Delta x \Delta y \Delta z \left(c(x, y, z, t + \Delta t) - c(x, y, z, t) \right). \quad (7.1.2)$$

The mass entering or leaving in the time interval Δt is the product of the mass flux, the surface area and Δt. The flux consists of two parts: the convective flux which is the product of the concentration and the component of the velocity perpendicular to the surface, and the diffusive flux. For the rear surface at $x - \Delta x/2$, the flux *entering* the volume in the $+x$ direction is $(cv_x + j_x)|_{x-\Delta x/2}$. The flux *leaving* the differential volume in the $+x$ at the surface at $x + \Delta x/2$ is $(cv_x + j_x)|_{x+\Delta x/2}$. Here, v_x and j_x are the fluid velocity and the diffusion flux in the x direction. The mass entering/leaving the differential volume through the surfaces is the product of the flux perpendicular to the surface, surface area and the time interval Δt,

$$
\begin{aligned}
\text{Mass entering rear surface at } x - \Delta x/2 &= (cv_x + j_x)|_{(x-\Delta x/2, y, z)} \, \Delta y \Delta z \Delta t, \\
\text{Mass leaving front surface at } x + \Delta x/2 &= (cv_x + j_x)|_{(x+\Delta x/2, y, z)} \, \Delta y \Delta z \Delta t, \\
\text{Mass entering left surface at } y - \Delta y/2 &= (cv_y + j_y)|_{(x, y-\Delta y/2, z)} \, \Delta x \Delta z \Delta t, \\
\text{Mass leaving right surface at } y + \Delta y/2 &= (cv_y + j_y)|_{(x, y+\Delta y/2, z)} \, \Delta x \Delta z \Delta t, \\
\text{Mass entering bottom surface at } z - \Delta z/2 &= (cv_z + j_z)|_{(x, y, z-\Delta z/2)} \, \Delta x \Delta y \Delta t, \\
\text{Mass leaving top surface at } z + \Delta z/2 &= (cv_z + j_z)|_{(x, y, z+\Delta z/2)} \, \Delta x \Delta y \Delta t.
\end{aligned}
$$
$$(7.1.3)$$

The last term on the right side of Eq. 7.1.1 is the production of mass in the volume ΔV due to, for example, a chemical reaction. If the production per unit volume per unit time is $\mathcal{S}$, the mass produced in the volume ΔV in the time Δt is $\mathcal{S}(x, y, z) \times \Delta V \times \Delta t$,

$$
\begin{pmatrix} \text{Production of mass} \\ \text{in differential volume} \end{pmatrix} = \mathcal{S}(x, y, z, t) \Delta x \Delta y \Delta z \Delta t. \qquad (7.1.4)
$$

This could be positive or negative depending on whether the mass is produced or consumed in the chemical reaction. Substituting Eqs. 7.1.2–7.1.4 into Eq. 7.1.1, the mass conservation equation is,

$$
\begin{aligned}
\Delta x \Delta y \Delta z \, &(c(x, y, z, t + \Delta t) - c(x, y, z, t)) \\
= \Delta t \, \Big(&(cv_x + j_x)|_{(x-\Delta x/2, y, z)} \, \Delta y \Delta z - (cv_x + j_x)|_{(x+\Delta x/2, y, z)} \, \Delta y \Delta z \\
&+ (cv_y + j_y)|_{(x, y-\Delta y/2, z)} \, \Delta x \Delta z - (cv_y + j_y)|_{(x, y+\Delta y/2, z)} \, \Delta x \Delta z \\
&+ (cv_z + j_z)|_{(x, y, z-\Delta z/2)} \, \Delta x \Delta y - (cv_z + j_z)|_{(x, y, z+\Delta z/2)} \, \Delta x \Delta y \Big) \\
&+ \mathcal{S}(x, y, z, t) \Delta x \Delta y \Delta z \Delta t.
\end{aligned}
$$
$$(7.1.5)$$

The difference equation for the concentration field is obtained by dividing the above balance equation by $\Delta x \Delta y \Delta z \Delta t$,

$$\frac{(c(x,y,z,t+\Delta t) - c(x,y,z,t))}{\Delta t} = \frac{(cv_x + j_x)|_{(x-\Delta x/2,y,z)} - (cv_x + j_x)|_{(x+\Delta x/2,y,z)}}{\Delta x}$$
$$+ \frac{(cv_y + j_y)|_{(x,y-\Delta y/2,z)} - (cv_y + j_y)|_{(x,y+\Delta y/2,z)}}{\Delta y}$$
$$+ \frac{(cv_z + j_z)|_{(x,y,z-\Delta z/2)} - (cv_z + j_z)|_{(x,y,z+\Delta z/2)}}{\Delta z}$$
$$+ \mathcal{S}(x,y,z,t). \tag{7.1.6}$$

The differential equation for mass conservation is obtained by taking the limit $\Delta x \to 0$, $\Delta y \to 0$, $\Delta z \to 0$ and $\Delta t \to 0$,

$$\frac{\partial c}{\partial t} = -\frac{\partial(cv_x + j_x)}{\partial x} - \frac{\partial(cv_y + j_y)}{\partial y} - \frac{\partial(cv_z + j_z)}{\partial z} + \mathcal{S}. \tag{7.1.7}$$

Recall that while calculating the partial derivatives with respect to one co-ordinate, the other co-ordinates are kept fixed. While taking the partial derivative with respect to time on the left side of Eq. 7.1.7, all the spatial co-ordinates are kept fixed. Similarly, for the partial derivative with respect to one of the co-ordinates on the right side of Eq. 7.1.7, the other two spatial co-ordinates and time are fixed. Also note the negative sign on the first three terms on the right side of Eq. 7.1.7. This is because the partial derivative with respect to x is defined as

$$\frac{\partial(cv_x + j_x)}{\partial x} = \lim_{\Delta x \to 0} \frac{(cv_x + j_x)|_{(x+\Delta x/2,y,z)} - (cv_x + j_x)|_{(x-\Delta x/2,y,z)}}{\Delta x}, \tag{7.1.8}$$

and the first term on the right side of Eq. 7.1.6 is the negative of the above partial derivative.

Eq. 7.1.7 is conventionally written with the time derivative and convection terms on the left, and the diffusion terms on the right,

$$\frac{\partial c}{\partial t} + \frac{\partial(cv_x)}{\partial x} + \frac{\partial(cv_y)}{\partial y} + \frac{\partial(cv_z)}{\partial z} = -\frac{\partial j_x}{\partial x} - \frac{\partial j_y}{\partial y} - \frac{\partial j_z}{\partial z} + \mathcal{S}. \tag{7.1.9}$$

The mass flux is related to the concentration variations by Fick's law for diffusion,

$$j_x = -\mathcal{D}\frac{\partial c}{\partial x}, \quad j_y = -\mathcal{D}\frac{\partial c}{\partial y}, \quad j_z = -\mathcal{D}\frac{\partial c}{\partial z}. \tag{7.1.10}$$

When the components of the flux, Eq. 7.1.10, are substituted into the conservation equation, Eq. 7.1.9, we obtain

$$\frac{\partial c}{\partial t} + \frac{\partial (cv_x)}{\partial x} + \frac{\partial (cv_y)}{\partial y} + \frac{\partial (cv_z)}{\partial z} = \frac{\partial}{\partial x}\left(\mathcal{D}\frac{\partial c}{\partial x}\right) + \frac{\partial}{\partial y}\left(\mathcal{D}\frac{\partial c}{\partial y}\right) + \frac{\partial}{\partial z}\left(\mathcal{D}\frac{\partial c}{\partial z}\right) + \mathcal{S}.$$

(7.1.11)

If the diffusion coefficient $\mathcal{D}$ is independent of position, the conservation equation reduces to,

$$\frac{\partial c}{\partial t} + \frac{\partial (cv_x)}{\partial x} + \frac{\partial (cv_y)}{\partial y} + \frac{\partial (cv_z)}{\partial z} = \mathcal{D}\left(\frac{\partial^2 c}{\partial x^2} + \frac{\partial^2 c}{\partial y^2} + \frac{\partial^2 c}{\partial z^2}\right) + \mathcal{S}. \quad (7.1.12)$$

The fluid velocity and mass flux can be expressed as vectors using a set of basis vectors for specific co-ordinate systems. In the Cartesian co-ordinate system, the unit vectors along the x, y and z directions are defined as $\boldsymbol{e}_x$, $\boldsymbol{e}_y$ and $\boldsymbol{e}_z$, respectively, as shown in Fig. 7.1(b). These vectors are of unit length—that is, $\boldsymbol{e}_x \cdot \boldsymbol{e}_x = \boldsymbol{e}_y \cdot \boldsymbol{e}_y = \boldsymbol{e}_z \cdot \boldsymbol{e}_z = 1$—and the vectors are perpendicular to each other—that is, $\boldsymbol{e}_x \cdot \boldsymbol{e}_y = \boldsymbol{e}_y \cdot \boldsymbol{e}_z = \boldsymbol{e}_x \cdot \boldsymbol{e}_z = 0$. Here, bold symbols are used to denote vectors, and the dot product of two vectors $\boldsymbol{A} \cdot \boldsymbol{B} = |\boldsymbol{A}||\boldsymbol{B}|\cos(\theta_{AB})$, where $|\boldsymbol{A}|$ and $|\boldsymbol{B}|$ are the magnitudes (lengths) of the two vectors, and θ_{AB} is the angle between them. The directions of the unit vectors in a Cartesian co-ordinate system are independent of position. The velocity and the flux vectors are,

$$\boldsymbol{v} = v_x \boldsymbol{e}_x + v_y \boldsymbol{e}_y + v_z \boldsymbol{e}_z,$$
$$\boldsymbol{j} = j_x \boldsymbol{e}_x + j_y \boldsymbol{e}_y + j_z \boldsymbol{e}_z. \quad (7.1.13)$$

Fick's law, Eq. 7.1.10, is written in vector form,

$$\boxed{\boldsymbol{j} = -\mathcal{D}\boldsymbol{\nabla} c,} \quad (7.1.14)$$

where the 'gradient' of the concentration in a Cartesian co-ordinate system is,

$$\boxed{\boldsymbol{\nabla} c = \boldsymbol{e}_x \frac{\partial c}{\partial x} + \boldsymbol{e}_y \frac{\partial c}{\partial y} + \boldsymbol{e}_z \frac{\partial c}{\partial z}.} \quad (7.1.15)$$

The definition of the gradient, its physical interpretation and the integral relation for the gradient are discussed in Appendix 7.A.

The conservation Eq. 7.1.9 can be written in a compact form by using the divergence operator,

$$\frac{\partial c}{\partial t} + \nabla \cdot (cv) = -\nabla \cdot j + \mathcal{S}. \qquad (7.1.16)$$

where the divergence of a vector, $\nabla \cdot j$, is defined in Cartesian co-ordinates as,

$$\nabla \cdot j = \left(e_x \frac{\partial}{\partial x} + e_y \frac{\partial}{\partial y} + e_z \frac{\partial}{\partial z} \right) \cdot (j_x e_x + j_y e_y + j_z e_z) = \frac{\partial j_x}{\partial x} + \frac{\partial j_y}{\partial y} + \frac{\partial j_z}{\partial z}. \qquad (7.1.17)$$

Note that the divergence acts on a vector, and the result is a scalar. The definition of the divergence, its physical interpretation and the integral relation for the divergence are discussed in Appendix 7.B.

The expression for the flux, Eq. 7.1.14, is substituted into the conservation equation to obtain a second order partial differential equation for the concentration,

$$\frac{\partial c}{\partial t} + \nabla \cdot (cv) = \nabla \cdot (\mathcal{D}\nabla c) + \mathcal{S}. \qquad (7.1.18)$$

If the diffusion coefficient $\mathcal{D}$ is a constant, Eq. 7.1.18 can be expressed using the 'Laplacian' operator $\nabla^2 = \nabla \cdot \nabla$,

$$\frac{\partial c}{\partial t} + \nabla \cdot (cv) = \mathcal{D}\nabla^2 c + \mathcal{S}, \qquad (7.1.19)$$

where the Laplacian operator ∇^2 is

$$\nabla^2 = \left(e_x \frac{\partial}{\partial x} + e_y \frac{\partial}{\partial y} + e_z \frac{\partial}{\partial z} \right) \cdot \left(e_x \frac{\partial}{\partial x} + e_y \frac{\partial}{\partial y} + e_z \frac{\partial}{\partial z} \right) = \frac{\partial^2}{\partial x^2} + \frac{\partial^2}{\partial y^2} + \frac{\partial^2}{\partial z^2}. \qquad (7.1.20)$$

Note that the Laplacian operator acts on a scalar, and the result is a scalar.

The conservation equation, Eq. 7.1.19, can be expressed in terms or the 'material' or 'substantial' derivative, which is the derivative in a reference frame moving with the same velocity as the fluid velocity v, as shown in Fig. 7.2(a). The volume of fluid located at x with components (x, y, z) at time t moves to a new position $(x + v\Delta t)$ with components $(x + v_x\Delta t, y + v_y\Delta t, z + v_z\Delta t)$ at time $t + \Delta t$, as shown in Fig. 7.2(b). The rate of change of concentration in the material volume element

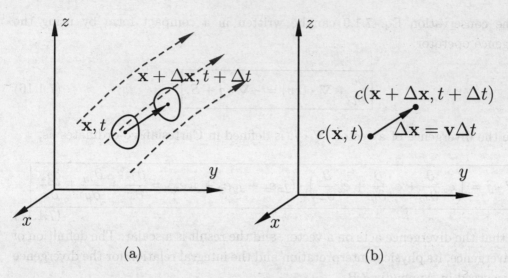

FIGURE 7.2. The motion of a 'material' volume element of the fluid (a), and the change in the concentration field in a reference frame moving with the velocity of the fluid (b).

is called the 'material derivative' or the 'substantial derivative',

$$\frac{Dc}{Dt} = \lim_{\Delta t \to 0} \frac{c(x + v_x \Delta t, y + v_y \Delta t, z + v_z \Delta t, t + \Delta t) - c(x, y, z, t)}{\Delta t}$$

$$= \frac{\partial c}{\partial t} + v_x \frac{\partial c}{\partial x} + v_y \frac{\partial c}{\partial y} + v_z \frac{\partial c}{\partial z}. \tag{7.1.21}$$

Chain rule for differentiation is used in the second step above to express the material derivative in terms of the spatial derivatives. The material derivative can be expressed using the gradient operator,

$$\boxed{\frac{Dc}{Dt} = \frac{\partial c}{\partial t} + \boldsymbol{v} \cdot \boldsymbol{\nabla} c.} \tag{7.1.22}$$

The conservation equation, Eq. 7.1.16, can be rewritten using the product rule for differentiation as,

$$\frac{\partial c}{\partial t} + \boldsymbol{v} \cdot \boldsymbol{\nabla} c + c(\boldsymbol{\nabla} \cdot \boldsymbol{v}) = -\boldsymbol{\nabla} \cdot \boldsymbol{j} + \mathcal{S}. \tag{7.1.23}$$

The sum of the first two terms on the left of Eq. 7.1.23 is the material derivative, and the mass conservation equation expressed in terms of the material derivative is,

$$\boxed{\frac{Dc}{Dt} + c(\boldsymbol{\nabla} \cdot \boldsymbol{v}) = -\boldsymbol{\nabla} \cdot \boldsymbol{j} + \mathcal{S}.} \tag{7.1.24}$$

7.1.2 Energy Conservation Equation

The conservation equation for the enthalpy density $(\rho C_p T)$ for processes at constant pressure, derived using procedures similar to that for the concentration Eq. 7.1.16, is

$$\frac{\partial(\rho C_p T)}{\partial t} + \boldsymbol{\nabla} \cdot (\boldsymbol{v} \rho C_p T) = -\boldsymbol{\nabla} \cdot \boldsymbol{q} + \mathcal{S}_e, \tag{7.1.25}$$

where $\boldsymbol{q}$ is the heat flux, and $\mathcal{S}_e$ is the rate of energy input per unit volume per unit time due to processes such as reactions or phase change. The heat flux is related to the temperature variation by Fourier's law for heat conduction,

$$\boldsymbol{q} = -k\boldsymbol{\nabla}T, \tag{7.1.26}$$

where k is the thermal conductivity.

If the density ρ and specific heat C_p are independent of position and time, Eq. 7.1.25 is divided by ρC_p, resulting in an equation of the same form as the diffusion Eq. 7.1.18,

$$\frac{\partial T}{\partial t} + \boldsymbol{\nabla} \cdot (\boldsymbol{v}T) = \boldsymbol{\nabla} \cdot (\alpha \boldsymbol{\nabla} T) + \frac{\mathcal{S}_e}{\rho C_p}, \tag{7.1.27}$$

where $\alpha = (k/\rho C_p)$ is the thermal diffusivity. If the thermal diffusivity is a constant, the energy conservation equation is of the same form as Eq. 7.1.19,

$$\frac{\partial T}{\partial t} + \boldsymbol{\nabla} \cdot (\boldsymbol{v}T) = \alpha \boldsymbol{\nabla}^2 T + \frac{\mathcal{S}_e}{\rho C_p}. \tag{7.1.28}$$

Summary (7.1)

1. The mass conservation equation for the concentration field c is Eq. 7.1.16, where the divergence of the flux $\boldsymbol{j}$ in Cartesian co-ordinates is defined in Eq. 7.1.17.

2. The flux is related to the concentration gradient by Fick's law, Eq. 7.1.14, where $\mathcal{D}$ is the diffusion coefficient, and the gradient of the concentration in Cartesian co-ordinates is defined in Eq. 7.1.15. When Eq. 7.1.14 is substituted for the flux, the conservation equation is Eq. 7.1.18.

3. If the diffusion coefficient is a constant, the conservation equation is Eq. 7.1.19, where the Laplacian of the concentration $\boldsymbol{\nabla}^2 c$ is defined in Eq. 7.1.20.

4. The substantial or material derivative is defined in Eq. 7.1.22.

5. Expressed in terms of the material derivative, the mass conservation equation is Eq. 7.1.24.

7.2   Spherical Co-ordinate System

In the spherical coordinate system, a point is represented by three coordinates, the distance from the origin r, the azimuthal angle θ that the position vector makes with the z axis, and the meridional angle ϕ that the projection of the position vector in the $x - y$ plane makes with the x axis, as shown in Fig. 7.3(a). The coordinate r is always positive, since it is defined as the distance from the origin, and surfaces of constant r are spherical surfaces. The azimuthal angle θ varies over the range $0 \leq \theta \leq \pi$, where $\theta = 0$ corresponds to the $+z$ axis and $\theta = \pi$ corresponds to the $-z$ axis. Surfaces of constant θ are conical surfaces with subtended angle 2θ. The meridional angle ϕ varies from 0 to 2π, since the projection made by the position vector on the $x - y$ plane can rotate around the z axis by a maximum angle 2π. The surfaces of constant ϕ are planes rotated around the z axis.

The sectioning of a spherical surface using the angles (θ, ϕ) is best understood using the familiar example of the geographic co-ordinate system for sectioning the earth's surface based on latitudes and longitudes shown in Fig. 7.3(b). The origin of the co-ordinate system is the centre of the earth, and the $+z$ and $-z$ axes are directed from the centre to the north and south poles, respectively, along the axis of rotation. The radius r is approximately a constant on the earth's surface, and the elevation in the geographic co-ordinate system is defined as the distance from the surface in the radial direction. The latitude φ in the geographic co-ordinate system is the angle between the position vector and the equatorial plane; this angle varies from $+(\pi/2)$ at the north pole to 0 at the equator and $-(\pi/2)$ at the south pole, as shown in Fig. 7.3(b). In contrast, the azimuthal angle θ in a spherical co-ordinate system is defined as the angle made by the position vector with the $+z$ axis. Therefore, the azimuthal angle varies from 0 on the $+z$ axis to $(\pi/2)$ on the equator to π on the $-z$ axis. It is easily verified that the relation between the azimuthal angle in a spherical co-ordinate system and the latitude in a geographic co-ordinate system is $\theta = (\pi/2) - \varphi$. The meridians are the great semi-circles extending between the two poles, and the meridional angle ϕ is measured from the $+x$ axis increasing towards the $+y$ axis, as shown in Fig. 7.3(c). Thus, $\phi = 0$ along the $+x$ axis, $\phi = (\pi/2)$ along the $+y$ axis, $\phi = \pi$ along the $-x$ axis and $\phi = (3\pi/2)$ along the $-y$ axis.

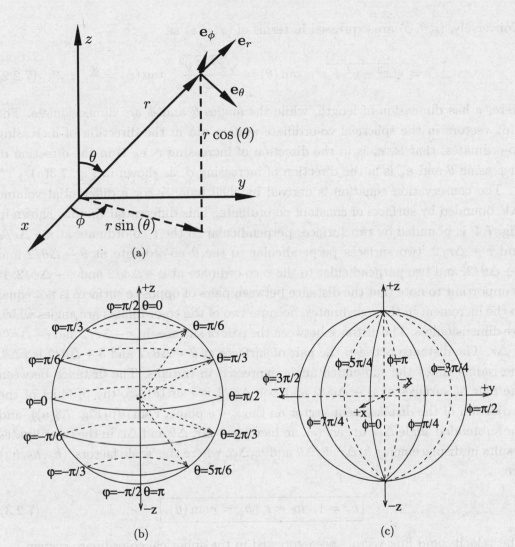

FIGURE 7.3. The spherical coordinate system (a), the azimuthal angle θ of a point on the sphere from the $+z$ axis compared with the latitude φ for the geographic co-ordinate system (b), and the meridional angle ϕ (c).

The spherical co-ordinate system is nearly identical to the geographic co-ordinate system, with the difference that the azimuthal angle is defined with respect to the $+z$ axis in the spherical co-ordinate system, whereas the latitude is defined relative to the equatorial plane in the geographic co-ordinate system.

In the spherical co-ordinate system, Fig. 7.3(a) shows that the z co-ordinate of a point is $r \cos(\theta)$, and the projection of the position vector on the $x - y$ plane is $r \sin(\theta)$. The coordinates x, y and z are expressed in terms of (r, θ, ϕ) as

$$x = r \sin(\theta) \cos(\phi), \quad y = r \sin(\theta) \sin(\phi), \quad z = r \cos(\theta). \qquad (7.2.1)$$

Conversely, (r, θ, ϕ) are expressed in terms of (x, y, z) as

$$r = \sqrt{x^2 + y^2 + z^2}, \ \ \tan(\theta) = \frac{\sqrt{x^2 + y^2}}{z}, \ \ \tan(\phi) = \frac{y}{x}. \qquad (7.2.2)$$

Here, r has dimension of length, while the angles θ and ϕ are dimensionless. The unit vectors in the spherical co-ordinate system are in the direction of increasing co-ordinates, that is, e_r is in the direction of increasing r, e_θ is in the direction of increasing θ and e_ϕ is in the direction of increasing ϕ, as shown in Fig. 7.3(a).

The conservation equation is derived by shell balance for a differential volume ΔV bounded by surfaces of constant co-ordinate. This differential volume, shown in Fig. 7.4, is bounded by two surfaces perpendicular to the r co-ordinate at $r - \Delta r/2$ and $r + \Delta r/2$, two surfaces perpendicular to the θ co-ordinate at $\theta - \Delta\theta/2$ and $\theta + \Delta\theta/2$, and two perpendicular to the ϕ co-ordinate at $\phi - \Delta\phi/2$ and $\phi + \Delta\phi/2$. It is important to note that the distance between pairs of opposite surfaces is not equal to the increment in the coordinates, because two of the co-ordinates are angles which are dimensionless. The distance between the pair of surfaces at $r - \Delta r/2$ and $r + \Delta r/2$ is Δr. The distance between the pair of surfaces at $\theta - \Delta\theta/2$ and $\theta + \Delta\theta/2$ is $r\Delta\theta$, the radius times the subtended angle expressed in radians. The distance between the pair of surfaces at $\phi - \Delta\phi/2$ and $\phi + \Delta\phi/2$ is $r\sin(\theta)\Delta\phi$, the product of the projection of the displacement vector on the $x - y$ plane, $r\sin(\theta)$ (Fig. 7.3(a)), and the subtended angle $\Delta\phi$. In general, an increment Δr, $\Delta\theta$ and $\Delta\phi$ in the co-ordinates results in displacements $h_r\Delta r$, $h_\theta\Delta\theta$ and $h_\phi\Delta\phi$, where the 'scale factors' (h_r, h_θ, h_ϕ) are,

$$\boxed{h_r = 1, \ \ h_\theta = r, \ \ h_\phi = r\sin(\theta).} \qquad (7.2.3)$$

The velocity and flux vectors are expressed in the spherical co-ordinate system,

$$v = v_r e_r + v_\theta e_\theta + v_\phi e_\phi,$$
$$j = j_r e_r + j_\theta e_\theta + j_\phi e_\phi. \qquad (7.2.4)$$

The volume ΔV is enclosed by six surfaces shown in Fig. 7.4 is,

$$\Delta V = h_r \Delta r \, h_\theta \Delta\theta \, h_\phi \Delta\phi = r^2 \sin(\theta) \, \Delta r \, \Delta\theta \, \Delta\phi. \qquad (7.2.5)$$

The surface areas of the six surfaces enclosing the differential volume, listed in Table 7.2, are products of the two orthogonal differential displacements perpendicular to the co-ordinate direction. For example, the surface area of the surface perpendicular to e_r at $r - \Delta r/2$ is the product of the displacements in the θ

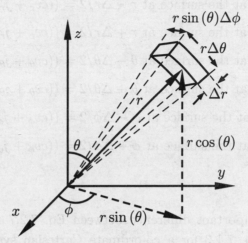

FIGURE 7.4. Differential volume for deriving the conservation equation in a spherical coordinate system.

TABLE 7.2. The surfaces and corresponding surface areas of the differential volume in a spherical co-ordinate system shown in Fig. 7.4.

Location	Perpendicular to	Surface area ΔS
$r - \Delta r/2$	e_r	$(h_\theta h_\phi)\big\|_{r-\Delta r/2}\Delta\theta\Delta\phi$
$r + \Delta r/2$	e_r	$(h_\theta h_\phi)\big\|_{r+\Delta r/2}\Delta\theta\Delta\phi$
$\theta - \Delta\theta/2$	e_θ	$(h_\phi h_r)\big\|_{\theta-\Delta\theta/2}\Delta\phi\Delta r$
$\theta + \Delta\theta/2$	e_θ	$(h_\phi h_r)\big\|_{\theta+\Delta\theta/2}\Delta\phi\Delta r$
$\phi - \Delta\phi/2$	e_ϕ	$(h_r h_\theta)\big\|_{\phi-\Delta\phi/2}\Delta r\Delta\theta$
$\phi + \Delta\phi/2$	e_ϕ	$(h_r h_\theta)\big\|_{\phi+\Delta\phi/2}\Delta r\Delta\theta$

and ϕ directions, $(h_\theta\Delta\theta)(h_\phi\Delta\phi)$. The scale factors h_θ and h_ϕ, Eq. 7.2.3, are functions of the co-ordinates; the former is only a function of r while the latter is a function of r and θ.

The balance equation, Eq. 7.1.1, for the volume ΔV in Fig. 7.4 can now be formulated. The change in mass in the time interval Δt is the change in concentration multiplied by the volume,

$$\begin{pmatrix}\text{Change of mass} \\ \text{in time } \Delta t\end{pmatrix} = h_r\Delta r\ h_\theta\Delta\theta\ h_\phi\Delta\phi\left(c(r,\theta,\phi,t+\Delta t) - c(r,\theta,\phi,t)\right). \quad (7.2.6)$$

The mass entering/leaving at a surface enclosing the differential volume is the product of the flux and the surface area,

Mass entering at the surface at $r - \Delta r/2 = ((cv_r + j_r)h_\theta\Delta\theta\, h_\phi\Delta\phi)|_{(r-\Delta r/2,\theta,\phi)}\,\Delta t,$

Mass leaving at the surface at $r + \Delta r/2 = ((cv_r + j_r)h_\theta\Delta\theta\, h_\phi\Delta\phi)|_{(r+\Delta r/2,\theta,\phi)}\,\Delta t,$

Mass entering at the surface at $\theta - \Delta\theta/2 = ((cv_\theta + j_\theta)h_r\Delta r\, h_\phi\Delta\phi)|_{(r,\theta-\Delta\theta/2,\phi)}\,\Delta t,$

Mass leaving at the surface at $\theta + \Delta\theta/2 = ((cv_\theta + j_\theta)h_r\Delta r\, h_\phi\Delta\phi)|_{(r,\theta+\Delta\theta/2,\phi)}\,\Delta t,$

Mass entering at the surface at $\phi - \Delta\phi/2 = ((cv_\phi + j_\phi)h_r\Delta r\, h_\theta\Delta\theta)|_{(r,\theta,\phi-\Delta\phi/2)}\,\Delta t,$

Mass leaving at the surface at $\phi + \Delta\phi/2 = ((cv_\phi + j_\phi)h_r\Delta r\, h_\theta\Delta\theta)|_{(r,\theta,\phi+\Delta\phi/2)}\,\Delta t.$

$$(7.2.7)$$

There is an important difference between Eq. 7.2.7 for the spherical co-ordinate system and Eq. 7.1.3 for a co-ordinate Cartesian system. The scale factors, and consequently the surface areas, in the spherical co-ordinate system do depend on the r and θ co-ordinates. For example, the surface area of the surface at $r + \Delta r/2$ is different from the surface area of the surface at $r - \Delta r/2$, because the surface area is proportional to $h_\theta h_\phi = r^2 \sin(\theta)$. In contrast, in the Cartesian co-ordinate system, the surface areas of opposing surfaces in Fig. 7.1(a) are equal.

The production of mass in the differential volume in time Δt is,

$$\left(\begin{array}{c}\text{Production of mass in}\\ \text{differential volume}\end{array}\right) = \mathcal{S}(r,\theta,\phi,t)h_r\Delta r\, h_\theta\Delta\theta\, h_\phi\Delta\phi\,\Delta t, \qquad (7.2.8)$$

where $\mathcal{S}$ is the mass produced due to reaction per unit area per unit time. Substituting Eqs. 7.2.6, 7.2.7 and 7.2.8 into the balance Eq. 7.1.1, we obtain the equation for the concentration,

$$(c(r,\theta,\phi,t+\Delta t) - c(r,\theta,\phi,t))\,h_r h_\theta h_\phi \Delta r\Delta\theta\Delta\phi$$

$$= \Delta t\left(((cv_r + j_r)h_\theta h_\phi\Delta\theta\Delta\phi)|_{r-\Delta r/2} - ((cv_r + j_r)h_\theta h_\phi\Delta\theta\Delta\phi)|_{r+\Delta r/2}\right.$$

$$+ ((cv_\theta + j_\theta)h_r h_\phi\Delta r\Delta\phi)|_{\theta-\Delta\theta/2} - ((cv_\theta + j_\theta)h_r h_\phi\Delta r\Delta\phi)|_{\theta+\Delta\theta/2}$$

$$\left. + ((cv_\phi + j_\phi)h_r h_\theta\Delta r\Delta\theta)|_{\phi-\Delta\phi/2} - ((cv_\phi + j_\phi)h_r h_\theta\Delta r\Delta\theta)|_{\phi+\Delta\phi/2}\right)$$

$$+ \mathcal{S}\,h_r h_\theta h_\phi \Delta r\Delta\theta\Delta\phi\Delta t. \qquad (7.2.9)$$

Eq. 7.2.9 is divided by $\Delta t \Delta V$ to obtain the difference equation,

$$
\begin{aligned}
&\frac{(c(r,\theta,\phi,t+\Delta t) - c(r,\theta,\phi,t))}{\Delta t} \\
&= \frac{((cv_r + j_r)h_\theta h_\phi)|_{r-\Delta r/2} - ((cv_r + j_r)h_\theta h_\phi)|_{r+\Delta r/2}}{h_r h_\theta h_\phi \Delta r} \\
&\quad + \frac{((cv_\theta + j_\theta)h_r h_\phi)|_{\theta-\Delta\theta/2} - ((cv_\theta + j_\theta)h_r h_\phi)|_{\theta+\Delta\theta/2}}{h_r h_\theta h_\phi \Delta \theta} \\
&\quad + \frac{((cv_\phi + j_\phi)h_r h_\theta)|_{\phi-\Delta\phi/2} - ((cv_\phi + j_\phi)h_r h_\theta)|_{\phi+\Delta\phi/2}}{h_r h_\theta h_\phi \Delta \phi} + \mathcal{S}.
\end{aligned}
$$

$$(7.2.10)$$

Taking the limit $\Delta r \to 0$, $\Delta \theta \to 0$, $\Delta \phi \to 0$ and $\Delta t \to 0$, the following partial differential equation is obtained for the concentration field,

$$
\begin{aligned}
&\frac{\partial c}{\partial t} + \frac{1}{h_r h_\theta h_\phi}\left(\frac{\partial(h_\theta h_\phi c v_r)}{\partial r} + \frac{\partial(h_\phi h_r c v_\theta)}{\partial \theta} + \frac{\partial(h_r h_\theta c v_\phi)}{\partial \phi}\right) \\
&= -\frac{1}{h_r h_\theta h_\phi}\left(\frac{\partial(h_\theta h_\phi j_r)}{\partial r} + \frac{\partial(h_\phi h_r j_\theta)}{\partial \theta} + \frac{\partial(h_r h_\theta j_\phi)}{\partial \phi}\right).
\end{aligned}
$$

$$(7.2.11)$$

As is conventional, the convective terms have been written on the left side of Eq. 7.2.11, and the diffusive terms on the right side. If the scale factors in Eq. 7.2.3, are substituted in Eq. 7.2.11, we obtain, after some simplification,

$$
\begin{aligned}
&\frac{\partial c}{\partial t} + \frac{1}{r^2}\frac{\partial(r^2 c v_r)}{\partial r} + \frac{1}{r\sin(\theta)}\frac{\partial(\sin(\theta)c v_\theta)}{\partial \theta} + \frac{1}{r\sin(\theta)}\frac{\partial(c v_\phi)}{\partial \phi} \\
&= -\frac{1}{r^2}\frac{\partial(r^2 j_r)}{\partial r} - \frac{1}{r\sin(\theta)}\frac{\partial(\sin(\theta)j_\theta)}{\partial \theta} - \frac{1}{r\sin(\theta)}\frac{\partial j_\phi}{\partial \phi}.
\end{aligned}
$$

$$(7.2.12)$$

The fluxes are now expressed in terms of the concentration variation using Fick's law for diffusion. In the Cartesian co-ordinate system, the fluxes were proportional to the partial derivatives with respect to the co-ordinates in Eq. 7.1.10. In a spherical co-ordinate system, two of the co-ordinates are angles, which are dimensionless. Therefore, the scale factors have to be included in the definition of the flux. For a concentration difference Δc between two surfaces perpendicular to the θ co-ordinate with azimuthal angles $\theta - \Delta\theta/2$ and $\theta + \Delta\theta/2$, the Fick's law of diffusion states that the flux is

$$j_\theta = -\mathcal{D}\frac{\Delta c}{h_\theta \Delta \theta}, \qquad (7.2.13)$$

because the separation between these two surfaces is $h_\theta \Delta \theta$. Taking the limit $\Delta \theta \to 0$, we obtain the flux in terms of the partial derivative,

$$j_\theta = -\frac{\mathcal{D}}{h_\theta}\frac{\partial c}{\partial \theta}. \qquad (7.2.14)$$

In a similar manner, the fluxes in the r and ϕ directions are

$$j_r = -\frac{\mathcal{D}}{h_r}\frac{\partial c}{\partial r}, \quad j_\phi = -\frac{\mathcal{D}}{h_\phi}\frac{\partial c}{\partial \phi}. \qquad (7.2.15)$$

When Eqs. 7.2.14–7.2.15 for the flux are substituted in Eq. 7.2.11, the mass conservation equation is obtained,

$$\frac{\partial c}{\partial t} + \frac{1}{h_r h_\theta h_\phi}\left(\frac{\partial(h_\theta h_\phi c v_r)}{\partial r} + \frac{\partial(h_r h_\phi c v_\theta)}{\partial \theta} + \frac{\partial(h_r h_\theta c v_\phi)}{\partial \phi}\right)$$
$$= \frac{1}{h_r h_\theta h_\phi}\left(\frac{\partial}{\partial r}\left(\frac{\mathcal{D}h_\theta h_\phi}{h_r}\frac{\partial c}{\partial r}\right) + \frac{\partial}{\partial \theta}\left(\frac{\mathcal{D}h_\phi h_r}{h_\theta}\frac{\partial c}{\partial \theta}\right) + \frac{\partial}{\partial \phi}\left(\frac{\mathcal{D}h_r h_\theta}{h_\phi}\frac{\partial c}{\partial \phi}\right)\right). \qquad (7.2.16)$$

If the the scale factors in Eq. 7.2.3 are substituted in Eq. 7.2.16, and the diffusion coefficient is considered to be a constant, we obtain after some simplification,

$$\frac{\partial c}{\partial t} + \frac{1}{r^2}\frac{\partial(r^2 c v_r)}{\partial r} + \frac{1}{r\sin(\theta)}\frac{\partial(\sin(\theta)c v_\theta)}{\partial \theta} + \frac{1}{r\sin(\theta)}\frac{\partial(c v_\phi)}{\partial \phi}$$
$$= \mathcal{D}\left(\frac{1}{r^2}\frac{\partial}{\partial r}\left(r^2\frac{\partial c}{\partial r}\right) + \frac{1}{r^2\sin(\theta)}\frac{\partial}{\partial \theta}\left(\sin(\theta)\frac{\partial c}{\partial \theta}\right) + \frac{1}{r^2\sin(\theta)^2}\frac{\partial^2 c}{\partial \phi^2}\right). \qquad (7.2.17)$$

Eqs. 7.2.14–7.2.15 for Fick's law in spherical co-ordinates are the components of the vector form, Eq. 7.1.14, where the definition of the gradient is,

$$\boxed{\begin{aligned}\nabla c &= \frac{e_r}{h_r}\frac{\partial c}{\partial r} + \frac{e_\theta}{h_\theta}\frac{\partial c}{\partial \theta} + \frac{e_\phi}{h_\phi}\frac{\partial c}{\partial \phi}\\ &= e_r\frac{\partial c}{\partial r} + \frac{e_\theta}{r}\frac{\partial c}{\partial \theta} + \frac{e_\phi}{r\sin(\theta)}\frac{\partial c}{\partial \phi}.\end{aligned}} \qquad (7.2.18)$$

The mass conservation equation in spherical co-ordinates, Eq. 7.2.11 or 7.2.12, is identical to the vector form, Eq. 7.1.16, where the divergence operator defined as,

$$\boldsymbol{\nabla} \cdot \boldsymbol{j} = \frac{1}{h_r h_\theta h_\phi} \left(\frac{\partial (h_\theta h_\phi j_r)}{\partial r} + \frac{\partial (h_\phi h_r j_\theta)}{\partial \theta} + \frac{\partial (h_r h_\theta j_\phi)}{\partial \phi} \right)$$
$$= \frac{1}{r^2} \frac{\partial (r^2 j_r)}{\partial r} + \frac{1}{r \sin(\theta)} \frac{\partial (\sin(\theta) j_\theta)}{\partial \theta} + \frac{1}{r \sin(\theta)} \frac{\partial j_\phi}{\partial \phi}.$$

(7.2.19)

The mass conservation equation for a constant diffusivity, Eq. 7.2.17, is the same as the vector representation, Eq. 7.1.19, with the Laplacian defined in spherical co-ordinates as,

$$\boldsymbol{\nabla}^2 c = \frac{1}{h_r h_\theta h_\phi} \left(\frac{\partial}{\partial r} \left(\frac{h_\theta h_\phi}{h_r} \frac{\partial c}{\partial r} \right) + \frac{\partial}{\partial \theta} \left(\frac{h_\phi h_r}{h_\theta} \frac{\partial c}{\partial \theta} \right) + \frac{\partial}{\partial \phi} \left(\frac{h_r h_\theta}{h_\phi} \frac{\partial c}{\partial \phi} \right) \right)$$
$$= \frac{1}{r^2} \frac{\partial}{\partial r} \left(r^2 \frac{\partial c}{\partial r} \right) + \frac{1}{r^2 \sin(\theta)} \frac{\partial}{\partial \theta} \left(\sin(\theta) \frac{\partial c}{\partial \theta} \right) + \frac{1}{r^2 \sin(\theta)^2} \frac{\partial^2 c}{\partial \phi^2}.$$

(7.2.20)

Summary (7.2)

1. In an orthogonal curvilinear co-ordinate system with co-ordinates (x_a, x_b, x_c), unit vectors (e_a, e_b, e_c) and scale factors (h_a, h_b, h_c), the velocity and flux are expressed in vector form as,

$$\boldsymbol{v} = v_a \boldsymbol{e}_a + v_b \boldsymbol{e}_b + v_c \boldsymbol{e}_c,$$
$$\boldsymbol{j} = j_a \boldsymbol{e}_a + j_b \boldsymbol{e}_b + j_c \boldsymbol{e}_c,$$

(7.2.21)

where (v_a, v_b, v_c) and (j_a, j_b, j_c) are the components in the three orthogonal directions.

2. The mass conservation equation is Eq. 7.1.16, where the divergence operator defined as,

$$\boldsymbol{\nabla} \cdot \boldsymbol{j} = \frac{1}{h_a h_b h_c} \left(\frac{\partial (h_b h_c j_a)}{\partial x_a} + \frac{\partial (h_c h_a j_b)}{\partial x_b} + \frac{\partial (h_a h_b j_c)}{\partial x_c} \right).$$

(7.2.22)

3. The Fick's law for diffusion is of the form in Eq. 7.1.14, where the gradient operator is defined as,

$$\nabla c = \frac{e_a}{h_a}\frac{\partial c}{\partial x_a} + \frac{e_b}{h_b}\frac{\partial c}{\partial x_b} + \frac{e_c}{h_c}\frac{\partial c}{\partial x_c}. \qquad (7.2.23)$$

The conservation equation, Eq. 7.1.18, is obtained when Eq. 7.1.14 is substituted for the flux.

4. If the diffusion coefficient is a constant, the conservation equation is Eq. 7.1.19, where the Laplacian operator is,

$$\nabla^2 c = \frac{1}{h_a h_b h_c}\left(\frac{\partial}{\partial x_a}\left(\frac{h_b h_c}{h_a}\frac{\partial c}{\partial x_a}\right) + \frac{\partial}{\partial x_b}\left(\frac{h_c h_a}{h_b}\frac{\partial c}{\partial x_b}\right) + \frac{\partial}{\partial x_c}\left(\frac{h_a h_b}{h_c}\frac{\partial c}{\partial x_c}\right)\right). \qquad (7.2.24)$$

5. For the spherical co-ordinate system, the gradient, divergence and Laplacian can be determined from the scale factors, Eq. 7.2.3. The gradient, divergence and Laplacian are, respectively, Eqs. 7.2.18, 7.2.19 and 7.2.20.

7.3 Navier–Stokes Equations

7.3.1 Mass Conservation

In Sections 7.1 and 7.2, the mass conservation equation was derived for one component of the fluid mixture. Here, we consider the mass conservation equation for the total mass density ρ of the fluid. Since the fluid velocity v is the velocity of the centre of mass of the fluid, there is no diffusion relative to the centre of mass, and the flux of mass is entirely due to convection. Therefore, the conservation equation for the density ρ is the same as that for the concentration, Eq. 7.1.19, without the diffusive flux. The conservation equation for total mass does not contain a source because the total mass does not change when there is interconversion between different molecular species in a reaction. Therefore, the mass conservation equation for the fluid is,

$$\frac{\partial \rho}{\partial t} + \nabla \cdot (\rho v) = 0. \qquad (7.3.1)$$

The definition of the divergence in Cartesian and spherical co-ordinates, Eqs. 7.1.17 and 7.2.19, can be used for the mass conservation equation in the respective co-ordinate systems.

The mass conservation equation, Eq. 7.3.1, can be expressed using the material or substantial derivative, Eq. 7.1.22, as

$$\frac{D\rho}{Dt} + \rho \boldsymbol{\nabla} \cdot \boldsymbol{v} = 0. \tag{7.3.2}$$

For an 'incompressible' flow, the density is a constant, $(D\rho/Dt) = 0$, and the mass conservation equation reduces to,

$$\boxed{\boldsymbol{\nabla} \cdot \boldsymbol{v} = 0.} \tag{7.3.3}$$

Thus, the divergence of the velocity is zero throughout the domain in an incompressible flow.

Fluid flows are incompressible if the flow velocity is much smaller than the speed of sound in the fluid. The speed of sound in gases is typically hundreds of meters per second—the speed of sound in air at standard temperature and pressure is about 340 m/s, while that in hydrogen is about 1270 m/s. The speed of sound in liquids is larger—the speed of sound in water is approximately 1500 m/s. In most applications of interest, the flow velocities are much smaller than the speed of sound, and so the flows can be considered as incompressible. In this text, attention is restricted to incompressible flows.

Consider a two-dimensional incompressible flow, in which the velocity components (v_x, v_y) are functions of the co-ordinates (x, y). The components of the velocity can be expressed in terms of the stream function ψ,

$$v_x = \frac{\partial \psi}{\partial y}, \quad v_y = -\frac{\partial \psi}{\partial x}. \tag{7.3.4}$$

By construction, the above velocity components satisfy the mass conservation equation, Eq. 7.3.3, in two dimensions, $(\partial v_x/\partial x) + (\partial v_y/\partial y) = 0$. Therefore, both components of the velocity are expressed as derivatives of one scalar function ψ.

The physical significance of the stream function is as follows. The difference in the stream function $\psi_B - \psi_A$ between two locations A and B is the flow rate (per unit length perpendicular to the plane of flow) between A and B. Consider the differential displacement $\Delta \boldsymbol{x}$ along the path from A to B shown in Fig. 7.5. The displacement $\Delta \boldsymbol{x}$ can be written as $\boldsymbol{t}\Delta s$, where Δs is the scalar displacement and $\boldsymbol{t}$ is the unit tangent to the path. The change in the stream function over the distance Δs is,

$$\Delta \psi = \Delta s \left(t_x \frac{\partial \psi}{\partial x} + t_y \frac{\partial \psi}{\partial y} \right), \tag{7.3.5}$$

where t_x and t_y are the components of the unit tangent to the path, and the differential displacements are $\Delta x = t_x \Delta s$ and $\Delta y = t_y \Delta s$. The relation between

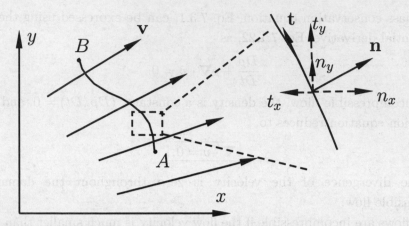

FIGURE 7.5. The flow across a surface between points A to B, and the unit tangent and unit normal at a point along the surface.

the components of the unit tangent t and the unit normal n to the surface are shown in Fig. 7.5,

$$n_x = t_y, \quad n_y = -t_x. \qquad (7.3.6)$$

Here, it is assumed that the rotation from the unit normal to the unit tangent is in the anticlockwise direction, or $n \times t = e_z$ in accordance with the right-hand rule, where e_z is in the $+z$ direction perpendicular to the plane in a right-handed co-ordinate system. The components of the unit tangent are expressed in terms of those of the unit normal using Eq. 7.3.6, and the expressions in Eq. 7.3.4 are used for the derivatives of the stream function in terms of the velocity,

$$\Delta\psi = \Delta s \left[(-v_y)(-n_y) + (v_x)(n_x) \right] = \Delta s \, n \cdot v. \qquad (7.3.7)$$

The change in the stream function between the locations A and B is,

$$\psi_B - \psi_A = \int_{x_A}^{x_B} ds \, n \cdot v. \qquad (7.3.8)$$

The right side of Eq. 7.3.8 is the total flow rate across the the contour from A to B per unit length perpendicular to the plane of flow. Since the left side of Eq. 7.3.8 is independent of the path from A to B, the flow rate across any path from A to B is the same.

The velocity can be expressed in terms of the stream function for a two-dimensional polar (r, ϕ) co-ordinate system, where r is the distance from the origin and ϕ is the polar angle made by the displacement vector with the x axis. The expressions for the components of the velocity are,

$$v_r = \frac{1}{r}\frac{\partial \psi}{\partial \phi}, \quad v_\phi = -\frac{\partial \psi}{\partial r}. \tag{7.3.9}$$

The expression for an axisymmetric flow in a spherical co-ordinate system (r, θ, ϕ), where the flow is independent of the meridional angle ϕ is

$$v_r = \frac{1}{r^2 \sin(\theta)}\frac{\partial \psi}{\partial \theta}, \quad v_\theta = -\frac{1}{r \sin(\theta)}\frac{\partial \psi}{\partial r}. \tag{7.3.10}$$

7.3.2 Stress Tensor

In the case of mass and energy transfer, the quantity being transported is a scalar. The flux of the quantity is a vector, which is oriented in the direction of transport. In the case of momentum transfer, the quantity being transported is a vector, which has both magnitude and direction. Therefore, the flux of this quantity has two directions associated with it: the first is the direction of momentum, and the second is the direction of the transport of momentum. In Chapter 2, the expression for the shear stress τ_{xz} contained two subscripts – the first (x) is the direction of the momentum or the flow velocity, and the second (z) is the direction of transport. Other components of the stress can be defined in a similar manner.

The symbol $\boldsymbol{\sigma}$ is used here for the total stress which includes the pressure, to distinguish it from the symbol $\boldsymbol{\tau}$ for the shear stress. The component σ_{ij} of the stress tensor is defined as the force per unit area in the i direction acting at a surface whose outward unit normal is in the j direction, where i and j could be x, y or z. The flux vector was written in terms of its components in a Cartesian co-ordinate system as $\boldsymbol{j} = j_x \boldsymbol{e}_x + j_y \boldsymbol{e}_y + j_z \boldsymbol{e}_z$. In a similar manner, the stress 'tensor' can be written in terms of its components in a Cartesian co-ordinate system as,

$$\boldsymbol{\sigma} = \boldsymbol{e}_x \sigma_{xx} \boldsymbol{e}_x + \boldsymbol{e}_x \sigma_{xy} \boldsymbol{e}_y + \boldsymbol{e}_x \sigma_{xz} \boldsymbol{e}_z + \ldots + \boldsymbol{e}_z \sigma_{zz} \boldsymbol{e}_z. \tag{7.3.11}$$

There are analogous expressions for the stress tensor in terms of its components in other co-ordinate systems. In these expressions, the unit vector on the left and the first subscript of $\boldsymbol{\sigma}$ is the direction of momentum, and the unit vector on the right and the second subscript of $\boldsymbol{\sigma}$ is the direction in which momentum is transported.

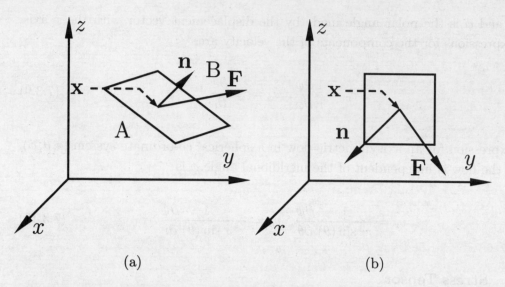

FIGURE 7.6. The force acting at a surface with unit normal n (a) and with unit normal e_x (b).

The stress tensor can also be written in matrix form,

$$\sigma = \begin{pmatrix} e_x & e_y & e_z \end{pmatrix} \begin{pmatrix} \sigma_{xx} & \sigma_{xy} & \sigma_{xz} \\ \sigma_{yx} & \sigma_{yy} & \sigma_{yz} \\ \sigma_{zx} & \sigma_{zy} & \sigma_{zz} \end{pmatrix} \begin{pmatrix} e_x \\ e_y \\ e_z \end{pmatrix}. \qquad (7.3.12)$$

The row matrix of unit vectors on the left and the column matrix on the right are usually omitted, and stress is written as a 3×3 square matrix.

Consider a surface of area ΔS centred at x which separates the fluid into two volumes, A and B, as shown in Fig. 7.6(a). The fluid on one side of the surface exerts a force on the fluid on the other side, which results in momentum transport across the surface. If we consider the momentum change in the fluid volume A on one side of the surface, the 'outward' unit normal to the surface n is directed into the fluid volume B, as shown in Fig. 7.6(a). The force acting on the surface is,

$$\begin{aligned} F(n) &= \Delta S(\sigma \cdot n) \\ &= \Delta S[e_x \left(\sigma_{xx} n_x + \sigma_{xy} n_y + \sigma_{xz} n_z \right) + e_y \left(\sigma_{yx} n_x + \sigma_{yy} n_y + \sigma_{yz} n_z \right) \\ &\quad + e_z \left(\sigma_{zx} n_x + \sigma_{zy} n_y + \sigma_{zz} n_z \right)], \end{aligned} \qquad (7.3.13)$$

where n_x, n_y and n_z are the components of the unit normal along the three co-ordinate directions. Here, the second index of the stress σ has been dotted with the unit normal to the surface, to obtain the force on the surface.

It is important to note that the force $F(n)$ on a surface within fluid is not uniquely specified at a location, since it depends on the surface orientation of the direction of the outward unit normal. If the outward unit normal is in the x direction, as shown in Fig. 7.6(b), the force exerted on the surface centred at the same point $\boldsymbol{x}$ is,

$$
\begin{aligned}
\boldsymbol{F}(\boldsymbol{e}_x) &= \Delta S(\boldsymbol{\sigma} \cdot \boldsymbol{e}_x) \\
&= \Delta S[\boldsymbol{e}_x \sigma_{xx} + \boldsymbol{e}_y \sigma_{yx} + \boldsymbol{e}_z \sigma_{zx}].
\end{aligned}
\tag{7.3.14}
$$

This force could, in general be different from that acting on a surface with unit normal $\boldsymbol{n}$. However, all the components of the stress tensor are uniquely specified at each point within the fluid.

From Eq. 7.3.13, it is evident that $\boldsymbol{F}(-\boldsymbol{n}) = -\boldsymbol{F}(\boldsymbol{n})$, that is, the direction of the force is reversed if the direction of the unit normal is reversed. The force $\boldsymbol{F}(\boldsymbol{n})$ acting on the surface with unit normal $\boldsymbol{n}$ is the force exerted on the volume A by the volume B. The force $\boldsymbol{F}(-\boldsymbol{n})$ is the force exerted on the volume B by the volume A, because $-\boldsymbol{n}$ is the outward unit normal to the volume B. These two forces are necessarily equal in magnitude and opposite in direction as a consequence of Newton's third law.

There is an important difference in the convention adopted for the flux vector and the stress tensor. The flux j_z, for example, is defined to be positive if mass is transported in the $+z$ direction, resulting in an accumulation of mass in the volume above the surface. In contrast, the stress tensor σ_{xz} is defined to be positive if a force in the $+x$ direction acts on a surface whose outward unit normal is in the $+z$ direction, resulting in an increase in the momentum of the volume below this surface. For this reason, there is a negative sign in the definition of the Fick's law, Eq. 7.1.14, whereas there is no negative sign in the definition of Newton's law for viscosity discussed later in Section 7.3.4.

The force acting at a surface can be separated into two components, one is the 'normal' force along the unit normal to the surface, and the second is the shear force acting parallel to the surface. The normal force is,

$$
F_n = \boldsymbol{n} \cdot \boldsymbol{F} \cdot \boldsymbol{n} = \Delta S \boldsymbol{n} \cdot \boldsymbol{\sigma} \cdot \boldsymbol{n}.
\tag{7.3.15}
$$

The force per unit area perpendicular to the surface $\boldsymbol{n} \cdot \boldsymbol{\sigma} \cdot \boldsymbol{n}$ is sometimes called the normal stress at the surface; this does not change when the direction of the unit normal is reversed. A positive normal stress exerts a tensile force, while a negative normal stress is compressive. The force parallel to the surface is the difference between the total force and the force normal to the surface,

$$
\boldsymbol{F}_t = \boldsymbol{F} - \boldsymbol{n} F_n = \Delta S[\boldsymbol{\sigma} \cdot \boldsymbol{n} - \boldsymbol{n}(\boldsymbol{n} \cdot \boldsymbol{\sigma} \cdot \boldsymbol{n})].
\tag{7.3.16}
$$

The force per unit area parallel to the surface is called the shear stress.

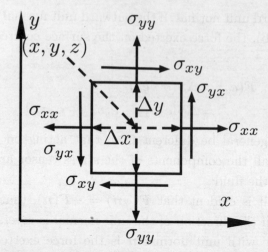

FIGURE 7.7. The shear and normal forces acting in the $x - y$ plane on the surfaces of a volume element of fluid of dimension $(\Delta x, \Delta y, \Delta z)$ about the point (x, y, z).

Consider the volume element of fluid of dimension $\Delta x, \Delta y, \Delta z$ about the point x, y, z shown in the $x - y$ plane in Fig. 7.7. The z axis is perpendicular to the plane. The normal stresses are tensile and directed along the outward normals if σ_{xx} and σ_{yy} are positive, and are compressive and directed along the inward normals if σ_{xx} and σ_{yy} are negative. The shear stress on the right face σ_{yx} is the force per unit area in the y direction, since the outward unit normal is in the x direction. On the left face, the shear stress σ_{yx} is in the negative y direction because the outward unit normal is in the $-x$ direction. The same convention applies for the top and bottom faces.

There is an important restriction on the form of the stress tensor, Eq. 7.3.12, due to angular momentum conservation. Consider the torque exerted in the z direction by the shear stresses on the differential volume in Fig. 7.7. In the right-hand convention, the torque is considered positive if the rotation is anticlockwise. The contribution to the torque in the z direction due to the force on the right face is the product of the tangential force $(\sigma_{yx}|_{x+\Delta x/2} \Delta y \Delta z)$ and the displacement from the axis of rotation $(\Delta x/2)$. The force tangential to the left face is $(-\sigma_{yx}|_{x-\Delta x/2} \Delta y \Delta z)$ since the unit normal is in the $-x$ direction, and the displacement from the centre of rotation is also in the $-x$ direction. On the top face, the tangential force acts in the clockwise direction relative to the centre of rotation if σ_{xy} is positive, and so the torque is the force $(-\sigma_{xy}|_{y+\Delta y/2} \Delta x \Delta z)$ times the displacement from the centre of rotation, $(\Delta y/2)$. On the bottom face, the directions of the force and the displacement from the centre of rotation are reversed relative to the top face. The four contributions are,

Contribution to torque from right face $= \sigma_{yx}|_{x+\Delta x/2} \Delta y \Delta z (\Delta x/2)$,

Contribution to torque from left face $= -\sigma_{yx}|_{x-\Delta x/2} \Delta y \Delta z (-\Delta x/2)$,

Contribution to torque from top face $= -\sigma_{xy}|_{y+\Delta y/2} \Delta x \Delta z (\Delta y/2)$,

Contribution to torque from left face $= \sigma_{xy}|_{y-\Delta y/2} \Delta x \Delta z (-\Delta y/2)$. (7.3.17)

The angular momentum conservation equation in the z direction is,

$$I_z \frac{d\Omega_z}{dt} = (\sigma_{yx}|_{x+\Delta x/2} + \sigma_{yx}|_{x-\Delta x/2})(\Delta y \Delta z)(\Delta x/2)$$
$$- (\sigma_{xy}|_{y+\Delta y/2} + \sigma_{xy}|_{y-\Delta y/2})(\Delta x \Delta z)(\Delta y/2), \qquad (7.3.18)$$

where Ω_z is the angular velocity and I_z is the moment of inertia of the fluid element around the z axis. The moment of inertia around the z axis is the integral of the distance of a differential volume from the axis, $x'^2 + y'^2$, times the mass of the differential volume, $\rho \, dx' \, dy' \, dz'$, integrated over the volume,

$$I_z = \rho \int_{-\Delta z/2}^{\Delta z/2} dz' \int_{-\Delta x/2}^{\Delta x/2} dx' \int_{-\Delta y/2}^{\Delta y/2} dy' (x'^2 + y'^2) = \frac{\rho(\Delta x^2 + \Delta y^2)\Delta x \Delta y \Delta z}{12}.$$
$$(7.3.19)$$

Eq. 7.3.19 is substituted into Eq. 7.3.18, and the equation is divided by $\Delta x \Delta y \Delta z$ to obtain,

$$\frac{\rho(\Delta x^2 + \Delta y^2)}{12} \frac{d\Omega_z}{dt} = \tfrac{1}{2}(\sigma_{yx}|_{x+\Delta x/2} + \sigma_{yx}|_{x-\Delta x/2})$$
$$- \tfrac{1}{2}(\sigma_{xy}|_{y+\Delta y/2} + \sigma_{xy}|_{y-\Delta y/2}). \qquad (7.3.20)$$

In the limit $\Delta x, \Delta y \to 0$, the left side of Eq. 7.3.20 tends to zero. The right side is $\sigma_{yx} - \sigma_{xy}$ at the location (x, y) in this limit. Therefore, the angular momentum balance is satisfied only if $\sigma_{xy} = \sigma_{yx}$. A similar calculation for the angular momentum balance around the x and y axes shows that $\sigma_{yz} = \sigma_{zy}$ and $\sigma_{zx} = \sigma_{xz}$. Therefore, the angular momentum balance condition is satisfied only if the stress tensor, Eq. 7.3.12, is symmetric.

There is a contribution to the stress due to the fluid pressure p which is non-zero even in the absence of flow. The pressure force exerted at a surface is perpendicular to the surface, and it is compressive, acting along the inward unit normal to the volume. The pressure is also independent of the orientation of the surface. The stress tensor due to the fluid pressure is,

$$\sigma^p = \begin{pmatrix} e_x \ e_y \ e_z \end{pmatrix} \begin{pmatrix} -p & 0 & 0 \\ 0 & -p & 0 \\ 0 & 0 & -p \end{pmatrix} \begin{pmatrix} e_x \\ e_y \\ e_z \end{pmatrix} = -e_x p e_x - e_y p e_y - e_z p e_z.$$

(7.3.21)

The diagonal terms are the negative of the pressure because the pressure acts along the inward unit normal to the surface, whereas the right unit vector in the stress tensor is the direction of the outward unit normal. It is easily verified that for a surface with unit vector n, the force per unit area along the outward unit normal to the surface $n \cdot \sigma^p \cdot n = -p$, and the force per unit area in any direction parallel to the surface is zero.

When there is flow, there is a viscous contribution to the stress due to velocity gradient. The form of the Newton's law for viscosity for a general flow field is more complicated than Fick's law, Eq. 7.1.14, because both the stress and the velocity gradient are second order tensors. The physical meaning of the velocity gradient, also called the rate of deformation tensor, is first discussed, followed by the statement of the Newton's law of viscosity in vector form.

7.3.3 Rate of Deformation Tensor

The rate of deformation tensor is the gradient of the velocity field. Formally, the rate of deformation tensor in a Cartesian co-ordinate system is defined as,

$$\nabla v = \left(e_x \frac{\partial}{\partial x} + e_y \frac{\partial}{\partial y} + e_z \frac{\partial}{\partial z} \right) (v_x e_x + v_y e_y + v_z e_z).$$

(7.3.22)

It is important to note that there is no dot product between the gradient and the velocity in the above expression. The rate of deformation tensor is a second order tensor, because it has two fundamental directions associated with it, the direction of the velocity and the direction in which there is a variation in the velocity. The rate of deformation tensor can be written in matrix form in a manner similar to the

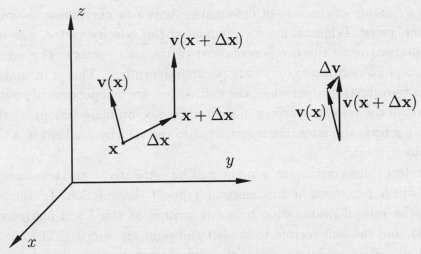

FIGURE 7.8. The difference in the fluid velocity Δv between two nearby locations $x + \Delta x$ and x.

stress tensor in Eq. 7.3.12,

$$
\nabla v = \begin{pmatrix} e_x & e_y & e_z \end{pmatrix} \begin{pmatrix} \dfrac{\partial v_x}{\partial x} & \dfrac{\partial v_y}{\partial x} & \dfrac{\partial v_z}{\partial x} \\[3mm] \dfrac{\partial v_x}{\partial y} & \dfrac{\partial v_y}{\partial y} & \dfrac{\partial v_z}{\partial y} \\[3mm] \dfrac{\partial v_x}{\partial z} & \dfrac{\partial v_y}{\partial z} & \dfrac{\partial v_z}{\partial z} \end{pmatrix} \begin{pmatrix} e_x \\ e_y \\ e_z \end{pmatrix}. \tag{7.3.23}
$$

The physical interpretation of the velocity gradient is similar to the concentration gradient. Consider two nearby locations x and $x + \Delta x$ shown in Fig. 7.8. The difference in the velocity between these two locations, $\Delta v = v(x + \Delta x) - v(x)$ is,

$$
\begin{aligned}
v(x + \Delta x) - v(x) &= \Delta x \cdot \nabla v \\
&= \left(\Delta x \frac{\partial v_x}{\partial x} + \Delta y \frac{\partial v_x}{\partial y} + \Delta z \frac{\partial v_x}{\partial z} \right) e_x \\
&\quad + \left(\Delta x \frac{\partial v_y}{\partial x} + \Delta y \frac{\partial v_y}{\partial y} + \Delta z \frac{\partial v_y}{\partial z} \right) e_y \\
&\quad + \left(\Delta x \frac{\partial v_z}{\partial x} + \Delta y \frac{\partial v_z}{\partial y} + \Delta z \frac{\partial v_z}{\partial z} \right) e_z. \tag{7.3.24}
\end{aligned}
$$

Note that there is a dot product between Δx and the gradient operator in Eq. 7.3.24, and the resultant vector is the difference in the velocity between the locations x and $x + \Delta x$, as shown on the right in Fig. 7.8.

The calculation of the rate of deformation tensor in curvilinear co-ordinates is more complicated. While taking the gradient of the velocity vector, it is necessary to take derivatives of the components and of the unit vectors. The unit vectors in curvilinear co-ordinate systems are position-dependent. This is in contrast to a Cartesian co-ordinate system where the unit vectors are independent of position. The calculation of the velocity gradient in a curvilinear co-ordinate system is beyond the scope of this text, and attention is restricted to the velocity gradient in a Cartesian co-ordinate system.

The rate of deformation at a point can be separated into three components, each of which represents a fundamental type of deformation. In the following analysis, the rate of deformation tensor is written as the 3×3 matrix shown in Eq. 7.3.23, and the unit vectors to the left and right are omitted. The 3×3 rate of deformation matrix can be written as the sum of a symmetric and an antisymmetric matrix,

$$\boldsymbol{\nabla v} = \boldsymbol{S} + \boldsymbol{A}, \tag{7.3.25}$$

where

$$\boldsymbol{S} = \frac{1}{2}\left(\boldsymbol{\nabla v} + (\boldsymbol{\nabla v})^T\right)$$

$$= \begin{pmatrix} \dfrac{\partial v_x}{\partial x} & \dfrac{1}{2}\left(\dfrac{\partial v_x}{\partial y} + \dfrac{\partial v_y}{\partial x}\right) & \dfrac{1}{2}\left(\dfrac{\partial v_x}{\partial z} + \dfrac{\partial v_z}{\partial x}\right) \\[2ex] \dfrac{1}{2}\left(\dfrac{\partial v_y}{\partial x} + \dfrac{\partial v_x}{\partial y}\right) & \dfrac{\partial v_y}{\partial y} & \dfrac{1}{2}\left(\dfrac{\partial v_y}{\partial z} + \dfrac{\partial v_z}{\partial y}\right) \\[2ex] \dfrac{1}{2}\left(\dfrac{\partial v_z}{\partial x} + \dfrac{\partial v_x}{\partial z}\right) & \dfrac{1}{2}\left(\dfrac{\partial v_z}{\partial y} + \dfrac{\partial v_y}{\partial z}\right) & \dfrac{\partial v_z}{\partial z} \end{pmatrix}, \tag{7.3.26}$$

$$\boldsymbol{A} = \frac{1}{2}\left(\boldsymbol{\nabla v} - (\boldsymbol{\nabla v})^T\right)$$

$$= \begin{pmatrix} 0 & \dfrac{1}{2}\left(\dfrac{\partial v_y}{\partial x} - \dfrac{\partial v_x}{\partial y}\right) & \dfrac{1}{2}\left(\dfrac{\partial v_z}{\partial x} - \dfrac{\partial v_x}{\partial z}\right) \\[2ex] \dfrac{1}{2}\left(\dfrac{\partial v_x}{\partial y} - \dfrac{\partial v_y}{\partial x}\right) & 0 & \dfrac{1}{2}\left(\dfrac{\partial v_z}{\partial y} - \dfrac{\partial v_y}{\partial z}\right) \\[2ex] \dfrac{1}{2}\left(\dfrac{\partial v_x}{\partial z} - \dfrac{\partial v_z}{\partial x}\right) & \dfrac{1}{2}\left(\dfrac{\partial v_y}{\partial z} - \dfrac{\partial v_z}{\partial y}\right) & 0 \end{pmatrix}, \tag{7.3.27}$$

Here, the superscript ^{T} indicates the transpose of the matrix, with the rows and columns interchanged. The trace of the symmetric matrix $\boldsymbol{S}$, which is the sum of the diagonal elements of S, is the divergence of the velocity $\boldsymbol{\nabla} \cdot \boldsymbol{v}$. Therefore, the symmetric part is expressed as the sum of two parts,

$$S = (1/3)(\boldsymbol{\nabla} \cdot \boldsymbol{v})\boldsymbol{I} + \boldsymbol{E}, \tag{7.3.28}$$

where $\boldsymbol{I}$ is the identity matrix, the first term on the right is the isotropic part of the rate of deformation tensor,

$$(1/3)(\boldsymbol{\nabla} \cdot \boldsymbol{v})\boldsymbol{I} = \begin{pmatrix} \frac{1}{3}\boldsymbol{\nabla} \cdot \boldsymbol{v} & 0 & 0 \\ 0 & \frac{1}{3}\boldsymbol{\nabla} \cdot \boldsymbol{v} & 0 \\ 0 & 0 & \frac{1}{3}\boldsymbol{\nabla} \cdot \boldsymbol{v} \end{pmatrix}, \tag{7.3.29}$$

and the second term on the right in Eq. 7.3.28 is the extensional strain,

$$\boldsymbol{E} = \tfrac{1}{2}(\boldsymbol{\nabla v} + (\boldsymbol{\nabla v})^T) - \tfrac{1}{3}(\boldsymbol{\nabla} \cdot \boldsymbol{v})\boldsymbol{I}$$
$$= \begin{pmatrix} \dfrac{\partial v_x}{\partial x} - \dfrac{1}{3}\boldsymbol{\nabla} \cdot \boldsymbol{v} & \dfrac{1}{2}\left(\dfrac{\partial v_x}{\partial y} + \dfrac{\partial v_y}{\partial x}\right) & \dfrac{1}{2}\left(\dfrac{\partial v_x}{\partial z} + \dfrac{\partial v_z}{\partial x}\right) \\ \dfrac{1}{2}\left(\dfrac{\partial v_y}{\partial x} + \dfrac{\partial v_x}{\partial y}\right) & \dfrac{\partial v_y}{\partial y} - \dfrac{1}{3}\boldsymbol{\nabla} \cdot \boldsymbol{v} & \dfrac{1}{2}\left(\dfrac{\partial v_y}{\partial z} + \dfrac{\partial v_z}{\partial y}\right) \\ \dfrac{1}{2}\left(\dfrac{\partial v_z}{\partial x} + \dfrac{\partial v_x}{\partial z}\right) & \dfrac{1}{2}\left(\dfrac{\partial v_z}{\partial y} + \dfrac{\partial v_y}{\partial z}\right) & \dfrac{\partial v_z}{\partial z} - \dfrac{1}{3}\boldsymbol{\nabla} \cdot \boldsymbol{v} \end{pmatrix}. \tag{7.3.30}$$

The extensional strain tensor $\boldsymbol{E}$ is symmetric and traceless, that is, the sum of the diagonal elements of $\boldsymbol{E}$ is zero. Thus, we have separated the rate of deformation tensor into three parts, the antisymmetric tensor $\boldsymbol{A}$, the symmetric traceless tensor $\boldsymbol{E}$ and the isotropic tensor $(1/3)(\boldsymbol{\nabla} \cdot \boldsymbol{v})\boldsymbol{I}$. The type of deformation due to each of these is explained in two dimensions for ease of visualisation.

The difference in velocity at the locations $(x + \Delta x, y + \Delta y)$ and (x, y) is,

$$\begin{pmatrix} \Delta v_x & \Delta v_y \end{pmatrix} = \begin{pmatrix} \Delta x & \Delta y \end{pmatrix}(\boldsymbol{\nabla v}) = \begin{pmatrix} \Delta x & \Delta y \end{pmatrix} \begin{pmatrix} \dfrac{\partial v_x}{\partial x} & \dfrac{\partial v_y}{\partial x} \\ \dfrac{\partial v_x}{\partial y} & \dfrac{\partial v_y}{\partial y} \end{pmatrix}. \tag{7.3.31}$$

In two dimensions, the isotropic, symmetric traceless and antisymmetric components of the rate of deformation tensor are,

$$
\nabla v = \begin{pmatrix} \dfrac{1}{2}\left(\dfrac{\partial v_x}{\partial x} + \dfrac{\partial v_y}{\partial y}\right) & 0 \\[3ex] 0 & \dfrac{1}{2}\left(\dfrac{\partial v_x}{\partial x} + \dfrac{\partial v_y}{\partial y}\right) \end{pmatrix}
$$

$$
+ \begin{pmatrix} \dfrac{\partial v_x}{\partial x} - \dfrac{1}{2}\left(\dfrac{\partial v_x}{\partial x} + \dfrac{\partial v_y}{\partial y}\right) & \dfrac{1}{2}\left(\dfrac{\partial v_x}{\partial y} + \dfrac{\partial v_y}{\partial x}\right) \\[3ex] \dfrac{1}{2}\left(\dfrac{\partial v_x}{\partial y} + \dfrac{\partial v_y}{\partial x}\right) & \dfrac{\partial v_y}{\partial y} - \dfrac{1}{2}\left(\dfrac{\partial v_x}{\partial x} + \dfrac{\partial v_y}{\partial y}\right) \end{pmatrix}
$$

$$
+ \begin{pmatrix} 0 & \dfrac{1}{2}\left(\dfrac{\partial v_y}{\partial x} - \dfrac{\partial v_x}{\partial y}\right) \\[3ex] \dfrac{1}{2}\left(\dfrac{\partial v_x}{\partial y} - \dfrac{\partial v_y}{\partial x}\right) & 0 \end{pmatrix}. \tag{7.3.32}
$$

General forms of $\frac{1}{2}(\nabla \cdot v)$ and A, and a specific form for E, are considered to illustrate the types of deformation in two dimensions,

$$
\nabla v = \tfrac{1}{2}\nabla \cdot v \begin{pmatrix} 1 & 0 \\ 0 & 1 \end{pmatrix} + \begin{pmatrix} E & 0 \\ 0 & -E \end{pmatrix} + \begin{pmatrix} 0 & A \\ -A & 0 \end{pmatrix}. \tag{7.3.33}
$$

The form of the symmetric traceless tensor, the second term on the right in Eq. 7.3.33, appears to be restricted to the special case $(\partial v_x/\partial y) + (\partial v_y/\partial x) = 0$. However, this representation is general: the symmetric traceless part of the rate of deformation tensor can always be reduced to the form in Eq. 7.3.33 by a suitable rotation of axes.

The deformation due to the isotropic part of the rate of deformation tensor, the first term on the right in Eq. 7.3.33, is shown in Fig. 7.9(a). The components of the velocity difference are,

$$
\Delta v_x = \frac{1}{2}(\nabla \cdot v)\Delta x,
$$

$$
\Delta v_y = \frac{1}{2}(\nabla \cdot v)\Delta y. \tag{7.3.34}
$$

The direction of the velocity difference at a fixed distance from (x, y) is shown by the arrows in Fig. 7.9(a). It is evident that the isotropic part of the rate of deformation tensor results in radial volumetric expansion/compression. The volume of a parcel

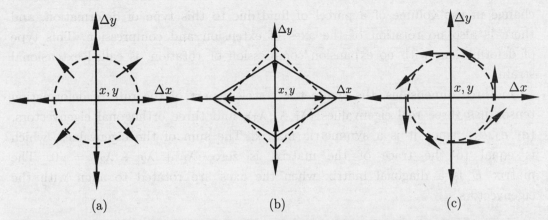

FIGURE 7.9. The velocity at the location $(x + \Delta x, y + \Delta y)$ relative to the location x, y due to the isotropic part (a), symmetric traceless part (b) and antisymmetric part (c) of the rate of deformation tensor.

of fluid at (x, y) increases with time if $\nabla \cdot v > 0$, and it decreases with time for $\nabla \cdot v < 0$. If the mass of a parcel of fluid is constant, volumetric expansion results in a decrease in the density, while volumetric compression results in an increase in the density. This is also inferred from the mass conservation equation expressed in terms of the substantial derivative, Eq. 7.3.2, which can be rewritten as,

$$\frac{1}{\rho}\frac{D\rho}{Dt} = - \nabla \cdot v. \tag{7.3.35}$$

The isotropic part of the rate of deformation tensor is non-zero only if the density varies in time. For an incompressible fluid, where the density is a constant, the divergence of the velocity is zero (Eq. 7.3.3), and the isotropic part of the rate of deformation tensor is zero.

The deformation due to the symmetric traceless part of the rate of deformation tensor, the second term on the right in Eq. 7.3.33, is shown in Fig. 7.9(b). The components of the relative velocity between the locations $(x + \Delta x, y + \Delta y)$ and (x, y) are

$$\Delta v_x = E\Delta x,$$
$$\Delta v_y = - E\Delta y. \tag{7.3.36}$$

The resultant velocity vectors are shown by the arrows in Fig. 7.9(b). Here, the flow is radially outward along the x axis and inward along the y axis if E is positive. The rectangular volume shown by the dashed line deforms to that shown by the solid line at a later instant, where the diagonal along the x axis increases in length and the diagonal along the y axis decreases in length. There is no net change in the volume of a parcel of fluid due to this type of deformation, and there is also no rotation of the axes of extension and compression. This type of deformation, with no expansion/compression or rotation, is called extensional strain.

In three dimensions, the symmetric traceless part of the rate of deformation tensor has three real eigenvalues $(\lambda_1, \lambda_2, \lambda_3)$, and three orthogonal eigenvectors, (e_1, e_2, e_3), since it is a symmetric matrix. The sum of the eigenvalues, which is equal to the trace of the matrix, is zero, $\lambda_1 + \lambda_2 + \lambda_3 = 0$. The matrix E is a diagonal matrix when the axes are rotated to align with the eigenvectors,

$$E = \begin{pmatrix} e_1 & e_2 & e_3 \end{pmatrix} \begin{pmatrix} \lambda_1 & 0 & 0 \\ 0 & \lambda_2 & 0 \\ 0 & 0 & \lambda_3 \end{pmatrix} \begin{pmatrix} e_1 \\ e_2 \\ e_3 \end{pmatrix}. \tag{7.3.37}$$

The orthogonal eigenvectors are the principal directions of extension/compression, and the eigenvalues are the rates of extension/compression along the direction of the eigenvectors. The deformation could be of three types, as shown in Fig. 7.10. If two of the eigenvalues are positive and one is negative, there is extension along two orthogonal axes and compression along the third, resulting in biaxial extension. This is shown in Fig. 7.10(a), where there is extension along the x and y axes, and compression along the z axis. If one eigenvalue is positive and two are negative, there is extension along one axis and compression along two axes, resulting in uniaxial extension. This is shown in Fig. 7.10(b), where there is compression along the x and y axes, and extension along the z axis. If one eigenvalue is zero, there is no extension/compression along the corresponding eigenvector, resulting is planar extension. This is shown in Fig. 7.10(c), where there is extension along the z axis, compression along the y axis and no deformation in the x direction.

The components of the relative velocity vector due to the antisymmetric part of the rate of deformation tensor in Eq. 7.3.33 are,

$$\Delta v_x = -A\Delta y,$$
$$\Delta v_y = A\Delta x. \tag{7.3.38}$$

The relative velocity vector, shown by the arrows in Fig. 7.9(c), results in a rotation around the axis perpendicular to the $x - y$ plane. The velocity is directed perpendicular to the position vector and is proportional to the distance from the

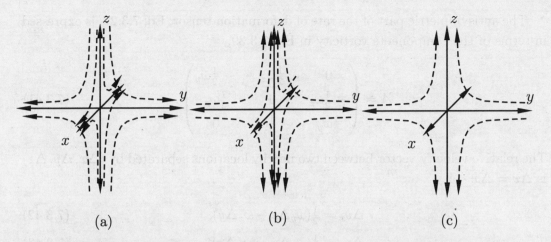

FIGURE 7.10. Biaxial extension (a), uniaxial extension (b), and planar extension (c).

centre of rotation. Therefore, the relative angular velocity (ratio of relative linear velocity and distance from centre) is a constant. This type of deformation is solid body rotation around the axis perpendicular to the plane of flow.

In three dimensions, the components of the antisymmetric part of the rate of deformation tensor are related to the 'vorticity' $\boldsymbol{\omega}$, which is the curl of the velocity,

$$\boldsymbol{\omega} = \boldsymbol{\nabla} \times \boldsymbol{v} = \begin{vmatrix} \boldsymbol{e}_x & \boldsymbol{e}_y & \boldsymbol{e}_z \\ \dfrac{\partial}{\partial x} & \dfrac{\partial}{\partial y} & \dfrac{\partial}{\partial z} \\ v_x & v_y & v_z \end{vmatrix}$$

$$= \boldsymbol{e}_x \left(\frac{\partial v_z}{\partial y} - \frac{\partial v_y}{\partial z} \right) + \boldsymbol{e}_y \left(\frac{\partial v_x}{\partial z} - \frac{\partial v_z}{\partial x} \right) + \boldsymbol{e}_z \left(\frac{\partial v_y}{\partial x} - \frac{\partial v_x}{\partial y} \right). \quad (7.3.39)$$

The generalisation of Eq. 7.3.39 for a curvilinear orthogonal co-ordinate system, such as the cylindrical or spherical co-ordinate system, is

$$\boldsymbol{\nabla} \times \boldsymbol{v} = \frac{1}{h_a h_b h_c} \begin{vmatrix} h_a \boldsymbol{e}_a & h_b \boldsymbol{e}_b & h_c \boldsymbol{e}_c \\ \dfrac{\partial}{\partial x_a} & \dfrac{\partial}{\partial x_b} & \dfrac{\partial}{\partial x_c} \\ h_a v_a & h_b v_b & h_c v_c \end{vmatrix}, \quad (7.3.40)$$

where x_a, x_b, x_c are the co-ordinates and h_a, h_b, h_c are the scale factors. The curl is defined in Appendix 7.C, and the Stokes' theorem relating the surface integral of the curl of a vector and the line integral of the vector around the perimeter of the surface is derived.

The antisymmetric part of the rate of deformation tensor, Eq. 7.3.27, is expressed in terms of the components vorticity in Eq. 7.3.39,

$$A = \begin{pmatrix} 0 & \frac{1}{2}\omega_z & -\frac{1}{2}\omega_y \\ -\frac{1}{2}\omega_z & 0 & \frac{1}{2}\omega_x \\ \frac{1}{2}\omega_y & -\frac{1}{2}\omega_x & 0 \end{pmatrix}. \tag{7.3.41}$$

The relative velocity vector between two nearby locations separated by $(\Delta x, \Delta y, \Delta z)$ is $\Delta v = \Delta x \cdot A$,

$$\Delta v_x = \frac{1}{2}(\omega_y \Delta z - \omega_z \Delta y), \tag{7.3.42}$$

$$\Delta v_y = \frac{1}{2}(\omega_z \Delta x - \omega_x \Delta z), \tag{7.3.43}$$

$$\Delta v_z = \frac{1}{2}(\omega_x \Delta y - \omega_y \Delta x). \tag{7.3.44}$$

Therefore, the relative velocity vector is

$$\Delta v = \frac{1}{2}\omega \times \Delta x. \tag{7.3.45}$$

Eq. 7.3.45 is the velocity relative to the axis for solid body rotation with angular velocity equal to $\frac{1}{2}\omega$. Therefore, the antisymmetric part of the rate of deformation tensor causes a solid body rotation with angular velocity equal to one half of the vorticity, and with axis of rotation along the vorticity vector.

It is important to understand that the decomposition shown in Fig. 7.9 is 'frame-invariant'. If the co-ordinate axes (x, y, z) are rotated, the components of the rate of deformation tensor (Eq. 7.3.23) will change. However, the flow patterns shown in Fig. 7.9 will not change. Similarly, if the velocity is expressed in a spherical instead of a Cartesian co-ordinate system, the components of the rate of deformation tensor change, but the decomposition shown in Fig. 7.9 is unchanged.

EXAMPLE 7.3.1: Determine the symmetric and antisymmetric components of the rate of deformation tensor, the vorticity, the principal axes and the rates of extension/compression for the extensional component for the flow in a channel of height h shown in Fig. 6.20 in Exercise 6.9 (Chapter 6). The velocity in the x direction is

$$v_x = \frac{6v_{av}(zh - z^2)}{h^2}, \tag{7.3.46}$$

where h is the channel height.

Solution: The rate of deformation tensor, Eq. 7.3.23, is

$$\boldsymbol{\nabla v} = \begin{pmatrix} 0 & 0 & 0 \\ 0 & 0 & 0 \\ \dfrac{6v_{av}(h-2z)}{h^2} & 0 & 0 \end{pmatrix}. \tag{7.3.47}$$

The symmetric and antisymmetric part of the rate of deformation tensor are

$$\tfrac{1}{2}(\boldsymbol{\nabla v} + (\boldsymbol{\nabla v})^T) = \begin{pmatrix} 0 & 0 & \dfrac{3v_{av}(h-2z)}{h^2} \\ 0 & 0 & 0 \\ \dfrac{3v_{av}(h-2z)}{h^2} & 0 & 0 \end{pmatrix}, \tag{7.3.48}$$

$$\tfrac{1}{2}(\boldsymbol{\nabla v} - (\boldsymbol{\nabla v})^T) = \begin{pmatrix} 0 & 0 & -\dfrac{3v_{av}(h-2z)}{h^2} \\ 0 & 0 & 0 \\ \dfrac{3v_{av}(h-2z)}{h^2} & 0 & 0 \end{pmatrix}. \tag{7.3.49}$$

The eigenvalues λ of the symmetric part of the rate of deformation tensor are determined from the equation,

$$|\tfrac{1}{2}(\boldsymbol{\nabla v} + (\boldsymbol{\nabla v})^T) - \lambda \boldsymbol{I}| = \begin{vmatrix} -\lambda & 0 & \frac{3v_{av}(h-2z)}{h^2} \\ 0 & -\lambda & 0 \\ \frac{3v_{av}(h-2z)}{h^2} & 0 & -\lambda \end{vmatrix} = 0, \tag{7.3.50}$$

where $\boldsymbol{I}$ is the identity matrix. The solutions for the eigenvalues are

$$\lambda_1 = 0, \lambda_2 = \frac{3v_{av}(h-2z)}{h^2}, \lambda_3 = -\frac{3v_{av}(h-2z)}{h^2}, \tag{7.3.51}$$

and the corresponding eigenvectors are,

$$e_1 = e_y, e_2 = \frac{e_x + e_z}{\sqrt{2}}, e_3 = \frac{e_x - e_z}{\sqrt{2}}. \tag{7.3.52}$$

Thus, this is a planar extensional flow with no extension/compression along the y axis. There is extension along the direction rotated $+(\pi/4)$ from the x axis, and compression along the direction rotated $-(\pi/4)$ from the x axis for $h - 2z > 0$. The extensional and compressional axes are interchanged for $h - 2z < 0$.

The vorticity is determined by comparing the antisymmetric matrices, Eqs. 7.3.49 and 7.3.41,

$$\omega_x = 0, \omega_y = \frac{6v_{av}(h - 2z)}{h^2}, \omega_z = 0. \tag{7.3.53}$$

□

7.3.4 Constitutive Relation

Newton's law of viscosity is a relation between the stress tensor and the rate of deformation tensor. While this is often expressed as a relation between one component of the stress and strain rate for simple unidirectional flows, the relation for complex three-dimensional flows can be correctly expressed only in tensor form. Newton's law of viscosity is obtained if we assume that the stress is a linear function of the rate of deformation tensor. In Section 7.3.3, the rate of deformation tensor was expressed as the sum of the the radial expansion/compression, extensional strain and solid body rotation. The antisymmetric part of the rate of deformation tensor results in solid body rotation. When an object is rotated, the distance between material points within the object does not change, and there is no internal deformation. Therefore, the stress does not depend on the antisymmetric part of the rate of deformation tensor. The stress depends on the symmetric and isotropic parts, in addition to the pressure which is present even in the absence of flow,

$$\boldsymbol{\sigma} = -p\boldsymbol{I} + 2\mu\boldsymbol{E} + \mu_b(\boldsymbol{\nabla} \cdot \boldsymbol{v})\boldsymbol{I}, \tag{7.3.54}$$

where $\boldsymbol{E}$ is given in Eq. 7.3.30. Here, the first term on the right is due to the pressure, Eq. 7.3.21, the second term proportional to the viscosity μ is due to extensional strain, and the third term proportional to the bulk viscosity μ_b is due to the radial expansion/compression. The momentum conservation equation is,

$$\frac{\partial(\rho\boldsymbol{v})}{\partial t} + \boldsymbol{\nabla} \cdot (\rho\boldsymbol{v}\boldsymbol{v}) = \boldsymbol{\nabla} \cdot \boldsymbol{\sigma} + \boldsymbol{f}, \tag{7.3.55}$$

where $\boldsymbol{f}$ is the body force density. The terms on the left are the analogues of the time derivative and the convective term in Eq. 7.1.19. The left side of the momentum conservation equation is simplified using product rule for differentiation,

$$\frac{\partial(\rho \boldsymbol{v})}{\partial t} + \boldsymbol{\nabla} \cdot (\rho \boldsymbol{v}\boldsymbol{v}) = \rho \frac{\partial \boldsymbol{v}}{\partial t} + \boldsymbol{v} \frac{\partial \rho}{\partial t} + \boldsymbol{v} \boldsymbol{\nabla} \cdot (\rho \boldsymbol{v}) + \rho \boldsymbol{v} \cdot \boldsymbol{\nabla} \boldsymbol{v}$$

$$= \rho \left(\frac{\partial \boldsymbol{v}}{\partial t} + \boldsymbol{v} \cdot \boldsymbol{\nabla} \boldsymbol{v} \right). \tag{7.3.56}$$

Here, the second and third terms on the right side in the first line in Eq. 7.3.56 sum to zero from the mass conservation Eq. 7.3.1.

Eqs. 7.3.56, 7.3.54, and 7.3.30 are substituted into Eq. 7.3.55,

$$\rho \left(\frac{\partial \boldsymbol{v}}{\partial t} + \boldsymbol{v} \cdot \boldsymbol{\nabla} \boldsymbol{v} \right) = -\boldsymbol{\nabla} p + \boldsymbol{\nabla} \cdot [\mu (\boldsymbol{\nabla} \boldsymbol{v} + (\boldsymbol{\nabla} \boldsymbol{v})^T - \tfrac{2}{3} (\boldsymbol{\nabla} \cdot \boldsymbol{v}) \boldsymbol{I})]$$

$$+ \boldsymbol{\nabla} [\mu_b (\boldsymbol{\nabla} \cdot \boldsymbol{v})] + \boldsymbol{f}, \tag{7.3.57}$$

where the superscript ^{T} denotes the transpose. Here, the first term on the right is $-\boldsymbol{\nabla} \cdot (\boldsymbol{I} p) = -\boldsymbol{\nabla} p$, the gradient of the pressure. The second and third terms on the right is the divergence of the shear stress, and the last term on the right is the body force.

For an incompressible flow, $\boldsymbol{\nabla} \cdot \boldsymbol{v}$ is zero and the stress tensor is

$$\boxed{\sigma = -p\boldsymbol{I} + 2\mu \boldsymbol{E},} \tag{7.3.58}$$

where the extensional strain tensor is,

$$\boldsymbol{E} = \tfrac{1}{2} (\boldsymbol{\nabla} \boldsymbol{v} + (\boldsymbol{\nabla} \boldsymbol{v})^T). \tag{7.3.59}$$

The momentum conservation equation for an incompressible flow is obtained by setting $\boldsymbol{\nabla} \cdot \boldsymbol{v} = 0$ in Eq. 7.3.57,

$$\rho \left(\frac{\partial \boldsymbol{v}}{\partial t} + \boldsymbol{v} \cdot \boldsymbol{\nabla} \boldsymbol{v} \right) = -\boldsymbol{\nabla} p + \boldsymbol{\nabla} \cdot [\mu (\boldsymbol{\nabla} \boldsymbol{v} + (\boldsymbol{\nabla} \boldsymbol{v})^T)] + \boldsymbol{f}. \tag{7.3.60}$$

For an incompressible flow with constant viscosity, the momentum conservation equation reduces to,

$$\rho \left(\frac{\partial \boldsymbol{v}}{\partial t} + \boldsymbol{v} \cdot \boldsymbol{\nabla} \boldsymbol{v} \right) = -\boldsymbol{\nabla} p + \mu \boldsymbol{\nabla}^2 \boldsymbol{v} + \mu \boldsymbol{\nabla} (\boldsymbol{\nabla} \cdot \boldsymbol{v}) + \boldsymbol{f}. \tag{7.3.61}$$

The second term on the right is the divergence of the velocity gradient, where the dot product is between the divergence and the gradient operators, resulting in the Laplacian of the velocity. The third term on the right is the divergence of the transpose of the velocity gradient, where the dot product is between the divergence and the velocity. For an incompressible flow, the divergence of the velocity in the third term on the right in Eq. 7.3.61 is zero, and the momentum conservation equations is,

$$\rho\left(\frac{\partial v}{\partial t} + v \cdot \nabla v\right) = -\nabla p + \mu \nabla^2 v + f. \qquad (7.3.62)$$

Eqs. 7.3.3 and 7.3.62 are the Navier–Stokes mass and momentum equations for an incompressible flow.

7.3.5   Potential Flow

The Bernoulli equation for a potential flow, discussed in Chapter 6, is derived as follows. The flow is irrotational if the curl of the velocity is zero, $\nabla \times v = 0$. The velocity can then be expressed as the gradient of a potential, $v = \nabla \phi_v$, because the curl of the gradient of a scalar is necessarily zero (Exercise 7.11). The mass conservation equation is,

$$\nabla \cdot v = \nabla^2 \phi_v = 0. \qquad (7.3.63)$$

For an irrotational flow, the rate of deformation tensor is symmetric, and it can be shown that $v \cdot \nabla v = \nabla(\frac{1}{2}v^2)$ (Exercise 7.12). For an inviscid flow, the viscous term on the right side of Eq. 7.3.62 is neglected. The momentum conservation equation then becomes,

$$\rho\left(\frac{\partial(\nabla \phi_v)}{\partial t} + \nabla(\tfrac{1}{2}v^2)\right) = -\nabla p + f. \qquad (7.3.64)$$

If the body force density is expressed as $= -\nabla \Phi$, where Φ is the potential energy density, Eq. 7.3.64 simplifies as,

$$\nabla\left(\rho\frac{\partial \phi_v}{\partial t} + \tfrac{1}{2}\rho v^2 + p + \Phi\right) = 0. \qquad (7.3.65)$$

If the gradient of a scalar function is zero at all locations in the flow, then the scalar function is necessarily a constant,

$$\rho\frac{\partial \phi_v}{\partial t} + \tfrac{1}{2}\rho v^2 + p + \Phi = p_0. \qquad (7.3.66)$$

Eq. 7.3.66 is the Bernoulli equation for a time-dependent potential flow. The time derivative of the potential is zero for a steady flow, and the Bernoulli equation reduces to Eq. 6.1.1 in Chapter 6, where the potential energy density is $\Phi = \rho g z$.

Summary (7.3)

1. The mass conservation equation is Eq. 7.3.3—that is, the divergence of the velocity is zero in an incompressible flow.

2. The force per unit area exerted at a surface with outward unit normal n can be written as $\sigma \cdot n$, where σ is the second order stress tensor which has the general form $\sigma = \sum_{i=x,y,z} \sum_{j=x,y,z} e_i \sigma_{ij} e_j$. Here, the first index i is the direction of the force, and the second index j is the direction of transport or the direction of the unit normal to the surface across which momentum is transported. Whereas the force on a surface depends on the position and orientation of the surface, the stress tensor is a single-valued function of position.

3. Angular momentum conservation requires that the stress tensor is symmetric.

4. In the absence of flow, the stress tensor, Eq. 7.3.21, is $-pI$, where p is the pressure and I is the identity tensor.

5. The relative velocity of points separated by an infinitesimal distance Δx in a fluid is $\Delta x \cdot \nabla v$, where ∇v (Eq. 7.3.23) is the velocity gradient or rate of deformation tensor.

6. The rate of deformation tensor is separated into isotropic, symmetric traceless and antisymmetric parts.

7. The isotropic part of the rate of deformation tensor, Eq. 7.3.29, is radial expansion or compression. The trace of the rate of deformation tensor is equal to the divergence of the velocity $\nabla \cdot v$. This is zero for an incompressible flow.

8. The antisymmetric part of the rate of deformation tensor, Eq. 7.3.27, is solid body rotation. The angular velocity is equal to one half of the vorticity $\frac{1}{2}\omega$, and axis of rotation is along the vorticity vector ω, where $\omega = \nabla \times v$ is the curl of the velocity.

9. The symmetric traceless part of the rate of deformation tensor, Eq. 7.3.30, is extensional strain. The eigenvectors of the matrix are the principal axes of expansion/compression. There is expansion along the axes with positive eigenvalues, and compression along the axes with negative eigenvalues. There is no net change in volume, and no rotation of the axes, due to the symmetric traceless part of the rate of deformation tensor.

10. The constitutive relation for the stress tensor for an incompressible flow of a Newtonian fluid, Eq. 7.3.58, comprises an isotropic component due to the pressure and a symmetric traceless component due to the symmetric traceless part of the rate of deformation tensor.

11. The Navier–Stokes momentum balance equation is Eq. 7.3.62.

12. For a potential flow which is incompressible, inviscid and irrotational, the velocity can be expressed as the gradient of a potential, $v = \nabla \phi_v$. The mass conservation equation reduces to a Laplace equation, Eq. 7.3.63, for the velocity potential, and the momentum conservation equation is the unsteady Bernoulli equation, Eq. 7.3.66.

Exercises

EXERCISE 7.1 The cylindrical coordinate system consists of the coordinates (r, ϕ, z), where r is the distance from the z axis, and ϕ is the angle made by the projection of the position vector on the $x - y$ plane with the x axis, as shown in Fig. 7.11. For this coordinate system,

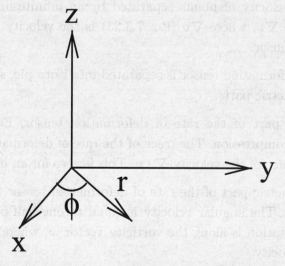

FIGURE 7.11. Cylindrical coordinate system.

a) Determine the coordinates (x, y, z) in terms of (r, ϕ, z), and the coordinates (r, ϕ, z) in terms of (x, y, z). How are the unit vectors (e_r, e_ϕ, e_z) related to (e_x, e_y, e_z)?

b) Write down the conservation equation for the concentration field for the appropriate differential volume in cylindrical coordinates. What is the divergence operator $\nabla \cdot$ in this coordinate system?

c) Express the flux in terms of the gradient of concentration in the cylindrical coordinate system. What is the Laplacian operator ∇^2 in this coordinate system?

EXERCISE 7.2 A toroidal co-ordinate system is used for analysing the flow in a curved tube which has inner radius R_t and radius of curvature of the centreline of the tube R_c about an axis as shown in Fig. 7.12. The three co-ordinates are the radius r which is the distance from the centreline of the tube, the polar angle ϕ which is the angle made by the radius vector with the x axis in the cross section of the tube, and the angle θ around the axis of curvature of the tube, as shown in Fig. 7.12. The unit vectors e_r, e_ϕ and e_θ form an orthogonal co-ordinate system at each point but the direction of the unit vectors change with location. The distance Δs between two locations (r, ϕ, θ) and $(r + \Delta r, \phi + \Delta \phi, \theta + \Delta \theta)$ is given by,

$$(\Delta s)^2 = (\Delta r)^2 + r^2(\Delta \phi)^2 + (R_c + r \cos(\phi))^2 (\Delta \theta)^2.$$

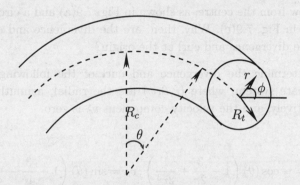

FIGURE 7.12. The toroidal co-ordinate system.

a) What is the differential volume element in this co-ordinate system?
b) What is the expression for the gradient of a scalar T in the toroidal co-ordinate system?
c) What is the expression for the divergence of a vector v?
d) What is the expression for the Laplacian in this co-ordinate system?

EXERCISE 7.3 Derive the expressions for the curl of a vector in cylindrical and spherical co-ordinate systems using Eq. 7.3.40.

EXERCISE 7.4 Verify that the mass conservation equation is satisfied if the velocities are expressed in terms of the stream functions, Eqs. 7.3.9 and 7.3.10, in polar and spherical co-ordinates, respectively. Verify the relation between the vorticity and the stream function in Cartesian co-ordinates,

$$\omega_z = -\nabla^2 \psi,$$

and in polar co-ordinates for a flow in the $r - \phi$ plane,

$$\omega_x = -\nabla^2 \psi.$$

EXERCISE 7.5 Determine the divergence and the curl of the following velocity fields in a cylindrical co-ordinate system, where (r, ϕ, x) are the radial, polar and axial co-ordinates.

a)

$$v_r = 0, v_\phi = 0, v_x = (1 - r^2).$$

b)

$$v_r = \frac{\cos(n\phi)}{r^{n+1}}, v_\phi = \frac{\sin(n\phi)}{r^{n+1}}, v_x = 0.$$

c) Show that the divergence and curl of the velocity field

$$v_r = \frac{m}{2\pi r}, v_\phi = \frac{\Gamma}{2\pi r}, v_x = 0,$$

are zero. Sketch the velocity field. The velocity field seems to have a radial component, which is a net flow from the centre, as shown in Fig. 7.9(a) and a circulation around the origin, as shown in Fig. 7.9(c). Why, then, are the divergence and curl of the velocity zero? What is the divergence and curl at the origin?

EXERCISE 7.6 Determine the divergence and curl of the following velocity fields in a spherical co-ordinate system, where (r, θ, ϕ) are the radial, azimuthal and meridional co-ordinates, respectively, and the velocity component v_ϕ is zero.

a)

$$v_r = -\cos(\theta)\left(1 - \frac{3}{2r} + \frac{1}{2r^3}\right), v_\theta = \sin(\theta)\left(1 - \frac{3}{4r} - \frac{1}{4r^3}\right).$$

b)

$$v_r = -\left(1 - \frac{1}{r}\right)\cos(\theta), v_\theta = \left(1 - \frac{1}{2r}\right)\sin(\theta).$$

c)

$$v_r = \cos(\theta)\left(1 - \frac{1}{r^3}\right), v_\theta = -\sin(\theta)\left(1 + \frac{1}{2r^3}\right).$$

EXERCISE 7.7 The stress tensor at a point in a fluid, in N/m^2, is

$$\mathbf{T} = \begin{pmatrix} \mathbf{e}_x & \mathbf{e}_y & \mathbf{e}_z \end{pmatrix} \begin{pmatrix} 1 & 3 & 2 \\ 3 & 1 & 4 \\ 2 & 4 & 1 \end{pmatrix} \begin{pmatrix} \mathbf{e}_x \\ \mathbf{e}_y \\ \mathbf{e}_z \end{pmatrix}.$$

Here, the first unit vector is the direction of the force, and the second is the direction of the unit normal to the surface. What is the force per unit area, in N/m^2, acting parallel and perpendicular to the surface with outward unit normal, $\mathbf{n} = 0.8\mathbf{e}_x + 0.31\mathbf{e}_y + 0.52\mathbf{e}_z$?

EXERCISE 7.8 In Section 7.3.2, a torque balance was used to show that the stress tensor is symmetric if there are no torques exerted on the fluid. If there is a body torque density (torque per unit volume), L, exerted on the fluid, such that the torque exerted on each differential volume ΔV is $L\Delta V$, how is the antisymmetric part of the stress tensor related to the components of L from angular momentum balance?

EXERCISE 7.9 Determine the symmetric and antisymmetric part of the rate of deformation tensor, the extension rates, the principal axes for the symmetric part and the vorticity for the flow in a pipe with velocity profile

$$v_x = 2v_{av}\left(1 - \left(\frac{r}{R}\right)^2\right),$$

where r and x are the radial and axial co-ordinates in a cylindrical co-ordinate system, R is the pipe radius and v_{av} is the average velocity. Use a Cartesian co-ordinate system.

EXERCISE 7.10 What are the two components of the momentum conservation equations, Eq. 7.3.62, expressed in terms of the components of the velocity in a two-dimensional Cartesian co-ordinate system? Express these equations in terms of the stream function, Eq. 7.3.4. Eliminate the pressure by taking

$$\frac{\partial(x \text{ momentum equation})}{\partial y} - \frac{\partial(y \text{ momentum equation})}{\partial x},$$

to obtain an equation for the stream function. Obtain an equation for the pressure by taking,

$$\frac{\partial(x \text{ momentum equation})}{\partial x} + \frac{\partial(y \text{ momentum equation})}{\partial y}.$$

EXERCISE 7.11 Using the definition Eq. 7.3.40 for the curl, show that $\nabla \times \nabla\phi = 0$ for any scalar field ϕ in a general co-ordinate system. What is the physical interpretation of this result?

EXERCISE 7.12 If the flow is irrotational and the rate of deformation tensor is symmetric, show that

$$v \cdot \nabla v = \nabla(\tfrac{1}{2}v^2).$$

Expand the left side in component form for one of the components, and show that it is equal to the right side for the same component.

Appendix

7.A   Gradient

Consider a scalar field, such as the concentration field, where the concentration $c(x)$ is a continuously varying function of the position x. The difference in concentration between two nearby locations, $\boldsymbol{x}$ and $\boldsymbol{x} + \Delta \boldsymbol{x}$, shown in Fig. 7.13(a), can be expressed in the Cartesian co-ordinate system as

$$c(\boldsymbol{x} + \Delta \boldsymbol{x}, t) - c(\boldsymbol{x}, t) = c(x + \Delta x, y + \Delta y, z + \Delta z, t) - c(x, y, z, t)$$

$$= \Delta x \frac{\partial c}{\partial x} + \Delta y \frac{\partial c}{\partial y} + \Delta z \frac{\partial c}{\partial z}. \tag{7.A.1}$$

Eq. 7.A.1 can be rewritten in vector notation,

$$c(\boldsymbol{x} + \Delta \boldsymbol{x}, t) - c(\boldsymbol{x}, t) = \Delta \boldsymbol{x} \cdot \boldsymbol{\nabla} c. \tag{7.A.2}$$

where $\Delta \boldsymbol{x} = \Delta x \, \boldsymbol{e}_x + \Delta y \, \boldsymbol{e}_y + \Delta z \, \boldsymbol{e}_z$, and $\boldsymbol{\nabla} c$ is given by Eq. 7.1.15. Here, we have considered the limit $|\Delta \boldsymbol{x}| \to 0$ so that the Taylor series expansion in Eq. 7.A.2 is approximated by the first term.

It is important to note that the gradient of the concentration field (or any scalar field) defined in Eq. 7.A.2 is a function of the location $\boldsymbol{x}$ and is independent of $\Delta \boldsymbol{x}$ for $|\Delta \boldsymbol{x}| \to 0$. The components of the gradient vector do depend on the choice of co-ordinate system, but the gradient vector has a unique magnitude and direction which is independent of the co-ordinate system.

The physical significance of the gradient is as follows. Consider the $\boldsymbol{\nabla} c$ vector at a location $\boldsymbol{x}$, shown in Fig. 7.13(a). If we travel equal distances $|\Delta \boldsymbol{x}|$ along different directions, the change in concentration will depend on the direction travelled. The change in concentration is largest when the displacement vector $\Delta \boldsymbol{x}$ is in the same direction as $\boldsymbol{\nabla} c$. This implies that $\boldsymbol{\nabla} c$ is in the direction of maximum variation in the concentration. There is no change in the concentration if $\Delta \boldsymbol{x}$ and $\boldsymbol{\nabla} c$ are perpendicular to each other. Therefore, $\boldsymbol{\nabla} c$ is perpendicular to the surface S of constant concentration, as shown in Fig. 7.13(a). Thus, the concentration gradient at a point is a vector along the direction of maximum variation in the concentration, and is perpendicular to the surface of constant concentration.

The gradient is a vector derivative operator—that is, its components are the spatial derivatives of the scalar concentration field. An integral relation can be derived for the gradient operator, which is independent of the co-ordinate system used, as follows. Consider two points A and B with position vectors $\boldsymbol{x}_A$ and $\boldsymbol{x}_B$ shown in Fig. 7.13(b). The change in the concentration between these two points, $c_B - c_A$, can be expressed as the integral of the gradient along any path connecting these two points,

$$c_B - c_A = \int_A^B \mathrm{d}\boldsymbol{x} \cdot \boldsymbol{\nabla} c. \tag{7.A.3}$$

Here, $\mathrm{d}\boldsymbol{x}$ is the infinitesimal vector displacement along a path between A and B, as shown in Fig. 7.13, and $\boldsymbol{\nabla} c$ is the gradient of the concentration at the point $\boldsymbol{x}$.

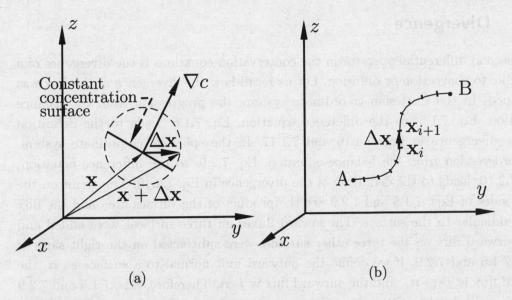

(a)   (b)

FIGURE 7.13. The gradient of the concentration field at a point (a), and the path between two points A and B for deriving the integral relation for the gradient (b). In (a), ∇c is the concentration gradient vector at the position x. For displacements of equal magnitude $|\Delta x|$ from the point x shown by the sphere, the change in concentration is maximum when $|\Delta x|$ is in the same direction as Δc. The change in concentration is zero on the surface perpendicular to ∇c.

To derive Eq. 7.A.3, the path between the locations A and B is divided into N infinitesimal displacements, $\Delta x_1 = x_1 - x_A$, $\Delta x_2 = x_2 - x_1, \ldots, \Delta x_N = x_B - x_{N-1}$. The change in concentration for each of these infinitesimal displacements is given by Eq. 7.A.2,

$$c(x_1) - c(x_A) = \Delta x_1 \cdot \nabla c|_{x=x_1},$$
$$c(x_2) - c(x_1) = \Delta x_2 \cdot \nabla c|_{x=x_2},$$
$$\cdots \cdots$$
$$c(x_B) - c(x_{N-1}) = \Delta x_N \cdot \nabla c|_{x=x_B}. \tag{7.A.4}$$

If all of the above equations are added, the result is,

$$c(x_B) - c(x_A) = \sum_{i=1}^{N} \Delta x_i \cdot \nabla c|_{x=x_i} \tag{7.A.5}$$

In the limit $|\Delta x_1|, |\Delta x_2|, \ldots, |\Delta x_N| \to 0$, the summation in Eq. 7.A.5 is replaced by the path integral, Eq. 7.A.3, along the path between A and B.

Eq. 7.A.3 is valid for any path between the end points x_A and x_B. This implies that the integral of ∇c is the same along any path between the same two end points. This also implies that if a path starts at a point x_A and returns to the same point, the integral of ∇c on this closed path has to be zero.

7.B   Divergence

The second differential operator in the conservation equations is the divergence of a
flux due to convection or diffusion. Let us recall how the divergence of the flux was
obtained. In the Cartesian co-ordinate system, the progression from the balance
equation, Eq. 7.1.5, to the difference equation, Eq. 7.1.6, leads to the definition
of the divergence in Eqs. 7.1.16 and 7.1.17. In the spherical co-ordinate system,
the progression from the balance equation, Eq. 7.2.9, to the difference equation,
Eq. 7.2.10, leads to the definition of the divergence in Eq. 7.2.19. The terms on the
right sides of Eqs. 7.1.5 and 7.2.9 are the product of the surface area and the flux
perpendicular to the surface. The inward fluxes on three surfaces were added and
the outward flux on the three other surfaces were subtracted on the right sides of
Eqs. 7.1.5 and 7.2.9. If we define the outward unit normal to a surface as n, the
inward flux is $j \cdot (-n)$, and the outward flux is $j \cdot n$. Therefore, Eqs. 7.1.5 and 7.2.9
can be equivalently written as

$$\Delta V \left(c(x, y, z, t + \Delta t) - c(x, y, z, t) \right)$$

$$= - \Delta t \sum_{I=1}^{6} (\Delta S_I n_I \cdot (cv + j)) + \mathcal{S}(x, y, z, t) \Delta V \Delta t, \qquad (7.B.1)$$

where the summation is carried out over all the six surfaces enclosing the differential
volume, S_I is the surface area of surface I, and n_I is the outward unit normal to the
surface I. Eq. 7.1.5 (or 7.2.9) was then divided by $\Delta V \Delta t$ to obtain the difference
Eq. 7.1.6 (or 7.2.10),

$$\frac{(c(x, y, z, t + \Delta t) - c(x, y, z, t))}{\Delta t}$$

$$= - \frac{1}{\Delta V} \sum_{I=1}^{6} (\Delta S_I n_I \cdot (cv + j)) + \mathcal{S}(x, y, z, t). \qquad (7.B.2)$$

By comparing Eq. 7.B.2 and the conservation Eq. 7.1.16, it is clear that the definition
of the divergence is,

$$\nabla \cdot (cv + j) = \frac{1}{\Delta V} \sum_{I=1}^{6} (\Delta S_I n_I \cdot (cv + j)). \qquad (7.B.3)$$

The summation in Eq. 7.B.3 is for a set of plane surfaces surrounding a differential
volume. For a smooth surface S of arbitrary shape surrounding a differential volume

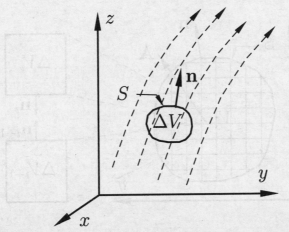

FIGURE 7.14. The definition of the divergence, Eq. 7.B.5, for a vector field A, depicted by the dashed arrows, applied for a volume of arbitrary shape ΔV surrounded by a smooth surface S with outward unit normal n.

ΔV shown in Fig. 7.14, the generalisation of the definition, Eq. 7.B.3, is

$$\nabla \cdot (cv + j) = \lim_{\Delta V \to 0} \frac{1}{\Delta V} \int_S \mathrm{d}S \, n \cdot (cv + j). \qquad (7.B.4)$$

For a vector field A, the divergence at a point is defined as

$$\nabla \cdot A = \lim_{\Delta V \to 0} \frac{1}{\Delta V} \int_S \mathrm{d}S \, n \cdot A. \qquad (7.B.5)$$

As the volume ΔV in Eq. 7.B.5 decreases to zero, the divergence of the vector at a location tends to a unique scalar value which is independent of the volume ΔV.

The physical interpretation of the divergence is, literally, the divergence of the vector field at a point. Consider the vector field, with value $v(x)$ at the location x, as shown in Fig. 7.9(a). The difference $\Delta v = v(x + \Delta x) - v(x)$ is the difference in the vector between nearby locations $x + \Delta x$ and x. If the difference Δv is directed outward from the point x—that is, Δv diverges at the point x—then the divergence $\nabla \cdot v$ is positive. If the difference Δv is directed inward—that is Δv converges to the location x—the divergence is negative. If there is no net convergence or divergence of Δv at the point x, as shown in Fig. 7.9(b) and (c), the divergence is zero.

With this physical interpretation of divergence, the meaning of the conservation Eq. 7.1.16 becomes clear. When the divergence of the flux $\nabla \cdot j$ is positive, that is, when the flux diverges at a point, there is a net outflow of mass from a differential volume at this location, and the concentration decreases in time. When the divergence of the flux $\nabla \cdot j$ is negative, that is, the flux converges to a point,

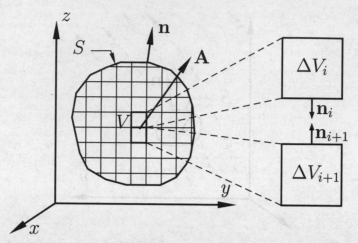

FIGURE 7.15. The derivation of the integral relation for the divergence of a vector $\boldsymbol{A}$ over a volume by dividing the volume V into infinitesimal sub-volumes ΔV_i and applying the definition of the divergence to the sub-volumes.

there is a net inflow of mass into a differential volume at this location, and the concentration at that location increases with time.

The expressions for the divergence in the Cartesian and spherical co-ordinate systems are Eqs. 7.1.17 and 7.2.19, respectively. In a general orthogonal co-ordinate system with co-ordinates (x_a, x_b, x_c) and scale factors (h_a, h_b, h_c), the expression for the divergence is Eq. 7.2.22.

The integral relation for the divergence states that for any macroscopic volume V surrounded by a surface S,

$$\int_V \mathrm{d}V (\boldsymbol{\nabla} \cdot \boldsymbol{A}) = \int_S \mathrm{d}S\, \boldsymbol{n} \cdot \boldsymbol{A} \qquad (7.B.6)$$

where $\boldsymbol{n}$ is the outward unit normal to the surface S. It should be emphasised that the integral relation, Eq. 7.B.6, is written for a macroscopic volume shown in Fig. 7.15, in contrast to the definition of the divergence in Eq. 7.B.5 which applies a differential volume. To prove Eq. 7.B.6, the volume V is divided into a large number of sub-volumes $\Delta V_1, \Delta V_2, \ldots, \Delta V_N$ with centres at the locations $\boldsymbol{x}_1, \boldsymbol{x}_2, \ldots, \boldsymbol{x}_N$. The left side of Eq. 7.B.6 is expressed as the summation over these differential volumes,

$$\int_V \mathrm{d}V (\boldsymbol{\nabla} \cdot \boldsymbol{A}) = \sum_{i=1}^{N} \Delta V_i (\boldsymbol{\nabla} \cdot \boldsymbol{A})\big|_{\boldsymbol{x}_i}. \qquad (7.B.7)$$

Eq. 7.B.5 applies to each of these sub-volumes,

$$\sum_{i=1}^{N} \Delta V_i (\boldsymbol{\nabla} \cdot \boldsymbol{A})\big|_{\boldsymbol{x}_i} = \sum_{i=1}^{N} \int_{S_i} \mathrm{d}S\, \boldsymbol{n} \cdot \boldsymbol{A}, \qquad (7.B.8)$$

where S_i is the surface surrounding sub-volume ΔV_i, and $\boldsymbol{n}$ is the outward unit normal to the surface. The surfaces enclosing the sub-volumes are of two types—the surface between two adjoining sub-volumes, and that which is a part of the surface S enclosing the entire volume V, as shown in Fig. 7.15. Consider the surface separating two typical adjoining sub-volumes, ΔV_i and ΔV_{i+1}. The value of the vector $\boldsymbol{A}$ is uniquely specified on the surface, but the outward unit normal $\boldsymbol{n}$ is opposite in direction for the two sub-volumes, as shown in Fig. 7.15. Therefore, the contributions to the surface integral on the right side of Eq. 7.B.8 sum to zero for the two adjacent sub-volumes. Similarly, the contributions to the surface integrals sum to zero for all internal surfaces between adjoining sub-volumes within the volume V. There is a non-zero contribution to the surface integral on the right side of Eq. 7.B.8 only due to the surfaces of the sub-volumes that are a part of the outer surface S enclosing the entire volume V. Therefore, the right side of Eq. 7.B.8 can be written as,

$$\sum_{i=1}^{N} \int_{S_i} \mathrm{d}S\, \boldsymbol{n} \cdot \boldsymbol{A} = \int_{S} \mathrm{d}S\, \boldsymbol{n} \cdot \boldsymbol{A}, \qquad (7.B.9)$$

where S is now the surface enclosing the volume V. This proves the integral relation, Eq. 7.B.6.

The divergence theorem, Eq. 7.B.6, has one important physical implication. In order to determine the volume integral of the divergence of a vector over a volume V, it is not necessary to know the value of the vector throughout the volume. It is sufficient to know the value of the vector on the surface enclosing the volume, since the right side of Eq. 7.B.6 depends only on the value of the vector on the surface S.

7.C   Curl

The curl of a vector field $\boldsymbol{A}$ at a point is defined by considering a differential volume ΔV around the point with surface S and outward unit normal $\boldsymbol{n}$ as shown in Fig. 7.14,

$$\mathrm{curl}(\boldsymbol{A}) = \lim_{\Delta V \to 0} \frac{\int \mathrm{d}S\, \boldsymbol{n} \times \boldsymbol{A}}{\Delta V}. \qquad (7.C.1)$$

In a Cartesian co-ordinate system, the cubic volume shown in Fig. 7.1(a) is considered. The surface integral on the right side of Eq. 7.C.1 is calculated on the six faces of the cube. On the left face at $(x, y - \Delta y/2, z)$, the surface area is $\Delta x \Delta z$, and the the unit normal is in the $-e_y$ direction. Therefore,

$$\int dS n \times A = \Delta x \Delta z (-e_y \times A) = \Delta x \Delta z (-A_z(x, y - \Delta y/2, z)e_x + A_x(x, y - \Delta y/2, z)e_z).$$
(7.C.2)

Similarly, on the right face at $(x, y + \Delta y/2, z)$, the surface area is $\Delta x \Delta z$, and the outward unit normal is in the $+e_y$ direction,

$$\int dS n \times A = \Delta x \Delta z e_y \times A = \Delta x \Delta z (A_z(x, y + \Delta y/2, z)e_x - A_x(x, y + \Delta y/2, z)e_z).$$
(7.C.3)

In a similar manner, the contributions to $\int dS\, n \times A$ from the six surfaces are,

Surface at $(x, y - \Delta y/2, z) = \Delta x \Delta z[-A_z(x, y - \Delta y/2, z)e_x + A_x(x, y - \Delta y/2, z)e_z],$

Surface at $(x, y + \Delta y/2, z) = \Delta x \Delta z[A_z(x, y + \Delta y/2, z)e_x - A_x(x, y + \Delta y/2, z)e_z],$

Surface at $(x - \Delta x/2, y, z) = \Delta y \Delta z[A_z(x - \Delta x/2, y, z)e_y - A_y(x - \Delta x/2, y, z)e_z],$

Surface at $(x + \Delta x/2, y, z) = \Delta y \Delta z[-A_z(x + \Delta x/2, y, z)e_y + A_y(x + \Delta x/2, y, z)e_z],$

Surface at $(x, y, z - \Delta z/2) = \Delta x \Delta y[-A_x(x, y, z - \Delta z/2)e_y + A_y(x, y, z - \Delta z/2)e_x],$

Surface at $(x, y, z + \Delta z/2) = \Delta x \Delta y[A_x(x, y, z + \Delta z/2)e_y - A_y(x, y, z + \Delta z/2)e_x].$
(7.C.4)

When the above contributions are added and divided by $\Delta V = \Delta x \Delta y \Delta z$, we obtain,

$$\begin{aligned}
\mathrm{curl}(A) = e_x &\left(\frac{A_z(x, y + \Delta y/2, z) - A_z(x, y - \Delta y/2, z)}{\Delta y} \right.\\
&\left. - \frac{A_y(x, y, z + \Delta z/2) - A_y(x, y, z - \Delta z/2)}{\Delta z} \right)\\
+ e_y &\left(\frac{A_x(x, y, z + \Delta z/2) - A_x(x, y, z - \Delta z/2)}{\Delta z} \right.\\
&\left. - \frac{A_z(x + \Delta x/2, y, z) - A_z(x - \Delta x/2, y, z)}{\Delta x} \right)\\
+ e_z &\left(\frac{A_y(x + \Delta x/2, y, z) - A_y(x - \Delta x/2, y, z)}{\Delta x} \right.\\
&\left. - \frac{A_x(x, y + \Delta y/2, z) - A_x(x, y - \Delta y/2, z)}{\Delta x} \right).
\end{aligned}$$
(7.C.5)

Taking the limit $\Delta x, \Delta y, \Delta z \to 0$, we obtain Eq. 7.3.39 for the cross product,

$$\mathrm{curl}(A) = e_x \left(\frac{\partial A_z}{\partial y} - \frac{\partial A_y}{\partial z} \right) + e_y \left(\frac{\partial A_x}{\partial z} - \frac{\partial A_z}{\partial x} \right) + e_z \left(\frac{\partial A_y}{\partial x} - \frac{\partial A_x}{\partial y} \right)$$
$$= \nabla \times A.$$
(7.C.6)

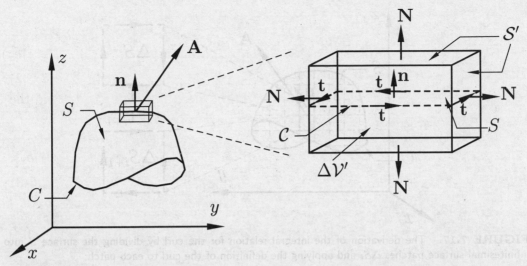

FIGURE 7.16. The cubic volume $\Delta \mathcal{V}'$ with six surfaces $\Delta \mathcal{S}'$ and unit normal N that straddles the surface S with unit volume n.

For an orthogonal co-ordinate system with co-ordinates (x_a, x_b, x_c) and scale factors (h_a, h_b, h_c), the expression for the curl of a vector is Eq. 7.3.40.

The integral relation for the curl of a vector, called the Stokes' theorem, relates the surface integral of $n \cdot \nabla \times A$ over a surface S with the line integral over the perimeter C of the surface of $A \cdot dx$, where dx is the differential displacement along the perimeter C as shown in Fig. 7.16,

$$\int dS\, n \cdot \nabla \times A = \oint_C A \cdot dx. \tag{7.C.7}$$

To derive this, we consider a cubic differential volume $\Delta \mathcal{V}'$ bounded by the surface $\mathcal{S}'$ of height Δh that straddles the surface, as shown in Fig. 7.16. The unit normal to the surface S is denoted n, while the unit normal to the surface $\mathcal{S}'$ of the cubic volume $\Delta \mathcal{V}'$ is denoted N. Eq. 7.C.1 is written for the volume $\Delta \mathcal{V}'$, multiplied by $\Delta \mathcal{V}'$ and dotted with n to obtain,

$$\Delta \mathcal{V}' n \cdot \mathrm{curl}(A) = \int_{\mathcal{S}'} d\mathcal{S}'\, n \cdot (N \times A). \tag{7.C.8}$$

The following simplifications are carried out in the above equation. The term on the right, $n \cdot (N \times A)$ is the triple product which can be equivalently written as $A \cdot (n \times N)$. This triple product is zero on the top and bottom surface, because n and N are parallel. Therefore, the integrand is non-zero only on the four surfaces of $\mathcal{S}'$ that intersect the surface S. On these surfaces, the differential surface area $d\mathcal{S}' = h\, dx$, where h is the height, and dx is the differential length on the contour

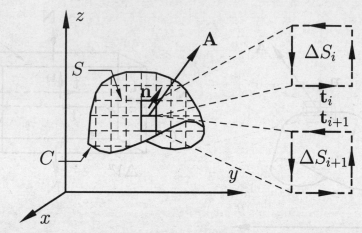

FIGURE 7.17. The derivation of the integral relation for the curl by dividing the surface S into infinitesimal surface patches ΔS_i and applying the definition of the curl to each patch.

$\mathcal{C}$ which is the intersection of S and $\mathcal{S}'$. The volume $\Delta \mathcal{V}' = h\,\Delta S$. The factor of h cancels on both sides of Eq. 7.C.8, resulting in,

$$\cancel{h}\,\Delta S\,\boldsymbol{n}\cdot\mathrm{curl}(\boldsymbol{A}) = \cancel{h}\oint_{\mathcal{C}}\mathrm{d}x\,\boldsymbol{A}\cdot(\boldsymbol{n}\times\boldsymbol{N}). \qquad (7.\mathrm{C}.9)$$

The cross product $\boldsymbol{n}\times\boldsymbol{N}$ is the tangent vector $\boldsymbol{t}$ along the contour $\mathcal{C}$, whose direction is given by the right-hand rule. Therefore, for the differential volume $\Delta\mathcal{V}'$, Eq. 7.C.9 is written as,

$$\Delta S\,\boldsymbol{n}\cdot\mathrm{curl}(\boldsymbol{A}) = \oint_{\mathcal{C}}\mathrm{d}x\,\boldsymbol{A}\cdot\boldsymbol{t} = \oint_{\mathcal{C}}\mathrm{d}\boldsymbol{x}\cdot\boldsymbol{A}, \qquad (7.\mathrm{C}.10)$$

where $\mathrm{d}\boldsymbol{x} = \boldsymbol{t}\,\mathrm{d}x$ is the differential displacement vector along the contour $\mathcal{C}$.

Eq. 7.C.10 applies for a differential patch of surface area, ΔS. The procedure for proving the integral relation, Eq. 7.C.7, which applies to the entire surface S, is similar to that for the integral relation for the divergence. The surface S is divided into patches $\Delta S_1, \Delta S_2, \ldots, \Delta S_N$, as shown in Fig. 7.17. The left side of Eq. 7.C.10 can be written in discrete form as,

$$\int \mathrm{d}S\,\boldsymbol{n}\cdot\boldsymbol{\nabla}\times\boldsymbol{A} = \sum_{i=1}^{N}\Delta S_i\,\boldsymbol{n}\cdot\boldsymbol{\nabla}\times\boldsymbol{A}. \qquad (7.\mathrm{C}.11)$$

Using Eq. 7.C.10, the right side of Eq. 7.C.11 is,

$$\int \mathrm{d}S\,\boldsymbol{n}\cdot\boldsymbol{\nabla}\times\boldsymbol{A} = \sum_{i=1}^{N}\oint \mathrm{d}\boldsymbol{x}_i\cdot\boldsymbol{A}. \qquad (7.\mathrm{C}.12)$$

The sum of the contour integrals on the right side of Eq. 7.C.12 is evaluated as follows. Consider two adjacent patches ΔS_i and ΔS_{i+1}, separated by a common boundary. The value of the vector $\boldsymbol{A}$ on this boundary is the same, but the direction of the tangent vector on this common boundary is in opposite directions, $\boldsymbol{t}_{i+1} = -\boldsymbol{t}_i$. Therefore, the sum of the contributions from this common boundary between ΔS_i and ΔS_{i+1} to the right side of Eq. 7.C.12 is zero. This sum is zero for all contours between adjacent patches. Therefore, the non-zero contribution to the right side of Eq. 7.C.12 is due to the integral over the perimeter C of the surface S, which is the right side of Eq. 7.C.7. This proves the the integral relation for the curl, Eq. 7.C.7, also called Stokes' theorem.

The following are the physical implications of the Stokes' theorem. The right side of Eq. 7.C.7 depends only on the value of $\boldsymbol{A}$ on the contour C. Therefore, the surface integral of $\boldsymbol{n} \cdot \text{curl}(\boldsymbol{A})$ is the same for all surfaces that have a common perimeter C. Since a closed surface has no perimeter, the surface integral of $\boldsymbol{n} \cdot \text{curl}(\boldsymbol{A})$ over a closed surface is necessarily equal to zero.

The physical significance of the integral relation for the curl is best explained by considering the curl of the velocity field. Eq. 7.C.7 for the velocity field $\boldsymbol{v}$ is,

$$\int \mathrm{d}S \, \boldsymbol{n} \cdot \boldsymbol{\omega} = \oint_C \boldsymbol{v} \cdot \mathrm{d}\boldsymbol{x}, \qquad (7.C.13)$$

where $\boldsymbol{\omega} = \boldsymbol{\nabla} \times \boldsymbol{v}$ is the curl of the velocity. The right side of 7.C.13, which is called the 'circulation' on the contour C, is the integral of the tangential component of the velocity along the contour. Eq. 7.C.13 states that the integral of the vorticity over a surface is equal to the circulation over the perimeter of the surface. A flow is considered 'irrotational' if the vorticity is zero everywhere, and there is no circulation on any contour within the flow.

8

Diffusion Equation

Convection can be neglected when the Peclet number is small, and the field variables are determined by solving a Poisson equation $\nabla^2 \Phi_{fv} + S = 0$ or a Laplace equation $\nabla^2 \Phi_{fv} = 0$, subject to boundary conditions, where Φ_{fv} and S are the field variable and the rate of production per unit volume, respectively. It is necessary to specify two boundary conditions in each co-ordinate to solve these equations. The separation of variables procedure is the general procedure to solve these problems in domains where the boundaries are surfaces of constant co-ordinate. This procedure was earlier used in Chapters 4 and 5 for unsteady one-dimensional transport problems.

The procedure for solving the heat conduction equation in Cartesian co-ordinates is illustrated in Section 8.1. The 'spherical harmonic' solution for the Laplace equation in spherical co-ordinates is derived using separation of variables in Section 8.2, first for an axisymmetric problem of the heat conduction in a composite, and then for a general three-dimensional configuration. There are two types of solutions, the 'growing harmonics' that increase proportional to a positive power of r, and the 'decaying harmonics' that decrease as a negative power of r, where r is distance from the origin in the spherical co-ordinate system.

An alternate interpretation of the decaying harmonic solutions of the Laplace equation as superpositions of point sources and sinks of heat is discussed in Section 8.3. It is shown that the each term in the spherical harmonic expansions is equivalent to a term obtained by the superposition of sources and sinks in a 'multipole expansion'. A physical interpretation of the growing harmonics is also provided.

The solution for a point source is extended to a distributed source in Section 8.4 by dividing the distributed source into a large number of point sources and taking the continuum limit. The Green's function procedure for a finite domain is illustrated by using image sources to satisfy the boundary conditions at planar surfaces.

8.1   Cartesian Co-ordinates

Consider the heat conduction in a rectangular block of length L and height H, in which the temperature is T_0 at $x = 0$ and $x = L$, T_A at $y = 0$ and T_B at $y = H$, as shown in Fig. 8.1. The temperature field is determined by solving the heat conduction equation in a Cartesian co-ordinate system,

$$\nabla^2 T = 0. \tag{8.1.1}$$

The separation of variables procedure is used to solve for the temperature field. Recall that in the separation of variables procedure, it is necessary to identify homogeneous boundary conditions in all directions except one. For the configuration shown in Fig. 8.1, it is convenient to define the reduced temperature, $T^* = T - T_0$, so that $T^* = 0$ at $x = 0, L$. The diffusion equation and boundary conditions expressed in terms of T^* are

$$\frac{\partial^2 T^*}{\partial x^2} + \frac{\partial^2 T^*}{\partial y^2} = 0, \tag{8.1.2}$$

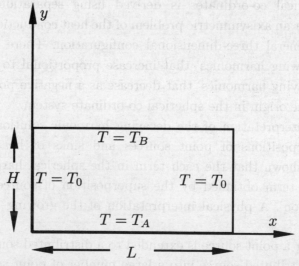

FIGURE 8.1.   The configuration and co-ordinate system for analysing the conduction in a rectangular block.

$$T^* = 0 \text{ at } x = 0 \ \& \ x = L, \tag{8.1.3}$$

$$T^* = T_A - T_0 \text{ at } y = 0, \tag{8.1.4}$$

$$T^* = T_B - T_0 \text{ at } y = H. \tag{8.1.5}$$

The substitution $T^* = X(x)Y(y)$ is made in Eq. 8.1.2, and the equation is divided by XY, to obtain,

$$\frac{1}{X}\frac{d^2 X}{dx^2} + \frac{1}{Y}\frac{d^2 Y}{dy^2} = 0. \tag{8.1.6}$$

In Eq. 8.1.6, the first term on the left is a function of x only, and the second is a function of y only; therefore, both have to be constants of equal magnitude and opposite sign. In Section 4.7, it was shown that homogeneous boundary conditions can be satisfied at both boundaries in the x direction only if,

$$\frac{1}{X}\frac{d^2 X}{dx^2} = -\frac{i^2 \pi^2}{L^2}, \tag{8.1.7}$$

where i is an integer. The solution for X,

$$X = \sin\left(\frac{i\pi x}{L}\right), \tag{8.1.8}$$

satisfies the homogeneous boundary conditions at $x = 0$ and $x = L$. From Eq. 8.1.6, the equation for Y is,

$$\frac{1}{Y}\frac{d^2 Y}{dx^2} = \frac{i^2 \pi^2}{L^2}, \tag{8.1.9}$$

and the solution is

$$Y = A_i e^{(i\pi y/L)} + B_i e^{(-i\pi y/L)}. \tag{8.1.10}$$

Therefore, the solution for T^* that satisfies the diffusion equation and the homogeneous boundary conditions in the x direction is,

$$T^* = \sum_{i=1}^{\infty} [A_i e^{(i\pi y/L)} + B_i e^{(-i\pi y/L)}] \sin\left(\frac{i\pi x}{L}\right). \tag{8.1.11}$$

The constants A_i and B_i are determined from the boundary conditions, Eqs. 8.1.4 and 8.1.5 at $y = 0$ and $y = H$,

$$T_A^* = \sum_{i=1}^{\infty} (A_i + B_i) \sin\left(\frac{i\pi x}{L}\right), \tag{8.1.12}$$

$$T_B^* = \sum_{i=1}^{\infty} (A_i e^{(i\pi H/L)} + B_i e^{(-i\pi H/L)}) \sin\left(\frac{i\pi x}{L}\right), \tag{8.1.13}$$

where $T_A^* = T_A - T_0$ and $T_B^* = T_B - T_0$.

The orthogonality relation, Eq. 4.7.25 in Section 4.7,

$$\int_0^L dx \, \sin\left(\frac{i\pi x}{L}\right) \sin\left(\frac{j\pi x}{L}\right) = \frac{L\delta_{ij}}{2},$$

(8.1.14)

is used to evaluate the constants A_i and B_i, where the Kronecker delta $\delta_{ij} = 1$ for $i = j$ and 0 for $i \neq j$. The left and right sides of Eqs. 8.1.12–8.1.13 are multiplied by $\sin\left(\frac{j\pi x}{L}\right)$ and integrated from 0 to L. The integral of $\sin\left(\frac{j\pi x}{L}\right)$ times the right side is evaluated using the expression,

$$\int_0^L dx \, \sin\left(\frac{j\pi x}{L}\right) = \frac{[1 - \cos(j\pi)]L}{j\pi} = \frac{[1 - (-1)^j]L}{j\pi}.$$

(8.1.15)

Multiplying Eqs. 8.1.12 and 8.1.13 by $\sin\left(\frac{j\pi x}{L}\right)$, integrating the resulting expressions from 0 to L, and using Eqs. 8.1.14 and 8.1.15, we obtain,

$$\frac{L[1 - (-1)^j]T_A^*}{j\pi} = \sum_{i=1}^{\infty} \frac{(A_i + B_i)L\delta_{ij}}{2} = \frac{(A_j + B_j)L}{2},$$

(8.1.16)

$$\frac{L[1 - (-1)^j]T_B^*}{j\pi} = \sum_{i=1}^{\infty} \frac{(A_i e^{(i\pi H/L)} + B_i e^{(-i\pi H/L)})L\delta_{ij}}{2}$$

$$= \frac{(A_j e^{(j\pi H/L)} + B_j e^{(-j\pi H/L)})L}{2}.$$

(8.1.17)

Eqs. 8.1.16 and 8.1.17 are solved to obtain,

$$A_j = \frac{2[1 - (-1)^j][T_B^* - T_A^* e^{(-j\pi H/L)}]}{j\pi[e^{(j\pi H/L)} - e^{(-j\pi H/L)}]},$$

(8.1.18)

$$B_j = \frac{2[1 - (-1)^j][T_A^* e^{(j\pi H/L)} - T_B^*]}{j\pi[e^{(j\pi H/L)} - e^{(-j\pi H/L)}]}.$$

(8.1.19)

The solution for the temperature field is,

$$T - T_0 = \sum_{i=1}^{\infty} \frac{2[1 - (-1)^i]}{i\pi[e^{(i\pi H/L)} - e^{(-i\pi H/L)}]} \left[[T_B^* - T_A^* e^{(-i\pi H/L)}]e^{(i\pi y/L)} \right.$$

$$\left. + [T_A^* e^{(i\pi H/L)} - T_B^*]e^{(-i\pi y/L)} \right] \sin\left(\frac{i\pi x}{L}\right).$$

(8.1.20)

In the preceding example, it was possible to obtain a solution by separation of variables because the boundary conditions in one direction, in this case x, could be reduced to homogeneous boundary conditions. The 'forcing' of the temperature field takes place in the other direction where the boundary conditions are not homogeneous. Even if the temperatures are different at $x = 0$ and $x = L$, it is possible to make the boundary conditions homogeneous in the x direction by defining a modified temperature that is the difference between the temperature T and a linear function of x, so that the modified temperature is zero at $x = 0$ and $x = L$. The modified temperature satisfies Eq. 8.1.2, because it differs from the actual temperature by a linear function of x.

The separation of variables procedure can also be used in cases where the boundary conditions are homogeneous on all boundaries, but there is forcing in the bulk of the flow by a force density. This is illustrated by the following example of the flow in a rectangular channel.

EXAMPLE 8.1.1: Determine the velocity profile for the steady and fully developed laminar flow of a Newtonian fluid in a rectangular channel of width W and height H driven by a pressure gradient, as shown in Fig. 8.2. The momentum conservation equation is,

$$-\frac{dp}{dx} + \mu\left(\frac{\partial^2 v_x}{\partial y^2} + \frac{\partial^2 v_x}{\partial z^2}\right) = 0, \tag{8.1.21}$$

where v_x, the stream-wise velocity in the x direction, is a function of y and z, and the pressure gradient (dp/dx) is a constant. The no-slip condition $v_x = 0$ applies at the four walls of the channel.

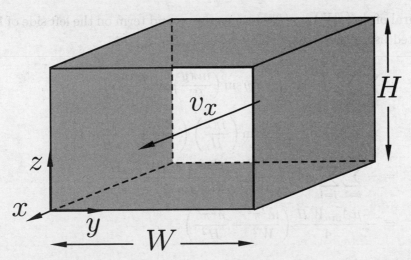

FIGURE 8.2. The configuration and co-ordinate system for analysing the pressure-driven flow in a rectangular channel.

Solution: Since the velocity is zero at the walls, the velocity can be expanded in a set of orthogonal basis functions which are zero at both walls. For the co-ordinate system in Fig. 8.2, the appropriate basis functions are $\sin\left(\frac{i\pi y}{W}\right)$ and $\sin\left(\frac{j\pi z}{H}\right)$ in the y and z directions, respectively, where i and j are integers. The general solution for the velocity field is of the form,

$$v_x = \sum_{i=1}^{\infty}\sum_{j=1}^{\infty} A_{ij} \sin\left(\frac{i\pi y}{W}\right)\sin\left(\frac{j\pi z}{H}\right). \qquad (8.1.22)$$

This solution is substituted into the momentum conservation equation, Eq. 8.1.21,

$$-\frac{dp}{dx} + \mu\sum_{i=1}^{\infty}\sum_{j=1}^{\infty} A_{ij}\left(-\frac{i^2\pi^2}{W^2} - \frac{j^2\pi^2}{H^2}\right)\sin\left(\frac{i\pi y}{W}\right)\sin\left(\frac{j\pi z}{H}\right) = 0. \quad (8.1.23)$$

In order to determine the coefficients A_{ij}, the orthogonality relation is applied in the y and z directions. Eq. 8.1.23 is multiplied by $\sin\left(\frac{m\pi y}{W}\right)\sin\left(\frac{n\pi z}{H}\right)$, and integrated from 0 to W in the y direction and 0 to H in the z direction. Here, m and n are integers. The integral of $\sin\left(\frac{m\pi y}{W}\right)\sin\left(\frac{n\pi z}{H}\right)$ times the first term on the left side of Eq. 8.1.23 is evaluated using Eq. 8.1.15,

$$-\int_0^H dz \int_0^W dy \frac{dp}{dx}\sin\left(\frac{m\pi y}{W}\right)\sin\left(\frac{n\pi z}{H}\right)$$
$$= -\left(\frac{[1-(-1)^m]W}{m\pi}\right)\left(\frac{[1-(-1)^n]H}{n\pi}\right)\frac{dp}{dx}. \qquad (8.1.24)$$

The integral of $\sin\left(\frac{m\pi y}{W}\right)\sin\left(\frac{n\pi z}{H}\right)$ times the second term on the left side of Eq. 8.1.23 is evaluated using the orthogonality relation, Eq. 8.1.14,

$$\sum_{i=1}^{\infty}\sum_{j=1}^{\infty}\int_0^H dz \int_0^W dy \sin\left(\frac{m\pi y}{W}\right)\sin\left(\frac{n\pi z}{H}\right)$$
$$\times\left[\mu A_{ij}\sin\left(\frac{i\pi y}{W}\right)\sin\left(\frac{j\pi z}{H}\right)\left(-\frac{i^2\pi^2}{W^2} - \frac{j^2\pi^2}{H^2}\right)\right]$$
$$= -\sum_{i=1}^{\infty}\sum_{j=1}^{\infty}\mu A_{ij}\left(\frac{i^2\pi^2}{W^2} + \frac{j^2\pi^2}{H^2}\right)\frac{\delta_{im}W}{2}\frac{\delta_{jn}H}{2}$$
$$= -\frac{\mu A_{mn}WH}{4}\left(\frac{m^2\pi^2}{W^2} + \frac{n^2\pi^2}{H^2}\right). \qquad (8.1.25)$$

The coefficients A_{mn} are determined from the condition that sum of the terms on the right side in Eqs. 8.1.24 and 8.1.25 is zero from the momentum conservation

Eq. 8.1.23,

$$A_{mn} = -\frac{4[1-(-1)^m][1-(-1)^n]}{\mu mn\pi^2}\left(\frac{m^2\pi^2}{W^2}+\frac{n^2\pi^2}{H^2}\right)^{-1}\frac{dp}{dx}. \qquad (8.1.26)$$

Note that the coefficients A_{mn} and the velocity v_x are positive if the pressure gradient is negative. The flow rate is

$$\dot{V} = \int_0^W dy \int_0^H dz\, v_x = \int_0^W dy \int_0^H dz \sum_{i=1}^\infty \sum_{j=1}^\infty A_{ij}\sin\left(\frac{i\pi y}{W}\right)\sin\left(\frac{j\pi z}{H}\right)$$

$$= \sum_{i=1}^\infty \sum_{j=1}^\infty \frac{A_{ij}[1-(-1)^i][1-(-1)^j]WH}{ij\pi^2}. \qquad (8.1.27)$$

Here, Eq. 8.1.15 has been used to simplify the integrals. The average flow velocity is the ratio of the flow rate and the area of cross section,

$$v_{av} = \frac{\dot{V}}{WH}$$

$$= \sum_{i=1}^\infty \sum_{j=1}^\infty \frac{A_{ij}[1-(-1)^i][1-(-1)^j]}{ij\pi^2}$$

$$= -\sum_{i=1}^\infty \sum_{j=1}^\infty \frac{4[1-(-1)^i]^2[1-(-1)^j]^2}{\mu\pi^4 i^2 j^2}\left(\frac{i^2\pi^2}{W^2}+\frac{j^2\pi^2}{H^2}\right)^{-1}\frac{dp}{dx}. \qquad (8.1.28)$$

$$\square$$

Summary (8.1)

1. It is necessary to identify homogeneous boundary conditions in all directions except the forcing direction to obtain a solution by separation of variables.

2. The orthogonal basis functions in the homogeneous directions in a Cartesian co-ordinate system are sine and cosine functions.

3. The spatial dependence in the homogeneous directions are expressed as an expansion in the basis functions, and the coefficients in the expansion are determined from the boundary conditions in the direction with inhomogeneous boundary conditions using the orthogonality relations.

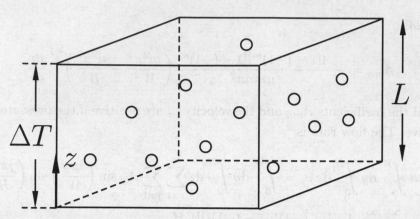

FIGURE 8.3. Effective conductivity of a composite material consisting of a matrix with conductivity k_m with spherical inclusions of radius R and conductivity k_p.

8.2 Spherical Co-ordinates

8.2.1 Effective Conductivity of a Composite

A composite material of thickness L consists of a matrix of thermal conductivity k_m containing randomly located particles of radius R and thermal conductivity k_p. A temperature difference ΔT is applied across the composite in the z direction, as shown in Fig. 8.3, and it is necessary to determine the average heat flux in the z direction, q_{av} as a function of the temperature gradient $T' = (\Delta T/L)$. It is assumed that the length L is large compared to the radius of the particles R, so that the material can be modelled as a continuum with an effective conductivity which is determined by the conduction of heat through the particles and the matrix,

$$q_{av} = -k_{eff}T'. \tag{8.2.1}$$

It is necessary to determine k_{eff} as a function of the conductivities of the particles and the matrix in the dilute limit where the temperature disturbance due to the presence of one particle does not affect the temperature field around another particle.

If there were no particles, the temperature gradient in the material would be uniform, and the conductivity would be equal to the conductivity of the matrix. The presence of the particles causes a disturbance to the temperature field around a particle, leading to a variation in the flux in the vicinity of the particle. The disturbance to the heat flux lines due to the presence of a particle in a matrix can be visualised as shown in Fig. 8.4. If the particle is more conducting than the matrix, $k_p > k_m$, the flux lines bend towards the particle as shown in Fig. 8.4(a). In the limiting case of a perfectly conducting particle, the heat flux lines are perpendicular to the surface of the particle. In contrast, if the particle is less conducting than

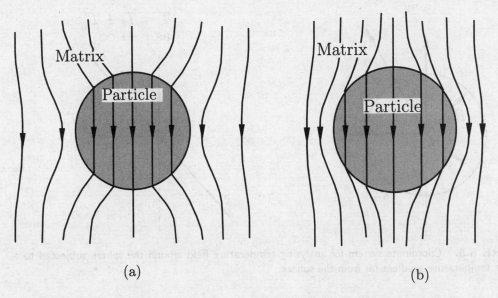

(a)   (b)

FIGURE 8.4. Schematic of the flux lines in the matrix and particles for $k_p > k_m$ (a) and $k_p < k_m$ (b).

the matrix, the flux lines bend around the particle, as shown in Fig. 8.4(b). In the limiting case of a perfectly insulating particle, there is no flux through the particle, and the flux lines are tangential to the particle surface. Thus, the contrast in the thermal conductivity causes a disturbance to the heat flux and the temperature field in the vicinity of the particle. Due to this, the effective conductivity of the composite is different from that of the matrix without particles.

Here, the 'dilute' limit is considered, where the volume fraction of particles is small, and the distance between particles is much larger than the particle radius. It is assumed that the temperature field around one particle is not influenced by the presence of other particles. Each particle is considered an isolated particle in a matrix of infinite extent subjected to a constant temperature gradient far from the particle. The effect of the disturbance to the heat flux due to all the particles is averaged, and divided by the applied temperature gradient $T' = (\Delta T / L)$, in order to determine the effective thermal conductivity.

The configuration for the single-particle problem, shown in Fig. 8.5, consists of a particle located at the origin of the spherical coordinate system. There is a uniform temperature gradient T' imposed in the z direction far from the particle. The temperature field far from the particle is $T_c + T'z$, where T_c is the temperature at $z = z_c$, and z_c is the location of the particle centre. It should be noted that the temperature T_c does not affect the disturbance to the flux around the particle, because the flux is proportional to the temperature gradient, and so it is sufficient

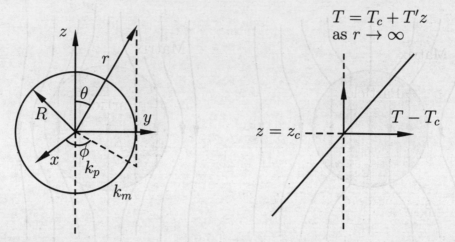

FIGURE 8.5. Coordinate system for analysing temperature field around the sphere subjected to a constant temperature gradient far from the sphere.

to consider the temperature field $T = T'z$ far from the particle. The boundary condition is,

$$T = T'z = T'r\cos(\theta) \text{ for } r \to \infty. \tag{8.2.2}$$

At steady state, the temperature fields in the particle T_p and in the matrix T_m satisfy the Laplace equation,

$$\nabla^2 T_p = 0, \quad \nabla^2 T_m = 0. \tag{8.2.3}$$

At the interface between the particle and the matrix, the temperature and the heat flux perpendicular to the surface are equal,

$$T_p = T_m, \tag{8.2.4}$$

$$-k_p \frac{\partial T_p}{\partial r} = -k_m \frac{\partial T_m}{\partial r} \text{ at } r = R. \tag{8.2.5}$$

It is evident the temperature field around the isolated particle, Fig. 8.5, is independent of the meridional angle ϕ. This is because the particle surface is spherically symmetric, and the boundary condition for the temperature field, Eq. 8.2.2, depends only on r and θ. Therefore, the temperature field depends only on (r, θ). Using the Laplacian, Eq. 7.2.20 in Chapter 7, the Laplace equation, Eq. 8.2.3, for the axisymmetric temperature field in the particle and the matrix is,

$$\frac{1}{r^2}\frac{\partial}{\partial r}\left(r^2\frac{\partial T}{\partial r}\right) + \frac{1}{r^2\sin(\theta)}\frac{\partial}{\partial\theta}\left(\sin(\theta)\frac{\partial T}{\partial\theta}\right) = 0. \tag{8.2.6}$$

The separation of variables procedure is used, where the temperature is expressed as $T = F(r)G(\theta)$ in Eq. 8.2.6. The resulting equation is multiplied by r^2 and divided by $F(r)G(\theta)$ to obtain,

$$\frac{1}{F(r)}\frac{d}{dr}\left(r^2\frac{dF}{dr}\right) + \frac{1}{G(\theta)\sin(\theta)}\frac{d}{d\theta}\left(\sin(\theta)\frac{dG}{d\theta}\right) = 0. \tag{8.2.7}$$

In the above equation, the first term on the left is a function of r and the second term is a function of θ; therefore, both terms are necessarily constants. Consider the second term on the left of Eq. 8.2.7,

$$\frac{1}{G(\theta)\sin(\theta)}\frac{d}{d\theta}\left(\sin(\theta)\frac{dG}{d\theta}\right) = -\beta_n, \tag{8.2.8}$$

where β_n, is a constant. Making the substitution $\xi = \cos(\theta)$ and $d\xi = -\sin(\theta)d\theta$, Eq. 8.2.8 can be simplified as,

$$\frac{d}{d\xi}\left((1-\xi^2)\frac{dG}{d\xi}\right) + \beta_n G = 0. \tag{8.2.9}$$

Here, the variable ξ varies in the range $-1 \le \xi \le 1$ corresponding to $0 \le \theta \le \pi$. Eq. 8.2.9 is called the Legendre equation, and the solutions of the equation are called the Legendre polynomials. It is shown in Appendix 8.A that the solutions of Eq. 8.2.9 are finite at $\xi = \pm 1$ if and only if the constant $\beta_n = n(n+1)$, where $n \ge 0$ is an integer. The solutions of the equation, denoted $P_n(\xi)$, are polynomials of order n, which are determined by the series expansion method in Appendix 8.A,

$$G(\theta) = P_n(\cos(\theta)). \tag{8.2.10}$$

The Legendre polynomials satisfy the orthogonality relation,

$$\int_{-1}^{1} d\xi\, P_n(\xi)P_k(\xi) = \int_0^\pi \sin(\theta)\, d\theta\, P_n(\cos(\theta))P_k(\cos(\theta)) = \frac{2\delta_{nk}}{2n+1}. \tag{8.2.11}$$

The integral in Eq. 8.2.11 is non-zero only for $n = k$, and zero otherwise.

The equation for $F(r)$ is obtained by substituting Eq. 8.2.8 and $\beta_n = n(n+1)$ in Eq. 8.2.7,

$$\frac{\mathrm{d}}{\mathrm{d}r}\left(r^2\frac{\mathrm{d}F}{\mathrm{d}r}\right) - n(n+1)F(r) = 0. \qquad (8.2.12)$$

This is solved to obtain,

$$F(r) = A_n r^n + \frac{B_n}{r^{n+1}}, \qquad (8.2.13)$$

where A_n and B_n are constants of integration. The general solution for the temperature field is the summation of FG for all possible values of n,

$$T = \sum_{n=0}^{\infty}\left(A_n r^n + \frac{B_n}{r^{n+1}}\right)P_n(\cos(\theta)). \qquad (8.2.14)$$

In the solution, Eq. 8.2.14, the terms proportional to r^n are called the growing harmonics, since these increase with r, while the terms proportional to $r^{-(n+1)}$ are called the decaying harmonics, since they decrease to zero for large r.

The series expansion in Eq. 8.2.14 is used to express the temperature field in the particles and matrix, since each satisfies the Laplace equation, Eq. 8.2.3,

$$T_p = \sum_{n=0}^{\infty}\left(A_n^p r^n + \frac{B_n^p}{r^{n+1}}\right)P_n(\cos(\theta)), \qquad (8.2.15)$$

$$T_m = \sum_{n=0}^{\infty}\left(A_n^m r^n + \frac{B_n^m}{r^{n+1}}\right)P_n(\cos(\theta)). \qquad (8.2.16)$$

The temperature field far from the particle is also expressed in terms of the Legendre polynomial solutions. The temperature field is $T = T'z = T'r\cos(\theta)$. It is shown that the function $\cos(\theta) = P_1(\cos(\theta))$ in Appendix 8.A, and therefore the boundary condition, Eq. 8.2.2 is,

$$T = T'rP_1(\cos(\theta)) \text{ for } r \to \infty. \qquad (8.2.17)$$

It should be noted that in the series expansion for the boundary condition in Eq. 8.2.17, there is only one coefficient for $n = 1$ that is non-zero, while all other coefficients are zero.

When Eqs. 8.2.15 and 8.2.16 are substituted into the boundary conditions, Eqs. 8.2.4 and 8.2.5, we obtain two relations containing the coefficients A_n^p, A_n^m, B_n^p and B_n^m at $r = R$ for each value of n. If the boundary condition for $r \to \infty$

(Eq. 8.2.17) is expanded in a series in Legendre polynomials, the coefficient is zero for all Legendre polynomials with $n \neq 1$. All the boundary conditions, Eqs. 8.2.4, 8.2.5 and 8.2.17, are satisfied for

$$A_n^p = B_n^p = A_n^m = B_n^m = 0 \text{ for } n \neq 1. \tag{8.2.18}$$

Since this is a linear problem with well-posed boundary conditions, it is guaranteed that the solution, Eq. 8.2.18, is unique. Therefore, the only possible non-zero coefficients in the expansion, Eqs. 8.2.15 and 8.2.16, are $A_1^p, B_1^p, A_1^m, B_1^m$. The solutions are,

$$T_p = A_1^p r P_1(\cos(\theta)) + \frac{B_1^p P_1(\cos(\theta))}{r^2}, \tag{8.2.19}$$

$$T_m = A_1^m r P_1(\cos(\theta)) + \frac{B_1^m P_1(\cos(\theta))}{r^2}. \tag{8.2.20}$$

The constant B_1^p is zero due to the requirement that the temperature in the particle T_p is finite at $r = 0$, and $A_1^m = T'$ in order to satisfy Eq. 8.2.17, the boundary condition in the limit $r \to \infty$. The constants A_1^p and B_1^m are determined from the boundary conditions, Eqs. 8.2.4 and 8.2.5 at the surface of the sphere $r = R$,

$$A_1^p R \cos(\theta) = T' R \cos(\theta) + \frac{B_1^m \cos(\theta)}{R^2}, \tag{8.2.21}$$

$$- k_p A_1^p \cos(\theta) = - k_m \left(T' \cos(\theta) - \frac{2 B_1^m \cos(\theta)}{R^3} \right). \tag{8.2.22}$$

The two simultaneous equations, Eqs. 8.2.21 and 8.2.22, are solved to obtain,

$$A_1^p = \frac{3 T'}{2 + k_R}, \tag{8.2.23}$$

$$B_1^m = \frac{(1 - k_R) R^3 T'}{2 + k_R}, \tag{8.2.24}$$

where $k_R = (k_p / k_m)$ is the ratio of the particle and matrix conductivities. The temperature fields in the particle and the matrix are,

$$T_p = \frac{3 T' r P_1(\cos(\theta))}{2 + k_R} = \frac{3 T' z}{2 + k_R}, \tag{8.2.25}$$

$$T_m = T' r P_1(\cos(\theta)) + \frac{(1 - k_R) R^3 T' P_1(\cos(\theta))}{(2 + k_R) r^2}. \tag{8.2.26}$$

Note that the temperature in the particle T_p is a linear function of $z = r \cos(\theta)$. Due to this, the flux lines in the particle are straight lines aligned in the $-z$ direction, as shown in Fig. 8.4.

The average heat flux is calculated by taking an average over the entire volume of the composite consisting of particles and matrix,

$$q_{av} = \frac{1}{V} \int dV\, q_z = -\frac{1}{V} \left(\int_{\text{matrix}} dV\, k_m (\nabla T) \cdot e_z + \int_{\text{particles}} dV\, k_p (\nabla T) \cdot e_z \right)$$

$$= -\frac{1}{V} \left(\underline{\int_{\text{matrix}} dV\, k_m (\nabla T) \cdot e_z + \int_{\text{particles}} dV\, k_m (\nabla T) \cdot e_z} \right.$$

$$\left. + \int_{\text{particles}} dV\, (k_p - k_m)(\nabla T) \cdot e_z \right)$$

$$= -\frac{1}{V} \left(\int_{\text{composite}} dV\, k_m (\nabla T) \cdot e_z + \int_{\text{particles}} dV\, (k_p - k_m)(\nabla T) \cdot e_z \right)$$

$$= -\frac{1}{V} \left(\int_{\text{composite}} dV\, k_m (\nabla T) \cdot e_z + N \int_{\text{1 particle}} dV\, (k_p - k_m)(\nabla T) \cdot e_z \right).$$

$$(8.2.27)$$

The integral of $k_p (\nabla T) \cdot e_z$ over the particles in the first line is expressed as the sum of two terms in the second line, the the integral of $k_m (\nabla T) \cdot e_z + (k_p - k_m)(\nabla T) \cdot e_z$. The two underlined terms in the second line on the right are added to obtain the first term on the right in the third line, which is the integral over the entire composite of $k_m (\nabla T) \cdot e_z$. This formulation clearly separates the excess particle contribution to the flux, the second term on the right, which is zero if the particle and matrix conductivities are equal. In the fourth line in Eq. 8.2.27, the sum of the integrals over over all particles is expressed as the number of particles N times the integral over one particle. This is valid in the dilute non-interacting limit, where the temperature field around one particle is not affected by the disturbance caused by other particles.

The first term on the right in the third line of Eq. 8.2.27 is the k_m times the volume average of the temperature gradient, which is necessarily equal to the imposed temperature gradient $T' e_z$, where $T' = (\Delta T / L)$. Since the temperature in the particle (Eq. 8.2.25) is a linear function of z, the temperature gradient in the particle is $(3T'/(2 + k_R))e_z$. Therefore, the average flux is,

$$q_{av} = -k_m T' - \frac{(k_p - k_m) N V_p}{V} \frac{3T'}{2 + k_R}$$

$$= -k_m T' \left(1 + \frac{3(k_R - 1)\Phi}{2 + k_R} \right). \qquad (8.2.28)$$

Here, V_p is the volume of one particle, and $\Phi = (N V_p / V)$ is the volume fraction of the particles.

Comparing Eqs. 8.2.28 and 8.2.1, we obtain the expression for the effective conductivity,

$$k_{eff} = k_m \left(1 + \frac{3(k_R - 1)\Phi}{2 + k_R} \right). \qquad (8.2.29)$$

As expected, the particle contribution to the effective conductivity is positive for $k_R > 1$ when the particle conductivity is greater than the matrix conductivity, and negative for $k_R < 1$ when the particle conductivity is less than the matrix conductivity. The particle contribution is independent of the particle radius, and is proportional to the volume fraction Φ of the particles in the dilute limit where the interaction between particles are neglected. This result is valid only when the particle volume fraction is small. As the particle volume fraction increases, it is necessary to incorporate interactions between particles. The effect of interactions is beyond the scope of this text.

8.2.2 Non-axisymmetric Temperature Fields

For a general non-axisymmetric temperature field, the dependence on the meridional angle ϕ cannot be neglected. Using Eq. 7.2.22 in Chapter 7 for the Laplacian, the Laplace equation for the temperature field is,

$$\frac{1}{r^2}\frac{\partial}{\partial r}\left(r^2\frac{\partial T}{\partial r}\right) + \frac{1}{r^2\sin(\theta)}\frac{\partial}{\partial \theta}\left(\sin(\theta)\frac{\partial T}{\partial \theta}\right) + \frac{1}{r^2\sin(\theta)^2}\frac{\partial^2 T}{\partial \phi^2} = 0. \qquad (8.2.30)$$

In the separation of variables procedure, the expression

$$T = F(r)G(\theta)H(\phi), \qquad (8.2.31)$$

for the temperature is substituted into Eq. 8.2.30, the resulting equation is divided by $F(r)G(\theta)H(\phi)$ and multiplied by $r^2 \sin(\theta)^2$, to obtain,

$$\sin(\theta)^2\left[\frac{1}{F}\frac{d}{dr}\left(r^2\frac{dF}{dr}\right) + \frac{1}{G}\frac{1}{\sin(\theta)}\frac{d}{d\theta}\left(\sin(\theta)\frac{dG}{d\theta}\right)\right] + \frac{1}{H}\frac{d^2H}{d\phi^2} = 0. \qquad (8.2.32)$$

Here, the first term on the left is a function of (r, θ), and the second term is only a function of ϕ. Therefore, the first and second terms are constants.

The equation for $H(\phi)$ is,

$$\frac{d^2 H}{d\phi^2} = \beta_m H, \qquad (8.2.33)$$

where β_m is a constant. The value of the constant β_m is fixed from the 'periodicity' condition in the ϕ direction shown in Fig. 8.6. Consider a point with co-ordinates

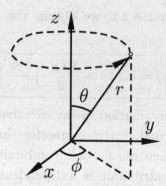

FIGURE 8.6. Periodicity in a spherical co-ordinate system. When the position vector is rotated around the z axis by an angle 2π, we return to the same physical location.

r, θ, ϕ. If the angle ϕ is increased by 2π, we go once around the z axis and return to the same physical point in space. Since the temperature field is a single valued function of position, the value of H has to be the same when we increase or decrease ϕ by $2p\pi$ where p is an integer. Therefore, the solution H of Eq. 8.2.33 is periodic in the ϕ co-ordinate with periodicity 2π. If β_m is positive, the solutions are exponential functions of ϕ which are not periodic in ϕ. If β_m is negative, the solutions of Eq. 8.2.33 are periodic with a period of 2π only if $\beta_m = -m^2$, where m is an integer. Thus, the periodicity condition fixes the value $\beta_m = -m^2$ in Eq. 8.2.33,

$$\frac{\mathrm{d}^2 H_m}{\mathrm{d}\phi^2} = -m^2 H_m, \tag{8.2.34}$$

and the solutions are,

$$H_m(\phi) = C_m \sin(m\phi) + D_m \cos(m\phi). \tag{8.2.35}$$

Here, m is a positive integer, because the cosine and sine solutions for $-m$ are ± 1 times those for $+m$, respectively. Conventionally, Eq. 8.2.35 is written in complex form,

$$H_m(\phi) = A_m \mathrm{e}^{(\imath m\phi)}. \tag{8.2.36}$$

Here, $\imath = \sqrt{-1}$, and m could be a positive or negative integer. Eq. 8.2.36 is entirely equivalent to Eq. 8.2.35, because the solutions in Eq. 8.2.35 can be written as linear combinations of H_m and H_{-m} in Eq. 8.2.36 using the substitution $\cos(m\phi) = \frac{1}{2}(\mathrm{e}^{(\imath m\phi)} + \mathrm{e}^{(-\imath m\phi)})$ and $\sin(m\phi) = \frac{1}{2\imath}(\mathrm{e}^{(\imath m\phi)} - \mathrm{e}^{(-\imath m\phi)})$. For the complex function H_m, the orthogonality relation is

$$\int_0^{2\pi} \mathrm{d}\phi \, H_m(\phi) H_l^*(\phi) = 2\pi \delta_{lm}, \tag{8.2.37}$$

where $H_l^* = H_{-l} = \mathrm{e}^{(-\imath l\phi)}$ is the complex conjugate of H_l.

Substituting $-m^2$ for the second term on the left in Eq. 8.2.32, and dividing the resulting equation by $\sin(\theta)^2$, we obtain

$$\frac{1}{F}\frac{\mathrm{d}}{\mathrm{d}r}\left(r^2\frac{\mathrm{d}F}{\mathrm{d}r}\right) + \left(\frac{1}{G\sin(\theta)}\frac{\mathrm{d}}{\mathrm{d}\theta}\left(\sin(\theta)\frac{\mathrm{d}G}{\mathrm{d}\theta}\right) - \frac{m^2}{\sin(\theta)^2}\right) = 0. \qquad (8.2.38)$$

Here, the first term on the left is a function of r and the second is a function of θ, and therefore, both are constants. The equation for $G(\theta)$, equivalent to Eq. 8.2.8 for an axisymmetric temperature field, is

$$\frac{1}{G\sin(\theta)}\frac{\mathrm{d}}{\mathrm{d}\theta}\left(\sin(\theta)\frac{\mathrm{d}G}{\mathrm{d}\theta}\right) - \frac{m^2}{\sin(\theta)^2} = -n(n+1). \qquad (8.2.39)$$

The condition for the existence of finite solutions of Eq. 8.2.39 at $\theta = 0, \pi$ is that m is an integer, $n \geq 0$ is an integer, and $|m| \leq n$. The requirement that n is an integer for the axisymmetric case is shown in Appendix 8.A. The solutions of Eq. 8.2.39 are called, associated Legendre polynomials of degree n and order m,

$$G(\theta) = P_n^m(\cos(\theta)). \qquad (8.2.40)$$

The orthogonality conditions for the associated Legendre polynomials, and the expression for the first few associated Legendre polynomials, are provided inAppendix 8.A. The orthogonality relation is

$$\int_{-1}^{1} \mathrm{d}x\, P_l^m(x)P_n^m(x) = \frac{2\delta_{ln}}{2l+1}\frac{(l+m)!}{(l-m)!}. \qquad (8.2.41)$$

Finally, the function $F(r)$ can be determined by substituting $-n(n+1)$ for the second term on the left side in Eq. 8.2.38, and multiplying the resulting equation by F,

$$\frac{\mathrm{d}}{\mathrm{d}r}\left(r^2\frac{\mathrm{d}F}{\mathrm{d}r}\right)\frac{\mathrm{d}^2F}{\mathrm{d}r^2} - n(n+1)F = 0. \qquad (8.2.42)$$

The solution of Eq. 8.2.42 is

$$F(r) = A_{nm}r^n + \frac{B_{nm}}{r^{n+1}}. \qquad (8.2.43)$$

Combining Eqs. 8.2.36, 8.2.40 and 8.2.43, the solution for the temperature field is of the form,

$$T = \left(A_{nm}r^n + \frac{B_{nm}}{r^{n+1}}\right) P_n^m(\cos(\theta))e^{(im\phi)}. \qquad (8.2.44)$$

It is convenient to express these solutions in the form of 'spherical harmonic' functions,

$$Y_n^m(\theta, \phi) = \sqrt{\frac{(2n+1)}{4\pi} \frac{(n-m)!}{(n+m)!}} P_n^m(\cos(\theta)) e^{(\imath m \phi)}. \qquad (8.2.45)$$

It should be noted that the spherical harmonic expansions are defined differently in different fields; for example, a factor $(-1)^m$ is included in the right side of Eq. 8.2.45 in quantum mechanics. Eqs. 8.2.37 and 8.2.41 are used to obtain the orthogonality relation for $Y_n^m(\theta, \phi)$,

$$\int_0^{2\pi} \mathrm{d}\phi \int_0^{\pi} \sin(\theta) \, \mathrm{d}\theta \, Y_n^m(\theta, \phi) Y_{n'}^{m'*}(\theta, \phi) = \int_0^{2\pi} \mathrm{d}\phi \int_0^{\pi} \sin(\theta) \mathrm{d}\theta \, Y_n^m(\theta, \phi) Y_{n'}^{-m'}(\theta, \phi)$$

$$= \delta_{nn'} \delta_{mm'}, \qquad (8.2.46)$$

where $Y_{n'}^{m'*}(\theta, \phi) = Y_{n'}^{-m'}(\theta, \phi)$ is the complex conjugate of $Y_{n'}^{m'}(\theta, \phi)$. In Eq. 8.2.45, the prefactor of $P_n^m(\cos(\theta)) e^{(\imath m \phi)}$ is chosen such that $Y_n^m(\theta, \phi)$ is orthonormal—that is, the scalar product is 1 for $n = n'$ and $m = m'$. The spherical harmonics for $n = 0, 1$ and 2 are provided in Table 8.1.

The solution, Eq. 8.2.44, for the temperature field is equivalently written with a redefinition of the coefficients A_{nm} and B_{nm} as,

$$T = \sum_{n=0}^{\infty} \sum_{m=-n}^{n} \left(A_{nm} r^n + \frac{B_{nm}}{r^{n+1}} \right) Y_n^m(\cos(\theta)). \qquad (8.2.47)$$

The integer n is positive, and for each value of n, the integer m varies from $-n$ to n. Therefore, there are $2n+1$ values of m for each n. For each value of n, the radial dependence of the temperature consists of two parts, the 'growing harmonic' which increases proportional to r^n for $r \to \infty$, and the 'decaying harmonic' which decreases proportional to $r^{-(n+1)}$ for $r \to \infty$. The coefficients A_{nm} and B_{nm} are determined from the boundary conditions for the specific problem under consideration. If we are solving the conduction equation within a finite domain, the origin of the co-ordinate system is within the domain. Since the decaying harmonics diverge as $r \to 0$, the coefficients of the decaying harmonics are set equal to zero. This was the reason B_1^p was set equal to zero in Eq. 8.2.19 for a spherical inclusion in a matrix. For the temperature field outside an object, the origin is usually placed within the object,

TABLE 8.1. The terms in the spherical harmonic expansion up to $n = 2$ in the spherical co-ordinate system (second column) and in the Cartesian co-ordinate system (third column). The third column is obtained using Eq. 7.2.1 in Chapter 7.

Y_0^0	$\sqrt{\dfrac{1}{4\pi}}$	$\sqrt{\dfrac{1}{4\pi}}$
Y_1^{-1}	$\sqrt{\dfrac{3}{8\pi}}\sin(\theta)e^{(-\imath\phi)}$	$\sqrt{\dfrac{3}{8\pi}}\dfrac{x-\imath y}{r}$
Y_1^0	$\sqrt{\dfrac{3}{4\pi}}\cos(\theta)$	$\sqrt{\dfrac{3}{4\pi}}\dfrac{z}{r}$
Y_1^1	$\sqrt{\dfrac{3}{8\pi}}\sin(\theta)e^{(\imath\phi)}$	$\sqrt{\dfrac{3}{8\pi}}\dfrac{x+\imath y}{r}$
Y_2^{-2}	$\sqrt{\dfrac{15}{32\pi}}\sin(\theta)^2 e^{(-2\imath\phi)}$	$\sqrt{\dfrac{15}{32\pi}}\dfrac{(x-\imath y)^2}{r^2}$
Y_2^{-1}	$\sqrt{\dfrac{15}{8\pi}}\sin(\theta)\cos(\theta)e^{(-\imath\phi)}$	$\sqrt{\dfrac{15}{8\pi}}\dfrac{(x-\imath y)z}{r^2}$
Y_2^0	$\sqrt{\dfrac{5}{16\pi}}(3\cos(\theta)^2-1)$	$\sqrt{\dfrac{5}{16\pi}}\dfrac{(3z^2-r^2)}{r^2}$
Y_2^1	$\sqrt{\dfrac{15}{8\pi}}\sin(\theta)\cos(\theta)e^{(\imath\phi)}$	$\sqrt{\dfrac{15}{8\pi}}\dfrac{(x+\imath y)z}{r^2}$
Y_2^2	$\sqrt{\dfrac{15}{32\pi}}\sin(\theta)^2 e^{(2\imath\phi)}$	$\sqrt{\dfrac{15}{32\pi}}\dfrac{(x+\imath y)^2}{r^2}$

and is therefore outside the fluid domain. Here, the coefficients of the decaying harmonics are non-zero, since these decrease to zero far from the object. The coefficients of the growing harmonics are non-zero only if they match an imposed temperature field at $r \to \infty$ which increases with r. This is the reason for setting $A_1^m = T'$ in Eq. 8.2.20 for a spherical particle with a linear temperature gradient far from the particle.

Summary (8.2)

1. The general solution of the Laplace equation in spherical co-ordinates, Eq. 8.2.30, is the spherical harmonic expansion, Eq. 8.2.47, and the basis functions $Y_n^m(\theta, \phi)$ satisfy the orthogonality relations, Eq. 8.2.46.

2. There are two types of solutions, growing harmonics which increase proportional to r^n for $r \to \infty$, and decaying harmonics which decrease proportional to $r^{-(n+1)}$ for $r \to \infty$.

3. For each value of n, there are $2n + 1$ terms in the expansion for $-n \leq m \leq n$.

4. To determine the field variable, the boundary conditions are expressed as the sum of the spherical harmonics, and the equation and boundary conditions for each harmonic are solved separately in order to determine the coefficients in the expansion.

5. The effective conductivity of a composite containing spherical particles of thermal conductivity k_p in a matrix of thermal conductivity k_m is Eq. 8.2.29 in the dilute limit, when the temperature disturbance due to one particle does not affect the temperature around another particle.

8.3   Multipole Expansions

8.3.1   Point Source

In Section 5.2.2 (Chapter 5), it was shown that the temperature field due to a sphere of radius R generating heat energy Q per unit time is,

$$T - T_\infty = \frac{Q}{4\pi k r}. \tag{8.3.1}$$

The solution for the temperature, Eq. 8.3.1, does not depend on the radius of the sphere, but only on the total amount of heat generated per unit time from the sphere. Therefore, the result is applicable even when the sphere is idealised as a 'point source' of zero radius, provided the total heat generated per unit time is Q. The solution, Eq. 8.3.1, is the $n = 0, m = 0$ term in the spherical harmonic expansion, Eq. 8.2.47, which is spherically symmetric (independent of θ and ϕ). Thus, the first term in the spherical harmonic expansion corresponds to a point source.

The temperature field due to a point source, Eq. 8.3.1, can be formally derived when the source is a Dirac delta function. The three-dimensional Dirac delta function, defined in Section 4.5 (Chapter 4), has the following properties,

$$\delta(\boldsymbol{x}) = 0 \text{ for } \boldsymbol{x} \neq 0, \tag{8.3.2}$$

$$\int_V dV\, \delta(\boldsymbol{x}) = 1, \tag{8.3.3}$$

and

$$\int_V dV\, \delta(\boldsymbol{x})\, g(\boldsymbol{x}) = g(0), \tag{8.3.4}$$

where $\boldsymbol{x} = x\boldsymbol{e}_x + y\boldsymbol{e}_y + z\boldsymbol{e}_z$ is the position vector, V is any volume that encompasses the origin $\boldsymbol{x} = 0$ and $g(\boldsymbol{x})$ is a function of $\boldsymbol{x}$. The conduction equation for the temperature field due to a point source generating energy per unit time Q is,

$$k\boldsymbol{\nabla} \cdot \boldsymbol{\nabla} T + Q\delta(\boldsymbol{x}) = 0. \tag{8.3.5}$$

When Eq. 8.3.5 is integrated over a volume V that includes the origin, we obtain,

$$k\int_V dV\, \boldsymbol{\nabla} \cdot \boldsymbol{\nabla} T = -\int_V dV\, \delta(\boldsymbol{x})Q = -Q. \tag{8.3.6}$$

The final step above follows from the property, Eq. 8.3.3, of the delta function. The divergence theorem (Appendix 7.B) is used to express the term on the left of Eq. 8.3.6 as a surface integral,

$$k\int_S dS\, \boldsymbol{n} \cdot \boldsymbol{\nabla} T = -Q, \tag{8.3.7}$$

where S is the surface of volume V, and $\boldsymbol{n}$ is the outward unit normal to the surface S. Consider a point source at the origin of a spherical co-ordinate system within a spherical surface of radius r with outward unit normal $\boldsymbol{n} = \boldsymbol{e}_r$ along the radial direction, as shown in Fig. 8.7(a). The term on the left of Eq. 8.3.7 is $4\pi r^2 k(\partial T/\partial r)$, and Eq. 8.3.7 is solved to obtain,

$$\frac{\partial T}{\partial r} = -\frac{Q}{4\pi k r^2}. \tag{8.3.8}$$

This equation is integrated subject to the boundary condition $T = T_\infty$ for $r \to \infty$, to obtain the solution, Eq. 8.3.1.

If the point source generating energy per unit time Q is at a location $\boldsymbol{x}_1$ instead of the origin, then the temperature field at the point $\boldsymbol{x}$ is, instead of Eq. 8.3.1,

$$T(\boldsymbol{x}) = \frac{Q}{4\pi k|\boldsymbol{x} - \boldsymbol{x}_1|} + T_\infty, \tag{8.3.9}$$

where $|\boldsymbol{x} - \boldsymbol{x}_1| = \sqrt{(x - x_1^2) + (y - y_1^2) + (z - z_1)^2}$ is the distance between the observation point $\boldsymbol{x}$ and the source point $\boldsymbol{x}_1$. The heat flux at the point $\boldsymbol{x}$ is

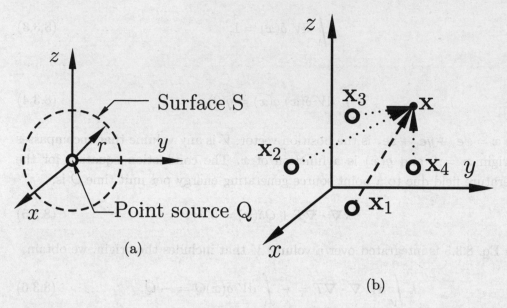

FIGURE 8.7. The temperature (a) due to a point source at the origin generating Q energy per unit time, and (b) due to multiple point sources generating $Q_1, Q_2, \ldots$ energy per unit time located at $x_1, x_2, \ldots$.

proportional to the gradient of the temperature,

$$
\begin{aligned}
\boldsymbol{q} &= -k\boldsymbol{\nabla}T = -k\left(\boldsymbol{e}_x\frac{\partial}{\partial x} + \boldsymbol{e}_y\frac{\partial}{\partial y} + \boldsymbol{e}_z\frac{\partial}{\partial z}\right)T \\
&= -\frac{Q}{4\pi}\left(\frac{-\boldsymbol{e}_x(x-x_1) - \boldsymbol{e}_y(y-y_1) - \boldsymbol{e}_z(z-z_1)}{((x-x_1)^2 + (y-y_1)^2 + (z-z_1)^2)^{3/2}}\right) \\
&= \frac{Q(\boldsymbol{x}-\boldsymbol{x}_1)}{4\pi|\boldsymbol{x}-\boldsymbol{x}_1|^3}.
\end{aligned}
\tag{8.3.10}
$$

If there are multiple point sources generating energy per unit time $Q_1, Q_2, \ldots, Q_N$ located at $\boldsymbol{x}_1, \boldsymbol{x}_2, \ldots, \boldsymbol{x}_N$, as shown in Fig. 8.7(b), the temperature at the observation point $\boldsymbol{x}$ is the summation of the temperatures due to the individual point sources,

$$
T(\boldsymbol{x}) = \sum_{i=1}^{N} \frac{Q_i}{4\pi k|\boldsymbol{x}-\boldsymbol{x}_i|} + T_\infty,
\tag{8.3.11}
$$

$$
\boldsymbol{q} = \sum_{i=1}^{N} \frac{Q_i(\boldsymbol{x}-\boldsymbol{x}_i)}{4\pi|\boldsymbol{x}-\boldsymbol{x}_i|^3}.
\tag{8.3.12}
$$

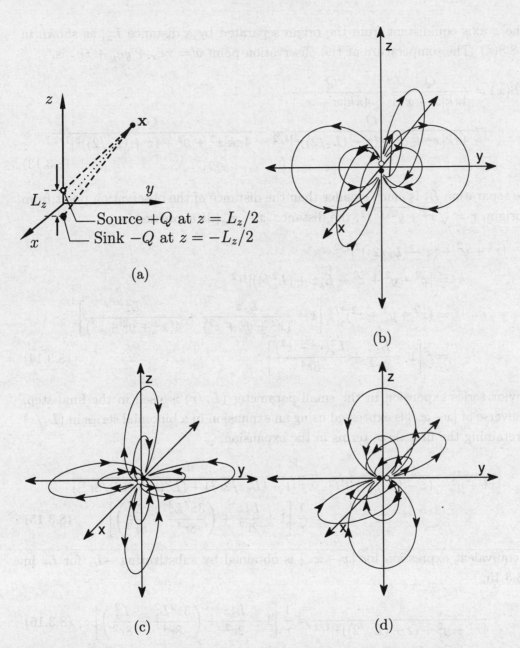

FIGURE 8.8. The configuration (a) for analysing the temperature field at the point $\boldsymbol{x}$ due to a point dipole, which is a combination of a source (open circle) and sink (filled circle) of equal magnitude Q separated by a distance $L \ll |\boldsymbol{x}|$ along the z axis; the flux lines due to a point dipole in the z direction (b), x direction (c) and y direction (d). The open circles are the sources and the filled circles are the sinks.

8.3.2 Point Dipole

Consider a source generating energy per unit time Q at the location $\boldsymbol{x}_+ = (L_z/2)\boldsymbol{e}_z$ and a sink absorbing energy per unit time Q at the point $\boldsymbol{x}_- = -(L_z/2)\boldsymbol{e}_z$, both on the z axis equidistant from the origin separated by a distance L_z, as shown in Fig. 8.8(a). The temperature at the observation point $\boldsymbol{x} = x\boldsymbol{e}_x + y\boldsymbol{e}_y + z\boldsymbol{e}_z$ is,

$$T(\boldsymbol{x}) = \frac{Q}{4\pi k|\boldsymbol{x} - \boldsymbol{x}_+|} - \frac{Q}{4\pi k|\boldsymbol{x} - \boldsymbol{x}_-|}$$
$$= \frac{Q}{4\pi k[x^2 + y^2 + (z - (L_z/2))^2]^{1/2}} - \frac{Q}{4\pi k[x^2 + y^2 + (z + (L_z/2))^2]^{1/2}}.$$
(8.3.13)

If the separation L_z is much smaller than the distance of the observation point from the origin, $r = \sqrt{x^2 + y^2 + z^2}$, the distance $|\boldsymbol{x} - \boldsymbol{x}_+|$ is expressed as,

$$[x^2 + y^2 + (z - (L_z/2))^2]^{1/2}$$
$$= [x^2 + y^2 + z^2 - L_z z + (L_z^2/4)]^{1/2}$$
$$= (x^2 + y^2 + z^2)^{1/2}\left[1 - \frac{L_z z}{(x^2 + y^2 + z^2)} + \frac{L_z^2}{4(x^2 + y^2 + z^2)}\right]^{1/2}$$
$$= r\left[1 - \frac{L_z z}{2r^2} + \frac{L_z^2(r^2 - z^2)}{8r^4}\right].$$
(8.3.14)

A Taylor series expansion in the small parameter (L_z/r) is used in the final step. The inverse of $|\boldsymbol{x} - \boldsymbol{x}_+|$ is expressed using an expansion in a binomial series in (L_z/r) and retaining the first three terms in the expansion,

$$\frac{1}{[x^2 + y^2 + (z - (L_z/2))^2]^{1/2}} = \frac{1}{r[1 - (L_z z/2r^2) + (L_z^2(r^2 - z^2)/8r^4)]}$$
$$= \frac{1}{r}\left[1 + \frac{L_z z}{2r^2} + \left(\frac{3z^2 L_z^2}{8r^4} - \frac{L_z^2}{8r^2}\right)\right].$$
(8.3.15)

The equivalent expression for $|\boldsymbol{x} - \boldsymbol{x}_-|$ is obtained by substituting $-L_z$ for L_z in Eq. 8.3.15,

$$\frac{1}{[x^2 + y^2 + (z + (L_z/2))^2]^{1/2}} = \frac{1}{r}\left[1 - \frac{L_z z}{2r^2} + \left(\frac{3z^2 L_z^2}{8r^4} - \frac{L_z^2}{8r^2}\right)\right].$$
(8.3.16)

These are substituted into Eq. 8.3.13 to obtain the expression for the temperature,

$$T(\boldsymbol{x}) = \frac{QL_z z}{4\pi kr^3} = \frac{QL_z \cos(\theta)}{4\pi kr^2}.$$
(8.3.17)

Here, the co-ordinate transformation, Eq. 7.2.1 in Chapter 7, is used to express $z = r\cos(\theta)$. Eq. 8.3.17 is the temperature field due to a point dipole aligned along the z axis.

To summarise, a point dipole aligned in the z direction is the combination of a source and sink emitting energy per unit volume $+Q$ and $-Q$ separated by an infinitesimal distance L_z, such that the 'dipole strength' QL_z is finite. The flux lines due to a point dipole, shown schematically in Fig. 8.8(b), emerge from the source and are absorbed into the sink. Eq. 8.3.17 shows that the temperature field due to the point dipole decreases proportional to r^{-2} for $r \to \infty$. This is a faster than the r^{-1} decay of the temperature field due to a point source, Eq. 8.3.1. Since the source and sink are of equal strength, there is no net heat generated by a point dipole, and the temperature field due to the point dipole does not contain a contribution proportional to r^{-1}.

Alignment along the x and y directions, shown in Fig. 8.8(c) and (d), respectively, are two other orthogonal alignments possible for the point dipole. The temperature field due to alignment along the x and y directions are, respectively,

$$T(x) = \frac{QL_x x}{4\pi k r^3} = \frac{QL_x \sin(\theta) \cos(\phi)}{4\pi k r^2}, \tag{8.3.18}$$

$$T(x) = \frac{QL_y y}{4\pi k r^3} = \frac{QL_y \sin(\theta) \sin(\phi)}{4\pi k r^2}. \tag{8.3.19}$$

Here, the transformation, Eq. 7.2.1 in Chapter 7, is used to relate the Cartesian and spherical co-ordinates. The temperature field due to a point dipole separated by distances (L_x, L_y, L_z) in the (x, y, z) directions is

$$\boxed{T(x) = \frac{QL_x x}{4\pi k r^3} + \frac{QL_y y}{4\pi k r^3} + \frac{QL_z z}{4\pi k r^3} = \frac{QL}{4\pi k} \cdot \left(\frac{x}{r^3}\right),} \tag{8.3.20}$$

where, $QL = Q(L_x e_x + L_y e_y + L_z e_z)$ is the vector dipole moment.

The point dipoles along the three co-ordinate directions are related to the three terms in the spherical harmonic expansion, Eq. 8.2.47 for $n = 1$ and $m = -1, 0, 1$. From Table 8.1, Eqs. 8.3.17–8.3.19 can be expressed in terms of the spherical harmonics,

$$\frac{QL_x x}{4\pi k r^3} = \frac{QL_x}{4\pi k} \sqrt{\frac{8\pi}{3}} \left(\frac{Y_1^1 + Y_1^{-1}}{2r^2}\right), \tag{8.3.21}$$

$$\frac{QL_y y}{4\pi k r^3} = \frac{QL_y}{4\pi k} \sqrt{\frac{8\pi}{3}} \left(\frac{Y_1^1 - Y_1^{-1}}{2ir^2}\right), \tag{8.3.22}$$

$$\frac{QL_z z}{4\pi k r^3} = \frac{QL_z}{4\pi k} \sqrt{\frac{4\pi}{3}} \left(\frac{Y_1^0}{r^2}\right). \tag{8.3.23}$$

Thus, the point dipole aligned in the z direction is the decaying harmonic solution for $n = 1, m = 0$, and the point dipoles in the x and y directions are linear combinations of the decaying harmonic solutions for $n = 1$ and $m = -1, 1$. The three decaying harmonics for $n = 1$ in Eq. 8.2.47 are generated by point dipoles aligned along the three co-ordinate directions.

There are also three growing harmonics for $n = 1$ in the spherical harmonic expansion, Eq. 8.2.47. These are of the form $rY_1^m(\theta, \phi)$, and from Table 8.1, it is easily seen that the three growing harmonics are just linear combinations of the co-ordinates x, y and z. Therefore, the most general solution for the temperature field due to the growing harmonics is

$$T(\boldsymbol{x}) = T_x'x + T_y'y + T_z'z,$$
$$= \boldsymbol{T}' \cdot \boldsymbol{x}, \tag{8.3.24}$$

where $\boldsymbol{T}'$ is the temperature gradient. These are generated by planar sources and sinks separated along the x, y and z co-ordinates, as shown in Fig. 8.9. When planar sources and sinks of infinite extent separated by a large distance, the field over a region much smaller than their separation is a growing harmonic. These are the inverse of the decaying harmonics which are generated by point sources and sinks of infinitesimal separation.

8.3.3   Point Quadrupole

A point quadrupole consists of two sources and two sinks of equal strength, arranged so that the net source is zero and the net dipole is zero. The five arrangements, corresponding to the $n = 2$ decaying harmonics proportional to r^{-3}, are shown in Fig. 8.10. Here, all sources emit energy per unit time Q, and all sinks absorb energy per unit time Q, with the exception of the sink at the origin for Φ_{z^2} which absorbs energy per unit time $2Q$. Pairs of sources are separated by distance L in all cases. The quadrupolar fields are determined by adding the fields due to individual sources, and using an expansion in the parameter (L/r) in the limit $L \ll r$—that is, the distance from the origin is much larger than the separation L. The procedure is similar to that adopted for the point dipole in Section 8.3.2, with the difference that we consider finite QL^2 (in contrast to finite QL for a point dipole). The source/sink locations and the quadrupolar fields in Cartesian and spherical co-ordinates are given in Table 8.2. The Φ_{z^2} quadrupole consists of two sources and one sink distributed along the z axis, and the resulting quadrupole field is axisymmetric. The quadrupole fields Φ_{xz} and Φ_{yz} and Φ_{xy} have sources and sinks along the diagonals in the $x - z$, $y - z$ and $x - y$ planes, respectively, while the field $\Phi_{x^2-y^2}$ has sources and sinks distributed along the x and y axes.

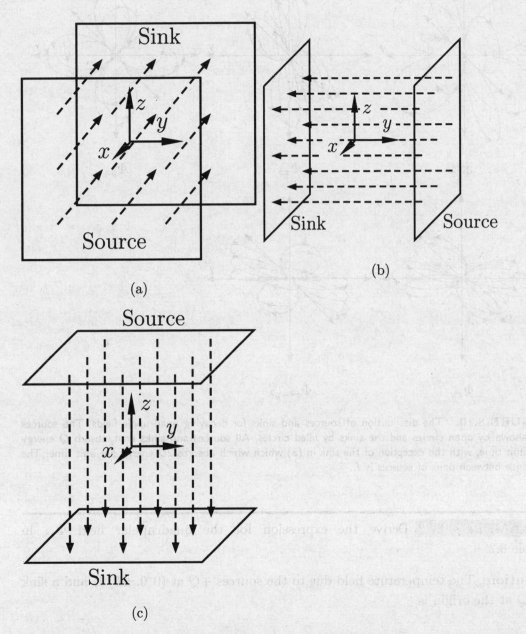

(a)

(b)

(c)

FIGURE 8.9. The infinite planar source source and sink and the flux lines corresponding to the growing harmonics in the x direction (a), y directions (b) and z direction (c).

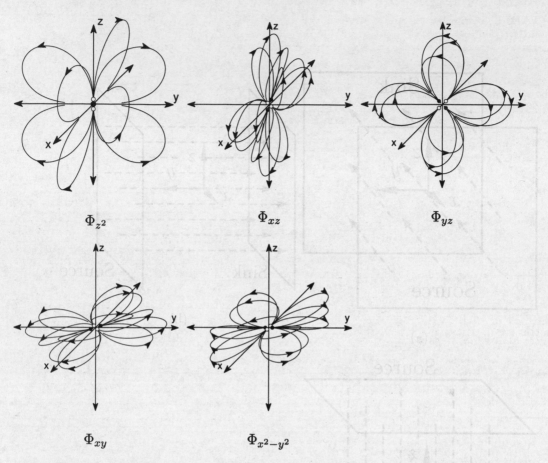

FIGURE 8.10. The distribution of sources and sinks for decaying quadrupole fields. The sources are shown by open circles and the sinks by filled circles. All sources and sinks emit/absorb Q energy per unit time, with the exception of the sink in (a) which which absorbs $2Q$ energy per unit time. The distance between pairs of sources is L.

EXAMPLE 8.3.1: Derive the expression for the quadrupolar field Φ_{z^2} in Table 8.2.

Solution: The temperature field due to the sources $+Q$ at $(0, 0, \pm L/2)$ and a sink $-2Q$ at the origin is,

$$T(\boldsymbol{x}) = \frac{Q}{4\pi k}\left[\frac{1}{\sqrt{x^2 + y^2 + (z - L/2)^2}} + \frac{1}{\sqrt{x^2 + y^2 + (z + L/2)^2}}\right.$$
$$\left. - \frac{2}{\sqrt{x^2 + y^2 + z^2}}\right].$$

$$(8.3.25)$$

TABLE 8.2. The source and sink locations that generate decaying quadrupole fields, and the fields in Cartesian and spherical co-ordinates. The distribution of sources and sinks is shown in Fig. 8.10.

	Source Locations (x,y,z)	Sink Locations (x,y,z)	Cartesian co-ordinates	Spherical co-ordinates
Φ_{z^2}	$(0,0,\pm\frac{L}{2})$	$(0,0,0)$	$\dfrac{QL^2(2z^2-x^2-y^2)}{16\pi(x^2+y^2+z^2)^{5/2}}$	$\sqrt{\dfrac{1}{80\pi}}\,\dfrac{QL^2 Y_2^0(\theta,\phi)}{r^3}$
Φ_{xz}	$(\pm\frac{L}{2\sqrt{2}},0,\pm\frac{L}{2\sqrt{2}})$	$(\pm\frac{L}{2\sqrt{2}},0,\mp\frac{L}{2\sqrt{2}})$	$\dfrac{3QL^2 xz}{8\pi(x^2+y^2+z^2)^{5/2}}$	$\sqrt{\dfrac{3}{160\pi}}\,\dfrac{QL^2[Y_2^1(\theta,\phi)+Y_2^{-1}(\theta,\phi)]}{r^3}$
Φ_{yz}	$(0,\pm\frac{L}{2\sqrt{2}},\pm\frac{L}{2\sqrt{2}})$	$(0,\pm\frac{L}{2\sqrt{2}},\mp\frac{L}{2\sqrt{2}})$	$\dfrac{3QL^2 yz}{8\pi(x^2+y^2+z^2)^{5/2}}$	$\sqrt{\dfrac{3}{160\pi}}\,\dfrac{QL^2[Y_2^1(\theta,\phi)-Y_2^{-1}(\theta,\phi)]}{ir^3}$
Φ_{xy}	$(\pm\frac{L}{2\sqrt{2}},\pm\frac{L}{2\sqrt{2}},0)$	$(\pm\frac{L}{2\sqrt{2}},\mp\frac{L}{2\sqrt{2}},0)$	$\dfrac{3QL^2 xy}{8\pi(x^2+y^2+z^2)^{5/2}}$	$\sqrt{\dfrac{3}{160\pi}}\,\dfrac{QL^2[Y_2^2(\theta,\phi)-Y_2^{-2}(\theta,\phi)]}{ir^3}$
$\Phi_{x^2-y^2}$	$(\pm\frac{L}{2},0,0)$	$(0,\pm\frac{L}{2},0)$	$\dfrac{3QL^2(x^2-y^2)}{16\pi(x^2+y^2+z^2)^{5/2}}$	$\sqrt{\dfrac{3}{160\pi}}\,\dfrac{QL^2[Y_2^2(\theta,\phi)+Y_2^{-2}(\theta,\phi)]}{r^3}$

Using Eqs. 8.3.15 and 8.3.16 for the first two terms within the square brackets on the right in Eq. 8.3.25, we obtain,

$$T(\boldsymbol{x}) = \frac{QL^2}{4\pi k}\left(\frac{3z^2-r^2}{4r^5}\right) = \frac{QL^2(2z^2-x^2-y^2)}{16\pi k(x^2+y^2+z^2)^{5/2}}. \tag{8.3.26}$$

$\square$

The five terms $n=2$ and $m=-2,-1,0,1,2$ in the solution, Eq. 8.2.47, are quadrupole terms, which are combinations of two sources and two sinks of equal magnitude arranged as shown in Fig. 8.10. The decaying quadrupole solutions decrease proportional to r^{-3} for $r \to \infty$; there is no contribution proportional to r^{-1} because the net source is zero, and no contribution proportional to r^{-2} because the net dipole moment is zero.

The growing quadrupole solutions increase as r increases. From Eq. 8.2.47, the growing harmonics are r^{2n+1} times the decaying harmonics. For $n=2$, the growing quadrupole fields are r^5 times the decaying quadrupole fields, and these increase proportional to r^2.

The term $n=p$ in the solution, Eq. 8.2.47, has $2p+1$ decaying harmonic solutions which decrease proportional to $r^{-(p+1)}$, and $2p+1$ growing harmonics which increase proportional to r^p for $r \to \infty$. The decaying harmonics are generated by a combination of 2^{p-1} sources and 2^{p-1} sinks of equal strength Q separated by distances L, in the limit $L \to 0$ and finite QL^p.

Summary (8.3)

1. The decaying harmonics in the solution of the Laplace equation in spherical co-ordinates, Eq. 8.2.47, can be equivalently expressed as the temperature field due to combinations of sources and sinks.

2. The term for $n = 0$ in Eq. 8.2.47 is due the temperature field due to a point source which emits a finite amount of heat per unit time. The temperature field, Eq. 8.3.1, decreases proportional to r^{-1}.

3. The three dipole terms for $n = 1$ in Eq. 8.2.47 are due to combinations of point sources and sinks of equal energy emitted/absorbed per unit time Q and infinitesimal separation L such that QL is finite. The temperature field, Eq. 8.3.20, decreases proportional to r^{-2}.

4. The five quadrupole terms for $n = 2$ in Eq. 8.2.47 are combinations of point sources and sinks of equal energy emitted/absorbed per unit time Q and infinitesimal separation L such that there is no net source, no net dipole, and QL^2, is finite. The temperature field, Table 8.2, decreases proportional to r^{-3}.

5. The growing harmonics for $n = 1$, Eq. 8.3.24, are the inversion of the point dipoles. The temperature field due to the $n = 1$ growing harmonics increases proportional to r.

6. The growing harmonics for $n = 2$, the inversion of the point quadrupoles, increase proportional to r^2.

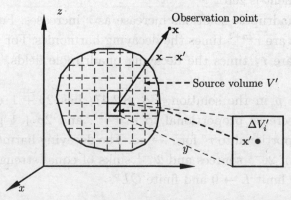

FIGURE 8.11. The configuration and co-ordinate system for analysing the temperature field due to a distributed source of heat.

8.4 Distributed Source

Consider a volumetric source that emits $Q_v(\boldsymbol{x})$ amount of heat per unit volume per unit time within the volume V', as shown in Fig. 8.11. The heat conduction equation is,

$$k\nabla^2 T + Q_v(\boldsymbol{x}) = 0, \qquad (8.4.1)$$

where k is the thermal conductivity. The heat conduction equation is solved by dividing the distributed source into a large number of differential volumes $\Delta V_1', \Delta V_2', \dots \Delta V_N'$ centred at the locations $\boldsymbol{x}_1', \boldsymbol{x}_2', \dots \boldsymbol{x}_n N'$. The amount of heat generated from each of these differential volumes per unit time is $Q_v(\boldsymbol{x}_1')\Delta V_1'$, $Q_v(\boldsymbol{x}_2')\Delta V_2'$, $\dots$, $Q_v(\boldsymbol{x}_N')\Delta V_N'$. The temperature at an observation point, $\boldsymbol{x}$, can be expressed as the sum of the temperature due to point sources located at the centres of these differential volumes,

$$T(\boldsymbol{x}) = \frac{1}{4\pi k} \sum_{i=1}^{N} \frac{Q_v(\boldsymbol{x}_i')\Delta V_i'}{|\boldsymbol{x} - \boldsymbol{x}_i'|}. \qquad (8.4.2)$$

Here, the primes are used to denote the source locations. In the limit $\Delta V_i' \to 0$, the above summation reduces to an integral,

$$T(\boldsymbol{x}) = \frac{1}{4\pi k} \int dV' \, \frac{Q_v(\boldsymbol{x}')}{|\boldsymbol{x} - \boldsymbol{x}'|}. \qquad (8.4.3)$$

Here, $\boldsymbol{x}$ is the observation location, $\boldsymbol{x}'$ is the source location, dV' is the differential volume $dx'\, dy'\, dz'$, $Q_v(\boldsymbol{x}')$ is the source density (energy emitted per unit volume per unit time) at the location $\boldsymbol{x}'$, and the integral in Eq. 8.4.3 is over the source locations. In Eq. 8.4.3, $|\boldsymbol{x} - \boldsymbol{x}'|$ is the distance between the source and observation points. Eq. 8.4.3 can be written as the convolution of the 'Green's function' $G(\boldsymbol{x}-\boldsymbol{x}')$ and the source strength $Q_v(\boldsymbol{x}')$,

$$\boxed{T(\boldsymbol{x}) = \int dV' \, G(\boldsymbol{x} - \boldsymbol{x}')Q_v(\boldsymbol{x}'),} \qquad (8.4.4)$$

where the Green's function for the unbounded domain is,

$$\boxed{G(\boldsymbol{x} - \boldsymbol{x}') = \frac{1}{4\pi k|\boldsymbol{x} - \boldsymbol{x}'|}.} \qquad (8.4.5)$$

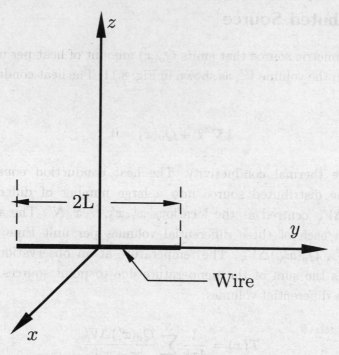

FIGURE 8.12.   The configuration and co-ordinate system for analysing the heating by a wire.

The heat flux at the location $\boldsymbol{x}$ is,

$$\boldsymbol{q}(\boldsymbol{x}) = -k\boldsymbol{\nabla}T = -k\boldsymbol{\nabla}\int \mathrm{d}V'\, G(\boldsymbol{x}-\boldsymbol{x}')Q_v(\boldsymbol{x}')$$

$$= -\frac{1}{4\pi}\boldsymbol{\nabla}\left(\int \mathrm{d}V'\,\frac{Q_v(\boldsymbol{x}')}{|\boldsymbol{x}-\boldsymbol{x}'|}\right). \tag{8.4.6}$$

The gradient in Eq. 8.4.6, which is with respect to the observation location $\boldsymbol{x}$, is simplified as shown in Eq. 8.3.10,

$$\boldsymbol{q}(\boldsymbol{x}) = \frac{1}{4\pi}\int \mathrm{d}V'\,\frac{(\boldsymbol{x}-\boldsymbol{x}')Q_v(\boldsymbol{x}')}{|\boldsymbol{x}-\boldsymbol{x}'|^3}. \tag{8.4.7}$$

EXAMPLE 8.4.1: A wire of length $2L$ immersed in a fluid generates heat at the rate of Q per unit length of the wire per unit time, as shown in Fig. 8.12. Determine the temperature field due to the wire. What is the approximation for the temperature field when the distance from the wire is much larger and much smaller than L?

Solution: A cylindrical coordinate system is used, where the axis is along the length of the wire, and the origin is at the centre of the wire. The wire is considered to be a line of infinitesimal thickness in the x and z directions, so that the energy emitted by the wire per unit volume per unit time is

$$Q_v(\pmb{x}) = Q\delta(x)\delta(z) \text{ for } -L < y < L. \tag{8.4.8}$$

Note that the delta functions $\delta(x)$ and $\delta(z)$ have dimension of $\mathcal{L}^{-1}$, and therefore, Q_v has the correct dimension of a volumetric source, which is the heat generated per unit volume per unit time. The temperature field due to the wire is,

$$
\begin{aligned}
T(\pmb{x}) &= \frac{Q}{4\pi k} \int_{-\infty}^{\infty} dx' \int_{-L}^{L} dy' \int_{-\infty}^{\infty} dz' \frac{Q\delta(x')\delta(z')}{((x-x')^2 + (y-y')^2 + (z-z')^2)^{1/2}} \\
&= \frac{Q}{4\pi k} \int_{-L}^{L} dy' \frac{1}{(x^2 + (y-y')^2 + z^2)^{1/2}} \\
&= \frac{Q}{4\pi k} \ln\left(\frac{L+y+\sqrt{r^2+(L+y)^2}}{-L+y+\sqrt{r^2+(y-L)^2}} \right), \tag{8.4.9}
\end{aligned}
$$

where $r = \sqrt{x^2 + z^2}$ is the distance from the y axis. In the second step above, the integrals of the delta functions in the x and z directions have been replaced by the integrands at $x = 0$ and $z = 0$.

The solution, Eq. 8.4.9, is examined as a function of r along the plane that bisects the wire, $y = 0$,

$$T(\pmb{x}) = \frac{Q}{4\pi k} \ln\left(\frac{L+\sqrt{r^2+L^2}}{-L+\sqrt{r^2+L^2}} \right). \tag{8.4.10}$$

In the limit $r \gg L$, an expansion in the small parameter (L/r) is used for the terms in solution, Eq. 8.4.10,

$$
\ln\left[\frac{\sqrt{r^2+L^2}+L}{\sqrt{r^2+L^2}-L} \right] = \ln\left[\frac{r\left(\sqrt{1+\frac{L^2}{r^2}} + \frac{L}{r} \right)}{r\left(\sqrt{1+\frac{L^2}{r^2}} - \frac{L}{r} \right)} \right] = \ln\left[\frac{1+\frac{L^2}{2r^2}+\frac{L}{r}}{1+\frac{L^2}{2r^2}-\frac{L}{r}} \right]
$$

$$
= \ln\left[1+\frac{L^2}{2r^2}+\frac{L}{r} \right] - \ln\left[1+\frac{L^2}{2r^2}-\frac{L}{r} \right] = \frac{2L}{r}. \tag{8.4.11}
$$

Here, the expansions $\sqrt{1+(L/r)^2} \approx 1 + \frac{1}{2}(L/r)^2$ and $\ln[1 \pm (L/r)] \approx \pm(L/r)$ have been used for $(L/r) \ll 1$. The above approximation is substituted in Eq. 8.4.10 to

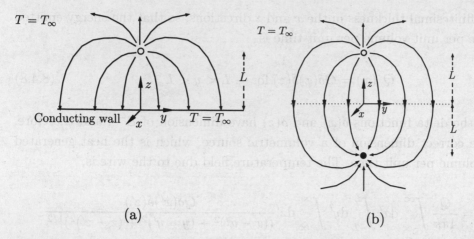

$T = T_\infty$

z

Conducting wall $\quad x \quad y \quad T = T_\infty$

$T = T_\infty$

L

z

$x \quad y$

L

L

(a)   (b)

FIGURE 8.13. A point source near a conducting wall (a), and a combination of a source and sink (b) satisfying the constant temperature condition at the wall location.

obtain the largest term in the expansion,

$$T(x) = \frac{2QL}{4\pi kr}. \qquad (8.4.12)$$

This is the solution for the temperature field due to a point source emitting $2QL$ energy per unit time, as expected when the distance from the source is large compared to L.

In the limit $r \ll L$ where the observation point is close to the wire, an expansion is used the parameter (r/L),

$$\ln\left[\frac{\sqrt{r^2 + L^2} + L}{\sqrt{r^2 + L^2} - L}\right] = \ln\left[\frac{L\left(\sqrt{1 + \frac{r^2}{L^2}} + 1\right)}{L\left(\sqrt{1 + \frac{r^2}{L^2}} - 1\right)}\right] = \ln\left[\frac{1 + \frac{r^2}{2L^2} + 1}{1 + \frac{r^2}{2L^2} - 1}\right] = \ln\left[\frac{4L^2}{r^2}\right].$$

$$(8.4.13)$$

The solution for the temperature field reduces to

$$T(x) = \frac{Q}{4\pi k}\ln\left(\frac{4L^2}{r^2}\right) = \frac{Q}{2\pi k}\ln\left(\frac{2L}{r}\right). \qquad (8.4.14)$$

The above solution is the temperature field due to a point source in two dimensions, Exercise problem 8.10.   □

8.4.1 Method of Images

Eq. 8.4.3 for the temperature field is applicable for a domain of infinite extent. Many practical problems involve finite domains where boundary conditions are to be applied at the boundaries. For simple geometries such as a planar boundary, the boundary conditions can be applied using the method of images. Consider a point source of strength Q located at $x_s = (0, 0, L)$ in a semi-infinite domain bounded by a surface at $z = 0$ with constant temperature, as shown in Fig. 8.13(a). For a constant temperature surface, there is no temperature variation along the surface, and therefore no flux in a direction parallel to the surface. The flux lines are necessarily perpendicular to the surface, as shown in Fig. 8.13(a). Since the surface is of infinite extent, the temperature at the surface is the same as that far from the source, T_∞.

The boundary condition at the surface is clearly not satisfied by Eqs. 8.3.9 and 8.3.10 for an infinite medium. However, the boundary condition is satisfied if we replace the finite domain by an infinite domain, in which there is a source emitting energy per unit time $+Q$ at $(0, 0, L)$, and an image source (sink) emitting energy per unit time $-Q$ at $x_I(0, 0, -L)$, as shown in Fig. 8.13(b). In this configuration, the flux lines are emitted from the source at $(0, 0, L)$ and are absorbed by the sink at $(0, 0, -L)$. At the mid-plane between the source and the sink, $z = 0$, the flux lines are necessarily perpendicular to the plane due to symmetry. Thus, the temperature field for the source near a perfectly conducting wall for the semi-infinite domain $z > 0$ in Fig. 8.13(a) is identical to that due to a source and a sink shown in Fig. 8.13(b),

$$T(x) - T_\infty = \frac{Q}{4\pi k |x - x_s|} - \frac{Q}{4\pi k |x - x_I|}, \tag{8.4.15}$$

where x_s is the source point and x_I is the image point. At any location on the surface $z = 0$, it is evident that $|x - x_s| = |x - x_I|$, and Eq. 8.4.15 predicts that $T(x) - T_\infty = 0$. Thus, the boundary condition at $z = 0$ has been satisfied by removing the surface and inserting an image emitting energy per unit time $-Q$ at $(0, 0, -L)$.

The method of images is quite easily extended to a distributed source $Q_v(x')$ near a conducting wall, as shown in Fig. 8.14. A distributed sink $-Q_v(x_I)$, which is a mirror image of the distributed source is placed symmetrically on the other side of the surface. The temperature field is given by Eq. 8.4.4, where the Green's function is

$$\boxed{G(x - x') = \frac{1}{4\pi k} \left(\frac{1}{|x - x'|} - \frac{1}{|x - x_I'|} \right).} \tag{8.4.16}$$

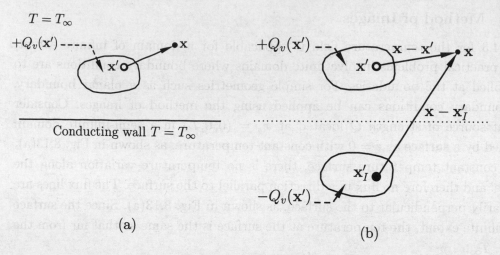

$T = T_\infty$

$+Q_v(\mathbf{x}')$

$\mathbf{x}'$

$\bullet\mathbf{x}$

$+Q_v(\mathbf{x}')$

$\mathbf{x}'$

$\mathbf{x} - \mathbf{x}'$

$\bullet\mathbf{x}$

$\mathbf{x} - \mathbf{x}_I'$

Conducting wall $T = T_\infty$

$\mathbf{x}_I'$

$-Q_v(\mathbf{x}')$

(a)   (b)

FIGURE 8.14. A distributed source near a conducting wall (a), and a combination of a source and sink (b) satisfying the constant temperature condition at the wall location.

The method of images can also be used for a source near an insulating wall, as shown in Fig. 8.15(a). In this case, the component of the flux perpendicular to the wall is zero, and so the flux lines are parallel to the wall. The zero normal flux condition is satisfied by removing the wall and placing a source emitting energy Q per unit time at the location $(0, 0, -L)$, as shown in Fig. 8.15(b). The temperature field is,

$$T(\mathbf{x}) - T_\infty = \frac{Q}{4\pi k |\mathbf{x} - \mathbf{x}_s|} + \frac{Q}{4\pi k |\mathbf{x} - \mathbf{x}_I|}. \qquad (8.4.17)$$

Since the sources at $(0, 0, L)$ and $(0, 0, -L)$ are emitting energy at equal rates, the flux perpendicular to the surface at $z = 0$ is zero by symmetry. From Eq. 8.4.7, the expression for the flux is,

$$\mathbf{q} = \frac{Q(\mathbf{x} - \mathbf{x}_s)}{4\pi k |\mathbf{x} - \mathbf{x}_s|^3} + \frac{Q(\mathbf{x} - \mathbf{x}_I)}{4\pi k |\mathbf{x} - \mathbf{x}_I|^3}. \qquad (8.4.18)$$

The flux perpendicular to the surface is

$$\mathbf{q} \cdot \mathbf{e}_z = \frac{Q(\mathbf{x} - \mathbf{x}_s) \cdot \mathbf{e}_z}{4\pi k |\mathbf{x} - \mathbf{x}_s|^3} + \frac{Q(\mathbf{x} - \mathbf{x}_I) \cdot \mathbf{e}_z}{4\pi k |\mathbf{x} - \mathbf{x}_I|^3}. \qquad (8.4.19)$$

From Fig. 8.15(b), the z components of $(\mathbf{x} - \mathbf{x}_s)$ and $(\mathbf{x} - \mathbf{x}_I)$ sum to zero for any point on the plane $z = 0$, and the magnitudes $|\mathbf{x} - \mathbf{x}_s|$ and $|\mathbf{x} - \mathbf{x}_I|$ are equal. Therefore, the flux perpendicular to the surface is zero. Thus, the combination of two sources in Fig. 8.15(b) satisfies the zero normal flux condition at the surface

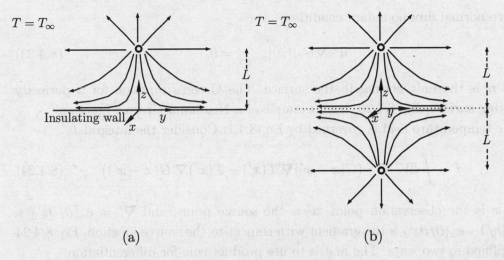

FIGURE 8.15. A point source near a perfectly insulating wall (a), and a combination of two sources (b) satisfying the zero normal flux condition at the wall location.

$z = 0$. For a distributed source near an insulating wall, the method of images can be extended to show that the temperature is given by Eq. 8.4.4, where the Green's function is,

$$G(\boldsymbol{x} - \boldsymbol{x}') = \frac{1}{4\pi k} \left(\frac{1}{|\boldsymbol{x} - \boldsymbol{x}'|} + \frac{1}{|\boldsymbol{x} - \boldsymbol{x}'_I|} \right). \qquad (8.4.20)$$

8.4.2 Green's Function Method

The Green's function method is a systematic method for solving the heat conduction Eq. 8.4.1 in a finite domain. The equation for the Green's function $G(\boldsymbol{x})$ is,

$$k\boldsymbol{\nabla}^2 G + \delta(\boldsymbol{x}) = 0, \qquad (8.4.21)$$

and boundary conditions are specified on the surface S. There are two types of Green's functions. The 'Dirichlet' Green's function $G_D(\boldsymbol{x})$ satisfies a constant temperature (Dirichlet) boundary condition on the surface,

$$G_D(\boldsymbol{x}_S) = 0. \qquad (8.4.22)$$

The Green's function for a perfectly conducting surface, Eq. 8.4.16, is an example of a Dirichlet Green's function. The Neumann boundary condition $G_N(\boldsymbol{x})$ satisfies the zero normal flux boundary condition,

$$n \cdot \nabla G_N(x)|_{x=x_S} = 0, \qquad (8.4.23)$$

where n is the unit normal to the surface. The Green's function for a perfectly insulating surface, Eq. 8.4.20, is an example of a Neumann Green's function.

The temperature field is governed by Eq. 8.4.1. Consider the integral I,

$$I = \int dV' \, \nabla' \cdot \left(G(x - x') \nabla' T(x') - T(x') \nabla' G(x - x') \right). \qquad (8.4.24)$$

Here, x is the observation point, x' is the source point, and $\nabla' = e_x(\partial/\partial x') + e_y(\partial/\partial y') + e_z(\partial/\partial z')$ is the gradient with respect to the source location. Eq. 8.4.24 is simplified in two ways. The first is to use product rule for differentiation,

$$I = \int dV' \left(G(x - x') \nabla'^2 T(x') - T(x') \nabla'^2 G(x - x') \right)$$
$$= \int dV' \left(-\frac{G(x - x') Q_v(x')}{k} + \frac{T(x') \delta(x' - x)}{k} \right). \qquad (8.4.25)$$

Eqs. 8.4.1 and 8.4.21 are used to simplify for the first and second terms on the right in the above equation. Eq. 8.4.25 is simplified by using the property of the delta function, Eq. 8.3.4,

$$I = -\int dV' \frac{G(x - x') Q_v(x')}{k} + \frac{T(x)}{k}. \qquad (8.4.26)$$

The divergence theorem is used to obtain a second simplification of Eq. 8.4.24,

$$I = \int dS' \, n' \cdot \left(G(x - x') \nabla' T(x') - T(x') \nabla' G(x - x') \right), \qquad (8.4.27)$$

where S' is the surface of the volume V', and n' is the outward unit normal to the surface. Since Eqs. 8.4.26 and 8.4.27 are two different simplifications of Eq. 8.4.24, the two are equal, and the equation for the temperature field is

$$\boxed{\begin{aligned} T(x) &= \int dV' \, G(x - x') Q_v(x') \\ &\quad + k \int dS' \, n' \cdot \left(G(x - x') \nabla' T(x') - T(x') \nabla' G(x - x') \right). \end{aligned}} \qquad (8.4.28)$$

Eq. 8.4.28 is the generalisation of Eq. 8.4.3 which includes the presence of surfaces on which the boundary conditions are specified. When there is a constant temperature (Dirichlet) boundary condition at the boundaries, this constant is subtracted from the temperature field so that $T(x')$ is zero at the boundaries in Eq. 8.4.28. The solution is simplified by choosing the Green's function so that $G(x - x')$ is also zero on the boundaries (Dirichlet Green's function), so that that the surface integral in Eq. 8.4.28 is zero. Instead, if zero flux (Neumann) boundary conditions are specified for the temperature field at the boundaries, $n' \cdot \nabla' T(x') = 0$, the surface integral in Eq. 8.4.28 is zero if the Neumann Green's function is chosen such that $n' \cdot \nabla' G(x - x') = 0$. Thus, Eq. 8.4.28 can be reduced to Eq. 8.4.3 by an appropriate choice of the Green's function in a finite domain.

Summary (8.4)

1. The temperature at an observation point, Eq. 8.4.4, due to a distributed source in an unbounded domain is expressed as an integral over the source domain of the source density (energy generated per unit volume per unit time) times the Green's function, Eq. 8.4.5.

2. In a bounded domain, the constant temperature/zero flux conditions at boundaries can be satisfied by removing the boundary and adding image sources/sinks of the appropriate magnitude at suitable locations so that the temperature field satisfies the conditions at the boundaries. For planar perfectly conducting and perfectly insulating walls, the Green's functions are Eqs. 8.4.16 and 8.4.20, respectively.

3. The general solution for the temperature field is Eq. 8.4.28; the solution is simplified by suitably choosing the Green's functions so that the surface integral on the right is zero.

Exercises

EXERCISE 8.1 Determine the temperature field at steady state for a rectangular block in which the temperature is fixed at three surfaces, and the fourth surface is insulated, as shown in Fig. 8.16.

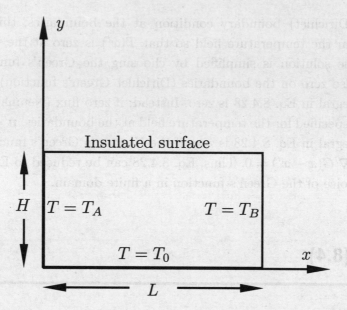

FIGURE 8.16.   Conduction in a rectangular block with one insulated surface.

EXERCISE 8.2 Determine the correlation for the friction factor for a square channel of side L using the solution for the average velocity derived in Example 8.1.1. The Reynolds number is defined as $\mathrm{Re} = (\rho v_{av} L/\mu)$ and the friction factor is $f = (dp/dx)(2\rho v_{av}^2/L)^{-1}$. Use a truncated expansion where only two terms each in the y and z directions are included.

EXERCISE 8.3 A cylindrical wedge has inner and outer radii R_i and R_o and subtended angle $(\pi/3)$, as shown in Fig. 8.17. The temperature on both the surfaces perpendicular to the ϕ co-ordinate at $\phi = 0$ and $\phi = \pi/3$ is T_s. The temperature on the surface perpendicular to the r direction at R_i is T_i, and that on the surface perpendicular to the r direction at R_o is T_o. The two-dimensional heat conduction equation is

$$\nabla^2 T = \frac{1}{r}\frac{\partial}{\partial r}\left(r\frac{\partial T}{\partial r}\right) + \frac{1}{r^2}\frac{\partial^2 T}{\partial \phi^2} = 0.$$

Determine the temperature field using separation of variables.

EXERCISE 8.4 Consider a fluid in a circular cylinder of radius R and height H, in which the top end cap rotates with angular velocity Ω, while the bottom end cap and the cylindrical surface are stationary, as shown in Fig. 8.18. Determine the velocity field if the only non-zero component of the velocity, v_ϕ in the polar direction, is a function of r and z. The equation for the velocity field is

$$\mu\left(\frac{\partial}{\partial r}\left(\frac{1}{r}\frac{\partial (rv_\phi)}{\partial r}\right) + \frac{\partial^2 v_\phi}{\partial z^2}\right) = 0.$$

EXERCISE 8.5 What is the number of independent decaying spherical harmonic solutions which decrease proportional to r^{-9} in three dimensions, where r is the distance from the origin?

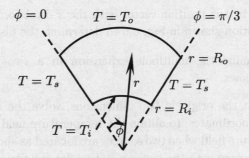

FIGURE 8.17. Heat conduction in a wedge.

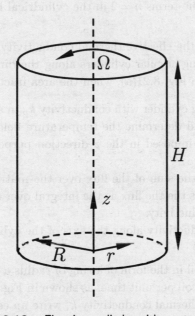

FIGURE 8.18. Flow in a cylinder with a rotating end cap.

EXERCISE 8.6 Derive the expressions for $\Phi_{x^2-y^2}$ and Φ_{xy} in Table 8.2.

EXERCISE 8.7 What are the growing quadrupole fields with the same angular dependence as the decaying quadrupole fields shown in Table 8.2? Sketch the field lines of these quadrupole fields, similar to those for the growing dipole fields in Fig. 8.9. Sketch these in the plane of the quadrupole fields where possible for ease of visualisation.

EXERCISE 8.8 Consider a particle of radius R and thermal conductivity k_p in a matrix with thermal conductivity k_m, shown in Fig. 8.5. Instead of a linear temperature gradient, the temperature far from the particle is $T = T''xy$ for $r \to \infty$. Determine the temperature disturbance due to the particle.

EXERCISE 8.9 Use separation of variables to solve the equation $k\nabla^2 T = 0$ in a two-dimensional polar co-ordinate system, (r, ϕ), where r is the distance from the origin and

ϕ is the angle subtended by the position vector with the x direction, as shown in Fig. 8.19. Use the periodicity condition shown in Fig. 8.6 to determine the discrete eigenvalues.

EXERCISE 8.10 Determine the multipole expansion in a two-dimensional cylindrical co-ordinate system as follows.

a) For a point source at the origin in two dimensions, solve the heat equation $k\nabla^2 T + Q\delta(x) = 0$ in polar coordinates, to obtain the temperature field.
b) What is the temperature field when two sources are located as shown in Figs. 8.20(a) and (b), and $L \ll r$? Compare with the terms $n = 1$ in the cylindrical harmonic expansion.
c) What is the temperature field when four sources are located as shown in Figs. 8.20(c) and (d)? Compare with the terms $n = 2$ in the cylindrical harmonic expansion.

EXERCISE 8.11 Determine the effective thermal conductivity for a composite consisting of dilute array of infinitely long circular cylinders along the direction perpendicular to the axis of the cylinders, shown in Fig. 8.21(a), when the area fraction of the cylinders is Φ.

a) Consider an infinitely long cylinder with conductivity k_p in a matrix of conductivity k_m, shown in Fig. 8.21(b), and determine the temperature field around the cylinder when a uniform gradient T' is imposed in the z direction perpendicular to the axis of the cylinder.
b) Express the heat flux as the sum of the flux over the matrix and the cylinders. When the array is dilute, express the the flux as the integral over one cylinder, and determine the effective thermal conductivity.
c) What is the effective conductivity along the axis of the cylinders?

EXERCISE 8.12 A heater coil in the form of a ring of radius a in the $x-y$ plane generates heat Q per unit length of the coil per unit time, as shown in Fig. 8.22. If the heater is placed in an unbounded medium of thermal conductivity k, write an equation for the temperature as a function of position in the medium. Plot the temperature as a function of position along

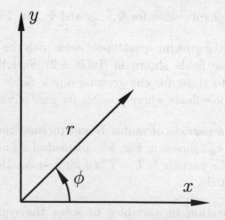

FIGURE 8.19. Polar co-ordinate system.

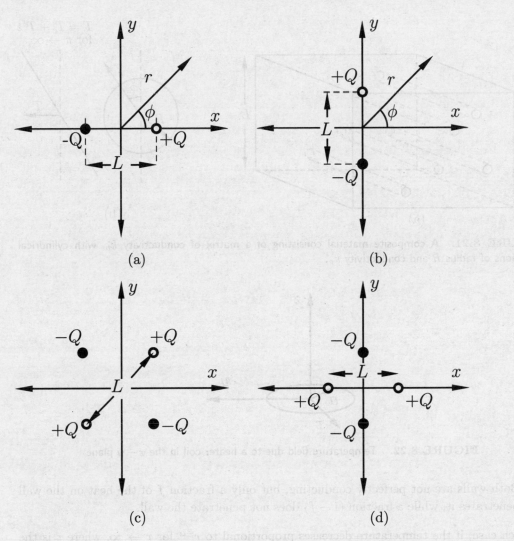

FIGURE 8.20. The distribution of sources and sinks for point dipoles ((a) and (b)), and point quadrupoles ((c) and (d)) in two dimensions.

the symmetry axis of the heater (z axis in the figure). Simplify the expressions for $z \ll a$ and $z \gg a$. What does the expression for $z \gg a$ correspond to?

EXERCISE 8.13 A point source of heat of strength Q (in units of heat energy per unit time) is placed at a distance L from two perpendicular walls in three dimensions, as shown in Fig. 8.23. Find the temperature profile for the following cases.

(a) Both walls are perfectly insulating.
(b) The horizontal wall has constant temperature and the vertical wall is perfectly insulating.

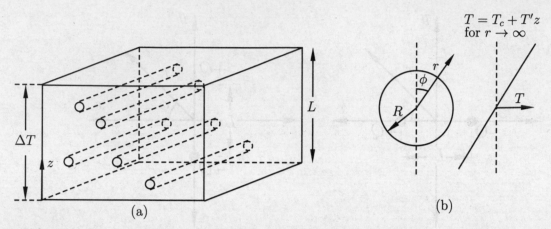

FIGURE 8.21. A composite material consisting of a matrix of conductivity k_m with cylindrical inclusions of radius R and conductivity k_p.

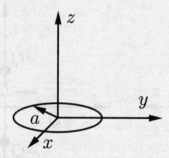

FIGURE 8.22. Temperature field due to a heater coil in the $x - y$ plane.

(c) Both walls are not perfectly conducting, but only a fraction f of the heat on the wall penetrates it, while a fraction $(1 - f)$ does not penetrate the wall.

In each case, if the temperature decreases proportional to $r^{-\alpha}$ for $r \to \infty$, where r is the distance from the origin, what is α?

EXERCISE 8.14 A point source generating energy Q per unit time is located at a distance r from the centre of a perfectly conducting sphere of radius R, as shown in Fig. 8.24(a). In order to satisfy the constant temperature boundary condition on the sphere, it is desired to use the method of images, where a point source generating energy Q' per unit time is located at a distance r', as shown in Fig. 8.24(b). What is the relation between (Q, r) and (Q', r')? *Hint:* Express locations on the surface of the sphere in a spherical co-ordinate system in which the axis is along the position vector of the source, so that the configuration is axisymmetric about this axis.

EXERCISE 8.15 Consider a particle of radius R moving with a velocity $v = v_x e_x + v_z e_z$ adjacent to a horizontal wall located at $z = 0$, as shown in Fig. 8.25(a). It is desired to replace this configuration by Fig. 8.25(b), consisting of a sphere and an image sphere, but

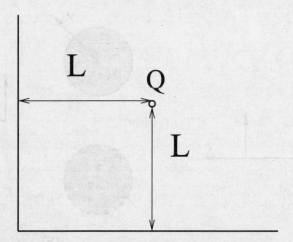

FIGURE 8.23. A point source located near two adjacent walls.

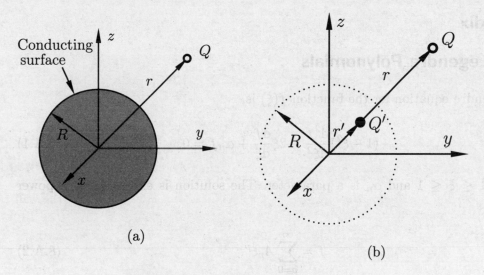

(a)

(b)

FIGURE 8.24. Method of images for a conducting sphere.

no wall. What should be the velocity of the image sphere, if the boundary condition at the wall is,

a) No-slip condition, $v_x = v_z = 0$.
b) Zero stress condition, $(dv_x/dz) = v_z = 0$.

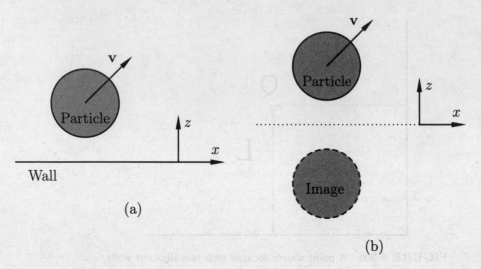

FIGURE 8.25. Method of images for a particle moving near a horizontal wall.

Appendix

8.A   Legendre Polynomials

The Legendre equation for the function $f(\xi)$ is,

$$(1 - \xi^2)\frac{\mathrm{d}^2 f}{\mathrm{d}\xi^2} - 2\xi\frac{\mathrm{d}f}{\mathrm{d}\xi} + \alpha_n f = 0, \tag{8.A.1}$$

where $-1 \leq \xi \leq 1$ and α_n is a parameter. The solution is expressed as a power series,

$$f = \sum_{p=0}^{\infty} A_p \xi^p. \tag{8.A.2}$$

The expansion 8.A.2 is substituted into the Legendre Eq. 8.A.1,

$$\sum_{p=0}^{\infty} p(p-1)A_p\xi^{p-2} - \sum_{p=0}^{\infty} p(p-1)A_p\xi^p - 2\sum_{p=0}^{\infty} pA_p\xi^p + \alpha_n \sum_{p=0}^{\infty} A_p\xi^p = 0. \tag{8.A.3}$$

The first term on the left in Eq. 8.A.3 is zero for $p = 0$ and $p = 1$. The transformation $p \to p + 2$ is made in the first term on the left, and Eq. 8.A.3 is simplified as,

$$\sum_{p=0}^{\infty} (p+2)(p+1)A_{p+2}\xi^p - \sum_{p=0}^{\infty} (p(p+1) - \alpha_n)A_p\xi^p = 0. \tag{8.A.4}$$

Equating coefficients of equal powers of ξ in Eq. 8.A.4, we obtain the recurrence relation for the coefficient A_{p+2} in terms of A_p,

$$A_{p+2} = \frac{(p(p+1) - \alpha_n)A_p}{(p+2)(p+1)}. \tag{8.A.5}$$

The recurrence relation can be used to determine the coefficients $A_2 \ldots$ in terms of the coefficients A_0 and A_1. It should be noted that the coefficients of the even powers of ξ are proportional to A_0, and the coefficients of the odd powers of ξ are related to A_1. Thus, there are two linearly independent solutions of the Legendre equation, Eq. 8.A.1, one containing odd powers of ξ and the other containing even powers of ξ.

If we examine Eq. 8.A.5 relating A_p and A_{p+2} for $p \gg 1$, it is evident that the ratio (A_{p+2}/A_p) tends to 1. Therefore, in Eq. 8.A.2, the coefficients of ξ^p for $p \gg 1$ approach a constant value. For $\xi = \pm 1$, this implies that the series does not converge. A finite solution for the expansion in Eq. 8.A.2 can be obtained only if $\alpha_n = n(n+1)$, where n is an integer. In this case, the coefficient $A_{n+2} = 0$ from the recurrence relation, Eq. 8.A.5. All higher coefficients $A_{n+4}, A_{n+6}, \ldots$ are also equal to zero from the recurrence relation. Thus, a finite series is obtained in which the largest term is proportional to ξ^n. This provides a condition $\alpha_n = n(n+1)$, where n is a positive integer, for the existence of finite solutions for the Legendre equation, Eq. 8.A.1. These are denoted the Legendre polynomials $P_n(\xi)$ or $P_n^0(\xi)$. The series for even n is an even function of ξ, and that for odd n is an odd function of ξ.

The convention, $P_n^0(\xi) = 1$ at $\xi = \pm 1$ for the even series, and $P_n^0(\xi) = \pm 1$ at $\xi = \pm 1$ for the odd series, are used to specify the normalisation for the Legendre polynomials. With this definition, the first few Legendre polynomials are,

$$
\begin{aligned}
P_0^0(\xi) &= 1, \\
P_1^0(\xi) &= \xi, \\
P_2^0(\xi) &= \tfrac{1}{2}(3\xi^2 - 1), \\
P_3^0(\xi) &= \tfrac{1}{2}(5\xi^3 - 3\xi), \\
P_4^0(\xi) &= \tfrac{1}{8}(35\xi^4 - 30\xi^2 + 3).
\end{aligned} \tag{8.A.6}
$$

The orthogonality relation for the Legendre polynomials is,

$$\int_{-1}^{1} d\xi \, P_n^0(\xi) P_{n'}^0(\xi) = \frac{2\delta_{nn'}}{2n+1}. \tag{8.A.7}$$

In heat conduction problems that are not axisymmetric, the associated Legendre equation is obtained by the separation of variables procedure,

$$(1-\xi^2)\frac{\mathrm{d}^2 f}{\mathrm{d}\xi^2} - 2\xi\frac{\mathrm{d}f}{\mathrm{d}\xi} + \left(n(n+1) - \frac{m^2}{1-\xi^2}\right)f = 0, \qquad (8.\text{A}.8)$$

In this case, we state without proof that the solutions of the associated Legendre equation are finite only if n is a positive integer, m is an integer and $|m| \leq n$. The solutions of the associated Legendre equations are denoted $P_n^m(\xi)$. For positive m these are expressed in terms of the Legendre polynomials as,

$$P_n^m(\xi) = (-1)^m(1-\xi^2)^{m/2}\frac{\mathrm{d}^m P_n^0(\xi)}{\mathrm{d}\xi^m}. \qquad (8.\text{A}.9)$$

and $P_n^{-m}(\xi)$ is related to $P_n^m(\xi)$,

$$P_n^{-m}(\xi) = (-1)^m\frac{(n-m)!}{(n+m)!}P_n^m(\xi). \qquad (8.\text{A}.10)$$

The orthogonality relation for the associated Legendre functions is,

$$\int_{-1}^{1}\mathrm{d}\xi\, P_n^m(\xi)P_{n'}^m(\xi) = \frac{2\delta_{nn'}}{2n+1}\frac{(n+m)!}{(n-m)!}. \qquad (8.\text{A}.11)$$

9

Forced Convection

Transport of heat/mass is enhanced by an externally generated flow past an object or a surface in 'forced convection'. Here, the flow is specified, and it is not affected by the change in temperature/concentration due to the heat/mass transfer. The known fluid velocity field is substituted into the convection–diffusion equation in order to determine the temperature/concentration field and the transport rate.

In the previous chapter, we examined the limit of low Peclet number, where transport due to convection is small compared that due to diffusion. There, the approach was to neglect convection altogether, and solve the diffusion equation. In the limit of high Peclet number, an equivalent approach would be to neglect diffusion altogether, and solve the convection equation to obtain the concentration/temperature fields. This approach is not correct for the following mathematical and physical reasons.

Mathematically, when the diffusion term is neglected, the convection–diffusion equation is reduced from a second order to a zeroth order differential equation in the cross-stream co-ordinate. The second order differential for the concentration/temperature field is well posed only if two boundary conditions are specified in each co-ordinate. When diffusion is neglected, the resulting zeroth order equation cannot satisfy both boundary conditions in the cross-stream co-ordinate specified for the original problem. Physically, when diffusion is neglected, there is transport due to convection only along fluid streamlines, and there is no transport across the streamlines. The concentration/temperature is a constant along streamlines in the flow. At bounding surfaces (the pipe surface in a heat exchanger, or particle surfaces in the case of suspended particles), there is no flow perpendicular to the surface. When we neglect diffusion, there is no flux across the surface. Therefore, we obtain the unphysical result that there is no mass/heat transfer across the surface.

A more sophisticated approach is required to obtain solutions for transport in strong convection, based on the following physical picture. In the limit of high Peclet number, mass or heat diffusing from a surface gets rapidly swept downstream due to the strong convection, and so the concentration/temperature variations are restricted to a thin 'boundary layer' close to the surface. Diffusion can be neglected in most of the flow domain, where transport takes place along the flow streamlines due to convection. However, it is necessary to incorporate diffusion in a boundary layer of thickness much smaller than the characteristic flow scale. The thickness of the layer is determined from the condition that convection and diffusion are comparable in this layer in the limit of large Peclet number—that is, $Pe \to \infty$. Stream-wise transport in the boundary layer is still primarily due to convection, but transport across streamlines is due to diffusion. The inclusion of cross-stream diffusive transport results in a second order differential equation in the cross-stream co-ordinate, and this enables us to satisfy the boundary conditions.

In this chapter, the boundary layer thickness is calculated by scaling the convection–diffusion equation for simple geometries such as a flat surface and a spherical particle/bubble. The simplified convection–diffusion equation for the temperature/concentration fields, which incorporates convection and cross-stream diffusion in the boundary layer, is solved using a similarity transform. The Nusselt/Sherwood number correlations are determined from the solutions for the temperature/concentration fields. As discussed in Section 2.3.2 (Chapter 2), there are two broad categories of boundary layer solutions, those for flow past a rigid surface with a no-slip boundary condition, and those for the flow past a mobile surface with a slip boundary condition. The Nusselt and Sherwood numbers scale as $Pe^{1/3}$ for the former, and as $Pe^{1/2}$ for the latter. The correlations for the flow past a rigid surface are derived in Section 9.1, and for a mobile interface in Section 9.2.

In Section 9.3, a microscopic model is analysed for the dispersion in a packed column. The particles in the column are modelled as a regular array of obstacles, and the spreading of a solute due to the flow around these obstacles is analysed. This provides an explanation for the result in Section 3.2.2 (Chapter 3), that the dispersion coefficient is proportional to the product of the size of the particles and the flow velocity. The axial dispersion in the flow through a pipe, called Taylor dispersion, is analysed in Section 9.4, and the Taylor dispersion coefficient is derived for a laminar flow.

9.1 Flow past a Rigid Surface

9.1.1 Flow Past a Heated Surface

The configuration consists of a flat plate in the $x - y$ plane in contact with a fluid which occupies $z > 0$, as shown in Fig. 9.1. The fluid temperature far from the plate is T_∞. The temperature of the plate is also T_∞ for $x < 0$, but the plate is heated with surface temperature T_w for $0 < x < L$. The velocity v_x in the x direction (parallel to the plate) is only a function of the z co-ordinate perpendicular to the plate. The specific form of the velocity profile is not important, since we shall be focusing on a thin boundary layer close to the surface. The Peclet number $\mathrm{Pe} = (v_c L / \alpha)$, based on the characteristic velocity v_c, length L and thermal diffusivity α, is large.

Since the only non-zero component of the velocity is v_x, the convection–diffusion equation at steady state is

$$v_x \frac{\partial T}{\partial x} = \alpha \left(\frac{\partial^2 T}{\partial x^2} + \frac{\partial^2 T}{\partial z^2} \right), \tag{9.1.1}$$

where α is the thermal diffusivity. The boundary conditions are,

$$T = T_w \quad \text{at} \ \ z = 0 \ \ \text{for} \ \ 0 < x < L, \tag{9.1.2}$$

$$T = T_\infty \quad \text{for} \ \ z \to \infty, \tag{9.1.3}$$

$$T = T_\infty \quad \text{at} \ \ x = 0 \ \ \text{for} \ \ z > 0. \tag{9.1.4}$$

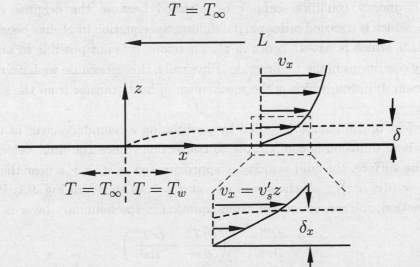

FIGURE 9.1. The configuration and co-ordinate system for analysing the heat transfer in the flow past a flat plate at high Peclet number.

Eq. 9.1.4 states that the temperature is T_∞ in the fluid at the start of the heating section, because the fluid has just come into contact with the surface and there has not been any heat transfer from the surface at this location.

The scaled temperature, velocity and spatial co-ordinates are defined as $T^* = (T - T_\infty)/(T_w - T_\infty)$, $v_x^* = (v_x/v_c)$, $x^* = (x/L)$ and $z^* = (z/L)$. The scaled convection–diffusion equation is,

$$\left(\frac{v_c L}{\alpha}\right) v_x^* \frac{\partial T^*}{\partial x^*} = \left(\frac{\partial^2 T^*}{\partial x^{*2}} + \frac{\partial T^*}{\partial z^{*2}}\right), \tag{9.1.5}$$

with boundary conditions

$$T^* = 1 \quad \text{at} \quad z^* = 0 \quad \text{for} \quad 0 < x^* < 1, \tag{9.1.6}$$

$$T^* = 0 \quad \text{for} \quad z^* \to \infty, \tag{9.1.7}$$

$$T^* = 0, \quad \text{at} \quad x^* = 0 \quad \text{for} \quad z^* > 0. \tag{9.1.8}$$

If the right side of Eq. 9.1.5 is neglected in comparison to the left side for $\text{Pe} = (v_c L/\alpha) \gg 1$, the convection–diffusion equation reduces to

$$v_x^* \frac{\partial T^*}{\partial x^*} = 0. \tag{9.1.9}$$

This implies that T^* is independent of the stream-wise co-ordinate x^*. Since the scaled temperature T^* is zero upstream of the heated section for $x^* < 0$, the solution of Eq. 9.1.9 is $T^* = 0$ everywhere. However, this solution does not satisfy the boundary conditions $T^* = 1$ at $z^* = 0$, Eq. 9.1.6.

The boundary condition cannot be satisfied because the original equation, Eq. 9.1.5, which is a second order partial differential equation in z^*, has been reduced to Eq. 9.1.9, which is zeroeth order in z^*. Therefore, it is not possible to satisfy two boundary conditions in the z^* direction. Physically, this is because we have neglected cross-stream diffusion, which is the mechanism of heat transfer from the surface to the fluid.

The effect of diffusion is included by postulating a boundary layer of thickness $\delta \ll L$ where diffusion is comparable to convection. Since the fluid velocity v_x is zero at the surface, the fluid velocity is approximated as $v_x = v_s' z$ near the surface, where $v_s' = (dv_x/dz)|_{z=0}$ is the strain rate at $z = 0$, as shown in Fig. 9.1. With this approximation, the convection–diffusion equation in the boundary layer is

$$v_s' z \frac{\partial T^*}{\partial x} = \alpha \left(\frac{\partial^2 T^*}{\partial x^2} + \frac{\partial^2 T^*}{\partial z^2}\right). \tag{9.1.10}$$

The scaled z co-ordinate is defined as $z^\dagger = (z/\delta)$, where the boundary layer thickness δ is to be determined by balancing convection and diffusion. The scaled co-ordinates are substituted into the convection–diffusion equation, Eq. 9.1.1,

$$\left(\frac{v_s' \delta z^\dagger}{L} \frac{\partial T^*}{\partial x^*} \right) = \frac{\alpha}{L^2} \frac{\partial^2 T^*}{\partial x^{*2}} + \frac{\alpha}{\delta^2} \frac{\partial^2 T^*}{\partial z^{\dagger 2}}. \tag{9.1.11}$$

Eq. 9.1.11 is divided by the coefficient of the second term on the right to obtain,

$$\left(\frac{v_s' \delta^3 z^\dagger}{\alpha L} \frac{\partial T^*}{\partial x^*} \right) = \frac{\delta^2}{L^2} \frac{\partial^2 T^*}{\partial x^{*2}} + \frac{\partial^2 T^*}{\partial z^{\dagger 2}}. \tag{9.1.12}$$

The stream-wise diffusion (first term on the right) in Eq. 9.1.12 is neglected, since this is smaller than the cross-stream diffusion (second term on the right) by a factor $(\delta/L)^2$ for $\delta \ll L$. There is a balance between stream-wise convection, the term on the left, and cross-stream diffusion, the second term on the right, in Eq. 9.1.12.

The simplified equation for the temperature field, which includes stream-wise convection and cross-stream diffusion, is

$$v_s' z \frac{\partial T^*}{\partial x} = \alpha \left(\frac{\partial^2 T^*}{\partial z^2} \right). \tag{9.1.13}$$

A similarity solution for Eq. 9.1.13 can be obtained based on the following simplification. When stream-wise diffusion is neglected, the temperature at a location is determined by convection from regions upstream of this location, and is not affected by the temperature distribution downstream of this location. For example, at a point (x, z) in the heated region $(0 < x < L)$, the temperature will depend only on the distance of x from the beginning of the heated section and the distance z from the surface, but not on the total length L of the heated section. This is because convection transports heat in the downstream direction, and we have neglected stream-wise diffusion which could transport heat upstream and downstream. Therefore, the total length L is not a relevant parameter at the location x, and it is appropriate to define the boundary layer thickness $\delta_x(x)$ which depends on the co-ordinate x.

The similarity variable for Eq. 9.1.13 is defined as

$$\eta = \frac{z}{\delta_x(x)}, \tag{9.1.14}$$

where $\delta_x(x)$, the boundary layer thickness at the location x, is defined in such a way that after transforming the independent variables from (x, z) to η, the resulting equation depends only on η. The temperature T^* is then only a function of the similarity variable η.

The distinct roles for the scaled similarity variable η and the boundary layer thickness δ_x in the analysis are as follows. The boundary layer thickness δ_x is a function of the Peclet number and the stream-wise co-ordinate x, and the boundary layer thickness decreases as the Peclet number increases. The boundary layer thickness can be understood as the inverse of the factor by which the actual distance from the surface has to be magnified (as shown in Fig. 9.1) at each stream-wise location x and Peclet number, so that the solution obtained in terms of the similarity variable η is independent of Pe and x. The boundary layer thickness is chosen such that, when the convection–diffusion equation is expressed in terms of the similarity variable $\eta = (z/\delta_x(x))$, the equation does not depend independently on x or Pe, but only on the combination η. The thickness of the boundary layer δ_x decreases to zero for Pe $\rightarrow \infty$, but η remains $O(1)$ in the boundary layer where the thermal diffusion from the surface causes a disturbance to the temperature.

The derivatives of the temperature field are expressed in terms of η using differentiation by chain rule,

$$\frac{\partial T^*}{\partial z} = \frac{\partial \eta}{\partial z}\frac{\mathrm{d}T^*}{\mathrm{d}\eta} = \frac{1}{\delta_x}\frac{\mathrm{d}T^*}{\mathrm{d}\eta},$$

$$\frac{\partial^2 T^*}{\partial z^2} = \frac{\partial \eta}{\partial z}\frac{\mathrm{d}}{\mathrm{d}\eta}\left(\frac{\partial \eta}{\partial z}\frac{\mathrm{d}T^*}{\mathrm{d}\eta}\right) = \frac{1}{\delta_x^2}\frac{\mathrm{d}^2 T^*}{\mathrm{d}\eta^2},$$

$$\frac{\partial T^*}{\partial x} = \frac{\partial \eta}{\partial x}\frac{\mathrm{d}T^*}{\mathrm{d}\eta} = -\frac{z}{\delta_x^2}\frac{\mathrm{d}\delta_x}{\mathrm{d}x}\frac{\mathrm{d}T^*}{\mathrm{d}\eta} = -\frac{\eta}{\delta_x}\frac{\mathrm{d}\delta_x}{\mathrm{d}x}\frac{\mathrm{d}T^*}{\mathrm{d}\eta}. \qquad (9.1.15)$$

The above expressions are substituted into the convection–diffusion equation, Eq. 9.1.13, to obtain,

$$v_s'\eta\delta_x\left(-\frac{\eta}{\delta_x}\frac{\mathrm{d}\delta_x}{\mathrm{d}x}\frac{\mathrm{d}T^*}{\mathrm{d}\eta}\right) = \frac{\alpha}{\delta_x^2}\frac{\mathrm{d}^2 T^*}{\mathrm{d}\eta^2}. \qquad (9.1.16)$$

The boundary conditions, Eqs. 9.1.2–9.1.4, expressed in terms of the similarity variable η, are

$$T^* = 1 \text{ at } z = 0 \,\&\, x > 0 \implies \boxed{T^* = 1 \text{ at } \eta = 0,} \qquad (9.1.17)$$

$$T^* = 0 \text{ for } z \rightarrow \infty \implies \boxed{T^* = 0 \text{ for } \eta \rightarrow \infty,} \qquad (9.1.18)$$

$$T^* = 0 \text{ for } x = 0 \,\&\, z > 0 \implies T^* = 0 \text{ for } \eta \rightarrow \infty. \qquad (9.1.19)$$

In Eq. 9.1.19, we have assumed that $\delta_x = 0$ at $x = 0$—that is, the boundary layer thickness is zero at the start of the heating section.

The convection–diffusion Eq. 9.1.13 is a second order equation in the z co-ordinate and a first order equation in the x co-ordinate. This requires two boundary conditions, Eqs. 9.1.17 and 9.1.18, in the z direction, and one 'initial' condition, Eq. 9.1.19 which states that the scaled temperature is zero for all z at the upstream edge of the heated section. When the similarity transform is applied, Eq. 9.1.16 is a second order equation in η, which requires two boundary conditions in the η co-ordinate. As required, we find that two boundary conditions, Eqs. 9.1.18 and 9.1.19, turn out to be identical when expressed in terms of η, so that there are only two independent boundary conditions for Eq. 9.1.16. This consistency requirement for the boundary conditions has to be satisfied in order to apply the similarity transform.

Eq. 9.1.16 is multiplied by (δ_x^2/α) to obtain,

$$-\left(\frac{v_s' \delta_x^2}{\alpha}\frac{d\delta_x}{dx}\right)\eta^2\frac{dT^*}{d\eta} = \frac{d^2T^*}{d\eta^2}. \qquad (9.1.20)$$

In the above equation, the left side is a function of x and η, whereas the right side is only a function of η. A similarity solution can be obtained only if the term in brackets on the left side of Eq. 9.1.20 is independent of x,

$$\boxed{\frac{v_s' \delta_x^2}{\alpha}\frac{d\delta_x}{dx} = C_\delta,} \qquad (9.1.21)$$

where C_δ is a constant. It turns out that the final result for the temperature field is independent of the C_δ. The solution of Eq. 9.1.20 expressed in terms of η does depend on the value of C_δ, but the final solution in terms of the original variables (x, z) does not depend on C_δ. In the present problem, we retain an unspecified constant C_δ on the right side of Eq. 9.1.21, in order to show that the final solution is independent of the value of C_δ. In subsequent problems, we will set the value of C_δ equal to 1 without loss of generality.

Eq. 9.1.21 is solved to obtain,

$$\delta_x = ((3C_\delta x\alpha/v_s') + C)^{1/3}. \qquad (9.1.22)$$

The constant of integration C is fixed by the condition that $\delta_x = 0$ at the upstream edge of the heated surface, $x = 0$, since there is no heating for $x < 0$,

$$\delta_x = 0 \text{ at } x = 0. \qquad (9.1.23)$$

Eq. 9.1.23 is satisfied if $C = 0$ in Eq. 9.1.22, and the solution for δ_x is,

$$\delta_x = \left(\frac{3C_\delta x\alpha}{v_s'}\right)^{1/3} = \left(\frac{3C_\delta\alpha}{x^2 v_s'}\right)^{1/3} x = \left(\frac{3C_\delta}{\mathrm{Pe}_{v_s'x}}\right)^{1/3} x, \qquad (9.1.24)$$

where $\mathrm{Pe}_{v'_s x} = (x^2 v'_s / \alpha)$ is the Peclet number based on the strain rate at the surface v'_s and the distance x from the upstream edge of the heated surface. Eq. 9.1.24 shows that the boundary layer thickness increases proportional to $x^{1/3}$.

When Eq. 9.1.21 for the boundary layer thickness is substituted into Eq. 9.1.20, the equation for the temperature field in terms of η is

$$-C_\delta \eta^2 \frac{\mathrm{d}T^*}{\mathrm{d}\eta} = \frac{\mathrm{d}^2 T^*}{\mathrm{d}\eta^2}. \qquad (9.1.25)$$

The above equation is reduced to a first order differential equation in η by the substitution $T^{*\prime} = (\mathrm{d}T^*/\mathrm{d}\eta)$,

$$\frac{\mathrm{d}T^{*\prime}}{\mathrm{d}\eta} = -C_\delta \eta^2 T^{*\prime}. \qquad (9.1.26)$$

This equation is divided by $T^{*\prime}$, integrated one time with respect to η, and exponentiated to obtain,

$$T^{*\prime} = \frac{\mathrm{d}T^*}{\mathrm{d}\eta} = C_1 \mathrm{e}^{(-C_\delta \eta^3 / 3)}. \qquad (9.1.27)$$

The expression for T^* is obtained by integrating the above equation,

$$T^* = C_1 \int_0^\eta \mathrm{d}\eta' \, \mathrm{e}^{(-C_\delta \eta'^3 / 3)} + C_2, \qquad (9.1.28)$$

where C_1 and C_2 are constants of integration. The boundary conditions, Eqs. 9.1.17–9.2.18, are satisfied if the constants are,

$$C_1 = -\frac{1}{\int_0^\infty \mathrm{d}\eta' \, \mathrm{e}^{(-C_\delta \eta'^3 / 3)}} = -\frac{3^{2/3} C_\delta^{1/3}}{\Gamma(1/3)},$$
$$C_2 = 1, \qquad (9.1.29)$$

where Γ is the Gamma function. More details regarding the Gamma function and its properties, and the integral in the expression for C_1, are provided in the Appendix 9.A.

The solution for the temperature field is obtained by substituting the constants (Eq. 9.1.29) in Eq. 9.1.28,

$$T^* = 1 - \frac{\int_0^\eta \mathrm{d}\eta' \, \mathrm{e}^{(-C_\delta \eta'^3 / 3)}}{\int_0^\infty \mathrm{d}\eta' \, \mathrm{e}^{(-C_\delta \eta'^3 / 3)}}$$
$$= 1 - \frac{3^{2/3} C_\delta^{1/3}}{\Gamma(1/3)} \int_0^{z/(3 C_\delta x \alpha / v'_s)^{1/3}} \mathrm{d}\eta' \, \mathrm{e}^{(-C_\delta \eta'^3 / 3)}. \qquad (9.1.30)$$

In the second line above, we have substituted the expression, Eq. 9.1.14, for η in the upper limit of integration, with δ_x given by Eq. 9.1.24.

It can be shown that the solution, Eq. 9.1.30, is independent of the value of C_δ, by defining a new variable of integration $\eta^* = C_\delta^{1/3}\eta'$. The substitutions $d\eta' = d\eta^*/C_\delta^{1/3}$ are made in the integral in Eq. 9.1.30, and the upper limit of integration is multiplied by $C_\delta^{1/3}$, to obtain the similarity solution,

$$T^* = 1 - \frac{3^{2/3}}{\Gamma(1/3)} \int_0^{z/(3x\alpha/v_s')^{1/3}} d\eta^* e^{(-\eta^{*3}/3)}. \qquad (9.1.31)$$

It is evident that the above solution, expressed in terms of the physical co-ordinates, is independent of the value of C_δ. Therefore, we can set C_δ to any finite value without loss of generality.

The heat flux in the z direction at the surface, $q_z|_{z=0}$, can now be calculated,

$$q_z|_{z=0} = -k\left.\frac{dT}{dz}\right|_{z=0} = -\frac{k(T_w - T_\infty)}{\delta_x}\left.\frac{dT^*}{d\eta}\right|_{\eta=0} = -\frac{k(T_w - T_\infty)C_1}{\delta_x}$$

$$= \frac{k(T_w - T_\infty)}{\delta_x}\frac{3^{2/3}C_\delta^{1/3}}{\Gamma(1/3)}. \qquad (9.1.32)$$

Here, $(dT^*/d\eta)|_{\eta=0}$ is determined from Eq. 9.1.27, and Eq. 9.1.29 is substituted for C_1. Substituting Eq. 9.1.24 for δ_x in Eq. 9.1.32, we obtain

$$q_z|_{z=0} = \frac{k(T_w - T_\infty)}{(3x\alpha/v_s')^{1/3}}\frac{3^{2/3}}{\Gamma(1/3)}. \qquad (9.1.33)$$

As expected, Eq. 9.1.33 for the heat flux is independent of the constant C_δ in Eq. 9.1.21.

The average heat flux q_{av} is,

$$q_{av} = \frac{1}{L}\int_0^L dx\ q_z|_{z=0} = \frac{(T_w - T_\infty)}{(\alpha L/v_s')^{1/3}}\frac{3^{4/3}}{2\Gamma(1/3)} = \frac{0.81k(T_w - T_\infty)}{(\alpha L/v_s')^{1/3}}. \qquad (9.1.34)$$

The dimensionless Nusselt number is,

$$\mathrm{Nu} = \frac{q_{av}L}{k(T_w - T_\infty)} = 0.81\,\mathrm{Pe}_{v_s'L}^{1/3}, \qquad (9.1.35)$$

where $\mathrm{Pe}_{v_s'L} = (v_s'L^2/\alpha)$ is the Peclet number based on the wall strain rate and the length of the heated section.

It is important to note that the Nusselt number is related to the Peclet number based on the strain rate at the wall v_s', and not to the characteristic velocity v_c.

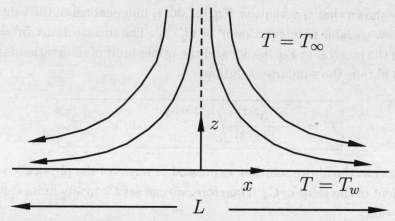

FIGURE 9.2. The configuration and co-ordinate system for the high Peclet number heat transfer in the stagnation point flow of a fluid towards a surface.

This is because convection and diffusion are comparable within a thin boundary layer at the surface. Since the thickness of the boundary layer is much smaller than the characteristic flow scale and the velocity is zero at this surface, the velocity can be approximated as $v_x = v'_s z$. The Peclet number is based on the total length of the length L of the heated section, because the system is considered of infinite extent in the cross-stream direction. The correlation applies for a system of finite extent in the cross-stream direction, provided the cross-stream length scale is large compared to the boundary layer thickness.

The similarity transform can be used even when the flow is two-dimensional, as illustrated in the following example of the stagnation point flow.

EXAMPLE 9.1.1: Consider the 'stagnation point' flow of a fluid incident on a rigid flat plate of length L, shown in Fig. 9.2. When there is a no-slip condition at the surface, the velocity close to the surface can be approximated as,

$$v_x = Kxz, \quad v_z = -\tfrac{1}{2}Kz^2, \tag{9.1.36}$$

where K is a constant. The temperature far from the plate is T_∞, and the plate is at a higher temperature T_w. What is the Peclet number for the heat transfer from the surface? Determine the Nusselt number correlation for high Peclet number.

Solution: The scaled temperature is $T^* = (T - T_\infty)/(T_w - T_\infty)$. The convection–diffusion equation expressed in terms of the scaled temperature is

$$v_x \frac{\partial T^*}{\partial x} + v_z \frac{\partial T^*}{\partial z} = \alpha \frac{\partial^2 T^*}{\partial z^2}. \tag{9.1.37}$$

Here, stream-wise diffusion has been neglected in comparison to cross-stream diffusion at high Peclet number. Substituting Eq. 9.1.36 for the velocity field, the convection–diffusion equation is

$$Kxz\frac{\partial T^*}{\partial x} - \frac{Kz^2}{2}\frac{\partial T^*}{\partial z} = \alpha\frac{\partial^2 T^*}{\partial z^2}. \tag{9.1.38}$$

The temperature boundary conditions are,

$$T^* = 0 \text{ for } z \to \infty, \tag{9.1.39}$$
$$T^* = 1 \text{ at } z = 0. \tag{9.1.40}$$

The condition at $x = 0$ is not obvious at present, but it will become clear when we determine the boundary layer thickness.

The similarity variable is defined as $\eta = (z/\delta_x)$. As in the case of the flow past a flat plate, we assume that the boundary layer thickness δ_x depends only on the distance x from the stagnation point, and not on the total length of the plate L. Substituting Eq. 9.1.15 for the spatial derivatives, the convection diffusion equation, Eq. 9.1.38 is

$$Kxz\left(-\frac{\eta}{\delta_x}\frac{\mathrm{d}\delta_x}{\mathrm{d}x}\right)\frac{\mathrm{d}T^*}{\mathrm{d}\eta} - \frac{Kz^2}{2\delta_x}\frac{\mathrm{d}T^*}{\mathrm{d}\eta} = \frac{\alpha}{\delta_x^2}\frac{\mathrm{d}^2 T^*}{\mathrm{d}\eta^2}. \tag{9.1.41}$$

The above equation is multiplied by (δ_x^2/α), and the substitution $z = \delta_x\eta$ is made, to obtain

$$-\left[\frac{K}{\alpha}\left(x\delta_x^2\frac{\mathrm{d}\delta_x}{\mathrm{d}x} + \frac{\delta_x^3}{2}\right)\right]\eta^2\frac{\mathrm{d}T^*}{\mathrm{d}\eta} = \frac{\mathrm{d}^2 T^*}{\mathrm{d}\eta^2}. \tag{9.1.42}$$

A similarity solution can be obtained if the term in the square brackets on the left is a constant, and the constant can be set equal to 1 without loss of generality. The equation for the boundary layer thickness is

$$\boxed{\frac{K}{\alpha}\left(x\delta_x^2\frac{\mathrm{d}\delta_x}{\mathrm{d}x} + \frac{\delta_x^3}{2}\right) = 1,} \tag{9.1.43}$$

and the equation for the temperature field in terms of the similarity variable η is the same as Eq. 9.1.25 for the flow past a flat plate with $C_\delta = 1$.

Eq. 9.1.43 is multiplied by $(3\alpha/K)$ to obtain a first order inhomogeneous linear differential equation for δ_x^3,

$$\frac{\mathrm{d}\delta_x^3}{\mathrm{d}x} + \frac{3\delta_x^3}{2x} = \frac{3\alpha}{Kx}. \tag{9.1.44}$$

The procedure for solving this equation is summarised in Appendix 9.B. The integrating factor for this equation is

$$\mathrm{IF} = \mathrm{e}^{\left(\int \mathrm{d}x\, \frac{3}{2x}\right)} = \mathrm{e}^{\left(\frac{3\ln(x)}{2}\right)} = x^{3/2}. \tag{9.1.45}$$

The solution for δ_x^3 is expressed in terms of the integrating factor,

$$\delta_x^3 = \frac{C}{x^{3/2}} + \frac{1}{x^{3/2}} \int_0^x \mathrm{d}x'\, x'^{3/2} \frac{3\alpha}{Kx'} = \frac{C}{x^{3/2}} + \frac{2\alpha}{K}. \tag{9.1.46}$$

The constant C is determined from the condition that the boundary layer thickness is finite at $x = 0$. The streamline $x = 0$ is the 'stagnation streamline', shown by the dashed vertical line in Fig. 9.2, where the flow is towards the surface. When there is strong convection towards the surface, it is unphysical for the boundary layer thickness to be infinite at $x = 0$. Therefore, $C = 0$ in Eq. 9.1.46, and the boundary layer thickness $\delta_x = (2\alpha/K)^{1/3}$ is independent of the downstream distance x.

The similarity equation for the temperature field, Eq. 9.1.25, is solved to obtain the Eq. 9.1.32 for the heat flux at the surface. Substituting the expression $\delta_x = (2\alpha/K)^{1/3}$ into Eq. 9.1.32, the heat flux is

$$q_z|_{z=0} = \frac{k(T_w - T_\infty)}{(2\alpha/K)^{1/3}} \frac{3^{2/3}}{\Gamma(1/3)} = \frac{0.62k(T_w - T_\infty)}{(\alpha/K)^{1/3}}. \tag{9.1.47}$$

Since the heat flux is independent of the x, the average flux is equal to q_z at $z = 0$. The Nusselt number is

$$\boxed{\mathrm{Nu} = \frac{q_z|_{z=0}}{k(T_w - T_\infty)/L} = 0.62 \left(\frac{KL^3}{\alpha}\right)^{1/3} = 0.62\, \mathrm{Pe}_{KL}^{1/3},} \tag{9.1.48}$$

where Pe_{KL} is the Peclet number based on the length L and the parameter K in Eq. 9.1.36 for the velocity.   □

Eq. 9.1.34 for the average heat flux is based on the strain rate at the surface v_s' and the total length of the heated section L. In practical problems, it is necessary to relate the wall strain rate to the characteristic velocity, in order to obtain a Nusselt number correlation which depends on the characteristic velocity and length. This is illustrated in the following section for the flow in a pipe.

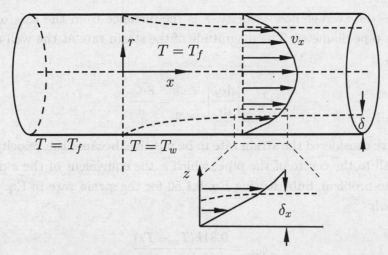

FIGURE 9.3. The configuration and co-ordinate system for analysing the thermal boundary layer at the wall of a pipe at high Peclet number.

9.1.2 Flow in a Pipe

Consider the pressure-driven laminar flow in a pipe of length L and diameter d shown in Fig. 9.3. The fluid at the entrance has a uniform temperature T_f, while the wall of the pipe has temperature T_w. Due to conduction from the surface, there is a boundary layer at the surface of the pipe which grows with downstream distance, until the boundary layer becomes comparable to the radius. When the boundary layer thickness is small compared to the radius, curvature effects are negligible in the boundary layer. The temperature field can be approximated as that near the flat plate analysed in the previous section if curvature effects are neglected. We first obtain the average heat flux with the 'thin boundary layer' approximation (boundary layer thickness small compared to the radius), and then examine the pipe length for which the thin boundary layer approximation is valid.

To analyse the temperature field in the boundary layer, it is convenient to define a co-ordinate z which is the distance from the pipe wall. The configuration and cylindrical co-ordinate system are shown in Fig. 9.3. When the curvature is neglected, the configuration is identical to that for the flow past a flat plate in Fig. 9.1. The average heat flux is expressed as a function of the strain rate at the wall in Eq. 9.1.34.

The parabolic velocity profile for the fully developed laminar flow was derived in Section 6.2 (Chapter 6),

$$v_x = 2v_{av} \left(1 - \frac{4r^2}{d^2} \right). \tag{9.1.49}$$

where v_{av} is the average flow velocity, r is the distance from the axis of the pipe, and d is the pipe diameter. The magnitude of the strain rate at the wall of the pipe $r = (d/2)$ is

$$v_s' = \left| \frac{\mathrm{d}v_x}{\mathrm{d}r} \right|_{r=d/2} = \frac{8v_{av}}{d}. \qquad (9.1.50)$$

Here, we have considered the strain rate to be positive, because the velocity increases from the wall to the centre of the pipe, which is the equivalent of the z direction in the flat plate problem. Substituting Eq. 9.1.50 for the strain rate in Eq. 9.1.34, the average flux is

$$q_{av} = \frac{0.81k(T_w - T_f)}{(\alpha L d/8v_{av})^{1/3}}. \qquad (9.1.51)$$

The Nusselt number correlation is

$$\mathrm{Nu} = \frac{q_{av}}{k(T_w - T_f)/d} = 1.62 \left(\frac{v_{av}d^2}{\alpha L} \right)^{1/3} = 1.62\,\mathrm{Pe}^{1/3}(d/L)^{1/3}, \qquad (9.1.52)$$

where $\mathrm{Pe} = (v_{av}d/\alpha)$ is the Peclet number based on the mean velocity and the pipe diameter. The correlation, Eq. 9.1.52, is numerically accurate to within 15% of the correlation, Eq. 2.3.4 in Chapter 2 for the laminar flow. The difference in the coefficient can be attributed to the pipe curvature, as well as the developing flow in the entrance section of the pipe which is not parabolic.

Next, we consider the conditions under which the 'thin boundary layer' approximation is valid. The boundary layer for the flow past the flat plate increases as $\delta_x = (3\alpha L/v_s')^{1/3}$. Since the strain rate at the wall is $(8v_{av}/d)$ (Eq. 9.1.50), the boundary layer thickness increases as $\delta_x = (3\alpha L d/8v_{av})^{1/3}$. The ratio of the boundary layer thickness and the pipe radius is

$$\frac{\delta_x}{(d/2)} = \left(\frac{3\alpha L}{d^2 v_{av}} \right)^{1/3} = \mathrm{Pe}^{-1/3}(3L/d)^{1/3}. \qquad (9.1.53)$$

Therefore, the 'thin boundary layer' approximation is valid when the above ratio is small,

$$\frac{L}{d} \ll \frac{\mathrm{Pe}}{3}. \qquad (9.1.54)$$

The 'thin boundary layer approximation' is valid in many practical situations, as illustrated by the following example.

EXAMPLE 9.1.2: Consider the flow of water with average velocity 1 cm/s in a pipe of diameter 2 cm. What is the length of the pipe for which the boundary layer thickness is comparable to the pipe radius? The thermal diffusivity of water is $1.5 \times 10^{-7} \text{ m}^2/\text{s}$.

Solution: The Peclet number $(v_{av}d/\alpha) = 1.33 \times 10^3$. The boundary layer thickness is comparable to the radius for $(L/d) \sim (\text{Pe}/3) \sim 4.44 \times 10^2$. Thus, the pipe radius and boundary layer thickness are comparable for $L \sim 8.9$ m. $\qquad\square$

EXAMPLE 9.1.3: The velocity profile for the flow of a power-law fluid in a pipe was derived in Example 6.2.2 in Chapter 6. What is the ratio of the boundary layer thickness and the pipe radius at a downstream distance x from the start of the heating section? Determine the Nusselt number correlation for a pipe of length L.

Solution: From Example 6.2.2, the average velocity (Eq. 6.2.36) and the strain rate at the wall (obtained by differentiating Eq. 6.2.35 with respect to r) are expressed as functions of the pressure gradient,

$$v_{av} = \left(-\frac{1}{2\kappa}\frac{dp}{dx}\right)^{1/n} \frac{n}{3n+1} \left(\frac{d}{2}\right)^{((n+1)/n)}, \qquad (9.1.55)$$

$$\left|\frac{dv_x}{dr}\right|_{r=d/2} = \left(-\frac{1}{2\kappa}\frac{dp}{dx}\right)^{1/n} \left(\frac{d}{2}\right)^{1/n}. \qquad (9.1.56)$$

The velocity and strain rate are expressed in terms of the diameter d instead of the radius in the above expressions. The strain rate is expressed in terms of the average velocity as,

$$\left|\frac{dv_x}{dr}\right|_{r=d/2} = \frac{2(3n+1)v_{av}}{nd}. \qquad (9.1.57)$$

The boundary layer thickness is determined by substituting the wall strain rate into Eq. 9.1.24,

$$\delta_x = \left(\frac{3\alpha x}{v_s'}\right)^{1/3} = \left(\frac{3\alpha n x d}{2(3n+1)v_{av}}\right)^{1/3}. \qquad (9.1.58)$$

The ratio of the boundary layer thickness and the pipe radius is

$$\frac{\delta_x}{d/2} = \left(\frac{12\alpha n x}{(3n+1)d^2 v_{av}}\right)^{1/3}. \qquad (9.1.59)$$

To determine the Nusselt number correlation, Eq. 9.1.57 for the strain rate is substituted into Eq. 9.1.34 for the average heat flux,

$$q_{av} = \frac{0.81k(T_w - T_\infty)}{(\alpha L)^{1/3}} \left(\frac{2(3n+1)v_{av}}{nd} \right)^{1/3}. \tag{9.1.60}$$

The Nusselt number is

$$\text{Nu} = \frac{q_{av}d}{k(T_w - T_\infty)} = 0.81 \, \text{Pe}^{1/3} \left(\frac{2(3n+1)}{n} \right)^{1/3} \left(\frac{d}{L} \right)^{1/3}, \tag{9.1.61}$$

where the Peclet number is $(v_{av}d/\alpha)$. It is easy to confirm that the above equation reduces to Eq. 9.1.52 for a $n = 1$, which is a Newtonian fluid. □

The Nusselt number correlation for a turbulent flow, Eq. 2.3.6 in Chapter 2, though empirical, is in agreement with experimental results for Reynolds number greater than about 10,000. This correlation cannot be derived in a manner similar to the correlation for a laminar flow, due to the complexity of the structure of a turbulent flow discussed in Section 6.3 (Chapter 6). The velocity profile is linear within the viscous sub-layer close to the wall of extent about $(5\nu/v_*)$, where ν is the kinematic viscosity, $v_* = \sqrt{\tau_w/\rho}$ is the friction velocity, τ_w and ρ are the wall shear stress and the density, respectively. If the boundary layer thickness is small compared to the thickness of the viscous sub-layer, the velocity can be approximated as the strain rate times the distance from the wall, and the Nusselt number can be related to the friction Reynolds number; this is considered in Exercise 9.3. However, this condition on the boundary layer thickness is restrictive, and cannot be applied in many practical situations.

9.1.3 Diffusion from a Solid Particle

Consider fluid flow around a stationary spherical particle of radius R shown in Fig. 9.4. Far from the particle, there is a constant free-stream fluid velocity of magnitude v_{fs} in the $-x$ direction, while the fluid velocity at the surface of the particle is zero due to the no-slip boundary condition. The temperature at the surface of the sphere is T_0, while the temperature far from the sphere in the incoming fluid stream is T_∞. We would like to determine the heat flux and the Nusselt number correlation in the limit of high Peclet number $\text{Pe} = (v_{fs}d/\alpha) \gg 1$ based on the sphere diameter d and the fluid velocity relative to the sphere v_{fs}.

Based on the symmetry of the configuration, it is appropriate to use a spherical (r, θ, ϕ) co-ordinate system, in which the radius r is the distance from the centre of the sphere, θ is the angle made by the position vector with the x axis and ϕ is the meridional angle around the x axis. Since the configuration is axisymmetric around the x axis, there is no dependence on the angle ϕ. The velocity has components v_r, v_θ in the r, θ directions, respectively, and the velocity and temperature fields depend only on (r, θ). The convection–diffusion equation at steady state for an incompressible flow is,[1]

$$v_r \frac{\partial T}{\partial r} + \frac{v_\theta}{r} \frac{\partial T}{\partial \theta} = \alpha \left(\frac{1}{r^2} \frac{\partial}{\partial r} \left(r^2 \frac{\partial T}{\partial r} \right) + \frac{1}{r^2 \sin(\theta)} \frac{\partial}{\partial \theta} \left(\sin(\theta) \frac{\partial T}{\partial \theta} \right) \right), \qquad (9.1.62)$$

with boundary conditions

$$T = T_0 \text{ at } r = R, \qquad (9.1.63)$$

$$T = T_\infty \text{ for } r \to \infty. \qquad (9.1.64)$$

The scaled temperature and velocity are defined as $T^* = (T - T_\infty)/(T_0 - T_\infty)$, $v_r^* = (v_r/v_{fs})$ and $v_\theta^* = (v_\theta/v_{fs})$, and the scaled radius is $r^* = (r/R)$. The convection–diffusion equation and the boundary conditions expressed in terms of

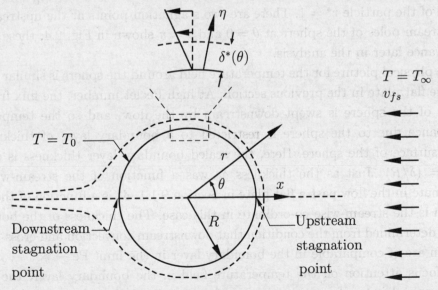

FIGURE 9.4. The configuration and co-ordinate system for analysing the thermal boundary layer in the flow past a spherical particle at high Peclet number.

---

[1]The convection term $\boldsymbol{\nabla} \cdot (\boldsymbol{v}T)$ in the convection–diffusion equation, Eq. 7.1.28, is expressed using product rule as $\boldsymbol{v} \cdot \boldsymbol{\nabla} T + T \boldsymbol{\nabla} \cdot \boldsymbol{v}$. The term $T \boldsymbol{\nabla} \cdot \boldsymbol{v}$ is zero for an incompressible flow, and the convection term is $\boldsymbol{v} \cdot \boldsymbol{\nabla} T$.

scaled variables are

$$\frac{\text{Pe}}{2}\left(v_r^*\frac{\partial T^*}{\partial r^*}+\frac{v_\theta^*}{r^*}\frac{\partial T^*}{\partial\theta}\right)=\left(\frac{1}{r^{*2}}\frac{\partial}{\partial r^*}\left(r^{*2}\frac{\partial T^*}{\partial r^*}\right)+\frac{1}{r^{*2}\sin(\theta)}\frac{\partial}{\partial\theta}\left(\sin(\theta)\frac{\partial T^*}{\partial\theta}\right)\right),$$

$$\text{(9.1.65)}$$

$$T^*=1 \text{ at } r^*=1, \tag{9.1.66}$$

$$T^*=0 \text{ for } r^*\to\infty, \tag{9.1.67}$$

where $\text{Pe} = (v_{fs}d/\alpha) = (2v_{fs}R/\alpha)$ is the Peclet number based on the sphere diameter d and the constant flow velocity v_{fs} far from the sphere.

It is necessary to specify the velocity field around the sphere in order to obtain the temperature field. The velocity components for the laminar flow around the sphere in the co-ordinate system in Fig. 9.4 are

$$v_r^* = -\cos(\theta)\left(1-\frac{3}{2r^*}+\frac{1}{2r^{*3}}\right),\quad v_\theta^* = \sin(\theta)\left(1-\frac{3}{4r^*}-\frac{1}{4r^{*3}}\right). \tag{9.1.68}$$

The derivation of the velocity components, Eq. 9.1.68, is beyond the scope of this text, but it can be verified that these velocity profiles satisfy the incompressibility condition $\nabla\cdot v = 0$ and the no-slip boundary conditions, $v_r^* = 0, v_\theta^* = 0$ at the surface of the particle $r^* = 1$. There are two stagnation points at the upstream and downstream poles of the sphere at $\theta = 0$ and $\theta = \pi$ shown in Fig. 9.4; these will be of relevance later in the analysis.

The physical picture for the temperature field around the sphere is similar to that near the flat plate in the previous section. At high Peclet number, the flux from the surface of the sphere is swept downstream by the flow, and so the temperature disturbance due to the sphere is restricted to a boundary layer of thickness δ^* at the surface of the sphere. Here, the scaled boundary layer thickness is defined as $\delta^* = (\delta/R)$. Just as the thickness δ_x was a function of the stream-wise (x) co-ordinate in the flow past a flat plate in Section 9.1.1, δ^* is a function of the angle θ which is the stream-wise co-ordinate in this case. The thickness of the boundary layer is determined from the condition that downstream convection and cross-stream diffusion are of comparable in the boundary layer in the limit $\text{Pe}\to\infty$.

To focus attention on the temperature field in the boundary layer, the scaled cross-stream distance from the sphere surface is defined as

$$\eta = \frac{(r^*-1)}{\delta^*(\theta)}. \tag{9.1.69}$$

The similarity variable η is zero at the surface of the sphere, and it increases radially outwards from the surface. In the following analysis, the thin boundary layer approximation $\delta^* \ll 1$ (boundary layer thickness small compared to R) is used to simplify the equations. The expressions for the velocity components, Eq. 9.1.68, are too complicated to obtain an analytical solution for the convection–diffusion equation. Within the boundary layer, these are approximated using Taylor expansions in the small parameter $\delta^*(\theta)$. The similarity variable η (Eq. 9.1.69) is substituted into the expressions for the velocity components, Eq. 9.1.68,

$$v_r^* = -\left(1 - \frac{3}{2(1 + \delta^*\eta)} + \frac{1}{2(1 + \delta^*\eta)^3}\right)\cos(\theta),$$

$$v_\theta^* = \left(1 - \frac{3}{4(1 + \delta^*\eta)} - \frac{1}{4(1 + \delta^*\eta)^3}\right)\sin(\theta). \tag{9.1.70}$$

A Taylor series expansion is used in the small parameter δ^* within the boundary layer,

$$(1 + \delta^*\eta)^{-n} = 1 - n\delta^*\eta + \frac{n(n + 1)(\delta^*\eta)^2}{2} + \dots. \tag{9.1.71}$$

The above expansion is substituted into the expressions, Eq. 9.1.70, and the largest terms in the expansions are retained,

$$v_r^* = -\frac{3}{2}\delta^{*2}\eta^2\cos(\theta), \quad v_\theta^* = \frac{3}{2}\delta^*\eta\sin(\theta). \tag{9.1.72}$$

Note that the stream-wise velocity component along the surface, v_θ^*, is proportional to $\delta^*\eta$, while the cross-stream component v_r^* is proportional to $(\delta^*\eta)^2$. The stream-wise component is zero at the sphere surface $\eta = 0$ due to the no-slip condition, and this component increases proportional to the distance from the surface, as shown in Fig. 9.4. The cross-stream component increases proportional to the square of the distance from the surface due to the incompressibility condition, as discussed in Section 9.1.4.

The derivatives of the temperature with respect to r^* and θ are expressed in terms of η derivatives using differentiation by chain rule,

$$\frac{\partial T^*}{\partial r^*} = \frac{1}{\delta^*}\frac{dT^*}{d\eta}, \tag{9.1.73}$$

$$\frac{\partial T^*}{\partial \theta} = -\frac{(1 - r^*)}{\delta^{*2}}\frac{d\delta^*}{d\theta}\frac{dT^*}{d\eta} = -\frac{\eta}{\delta^*}\frac{d\delta^*}{d\theta}\frac{dT^*}{d\eta}. \tag{9.1.74}$$

The left and right sides of Eq. 9.1.65 are expressed in terms of the similarity variable η using the above relations in the the the limit $\delta^* \ll 1$,

$$v_r^* \frac{\partial T^*}{\partial r^*} + \frac{v_\theta^*}{r^*} \frac{\partial T^*}{\partial \theta}$$

$$= -\frac{3}{2} \delta^{*2} \eta^2 \cos(\theta) \frac{1}{\delta^*} \frac{dT^*}{d\eta} + \frac{3}{2} \frac{\delta^* \eta \sin(\theta)}{1 + \delta^* \eta} \left(-\frac{\eta}{\delta^*} \frac{d\delta^*}{d\theta} \frac{dT^*}{d\eta} \right)$$

$$= -\frac{3}{2} \left(\delta^* \cos(\theta) + \sin(\theta) \frac{d\delta^*}{d\theta} \right) \eta^2 \frac{dT^*}{d\eta}, \tag{9.1.75}$$

$$\frac{1}{r^{*2}} \frac{\partial}{\partial r^*} \left(r^{*2} \frac{\partial T^*}{\partial r^*} \right) + \frac{1}{r^{*2} \sin(\theta)} \frac{\partial}{\partial \theta} \left(\sin(\theta) \frac{\partial T^*}{\partial \theta} \right)$$

$$= \frac{1}{(1 + \delta^* \eta)^2} \frac{1}{\delta^*} \frac{d}{d\eta} \left((1 + \delta^* \eta)^2 \frac{1}{\delta^*} \frac{dT^*}{d\eta} \right)$$

$$+ \frac{1}{(1 + \delta^* \eta)^2 \sin(\theta)} \frac{\eta}{\delta^*} \frac{d\delta^*}{d\theta} \frac{d}{d\eta} \left(\sin(\theta) \frac{\eta}{\delta^*} \frac{d\delta^*}{d\theta} \frac{dT^*}{d\eta} \right)$$

$$= \frac{1}{\delta^{*2}} \frac{d^2 T^*}{d\eta^2} + \frac{\eta}{\delta^* \sin(\theta)} \frac{d\delta^*}{d\theta} \frac{d}{d\eta} \left(\cancel{\sin(\theta) \frac{\eta}{\delta^*} \frac{d\delta^*}{d\theta} \frac{dT^*}{d\eta}} \right). \tag{9.1.76}$$

As expected, the cross-stream diffusion term on the right side is $O(1/\delta^{*2})$ larger than the stream-wise diffusion term, due to the large gradient in the cross-stream direction in the boundary layer. Therefore, the stream-wise diffusion term is neglected in the final step in Eq. 9.1.76.

Equating the right sides of Eqs. 9.1.75 and $\alpha\times$ 9.1.76, and multiplying the resulting equation by (δ^{*2}/α), we obtain

$$-\left[\frac{3}{4} \text{Pe} \left(\delta^{*3} \cos(\theta) + \sin(\theta)\delta^{*2} \frac{d\delta^*}{d\theta} \right) \right] \eta^2 \frac{dT^*}{d\eta} = \frac{d^2 T^*}{d\eta^2}. \tag{9.1.77}$$

The boundary conditions are identical to Eqs. 9.1.17–9.1.18 when expressed in terms of the similarity variable η.

For a similarity solution, the term in the square brackets on the left side is a constant. In Section 9.1.1, it was shown that the constant can be set equal to 1 without loss of generality, and the equation for the boundary layer thickness δ^* becomes

$$\boxed{\left[\frac{3}{4} \text{Pe} \left(\delta^{*3} \cos(\theta) + \frac{\sin(\theta)}{3} \frac{d\delta^{*3}}{d\theta} \right) \right] = 1.} \tag{9.1.78}$$

When Eq. 9.1.78 is substituted in Eq. 9.1.77, the equation for the temperature field is the same as Eq. 9.1.25 with $C_\delta = 1$, and the boundary conditions are

Eqs. 9.1.17–9.1.18. Therefore, the solution for T^* as a function of η is Eq. 9.1.31. The radial derivative of the temperature at the surface of the particle is

$$\frac{dT^*}{dr^*}\bigg|_{r^*=1} = \frac{1}{\delta^*}\frac{dT^*}{d\eta}\bigg|_{\eta=0} = -\frac{3^{2/3}}{\delta^*(\theta)\Gamma(1/3)}. \tag{9.1.79}$$

To determine the boundary layer thickness, Eq. 9.1.78 is recast as a first order linear inhomogeneous differential equation for δ^{*3},

$$\frac{d\delta^{*3}}{d\theta} + \frac{3\delta^{*3}\cos(\theta)}{\sin(\theta)} = \frac{4}{\text{Pe}\sin(\theta)}. \tag{9.1.80}$$

The integrating factor for this differential equation (see Appendix 9.B) is

$$\text{IF} = e^{\left(\int d\theta \frac{3\cos(\theta)}{\sin(\theta)}\right)} = e^{(3\,\ln(\sin(\theta)))} = \sin(\theta)^3. \tag{9.1.81}$$

The solution of Eq. 9.1.80 is expressed in terms of the integrating factor,

$$\begin{aligned}
\delta^{*3} &= \frac{C}{\sin(\theta)^3} + \frac{1}{\sin(\theta)^3}\int_0^\theta d\theta'\frac{4}{\text{Pe}\sin(\theta')}\times\sin(\theta')^3 \\
&= \frac{C}{\sin(\theta)^3} + \frac{2\theta - \sin(2\theta)}{\text{Pe}\sin(\theta)^3},
\end{aligned} \tag{9.1.82}$$

where C is a constant of integration. The constant C in Eq. 9.1.82 is determined from the 'initial condition' in the θ co-ordinate as follows. First, note that the solution for δ^*, Eq. 9.1.82, which is proportional to $(1/\sin(\theta)^3)$, diverges at the upstream and downstream stagnation points $\theta = 0$ and $\theta = \pi$. By a suitable choice of the constant C, δ^* can be made finite either at the $\theta = 0$ or $\theta = \pi$, but not both. At the upstream stagnation point $\theta = 0$, the flow is incident on the sphere. At this point, there is a strong convection towards the sphere, and so we would expect the boundary layer thickness to be finite here. In order to obtain a finite δ^* at $\theta = 0$, it is necessary to set the constant $C = 0$, and the solution for the boundary layer thickness becomes

$$\delta^*(\theta) = \frac{(2\theta - \sin(2\theta))^{1/3}}{\text{Pe}^{1/3}\sin(\theta)}. \tag{9.1.83}$$

The boundary layer thickness, Eq. 9.1.83, diverges at the downstream stagnation point $\theta = \pi$. Physically, this is because the velocity v_θ^*, Eq. 9.1.70, goes to zero at this point, and there is no velocity tangential to the surface surface. There is a radial velocity downstream from this point, which convects heat downstream in a thin temperature 'wake' along the streamline from the downstream stagnation point, as shown in Fig. 9.4. In order to determine the temperature field in the wake, it is necessary to rescale the equations in a thin wake region about the downstream stagnation streamline $\theta = \pi$, and balance the convection and diffusion terms in the resulting equation. It turns out that the thickness of the wake region is $O(\text{Pe}^{-1/4})$ in the limit of high Peclet number. However, the heat flux in the wake region does not contribute significantly to the total heat flux from the sphere, or the Nusselt number correlation, and so the wake region is not examined in further detail.

Using the solution, Eq. 9.1.79, for the scaled temperature gradient at the surface and Eq. 9.1.83 for the boundary layer thickness, the heat flux from the surface is given by

$$\boxed{q_r|_{r=R} = -k \left.\frac{\partial T}{\partial r}\right|_{r=R} = -\frac{k(T_0 - T_\infty)}{R\delta^*(\theta)} \left.\frac{\mathrm{d}T^*}{\mathrm{d}\eta}\right|_{\eta=0}}$$

$$\boxed{= \frac{k(T_0 - T_\infty)3^{2/3}}{R\delta^*(\theta)\Gamma(1/3)}.} \tag{9.1.84}$$

The flux is a function of position θ on the surface only due to the θ dependence of the boundary layer thickness $\delta^*(\theta)$. The average heat flux is determined by integrating the heat flux over the surface in a spherical co-ordinate system where the differential area element on the surface is $R^2 \sin(\theta) d\theta d\phi$, and dividing by the total surface area is $4\pi R^2$,

$$q_{av} = \frac{1}{4\pi R^2} \int_0^{2\pi} R^2 \, d\phi \int_0^\pi \sin(\theta) \, d\theta \, q_r|_{r=R}$$

$$= \frac{k(T_0 - T_\infty)\text{Pe}^{1/3}3^{2/3}}{2R\Gamma(1/3)} \int_0^\pi d\theta \frac{\sin(\theta)^2}{(2\theta - \sin(2\theta))^{1/3}}. \tag{9.1.85}$$

Here, Eq. 9.1.83 is substituted for $\delta^*(\theta)$, and the value of the integral over the meridional angle ϕ is 2π, because q_r is independent of ϕ. The integral in Eq. 9.1.85 is evaluated by recognising that the numerator of the integrand can be written as $\frac{1}{2}(1 - \cos(2\theta))$, which is proportional to the derivative of the term $2\theta - \sin(2\theta)$ in the denominator,

$$q_{av} = \frac{k\text{Pe}^{1/3}(T_0 - T_\infty)3^{2/3}}{2R\Gamma(1/3)} \times \left.\frac{3(2\theta - \sin(2\theta))^{2/3}}{8}\right|_0^\pi$$

$$= \frac{3^{5/3}\pi^{2/3}k(T_0 - T_\infty)\text{Pe}^{1/3}}{2^{10/3}\Gamma(1/3)R} = \frac{0.496k(T_0 - T_\infty)\text{Pe}^{1/3}}{R}. \tag{9.1.86}$$

The Nusselt number for the flow around the sphere is given by

$$\boxed{\text{Nu} = \frac{q_{av}}{(k(T_0 - T_\infty)/d)} = 0.992\,\text{Pe}^{1/3}.} \tag{9.1.87}$$

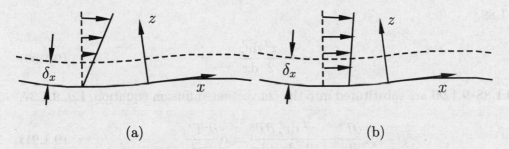

FIGURE 9.5. The configuration and co-ordinate system for a two-dimensional flow past a rigid surface (a) and mobile surface (b). It is assumed that the boundary layer thickness is small compared to the radius of curvature.

Here, the particle diameter is two times the radius, $d = 2R$.[2]

9.1.4 General Flow Fields

The analysis in the previous three sub-sections was carried out for specified velocity fields. This can be extended for a general flow past a surface, provided the boundary layer thickness is small compared to the radius of curvature of the surface, so that the surface can be considered locally flat. For a two-dimensional flow, a local co-ordinate system can be used at the surface, as shown in Fig. 9.5(a).

The tangential velocity is zero at the surface, and therefore the tangential velocity increases linearly from the surface,

$$v_x = v'_s(x) z, \qquad (9.1.88)$$

where z is the distance from the surface, and $v'_s(x) = (\partial v_x / \partial z)|_{z=0}$ is the fluid strain rate at the surface. The velocity field satisfies the incompressibility condition,

$$\frac{\partial v_x}{\partial x} + \frac{\partial v_z}{\partial z} = 0. \qquad (9.1.89)$$

When the boundary layer thickness is much smaller than the characteristic flow length scale, the cross-stream velocity v_z close to the surface is determined from

---

[2]In [2], the coefficient in Eq. 9.1.87 is 1.249, because the Peclet number is based on the particle radius and not the diameter.

Eq. 9.1.89,

$$v_z = -\frac{z^2}{2}\frac{\mathrm{d}v_s'}{\mathrm{d}x}.$$ (9.1.90)

Eqs. 9.1.88–9.1.90 are substituted into the convection–diffusion equation, Eq. 9.1.37,

$$v_s' z \frac{\partial T^*}{\partial x} - \frac{z^2}{2}\frac{\mathrm{d}v_s'}{\mathrm{d}x}\frac{\partial T^*}{\partial z} = \alpha\frac{\partial^2 T^*}{\partial z^2}.$$ (9.1.91)

Here, stream-wise diffusion is neglected in comparison to cross-stream diffusion for high Peclet number. The boundary conditions are given by Eqs. 9.1.17–9.1.18. The z co-ordinate is expressed in terms of the similarity variable, $z = \eta \delta_x(x)$, and the expressions in Eq. 9.1.15, are used or the spatial derivatives,

$$\left[v_s'\eta\delta_x\left(-\frac{\eta}{\delta_x}\frac{\mathrm{d}\delta_x}{\mathrm{d}x}\right) - \frac{\eta^2\delta_x}{2}\frac{\mathrm{d}v_s'}{\mathrm{d}x}\right]\frac{\mathrm{d}T^*}{\mathrm{d}\eta} = \frac{\alpha}{\delta_x^2}\frac{\mathrm{d}^2 T^*}{\mathrm{d}\eta^2}.$$ (9.1.92)

The above equation is multiplied by δ_x^2/α,

$$-\left[\frac{1}{\alpha}\left(v_s'(x)\delta_x^2\frac{\mathrm{d}\delta_x}{\mathrm{d}x} + \frac{\delta_x^3}{2}\frac{\mathrm{d}v_s'}{\mathrm{d}x}\right)\right]\eta^2\frac{\mathrm{d}T^*}{\mathrm{d}\eta} = \frac{\mathrm{d}^2 T^*}{\mathrm{d}\eta^2}.$$ (9.1.93)

For a similarity solution, the term in the square brackets is equal to a constant, which can be set equal to 1 without loss of generality. This results in Eq. 9.1.25 for the temperature field, which is solved to obtain Eq. 9.1.31. The heat flux at the surface is given by Eq. 9.1.33, where the strain rate at the surface v_s' depends on x for a general flow field.

From Eq. 9.1.93, the equation for δ_x is

$$\boxed{\left[\frac{1}{\alpha}\left(\frac{v_s'}{3}\frac{\mathrm{d}\delta_x^3}{\mathrm{d}x} + \frac{\delta_x^3}{2}\frac{\mathrm{d}v_s'}{\mathrm{d}x}\right)\right] = 1.}$$ (9.1.94)

This equation is recast as a first order inhomogeneous differential equation for δ_x^3,

$$\frac{\mathrm{d}\delta_x^3}{\mathrm{d}x} + \frac{3\delta_x^3}{2v_s'}\frac{\mathrm{d}v_s'}{\mathrm{d}x} = \frac{3\alpha}{v_s'}.$$ (9.1.95)

This equation is solved using the procedure in Appendix 9.B. The integrating factor for this equation is,

$$\mathrm{IF} = \mathrm{e}^{\left(\int \mathrm{d}x\,\frac{3}{2v_s'}\frac{\mathrm{d}v_s'}{\mathrm{d}x}\right)} = \mathrm{e}^{\left(\frac{3}{2}\ln(v_s')\right)} = (v_s')^{3/2}.$$ (9.1.96)

The inhomogeneous equation, Eq. 9.1.95, is solved using this integrating factor,

$$\delta_x^3 = \frac{1}{(v_s'(x))^{3/2}} \int_0^x dx' \frac{3\alpha}{v_s'(x')} \times (v_s'(x'))^{3/2} = \frac{3\alpha}{(v_s'(x))^{3/2}} \int_0^x dx' \sqrt{v_s'(x')}.$$

$$(9.1.97)$$

Here, it is assumed that $x = 0$ is the start of the heating section. It is easily verified that Eq. 9.1.97 reduces to Eq. 9.1.24 when $v_s'(x)$ is a constant, and to Eq. 9.1.46 for the stagnation point flow, $v_s'(x) = Kx$.

Eq. 9.1.97 for the boundary layer thickness is substituted into the equation for the heat flux, Eqs. 9.1.32, 9.1.47 and 9.1.84. The local flux is integrated over the length of the surface to determine the average flux, Eqs. 9.1.34 and 9.1.86. The Nusselt number is then calculated from the average flux.

Summary (9.1)

---

1. For high Peclet number transport, the temperature/concentration disturbance due to the surface is confined to a boundary layer of thickness δ_x much smaller than the characteristic length l_c in the stream-wise direction. There is a balance between convection and cross-stream diffusion in the boundary layer.

2. If l_c is the characteristic length scale for the flow, the stream-wise convection term in convection–diffusion Eq. 9.1.1 is $v_x(\partial T/\partial x) \sim (v_s'\delta_x T/l_c)$, where v_s' is the strain rate at the surface, while the cross-stream diffusion term is $(\alpha T/\delta_x^2)$. There is a balance between stream-wise convection and cross-stream diffusion for $\delta_x \sim (\alpha l_c/v_s')^{1/3}$, or $(\delta_x/l_c) \sim \text{Pe}_{v_s'l_c}^{-1/3}$, where the Peclet number $\text{Pe}_{v_s'l_c} = (v_s'l_c^2/\alpha)$ is based on the wall strain rate and the characteristic length.

3. When stream-wise diffusion is neglected, a similarity solution is obtained for the boundary layer thickness $\delta_x(x)$ based on the assumption that the temperature/concentration field depends only on the distance x from the start of the heating/contacting section.

4. The equation for the boundary layer thickness δ_x as a function of the stream-wise co-ordinate x is determined from the condition that the convection–diffusion equation, when expressed in terms of the similarity variable η (Eq. 9.1.14), does not explicitly depend on x.

5. The equations for the boundary layer thickness, Eqs. 9.1.21, 9.1.43, 9.1.78, and 9.1.94, do depend on geometry, and these are solved to determine the boundary layer thickness as a function of the stream-wise co-ordinate.

6. Eq. 9.1.25 for the η dependence of the temperature/ concentration field is independent of geometry, and the constant C_δ can be set equal to 1 without loss of generality. The boundary conditions, Eqs. 9.1.17–9.1.18, in the η co-ordinate are also independent of geometry. The solution for $T^*(\eta)$ is Eq. 9.1.31.

7. The local heat flux at the surface, Eq. 9.1.32 with $C_\delta = 1$ and Eq. 9.1.84, depends on geometry only through the dependence of the boundary layer thickness δ_x.

8. The flux is averaged over the stream-wise co-ordinate to determine the average flux and the Nusselt number correlation, Eq. 9.1.35, 9.1.48 and 9.1.87.

9. The boundary layer analysis for the flow past a flat surface is adapted for the flow through a pipe if the boundary layer thickness is much smaller than the pipe radius. This results in the correlation, Eq. 9.1.52, which has the same form as that for the laminar flow in a pipe.

9.2   Flow past a Mobile Surface

The Nusselt number correlation for the transport into a falling film was derived in Section 4.4.1 in Chapter 4, using the equivalence between the unsteady time evolution of the concentration profile for a stationary film and the spatial evolution in a fluid volume element moving with a constant velocity. This equivalence is valid for a unidirectional flow in which the stream-wise diffusion is neglected. The Sherwood number correlation, Eq. 4.4.25 or Eq. 4.4.26, can be formally derived using boundary layer analysis; this can be worked out in Exercise 9.5. Here, the Sherwood number correlation for the flow past a bubble is determined, and the similarity equation for the general flow past a mobile interface is derived.

9.2.1   Diffusion from a Gas Bubble

Consider the transport of a solute from a gas bubble of radius R into the surrounding liquid, where the liquid is moving with velocity v_{fs} relative to the bubble. The gas concentration at the surface is c_s, while that in the fluid far from the bubble surface

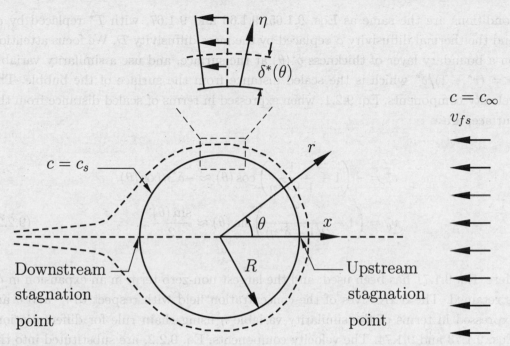

FIGURE 9.6. The configuration and co-ordinate system for analysing the diffusion from a gas bubble.

is c_∞. The average flux and the Sherwood number for the mass transfer from the bubble surface are determined as follows.

The configuration and spherical co-ordinate system for analysing the diffusion around a spherical gas bubble, shown in Fig. 9.6, is identical to that for the diffusion from the surface of a solid particle. The important difference between the flow around a particle and a gas bubble is the fluid velocity field. There is a slip velocity condition at the surface of the bubble, as shown in Fig. 9.6, due to the zero tangential stress condition on the liquid side at a liquid–gas interface. This is in contrast to the zero velocity condition at a solid surface shown in Fig. 9.4. The velocity components in the laminar flow past a spherical bubble, with a zero tangential stress condition applied at the surface of the bubble, and velocity $v_x = -v_{fs}$ far from the bubble, are

$$v_r = -v_{fs}\left(1 - \frac{R}{r}\right)\cos\left(\theta\right), \quad v_\theta = v_{fs}\left(1 - \frac{R}{2r}\right)\sin\left(\theta\right). \tag{9.2.1}$$

The solution procedure is very similar to that for the flow past a rigid particle, so only the important steps are described. The scaled concentration and velocity are defined as $c^* = (c - c_\infty)/(c_s - c_\infty)$, $v_r^* = (v_r/v_{fs})$ and $v_\theta^* = (v_\theta/v_{fs})$, and the scaled radius is $r^* = (r/R)$. The convection–diffusion equation and boundary conditions are the same as Eqs. 9.1.65, 9.1.66 and 9.1.67, with T^* replaced by c^* and the thermal diffusivity α replaced by the mass diffusivity $\mathcal{D}$. We focus attention on a boundary layer of thickness $\delta^*(\theta)$ at the surface, and use a similarity variable $\eta = (r^* - 1)/\delta^*$ which is the scaled distance from the surface of the bubble. The velocity components, Eq. 9.2.1, when expressed in terms of scaled distance from the surface, are

$$v_r^* = -\left(1 - \frac{1}{1 + \delta^*\eta}\right)\cos(\theta) \approx -\delta^*\eta\cos(\theta),$$

$$v_\theta^* = \left(1 - \frac{1}{2(1 + \delta^*\eta)}\right)\sin(\theta) \approx \frac{\sin(\theta)}{2}. \qquad (9.2.2)$$

Here, Eq. 9.1.71 has been used, and the largest non-zero term in an expansion in δ^* is retained. The derivatives of the concentration field with respect to r^* and θ are expressed in terms of the similarity variable η using chain rule for differentiation, Eqs. 9.1.73 and 9.1.74. The velocity components, Eq. 9.2.2, are substituted into the left side of the convection–diffusion Eq. 9.1.65,

$$v_r^* \frac{\partial c^*}{\partial r^*} + \frac{v_\theta^*}{r^*}\frac{\partial c^*}{\partial \theta}$$

$$= -\delta^*\eta\cos(\theta)\frac{1}{\delta^*(\theta)}\frac{dc^*}{d\eta} + \frac{\sin(\theta)}{2(1 + \delta^*(\theta)\eta)}\left(-\frac{\eta}{\delta^*(\theta)}\frac{d\delta^*}{d\theta}\frac{dc^*}{d\eta}\right)$$

$$\approx -\left(\cos(\theta) + \frac{\sin(\theta)}{2\delta^*}\frac{d\delta^*}{d\theta}\right)\eta\frac{dc^*}{d\eta}, \qquad (9.2.3)$$

and the right side of Eq. 9.1.65 is simplified as shown in Eq. 9.1.76, to obtain the boundary layer equation,

$$-\left[\frac{\mathrm{Pe}}{2}\left(\delta^{*2}\cos(\theta) + \frac{\delta^*\sin(\theta)}{2}\frac{d\delta^*}{d\theta}\right)\right]\eta\frac{dc^*}{d\eta} = \frac{d^2 c^*}{d\eta^2}, \qquad (9.2.4)$$

where $\mathrm{Pe} = (v_{fs}d/\mathcal{D})$ is the Peclet number based on the bubble diameter and the free stream velocity.

A similarity solution requires that the term in square brackets in Eq. 9.2.4 is a constant. The final solution does not depend on the value of the constant, and so we can set the constant equal to 1 without loss of generality. With this, the equation for the boundary layer thickness δ^* is

$$\boxed{\frac{Pe}{2}\left(\delta^{*2}\cos\left(\theta\right)+\frac{\sin\left(\theta\right)}{4}\frac{d\delta^{*2}}{d\theta}\right)=1.}$$ (9.2.5)

The similarity equation for the concentration field is Eq. 9.2.4, with the term in square brackets set equal to 1,

$$\boxed{-\eta\frac{dc^*}{d\eta}=\frac{d^2c^*}{d\eta^2}.}$$ (9.2.6)

The above equation is expressed in terms of the derivative, $c^{*\prime}=(dc^*/d\eta)$,

$$-\eta c^{*\prime}=\frac{dc^{*\prime}}{d\eta}.$$ (9.2.7)

The solution of the above equation is,

$$c^{*\prime}=\frac{dc^*}{d\eta}=C_1 e^{(-\eta^2/2)}.$$ (9.2.8)

The concentration field is determined by integrating the above equation with respect to η,

$$c^*=C_1\int_0^\eta d\eta'\,e^{(-\eta'^2/2)}+C_2$$ (9.2.9)

The constants of integration, C_1 and C_2, are determined from the boundary conditions, Eqs. 9.1.17 and 9.1.18,

$$C_1=-\frac{1}{\int_0^\infty d\eta' e^{(-\eta'^2/2)}}=-\sqrt{\frac{2}{\pi}},$$ (9.2.10)

$$C_2=1.$$ (9.2.11)

The integral in the denominator in the solution for C_1 in Eq. 9.2.10 is discussed in Appendix 9.A. The similarity solution for the scaled concentration field is

$$\boxed{c^*=1-\sqrt{\frac{2}{\pi}}\int_0^\eta d\eta'\,e^{(-\eta'^2/2)}.}$$ (9.2.12)

Eq. 9.2.5 for the boundary layer thickness is rewritten as a first order linear inhomogeneous equation for δ^{*2},

$$\frac{d\delta^{*2}}{d\theta}+\frac{4\delta^{*2}\cos\left(\theta\right)}{\sin\left(\theta\right)}=\frac{8}{Pe\sin\left(\theta\right)}.$$ (9.2.13)

This equation is solved as described in Appendix 9.B. The integrating factor for this equation is

$$\text{IF} = e^{\left(\int d\theta \frac{4\cos(\theta)}{\sin(\theta)}\right)} = e^{(4\,\ln(\sin(\theta)))} = \sin(\theta)^4. \qquad (9.2.14)$$

The solution for δ^{*2} is expressed in terms of the integrating factor,

$$\delta^{*2} = \frac{C}{\sin(\theta)^4} + \frac{1}{\sin(\theta)^4} \int_0^\theta d\theta' \frac{8}{\text{Pe}\sin(\theta')} \times \sin(\theta')^4. \qquad (9.2.15)$$

The trigonometric identity $\sin(\theta)^3 = \frac{1}{4}(3\sin(\theta) - \sin(3\theta))$ is used evaluate the integral in the above equation,

$$\delta^{*2} = \frac{C}{\sin(\theta)^4} + \frac{2[(\cos(3\theta)/3) - 3\cos(\theta) + (8/3)]}{\text{Pe}\sin(\theta)^4}. \qquad (9.2.16)$$

As in the case of the flow past a spherical particle in Section 9.1.3, we observe that the boundary layer thickness in Eq. 9.2.16 diverges at $\theta = 0$ and $\theta = \pi$. By a suitable choice of the constant C, it is possible to ensure that the boundary layer thickness is finite at one of these two points. Physically, the boundary layer thickness has to be finite at the upstream stagnation point $\theta = 0$, where the fluid is incident on the sphere. Using the series expansions for $\theta \to 0$,

$$\cos(\theta) = 1 - \frac{\theta^2}{2!} + \frac{\theta^4}{4!}, \quad \sin(\theta) = \theta - \frac{\theta^3}{3!}, \qquad (9.2.17)$$

it can be shown that the second term on the right side of Eq. 9.2.16 tends to a finite value, $(2/\text{Pe})$ for $\theta \to 0$. Therefore, the constant C is set equal to 0 so that the boundary layer thickness is finite at $\theta = 0$,

$$\boxed{\delta^* = \sqrt{\frac{2}{\text{Pe}}} \frac{1}{\sin(\theta)^2} \left(\frac{8}{3} - 3\cos(\theta) + \frac{\cos(3\theta)}{3}\right)^{1/2}.} \qquad (9.2.18)$$

The mass flux at the surface of the bubble is now determined using Fick's law for diffusion,

$$\boxed{j_r|_{r=R}} = -\mathcal{D}\left.\frac{\partial c}{\partial r}\right|_{r=R} = -\frac{\mathcal{D}(c_s - c_\infty)}{R}\left.\frac{\partial c^*}{\partial r^*}\right|_{r^*=1}$$

$$= -\frac{\mathcal{D}(c_s - c_\infty)}{R\delta^*(\theta)}\left.\frac{dc^*}{d\eta}\right|_{\eta=0} = \boxed{\frac{\mathcal{D}(c_s - c_\infty)}{R\delta^*(\theta)}\sqrt{\frac{2}{\pi}}.} \qquad (9.2.19)$$

Here, $(dc^*/d\eta)|_{\eta=0}$ is substituted from Eqs. 9.2.8 and 9.2.10. The average mass flux is

$$j_{av} = \frac{1}{4\pi R^2} \int_0^{2\pi} R^2 d\phi \int_0^\pi \sin(\theta)\,d\theta\; j_r|_{r=R} = \frac{1}{2} \int_0^\pi \sin(\theta)\,d\theta\; j_r|_{r=R}$$

$$= \frac{\mathcal{D}(c_s - c_\infty)}{2R} \sqrt{\frac{2}{\pi}} \sqrt{\frac{\mathrm{Pe}}{2}} \int_0^\pi \sin(\theta)\,d\theta\; \frac{\sin(\theta)^2}{(8/3 - 3\cos(\theta) + \cos(3\theta)/3)^{1/2}}. \quad (9.2.20)$$

The integral on the right side is evaluated by the substitution $w = (8/3 - 3\cos(\theta) + \cos(3\theta)/3)$, and $dw = (3\sin(\theta) - \sin(3\theta))d\theta = 4\sin(\theta)^3 d\theta$,

$$j_{av} = \frac{\mathcal{D}(c_s - c_\infty)\mathrm{Pe}^{1/2}}{2\sqrt{\pi}R} \times \frac{1}{2}\left(\frac{8}{3} - 3\cos(\theta) + \frac{\cos(3\theta)}{3}\right)^{1/2}\Bigg|_0^\pi$$

$$= \frac{\mathcal{D}(c_s - c_\infty)\mathrm{Pe}^{1/2}}{2\sqrt{\pi}R} \times \frac{2}{\sqrt{3}} = \frac{0.326\,\mathcal{D}(c_s - c_\infty)\mathrm{Pe}^{1/2}}{R}. \quad (9.2.21)$$

The correlation for the Sherwood number is,

$$\boxed{\mathrm{Sh} = \frac{j_{av}}{\mathcal{D}(c_s - c_\infty)/d} = 0.652\,\mathrm{Pe}^{1/2}.} \quad (9.2.22)$$

Here, the Sherwood number is expressed in terms of the particle diameter, $d = 2R$.[3]

9.2.2 General Flow Fields

The configuration and co-ordinate system for the flow past a mobile surface in two dimensions is shown in Fig. 9.5(b). The tangential velocity is non-zero, and the approximations for the velocity components for a two-dimensional flow which satisfy the incompressibility condition, Eq. 9.1.89, are

$$v_x = v_s(x), \quad v_z = -z\frac{dv_s}{dx}. \quad (9.2.23)$$

The expressions, Eq. 9.2.23, which retain only the largest terms in an expansion in small z, are valid when the distance from the surface is much smaller than the characteristic cross-stream length scale for the flow. The mass transfer equivalent of

---

[3]In [2], the coefficient in Eq. 9.2.22 is 0.921, because the Peclet number is based on the particle radius and not the diameter.

the convection–diffusion equation, Eq. 9.1.37, is

$$v_s \frac{\partial c^*}{\partial x} - z \frac{dv_s}{dx} \frac{\partial c^*}{\partial z} = \mathcal{D} \frac{\partial^2 c^*}{\partial z^2}. \tag{9.2.24}$$

The boundary conditions are $c^* = 1$ at the surface $z = 0$, and the scaled concentration decreases to zero far from the surface. Substituting the similarity variable $\eta = (z/\delta_x)$, using Eq. 9.1.15 to convert the spatial derivatives into η derivatives, and multiplying the resulting equation by $(\delta_x^2/\mathcal{D})$, we obtain,

$$-\left[\frac{1}{\mathcal{D}} \left(v_s \delta_x \frac{d\delta_x}{d\eta} + \delta_x^2 \frac{dv_s}{dx} \right) \right] \eta \frac{dc^*}{d\eta} = \frac{d^2 c^*}{d\eta^2}. \tag{9.2.25}$$

The term in the square brackets in Eq. 9.2.25 is a constant for a similarity solution; the constant is set equal to 1 without loss of generality. The equation for c^* then reduces to Eq. 9.2.6, and this is solved subject to the boundary conditions, Eqs. 9.1.17 and 9.1.18, to obtain the solution, Eq. 9.2.12. The equation for δ_x is

$$\frac{1}{\mathcal{D}} \left(\frac{v_s}{2} \frac{d\delta_x^2}{dx} + \delta_x^2 \frac{dv_s}{dx} \right) = 1. \tag{9.2.26}$$

This is recast as a first order inhomogeneous differential equation for δ_x^2,

$$\frac{d\delta_x^2}{dx} + \frac{2\delta_x^2}{v_s} \frac{dv_s}{dx} = \frac{2\mathcal{D}}{v_s}. \tag{9.2.27}$$

The solution procedure for this equation is summarised in Appendix 9.B. The integrating factor for this first order differential equation is

$$\text{IF} = e^{\left(\int dx \frac{2}{v_s} \frac{dv_s}{dx} \right)} = v_s^2. \tag{9.2.28}$$

This integrating factor is used to determine the solution of the inhomogeneous equation,

$$\delta_x^2 = \frac{1}{v_s(x)^2} \int_0^x dx' \, (2\mathcal{D}v_s(x')). \tag{9.2.29}$$

Here, it is assumed that the $x = 0$ is the start of the contacting section.

The local flux is calculated from Fick's law for diffusion,

$$\boxed{j_z} = -\mathcal{D} \frac{\partial c}{\partial z}\Big|_{z=0} = -\frac{\mathcal{D}(c_0 - c_\infty)}{\delta_x} \frac{dc^*}{d\eta}\Big|_{\eta=0} = \frac{\mathcal{D}(c_0 - c_\infty)}{\delta_x} \sqrt{\frac{2}{\pi}}. \tag{9.2.30}$$

Here, Eq. 9.2.12 has been used for $c^*(\eta)$. Eq. 9.2.29 is substituted into Eq. 9.2.30 to determine the local flux, and the average flux is then calculated as shown in the following example.

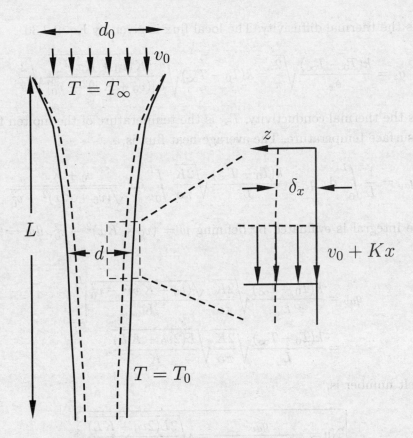

FIGURE 9.7. Heat transfer in a filament drawing process.

EXAMPLE 9.2.1: In a filament drawing process, molten polymer is drawn out from an orifice of diameter d_0 through cold air that freezes the polymer filament, as shown in Fig. 9.7. The flow of the polymer is extensional, with velocity $v_0 + Kx$ at a distance x from the orifice, where K is the extension rate. The diameter decreases with downstream distance. If the Peclet number for the heat transfer from the filament surface is large, and the thermal boundary layer thickness is small compared to the filament diameter, what is the expression for the Nusselt number for a filament of length L? Under what condition is the thermal boundary layer thickness much smaller than the filament diameter?

Solution: The boundary layer thickness is given by Eq. 9.2.29, where $v_s(x) = v_0 + Kx$,

$$\delta_x^2 = \frac{1}{(v_0 + Kx)^2} \int_0^x dx' \, [2\alpha(v_0 + Kx')] = \frac{\alpha[(v_0 + Kx)^2 - v_0^2]}{K(v_0 + Kx)^2}, \quad (9.2.31)$$

where α is the thermal diffusivity. The local flux is given by Eq. 9.2.30,

$$q_z = \frac{k(T_0 - T_\infty)}{\delta_x}\sqrt{\frac{2}{\pi}} = k(T_0 - T_\infty)\sqrt{\frac{K(v_0 + Kx)^2}{\alpha[(v_0 + Kx)^2 - v_0^2]}}\sqrt{\frac{2}{\pi}}, \quad (9.2.32)$$

where k is the thermal conductivity, T_∞ is the temperature of the molten fiber and T_0 is the surface temperature. The average heat flux is,

$$q_{av} = \frac{1}{L}\int_0^L dx\, q_z = \frac{k(T_0 - T_\infty)}{L}\sqrt{\frac{2K}{\pi\alpha}}\int_0^L dx\frac{v_0 + Kx}{\sqrt{(v_0 + Kx)^2 - v_0^2}}. \quad (9.2.33)$$

The above integral is evaluated by defining $w = (v_0 + Kx)^2 - v_0^2$, $dw = 2K(v_0 + Kx)dx$,

$$\begin{aligned}
q_{av} &= \frac{k(T_0 - T_\infty)}{L}\sqrt{\frac{2K}{\pi\alpha}}\left.\frac{\sqrt{(v_0 + Kx)^2 - v_0^2}}{K}\right|_{x=0}^{L} \\
&= \frac{k(T_0 - T_\infty)}{L}\sqrt{\frac{2K}{\pi\alpha}}\sqrt{\frac{L(2v_0 + KL)}{K}}. \quad (9.2.34)
\end{aligned}$$

The Nusselt number is,

$$\boxed{\mathrm{Nu} = \frac{q_{av}}{k(T_0 - T_\infty)/L} = \sqrt{\frac{2L(2v_0 + KL)}{\pi\alpha}}.} \quad (9.2.35)$$

We now examine the thin boundary layer approximation. The diameter of the filament at a downstream distance L is determined from the constant volumetric flow rate condition—that is, the product of the velocity v and the area of cross section $(\pi d^2/4)$ is a constant,

$$\frac{\pi d_0^2 v_0}{4} = \frac{\pi d^2(v_0 + KL)}{4}. \quad (9.2.36)$$

Therefore, the filament diameter at a downstream distance x is,

$$d^2 = \frac{d_0^2 v_0}{v_0 + KL}. \quad (9.2.37)$$

The square ratio of the boundary layer thickness (Eq. 9.2.31) and filament diameter is,

$$\frac{\delta_x^2}{d^2} = \frac{\alpha[(v_0 + KL)^2 - v_0^2]}{K(v_0 + KL)^2} \times \frac{v_0 + KL}{d_0^2 v_0} = \frac{\alpha L(2v_0 + KL)}{d_0^2 v_0(v_0 + KL)}. \quad (9.2.38)$$

This ratio is small compared to 1, and the boundary layer thickness is small compared to the filament diameter, for

$$\frac{\alpha L}{d_0^2 v_0} \ll 1 \text{ or } \frac{L}{d_0} \ll \frac{v_0 d_0}{\alpha}. \qquad (9.2.39)$$

□

Summary (9.2)

1. For a mobile surface, the tangential velocity at the surface is non-zero. For a flow with characteristic length l_c, the convection term close to the surface in the convection–diffusion equation, Eq. 9.1.1, is $v_s(\partial T/\partial x) \sim (v_s T/l_c)$, and the cross-stream diffusion term is $\alpha(\partial^2 T/\partial z^2) \sim (\alpha T/\delta_x^2)$, where v_s is the tangential velocity at the surface. Stream-wise convection and cross-stream diffusion are comparable for $\delta_x \sim (\alpha l_c/v_s)^{1/2}$, or $(\delta/l_c) \sim \mathrm{Pe}_{v_s l_c}^{-1/2}$, where the Peclet number $\mathrm{Pe}_{v_s l_c}$ is based on the surface velocity and characteristic length l_c.

2. An equation for the spatial evolution of the boundary layer thickness δ_x is obtained from the condition that the convection–diffusion equation, when expressed in terms of η, does not explicitly depend on x.

3. The general Eq. 9.2.6 for the η dependence of the temperature/ concentration field, and the boundary conditions, Eqs. 9.1.17 and 9.1.18, are independent of geometry. The solution is Eq. 9.2.12.

4. The equations for the boundary layer thickness, Eqs. 9.2.5 and 9.2.26, do depend on geometry, and these are solved to determine the boundary layer thickness as a function of the stream-wise co-ordinate, Eqs. 9.2.18 and 9.2.29.

5. The local flux at the surface, Eqs. 9.2.19 and 9.2.30, depends on geometry only through the dependence of the boundary layer thickness.

6. The flux is averaged over the stream-wise co-ordinate to determine the average flux and the Nusselt number correlations, Eqs. 9.2.22 and 9.2.35.

9.3   Flow through a Packed Column

Dispersion in a packed column was discussed in Section 3.2.2 (Chapter 3) and illustrated in Fig. 3.8. When the fluid flows around the particles in a packed column, nearby material points are separated by a distance comparable to the particle diameter when they travel past a particle. The only length and velocity scales that affect the dispersion coefficient are the particle diameter d and the velocity of the fluid through the interstitial spaces between the particles (v_s/ε), where v_s is the superficial velocity and ε is the void fraction. On this basis, the dispersion coefficient was estimated as (dv_s/ε). A microscopic model is used here to calculate the dispersion coefficient.

The dispersion process can be visualised by considering the fluid flow through a regular array of rectangular obstacles, with repeat length l, shown in Fig. 9.8. Successive rows of obstacles are displaced horizontally by half the repeat length, so that fluid has to follow a tortuous path as it passes through the array. Here, the average flow direction is x, and the cross-stream direction is z. At the each T junction, the vertical fluid stream bifurcates into two equal streams moving horizontally to the left and right. Two horizontal streams flowing around an obstacle meet at a $\perp$ junction, and mixing takes place due to cross-stream diffusion across the vertical channel/pore width. The concentration in the vertical stream at a T junction is the average of the concentrations in the horizontal streams at the $\perp$ junction below. Thus, diffusion across the width of the pore results in spreading of the mass across the much larger dimension of the obstacle or the particle in a packed column.

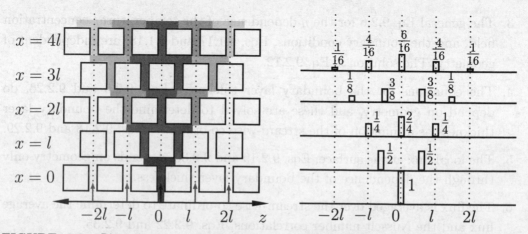

FIGURE 9.8. Dispersion in a regular array of staggered obstacles of repeat length l. The concentration in vertical streams at different x and z locations are shown on the right.

A solute of scaled concentration 1 is introduced in the first row at $x = 0, z = 0$ in figure 9.8. The concentration is $\frac{1}{2}$ at $z = \pm(l/2)$ in the second row $x = l$. In the third row at $x = 2l$, the concentration $\frac{1}{4}$ at $z = \pm l$ and $\frac{1}{2}$ at $z = 0$. The concentration in each row is calculated by splitting the vertical stream from the row below at the T junction, and taking the average of the horizontal streams at the $\perp$ junction. The concentrations in each successive row, shown on the right in Fig. 9.8, have an interesting pattern. The numerator is the term in the Pascal triangle, which is a triangular array of numbers where the number in a row is the sum of the nearest neighbours in the row below. The denominator is 2^n, where n is the row number starting with 0 at the first row. The terms are the binomial coefficients of the form $(C_m^n/2^n)$, where n is the row number and m is the column number starting from the left. It can also be shown that when n is large, the concentrations is a Gaussian function discussed in Section 4.5 (Chapter 4) and Appendix 4.A.

From the concentration values in Fig. 9.8, it is evident that the spread of the concentration is proportional to the number of rows crossed by the fluid. The spread in the concentration field can be formally expressed in terms of the mean square displacement MSD_z in the z direction,

$$\text{MSD}_z(x) = \frac{\sum_i c_i z_i^2}{\sum_i c_i}, \tag{9.3.1}$$

where c_i is the concentration in stream i at the location z_i. The denominator in Eq. 9.3.1 is 1 for every row, because the concentration is normalised by the value at the inlet. The mean square displacement is 0 at $x = 0$. In the second row $x = l$, the mean square displacement is $\frac{1}{2}(l/2)^2 + \frac{1}{2}(-l/2)^2 = (l^2/4)$. In a similar manner, the mean square displacement is $\text{MSD}_z = (nl^2/4)$ at $x = nl$; this can be worked out for the third, fourth and fifth rows in Exercise 9.11. Thus, when expressed in terms of x, the mean square displacement is

$$\text{MSD}_z = \frac{nl^2}{4} = \frac{xl}{4}. \tag{9.3.2}$$

The time required for the fluid to travel the distance x in the stream-wise direction is $t = (x/v) = (x\varepsilon/v_s)$, where v is the velocity of the fluid in the pores, v_s is the superficial velocity and ε is the void fraction. In Section 4.5 (Chapter 4), it is shown that the effective dispersion coefficient is,

$$\mathcal{D}_{eff} = \frac{\text{MSD}_z}{2t} = \frac{v_s l}{8\varepsilon}. \tag{9.3.3}$$

This simple calculation shows that the dispersion coefficient is proportional to the product of the flow velocity through the packed column and the characteristic length, which is l for the regular array in Fig. 9.8, and the particle diameter for a packed column.

Summary (9.3)

---

1. In the flow through packed columns, the fluid traverses tortuous paths between a densely packed assembly of particles. The width of these paths is typically much smaller than the diameter of the particles.

2. Nearby material elements in the fluid are separated by a distance comparable to the particle diameter as they travel around a particle.

3. Convection around particles followed by cross-stream mixing due to diffusion in the interstitial spaces between particles results in a dispersion coefficient proportional to the product of the fluid velocity (v_s/ε) and the particle diameter.

9.4   Taylor Dispersion

There is a significant enhancement in the transport along the axis of a pipe or channel in a laminar flow due to convection, in comparison to the transport due to molecular diffusion. The qualitative reason for this was explained in Section 3.2.3 (Chapter 3). Here, an analytical expression is derived for the dispersion coefficient for the laminar flow in a pipe of diameter d with average velocity v_{av}. There are three important assumptions in the analysis.

1. The Peclet number based on the pipe diameter d and the average flow velocity v_{av} is large, $\mathrm{Pe} = (v_{av}d/\mathcal{D}) \gg 1$, where $\mathcal{D}$ is the molecular diffusion coefficient.

2. At any downstream location, the concentration variation across the cross section is much smaller than the average concentration—that is, the concentration is nearly uniform across the cross section, but the relatively small variation in the concentration coupled with the variation in the stream-wise velocity could result in a stream-wise flux due to convection. The characteristic time for molecular diffusion across the cross section of the pipe is $(d^2/\mathcal{D})$, and the concentration is nearly constant across the cross section when the travel time t satisfies the condition $t \gg (d^2/\mathcal{D})$. The average downstream distance moved by the solute is proportional to $v_{av}t$, and therefore the analysis

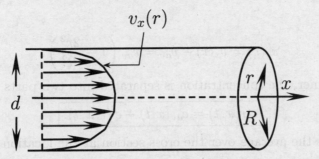

FIGURE 9.9. Configuration and co-ordinate system for analysing the Taylor dispersion due to a laminar flow in a pipe.

is valid when the downstream distance travelled L is larger than $(v_{av} d^2/\mathcal{D})$. The condition on the ratio of the distance travelled and the pipe diameter is $(L/d) \gg (v_{av} d/\mathcal{D}) \sim \text{Pe}$.

3. The length scale for the variation of the concentration in the x direction is large compared to the pipe radius. This is related to assumption 2 above. Since the travel time is much larger than $(d^2/\mathcal{D})$, the spread of the solute in the stream-wise direction, $\sqrt{\mathcal{D}t}$, is much larger than d.

With the above assumptions, it is possible to derive an analytical result for the dispersion coefficient for the average stream-wise flux in the pipe.

A fully developed laminar flow in a pipe of diameter d, shown in Fig. 9.9, is analysed using a cylindrical co-ordinate system where x is along the axis and r is in the radial direction. The parabolic velocity profile for the laminar flow in a pipe is,

$$v_x = 2v_{av} \left(1 - \left(\frac{r}{R}\right)^2\right), \tag{9.4.1}$$

where $R = (d/2)$ is the pipe radius. The above equation was derived in Section 6.2 (Chapter 6), where it was shown that v_{av}, the average velocity, is one half of the maximum velocity at the centre of the pipe. A concentration pulse is injected into the pipe, and the spreading of the concentration pulse is measured as it is convected downstream with the mean flow. The concentration diffusion equation for the unidirectional flow along the axis of the pipe is

$$\frac{\partial c}{\partial t} + v_x \frac{\partial c}{\partial x} = \mathcal{D} \left(\frac{1}{r} \frac{\partial}{\partial r} \left(r \frac{\partial c}{\partial r}\right) + \frac{\partial^2 c}{\partial x^2}\right). \tag{9.4.2}$$

The velocity is separated into two parts,

$$\boxed{v_x(r) = v_{av} + v'_x(r),} \tag{9.4.3}$$

where

$$v'_x(r) = v_x(r) - v_{av} = v_{av}\left(1 - \frac{2r^2}{R^2}\right). \qquad (9.4.4)$$

In a similar manner, the concentration is separated into two parts,

$$c(r, x, t) = c_{av}(x, t) + c'(r, x, t), \qquad (9.4.5)$$

where $c_{av}(x, t)$ is the average over the cross section at the location x and time t,

$$c_{av}(x, t) = \frac{1}{\pi R^2}\int_0^R 2\pi r\,dr\,c(r, x, t), \qquad (9.4.6)$$

and $c'(r, x, t)$ is the difference between the local concentration and the average over the cross section. The average over cross section of c' and v'_x are zero by definition,

$$\frac{1}{\pi R^2}\int_0^R 2\pi r\,dr\,c'(r, x, t) = \frac{1}{\pi R^2}\int_0^R 2\pi r\,dr\,v'_x(r, x, t) = 0. \qquad (9.4.7)$$

Instead of the complete description of the variation of the concentration in space and time in Eq. 9.4.2, we seek an equation for the effective flux $j_{av}(x, t)$ in the stream-wise x direction of the form,

$$j_{av}(x, t) = -\mathcal{D}_{eff}\frac{\partial c_{av}}{\partial x}, \qquad (9.4.8)$$

where $j_{av}(x, t)$ is the flux in the x direction averaged over the cross section of the pipe,

$$j_{av}(x, t) = \frac{1}{\pi R^2}\int_0^R 2\pi r\,dr\,j_x. \qquad (9.4.9)$$

The local flux j_x in a reference frame moving at the average velocity of the fluid consists of a convective and a diffusive part,

$$j_x = v_x c - \mathcal{D}\frac{\partial c}{\partial x} - c_{av}v_{av}$$

$$= \left[v_{av}c_{av} + v'_x c_{av} + v_{av}c' + v'_x c' - \mathcal{D}\frac{\partial c_{av}}{\partial x} - \mathcal{D}\frac{\partial c'}{\partial x}\right] - c_{av}v_{av}. \qquad (9.4.10)$$

Here, the convection of mass due to the average velocity , $c_{av}v_{av}$, has been subtracted from the total flux to determine the flux in a reference frame moving with the average velocity. When this local flux is averaged over the cross section, we obtain

$$j_{av} = \frac{1}{\pi R^2}\int_0^R 2\pi r\,dr\,v'_x c' - \mathcal{D}\frac{\partial c_{av}}{\partial x}. \qquad (9.4.11)$$

Eq. 9.4.10 was averaged to obtain Eq. 9.4.11 as follows. In the second term within the square brackets on the right in Eq.9.4.10, c_{av} is independent of r, and the average of v'_x over the cross section is zero from Eq. 9.4.7. In the third term in the square brackets on the right in Eq. 9.4.10, v_{av} is independent of r, and the average of c' over the cross section is zero from Eq. 9.4.7. In the sixth term within the square brackets on the right in Eq. 9.4.10, the average of c' over the cross section is zero at each location, and so the derivative is also zero. Therefore, only the fourth and fifth terms in the square brackets on the right side of Eq. 9.4.10 provide non-zero contributions to j_{av} in Eq. 9.4.11. It is now necessary to calculate the flux due to the first term on the right in Eq. 9.4.11, which is due to the the variations over the cross section of the product of the concentration and velocity fields. For this, $c'(r, x, t)$ has to be determined.

Eqs. 9.4.3 and 9.4.5 are substituted into the concentration Eq. 9.4.2,

$$\frac{\partial(c_{av} + c')}{\partial t} + (v_{av} + v'_x)\frac{\partial(c_{av} + c')}{\partial x} = \frac{\mathcal{D}}{r}\frac{\partial}{\partial r}\left(r\frac{\partial c'}{\partial r}\right) + \mathcal{D}\frac{\partial^2(c_{av} + c')}{\partial x^2}. \quad (9.4.12)$$

The next step is to take the average of Eq. 9.4.12 over the cross section of the pipe. In the first term on the left, the spatial averaging and the time derivative can be interchanged, since r and t are independent co-ordinates. The average over the cross section of c' is zero (Eq. 9.4.7), and c_{av} is independent of r. Therefore, the average over the cross section of the first term on the left is $(\partial c_{av}/\partial t)$. The average over the cross section of the first term on the right of the above equation is

$$\frac{\mathcal{D}}{\pi R^2}\int_0^R 2\pi r\, dr\, \frac{1}{r}\frac{\partial}{\partial r}\left(r\frac{\partial c'}{\partial r}\right) = \frac{2\mathcal{D}}{R^2}\,r\frac{\partial c'}{\partial r}\bigg|_0^R. \quad (9.4.13)$$

The right side of the above equation is zero at $r = 0$ and $r = R$, the latter because $\mathcal{D}(\partial c'/\partial r)$ is the flux at the pipe surface which has to be zero if the surface is impenetrable. When the second term on the right in Eq. 9.4.12 is averaged over the cross section, the derivative with respect to x and the integral with respect to r are interchanged because they are independent co-ordinates. Here, c_{av} is independent of r, and the average of c' is zero (Eq. 9.4.7), so the average over the cross section is $\mathcal{D}(\partial^2 c_{av}/\partial x^2)$. The average over the cross section of the second term on the left of Eq. 9.4.12 is

$$\frac{1}{\pi R^2}\int_0^R 2\pi r\, dr\, (v_{av} + v'_x)\frac{\partial(c_{av} + c')}{\partial x}$$

$$= \frac{2}{R^2}\int_0^R r\, dr\left[v_{av}\frac{\partial c_{av}}{\partial x} + v_{av}\frac{\partial c'}{\partial x} + v'_x\frac{\partial c_{av}}{\partial x} + v'_x\frac{\partial c'}{\partial x}\right]. \quad (9.4.14)$$

The average over the cross section of the second and third terms in the square brackets on the right in the above equation are zero, since these are linear in the fluctuation quantities v_x' or c' whose average is zero (Eq. 9.4.7). The average over the cross section of the first term in the square brackets on the right in Eq. 9.4.14 is $v_{av}(\partial c_{av}/\partial x)$, since v_{av} and c_{av} are independent of r. The average over the cross section of the last term in the square brackets on the right in Eq. 9.4.14 is not zero, and it is this term that gives rise to Taylor dispersion.

Based on the above simplifications, the average over the cross section of Eq. 9.4.12 is

$$\frac{\partial c_{av}}{\partial t} + v_{av}\frac{\partial c_{av}}{\partial x} + \frac{1}{\pi R^2}\int_0^R 2\pi r\,dr\left(v_x'\frac{\partial c'}{\partial x}\right) = \mathcal{D}\left(\frac{\partial^2 c_{av}}{\partial x^2}\right), \qquad (9.4.15)$$

Eq. 9.4.15 is subtracted from Eq. 9.4.12 to obtain an equation for c',

$$\frac{\partial c'}{\partial t} + v_x'\frac{\partial c_{av}}{\partial x} + v_{av}\frac{\partial c'}{\partial x} + v_x'\frac{\partial c'}{\partial x} - \frac{1}{\pi R^2}\int_0^R 2\pi r\,dr\left(v_x'\frac{\partial c'}{\partial x}\right) = \mathcal{D}\left(\frac{1}{r}\frac{\partial}{\partial r}\left(r\frac{\partial c'}{\partial r}\right) + \frac{\partial^2 c'}{\partial x^2}\right).$$
$$(9.4.16)$$

Two simplifications are made to solve Eq. 9.4.16. In accordance with assumption 2 at the beginning of this section, all terms containing the concentration c' are neglected in comparison to $v_x'(\partial c_{av}/\partial x)$ on the left side of Eq. 9.4.16. In assumption 3, the length scale for the variation in the x direction was considered to be much larger than the pipe radius. Therefore, the second derivative with respect to x on the right in Eq. 9.4.16 is much smaller than the radial derivative. With these simplifications, Eq. 9.4.16 reduces to,

$$\boxed{v_x'\frac{\partial c_{av}}{\partial x} = \frac{\mathcal{D}}{r}\frac{\partial}{\partial r}\left(r\frac{\partial c'}{\partial r}\right).} \qquad (9.4.17)$$

Since c_{av} is independent of r, Eq. 9.4.17 can be integrated two times with respect to r after substituting Eq. 9.4.4 for v_x',

$$c'(r,z,t) = \frac{v_{av}}{\mathcal{D}}\left(\frac{r^2}{4} - \frac{r^4}{8R^2}\right)\frac{\partial c_{av}}{\partial x} + C_1\ln(r) + C_2, \qquad (9.4.18)$$

where C_1 and C_2 are the constants of integration. The constant C_1 is set equal to zero to ensure that the concentration c' is finite at $r = 0$. The constant C_2 is determined from the integral condition, Eq. 9.4.7, that the average over the cross section of c' is zero,

$$C_2 = -\frac{v_{av}}{\mathcal{D}}\frac{R^2}{12}\frac{\partial c_{av}}{\partial x}. \tag{9.4.19}$$

The final expression for $c'(r, z, t)$ is

$$\boxed{c'(r, z, t) = \frac{v_{av}}{\mathcal{D}}\left(\frac{r^2}{4} - \frac{r^4}{8R^2} - \frac{R^2}{12}\right)\frac{\partial c_{av}}{\partial x}.} \tag{9.4.20}$$

Eq. 9.4.20 for c' is substituted into the equation for the average flux, Eq. 9.4.11, to obtain

$$\boxed{j_{av} = -\frac{v_{av}^2 R^2}{48\mathcal{D}}\frac{\partial c_{av}}{\partial x} - \mathcal{D}\frac{\partial c_{av}}{\partial x}.} \tag{9.4.21}$$

The flux due to Taylor dispersion is the first term on the right in Eq. 9.4.21. The effective dispersion coefficient can be written as the sum of the molecular diffusion coefficient and the Taylor dispersion coefficient,

$$\boxed{\mathcal{D}_{eff} = \mathcal{D} + \frac{v_{av}^2 R^2}{48\mathcal{D}} = \mathcal{D}\left(1 + \frac{v_{av}^2 R^2}{48\mathcal{D}^2}\right) = \mathcal{D}\left(1 + \frac{\text{Pe}^2}{192}\right).} \tag{9.4.22}$$

where $\text{Pe} = (v_{av}d/\mathcal{D})$ is the Peclet number based on the pipe diameter and the average velocity.

It is interesting to note that Taylor dispersion coefficient is inversely proportional to the molecular diffusion coefficient. The physical reason for this, explained in Section 3.2.3 (Chapter 3), is the stretching of a fluid element due to the parabolic velocity profile resulting in rapid axial spread of the solute. In flows of practical interest, axial transport is almost always due to Taylor dispersion in liquids, and for reasonably fast gas flows, as illustrated in the following example.

EXAMPLE 9.4.1: Consider the laminar flow of a fluid with velocity v_{av} in a pipe of diameter 1 cm. Determine the velocity for which the Taylor dispersion coefficient is equal to the molecular diffusion coefficient if the fluid is a liquid in which the solute has diffusion coefficient 10^{-9} m^2/s, and if the fluid is a gas in which the solute has diffusion coefficient 10^{-5} m^2/s.

Solution: The Taylor dispersion coefficient is equal to the molecular diffusion coefficient for

$$\mathcal{D} = \frac{v_{av}^2 d^2}{192 \mathcal{D}} \Rightarrow v_{av} = \frac{8\sqrt{3}\mathcal{D}}{d}. \tag{9.4.23}$$

The velocity is 1.39×10^{-6} m/s for a liquid with diffusion coefficient 10^{-9} m/s, and 1.39×10^{-2} m/s for a gas with diffusion coefficient 10^{-5} m^2/s. □

Summary (9.4)

1. Taylor dispersion is the axial dispersion in the flow in a pipe in the reference frame moving with the average fluid velocity at high Peclet number.

2. It is assumed that the concentration variation across the pipe is small compared to the average concentration, and the length scale for axial variation in the concentration is large compared to the diameter.

3. The velocity and concentration fields, Eqs. 9.4.3 and 9.4.5, are expressed as the sum of two parts, the average over the cross section, c_{av}, v_{av}, and the 'fluctuation' or difference between the local value and the average, c', v'. The averages over the cross section of the latter are zero, from Eq. 9.4.7.

4. There is a contribution to the average of the axial flux over the cross section due to the product of the velocity and concentration fluctuations, Eq. 9.4.11.

5. The velocity fluctuation is given by Eq. 9.4.4 for a parabolic flow, and the concentration fluctuation, Eq. 9.4.20, is determined from Eq. 9.4.17 for the concentration fluctuation where c' is neglected in comparison to c_{av} in the convection terms, and the stream-wise diffusion of c' is neglected in comparison to the cross-stream diffusion.

6. The expressions for v'_x and c' are substituted into Eq. 9.4.11, and the resulting contribution to the flux is the product of the Taylor dispersion coefficient and the axial gradient of c_{av}, Eq. 9.4.21.

7. The Taylor dispersion coefficient, Eq. 9.4.22, is proportional to $(v_{av}^2 d^2/\mathcal{D}) \sim \mathrm{Pe}^2 \mathcal{D}$, where d is the pipe diameter, and $\mathcal{D}$ is the molecular diffusion coefficient.

Exercises

EXERCISE 9.1 Consider a laminar flow with average velocity 1 m/s in a pipe of diameter 1 cm. Estimate the pipe length for which the concentration boundary layer thickness becomes comparable to the pipe radius for the flow of a liquid with mass diffusion coefficient 10^{-9} m^2/s and a gas with mass diffusion coefficient 10^{-5} m^2/s.

EXERCISE 9.2 For the heat transfer in a pipe, how does the total heat transfer rate depend volumetric flow rate $\dot{V}$, the pipe diameter d and length L for (a) a laminar flow and (b) a turbulent flow?

EXERCISE 9.3 For the turbulent flow in a pipe, the friction Reynolds number Re$_*$ was defined in Section 6.3 (Chapter 6) as $(v_* d/\nu)$, where the friction velocity $v_* = \sqrt{\tau_w/\rho}$, τ_w is the wall shear stress, ρ and ν are the density and kinematic viscosity, and d is the pipe diameter. There is a viscous wall layer at the wall of the pipe for $(w v_*/\nu) < 5$, where w is the distance from the wall of the pipe. For a high Peclet number flow, the temperature disturbance due to a heated wall is restricted to a thin boundary layer close to the wall. What is the appropriate Peclet number for this configuration? What is the condition for the 'thin boundary layer approximation', that the boundary layer thickness is much smaller than the thickness of the viscous wall layer? What is the correlation for the Nusselt number in this case? Express all answers in terms of the Prandtl number, the friction Reynolds number and the ratio (L/d), where L is the length of the pipe.

EXERCISE 9.4 The generalisation of the stagnation point flow in Example 9.1.1 is the flow in a corner shown in Fig. 9.10. Here, the velocity profiles close to the surface are approximated as,

$$v_x = K x^n z, \quad v_z = -\frac{n}{2} K x^{n-1} z^2,$$

where K is a constant, and n is related to the angle of the corner θ, $n = (\pi/2\theta)$. Determine the Nusselt number correlation for this flow.

EXERCISE 9.5 Derive the Sherwood number correlation, Eq. 4.4.25, using boundary layer analysis. Use Eq. 9.2.29 to determine boundary layer thickness.

EXERCISE 9.6 Consider cylindrical contactor of length L and diameter d, shown in Fig. 9.11, in which one gas component is absorbed into a falling film. When the volumetric flow rate of the liquid is $\dot{V}$, the total mass absorbed per unit time is $\dot{M}$. If the flow rate is increased to $2\dot{V}$, what is the total mass absorbed per unit time? Assume the film thickness is small compared to the diameter of the contactor, the liquid flow is laminar (Example 4.2.3) and the rate-limiting step is the diffusion of the solute into the liquid.

EXERCISE 9.7 Consider the stagnation point flow shown in Fig. 9.2, where the tangential velocity is non-zero at the surface. The components of the velocity in the two-dimensional

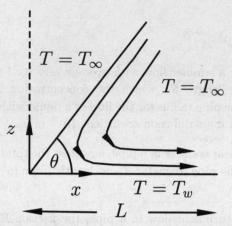

FIGURE 9.10. Configuration and co-ordinate system for analysing the high Peclet number flow in a corner with subtended angle θ.

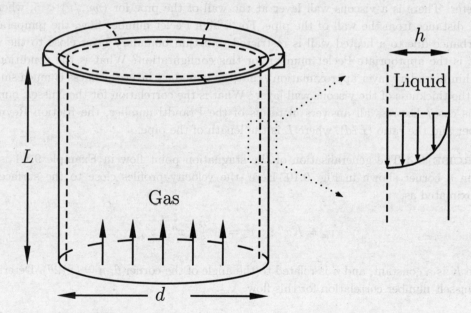

FIGURE 9.11. Absorption of a gas component into a liquid film in a cylindrical contactor.

$x - z$ co-ordinate system are

$$v_x = Kx, \quad v_z = -Kz,$$

where K is a constant. The surface is at a temperature T_w, while the fluid far from the surface has velocity T_∞. Solve the convection–diffusion equation to determine the temperature field and the heat flux from the surface. What is the Nusselt number correlation if the length of the surface is L?

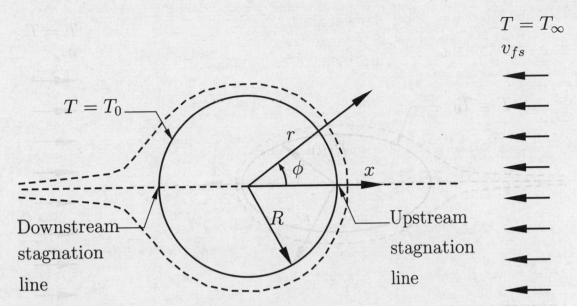

FIGURE 9.12. The configuration and co-ordinate system for analysing the high Peclet number heat transfer in the flow past a cylinder.

EXERCISE 9.8 Consider the laminar flow around a cylinder in two dimensions as shown in Fig. 9.12. The velocity field in the polar co-ordinate system is given by

$$v_r = -v_{fs}\cos(\phi)\left(1 - \frac{R^2}{r^2}\right), \quad v_\phi = v_{fs}\sin(\phi)\left(1 + \frac{R^2}{r^2}\right),$$

where R is the radius of the cylinder, r is the distance from the axis of the cylindrical co-ordinate system and ϕ is the polar angle. The cylinder surface is at a temperature T_0, while the temperature far from the cylinder is T_∞. Determine the Nusselt number correlation for high Peclet number, $(v_{fs}R/\alpha) \gg 1$, where α is the thermal diffusivity.

EXERCISE 9.9 For the potential flow around a bubble at high Reynolds number, the velocity field in the spherical co-ordinate system shown in Fig. 9.6 is

$$v_r = -v_{fs}\cos(\theta)\left(1 - \frac{R^3}{r^3}\right), \quad v_\theta = v_{fs}\sin(\theta)\left(1 + \frac{R^2}{2r^3}\right),$$

where R is bubble radius, and (r, θ) are the radial and azimuthal co-ordinates in the spherical co-ordinate system shown in Fig. 9.6. Determine the Sherwood number correlation for high Peclet number.

EXERCISE 9.10 In Sections 9.1.4 and 9.2.2, we had derived general expressions for the convection–diffusion equation in a Cartesian co-ordinate system for the two-dimensional flow past a solid surface, Eq. 9.1.93, and for the flow past a gas–liquid interface, Eq. 9.2.25. The equations for the boundary layer thickness, Eq. 9.1.97 for the flow past a solid surface and Eq. 9.2.29 for the liquid–gas interface, were also derived. Derive similar equations for the flow

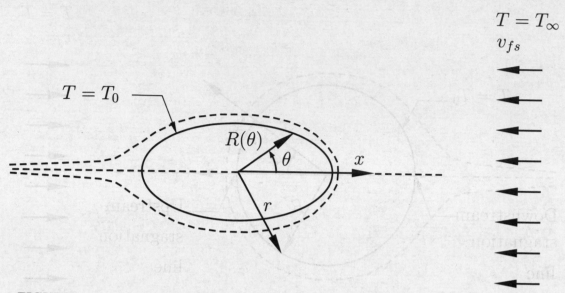

FIGURE 9.13. The configuration and co-ordinate system for a flow around an object for which the radius R is a function of the azimuthal angle θ in a spherical co-ordinate system with axis along the flow direction. The boundary layer thickness is small compared to the radius of curvature.

around an axisymmetric object in a spherical co-ordinate system shown in Fig. 9.13, where the radius of the object $R(\theta)$ is a function of the azimuthal angle. The mass conservation condition in the spherical co-ordinate system is

$$\frac{1}{r^2}\frac{\partial(r^2 v_r)}{\partial r} + \frac{1}{r\sin(\theta)}\frac{\partial(\sin(\theta)v_\theta)}{\partial\theta} = 0.$$

EXERCISE 9.11 Verify Eq. 9.3.2 for the third, fourth and fifth rows in Fig. 9.8. If instead of a square, the repeat unit in Fig. 9.8 is a rectangle of length l in the z direction and height h in the x direction, what is the dispersion coefficient?

EXERCISE 9.12 Consider the flow in the configuration shown in Fig. 9.8, where the branching is asymmetric. At each T junction, $\frac{3}{4}$ of the fluid flows to the right, and $\frac{1}{4}$ to the left. At each $\perp$ junction, the mixture in the vertical channel comprises $\frac{3}{4}$ of the fluid from the left and $\frac{1}{4}$ of the fluid from the right. For the first three rows above the entrance, calculate the mean location z_{mean} and the mean square displacement MSD_z defined as,

$$z_{mean} = \sum_i c_i z_i, \quad \mathrm{MSD}_z = \sum_i c_i z_i^2.$$

How does the spread, defined as $\mathrm{MSD}_z - z_{mean}^2$, increase with the number of rows of obstacles?

EXERCISE 9.13 Determine the Taylor dispersion coefficient for the laminar flow in a two-dimensional channel of height h, where the stream-wise velocity v_x is a function of the cross-stream distance z,

$$v_x = 6v_{av}\left(\frac{z}{h} - \frac{z^2}{h^2}\right).$$

EXERCISE 9.14 Determine the Taylor dispersion coefficient for the laminar flow in a pipe with a slip velocity v_s at the wall. The velocity profile is,

$$v_x = v_s + 2(v_{av} - v_s)\left(1 - \left(\frac{r}{R}\right)^2\right).$$

Appendix

9.A   Gamma Function

The Gamma function $\Gamma(x)$ is a generalisation of the factorial for a non-integer argument x,

$$\Gamma(x) = \int_0^\infty dt\, t^{x-1}e^{(-t)}, \tag{9.A.1}$$

where t is the variable of integration. When x is an integer, the above integral can be evaluated analytically by repeated integrations by parts,

$$\Gamma(x) = (x - 1)!, \tag{9.A.2}$$

where the factorial $(x - 1)! = (x - 1) \times (x - 2) \times \cdots \times 2 \times 1$. When x is not an integer, the function $\Gamma(x)$ has to be evaluated numerically. The recurrence relation, $\Gamma(x + 1) = x\Gamma(x)$ can be derived using integration by parts,

$$\Gamma(x + 1) = \int_0^\infty dt\, t^x e^{(-t)}$$

$$= -t^x\, e^{(-t)}\Big|_0^\infty + \int_0^\infty dt\, x\, t^{x-1}\, e^{(-t)} = x\Gamma(x). \tag{9.A.3}$$

The reason for the Gamma function in Eq. 9.1.30 is as follows. The substitution $w = (C_\delta\eta'^3/3)$ is made in the integral in the denominator in the expression for C_1 in Eq. 9.1.29,

$$\int_0^\infty d\eta'\, e^{(-C_\delta\eta'^3/3)} = \frac{1}{3^{2/3}C_\delta^{1/3}}\int_0^\infty dw\, w^{-2/3}e^{(-w)} = \frac{\Gamma(1/3)}{3^{2/3}C_\delta^{1/3}}. \tag{9.A.4}$$

In the first step above, the substitution $\eta' = (3/C_\delta)^{1/3}w^{1/3}$ and $d\eta' = (3/C_\delta)^{1/3}(1/3)\, w^{-2/3}\, dw$ has been made. This leads to the gamma function in Eq. 9.1.29, for C_1.

The definite integral in Eq. 9.2.10 can also be expressed as a Gamma function using the substitution $\eta'^2/2 = w$, and $dw = \eta' d\eta' = \sqrt{2w} d\eta'$,

$$\int_0^\infty d\eta' e^{(-\eta'^2/2)} = \int_0^\infty dw \frac{e^{(-w)}}{\sqrt{2w}} = \frac{\Gamma(\frac{1}{2})}{\sqrt{2}} = \sqrt{\frac{\pi}{2}}. \qquad (9.A.5)$$

The numerical value of $\Gamma(\frac{1}{2})$ is $\sqrt{\pi}$.

9.B   Integrating Factor

An inhomogeneous first order differential equation has the general form,

$$\frac{df}{dx} + P(x)f = Q(x). \qquad (9.B.1)$$

The integrating factor for this equation is the exponential of the antiderivative of $P(x)$,

$$IF(x) = e^{\left(\int dx P(x)\right)}. \qquad (9.B.2)$$

When Eq. 9.B.1 is multiplied by the integrating factor, we obtain

$$\frac{df}{dx} e^{\left(\int dx P(x)\right)} + P(x) f e^{\left(\int dx P(x)\right)} = Q(x) e^{\left(\int dx P(x)\right)}. \qquad (9.B.3)$$

The first two terms on the left side are combined,

$$\frac{d}{dx}\left(f e^{\left(\int dx P(x)\right)}\right) = Q(x) e^{\left(\int dx P(x)\right)}. \qquad (9.B.4)$$

This equation is integrated to obtain

$$f e^{\left(\int dx P(x)\right)} = C + \int_0^x dx' Q(x') e^{\left(\int dx' P(x')\right)}, \qquad (9.B.5)$$

where C is the constant of integration. The solution for f in terms of the integrating factor 9.B.2 is

$$f(x) = \frac{C}{IF(x)} + \frac{1}{IF(x)} \int_0^x dx' Q(x') IF(x'). \qquad (9.B.6)$$

Natural Convection

10

In the analysis of transport at high Peclet number in Chapter 9, it was assumed that the fluid velocity field is specified, and is not affected by the concentration or temperature variations. There are situations, especially in the case of heat transfer, where variations in temperature cause small variations in density, which results in flow in a gravitational field due to buoyancy. Examples of these flows range from circulation in the atmosphere to cooking by heating over a flame. In the former, air heated by the earth's surface rises and cold air higher up in the atmosphere descends due to buoyancy; in the latter, hotter and lighter fluid at the bottom rises due to buoyancy and is replaced by colder and heavier fluid at the top, resulting in significantly enhanced heat transfer.

The heat transfer due to natural convection from heated objects is considered here, and correlations are derived for the Nusselt number as a function of the Prandtl number and the Grashof number. The Prandtl number is the ratio of momentum and thermal diffusion. The Grashof number, defined in Section 2.4 (Chapter 2), is the square of the Reynolds number based on the characteristic fluid velocity generated by buoyancy. In order to determine the heat transfer rate, it is necessary to solve the coupled momentum and energy equations, the former for the velocity field due to temperature variations and the latter for the temperature field. The equations are too complex to solve analytically, and attention is restricted to scaling the equations to determine the relative magnitudes of convection, diffusion and buoyancy. We examine how the dimensionless groups emerge when the momentum and energy equations are scaled, and how these lead to correlations for the Nusselt number. The numerical coefficients in these correlations are not calculated here.

10.1　Boussinesq Equations

Consider a heated object with surface temperature T_0, in a ambient fluid with temperature T_∞ far from the object, as shown in Fig. 10.1 The fluid density is ρ_∞ far from the object, but the temperature variation causes a variation in the density near the object. This density variation results in a buoyancy force, which drives the flow. A specific shape is not considered while deriving the equations, but we assume that the characteristic dimension of the object is l_c. The objective is to determine the average heat flux q_{av}, or the Nusselt number $\mathrm{Nu} = q_{av}/(k(T_0 - T_\infty)/l_c)$.

A steady flow is considered, where the temperature and velocity fields are independent of time. The mass, momentum and energy equations are first simplified in vector notation, and then expressed in terms of the individual components. At

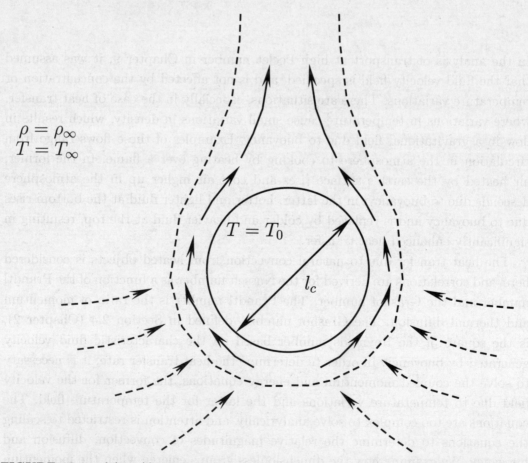

FIGURE 10.1. A hot object with surface temperature T_0 in an ambient fluid with density ρ_∞ and temperature T_∞ far from the object. Air close to the object which is hotter rises, resulting in a flow which draws in cold air from the ambient.

steady state, the energy conservation equation for the temperature T, Eq. 7.1.25, is

$$\boldsymbol{\nabla} \cdot (\rho C_p \boldsymbol{v} T) = \boldsymbol{\nabla} \cdot (k \boldsymbol{\nabla} T), \qquad (10.1.1)$$

where ρ, C_p and k are the density, specific heat and thermal conductivity, and $\boldsymbol{v}$ is the velocity field. The mass and momentum equations at steady state, Eqs. 7.3.1 and 7.3.57 in Chapter 7 are

$$\boldsymbol{\nabla} \cdot (\rho \boldsymbol{v}) = 0, \qquad (10.1.2)$$

$$\rho \boldsymbol{v} \cdot \boldsymbol{\nabla} \boldsymbol{v} = -\boldsymbol{\nabla} p + \boldsymbol{\nabla} \cdot [\mu(\boldsymbol{\nabla} \boldsymbol{v} + (\boldsymbol{\nabla} \boldsymbol{v})^T) + (\mu_b - \tfrac{2}{3}\mu)(\boldsymbol{\nabla} \cdot \boldsymbol{v})\boldsymbol{I}] + \rho \mathbf{g}, \qquad (10.1.3)$$

where $\mathbf{g}$ is the gravitational acceleration, μ is the shear viscosity and μ_b is the bulk viscosity. A simplification is made in the momentum conservation equation, Eq. 10.1.3, by recognising that there is no flow when there is no variation in density—that is, the velocity is zero when $\rho = \rho_\infty$. The momentum conservation equation, Eq. 10.1.3 reduces to,

$$0 = -\boldsymbol{\nabla} p_s + \rho_\infty \mathbf{g}, \qquad (10.1.4)$$

where p_s is the static pressure due to the weight of the fluid. When Eq. 10.1.4 is subtracted from Eq. 10.1.3, the momentum equation reduces to,

$$\rho \boldsymbol{v} \cdot \boldsymbol{\nabla} \boldsymbol{v} = -\boldsymbol{\nabla} p_d + \boldsymbol{\nabla} \cdot [\mu(\boldsymbol{\nabla} \boldsymbol{v} + (\boldsymbol{\nabla} \boldsymbol{v})^T) + (\mu_b - \tfrac{2}{3}\mu)(\boldsymbol{\nabla} \cdot \boldsymbol{v})\boldsymbol{I}] + (\rho - \rho_\infty)\mathbf{g}, \qquad (10.1.5)$$

where $p_d = p - p_s$ is the dynamic pressure. The above equation correctly captures the driving force for the fluid flow, which is the buoyancy force due to the difference between the local density and that far from the object.

The following approximations are made to order to derive the 'Boussinesq' equations for natural convection,

1. The density variation is a linear function of the temperature variation,

$$\boxed{\rho - \rho_\infty = \rho_\infty \beta(T - T_\infty),} \qquad (10.1.6)$$

where β is the coefficient of thermal expansion.

2. The density ρ is replaced by the constant density ρ_∞ in the mass conservation Eq. 10.1.2, in the inertial terms on the left side of the momentum equation, Eq. 10.1.5, in the viscous terms on the right side of the momentum conservation equation, Eq. 10.1.5, and the temperature equation, Eq. 10.1.1. Clearly, this is valid only when the density variation is small—that is, if $(\rho - \rho_\infty)/\rho_\infty = \beta \Delta T \ll 1$. The mass conservation equation, Eq. 10.1.2, reduces to

$$\boxed{\boldsymbol{\nabla} \cdot \boldsymbol{v} = 0.} \qquad (10.1.7)$$

3. The viscosity is considered a constant, $\mu = \mu_\infty$, where μ_∞ is the viscosity at the temperature T_∞, and the density is considered constant in the inertial term on the left in the momentum Eq. 10.1.3. The Boussinesq approximation, Eq. 10.1.6, is substituted for the body force due to density variation in the momentum equation, Eq. 10.1.5,

$$\rho_\infty v \cdot \nabla v = -\nabla p_d + \mu_\infty \nabla^2 v + \rho_\infty \beta (T - T_\infty) g. \qquad (10.1.8)$$

Note that we have set $\rho = \rho_\infty$ in the mass conservation Eq. 10.1.7, and in the inertial term on the left in the momentum conservation Eq. 10.1.8. We have *not* set $\rho = \rho_\infty$ in the body force. The density variation cannot be neglected in the body force on the right side of Eq. 10.1.8, because there would be no flow without the body force.

4. The variation in the thermal conductivity, density and specific heat are neglected in the energy balance equation, Eq. 10.1.1, the flow is considered incompressible (Eq. 10.1.7), and the thermal diffusivity is considered equal to that in the ambient,

$$v \cdot \nabla T = \alpha_\infty \nabla^2 T. \qquad (10.1.9)$$

EXAMPLE 10.1.1: Under what conditions are the constant density, viscosity and thermal conductivity approximations valid for an ideal gas of hard sphere molecules?

Solution: For an ideal gas at constant pressure p, the density is inversely proportional to the temperature,

$$\rho = \frac{pM}{RT}, \qquad (10.1.10)$$

where M is the molecular weight, R is the ideal gas constant and T is the temperature. For small variations in temperature, the variation in density can be estimated as follows,

$$\frac{\Delta \rho}{\rho_\infty} = \frac{1}{\rho}\frac{d\rho}{dT}\bigg|_{\rho=\rho_\infty} \Delta T = \frac{d(\ln(\rho))}{dT}\bigg|_{\rho=\rho_\infty} \Delta T. \qquad (10.1.11)$$

Since the density is inversely proportional to the temperature, $(d(\ln(\rho))/dT) = -(1/T)$,

$$\frac{\Delta \rho}{\rho_\infty} = -\frac{\Delta T}{T_\infty}. \qquad (10.1.12)$$

Therefore, the constant density approximation is valid for

$$\left| \frac{\Delta T}{T_\infty} \right| \ll 1. \tag{10.1.13}$$

The viscosity and thermal conductivity depend only on the absolute temperature, and are independent of the density in an ideal gas. The Eqs. 3.1.35 and 3.1.25 in Chapter 3 are used to estimate the variations in viscosity and thermal conductivity in an ideal gas of hard sphere molecules. Since the viscosity and thermal conductivity are proportional to $T^{1/2}$, for small variations in temperature,

$$\frac{\Delta \mu}{\mu_\infty} = \frac{1}{\mu} \frac{d\mu}{dT} \bigg|_{T=T_\infty} \Delta T = \frac{d(\ln(\mu))}{dT} \bigg|_{T=T_\infty} \Delta T = \frac{1}{2} \frac{\Delta T}{T_\infty}, \tag{10.1.14}$$

and a similar expression for the thermal conductivity. Thus, the Boussinesq approximation can be applied if Eq. 10.1.13 is satisfied—that is, if the temperature variation is small compared to the absolute temperature. □

EXAMPLE 10.1.2: Examine the validity of the Boussinesq approximation for a hot object at temperature 60°C placed in a water bath at temperature 20°C.

- The density of water at atmospheric pressure varies from 998.2 kg/m^3 at 20°C to 983.2 kg/m^3 at 60°C. The percentage variation in density is about 1.5%.

- The viscosity of water varies from 10^{-3} kg/m/s at 20°C to 0.467×10^{-3} kg/m/s at 60°C. Therefore, the percentage variation in viscosity is 53%.

- The thermal conductivity of water varies from 0.598 W/m/°C at 20°C to 0.654 W/m/°C at 60°C. The percentage variation in the thermal conductivity is about 9.3%.

- The specific heat of water at 20°C is 4.182×10^3 J/kg/°C, and at 60° is 4.184×10^3 J/kg/°C. The percentage variation in the specific heat is 0.05%.

In summary, the variation in the specific heat and density is negligible, and the variation in thermal conductivity is small. There is a significant variation in the viscosity, and therefore the Boussinesq approximation is not valid in this case. □

Summary (10.1)

1. When a heated object is placed in a cold ambient fluid, the fluid near the surface gets heated. Due to buoyancy, the hotter fluid close to the surface rises, and the colder fluid further away flows in. The heat transfer due to buoyancy-driven flow is called natural convection heat transfer.

2. The Boussinesq equations, Eqs. 10.1.7, 10.1.8 and 10.1.9, are the coupled momentum and energy balance equations for heat transfer. The following approximations are made:

   (a) The fluid viscosity, thermal conductivity and specific heat are considered independent of temperature, and are assumed to be the same as those for the ambient fluid.

   (b) The fluid density is considered the same as that for the ambient fluid in the mass and energy conservation equations, and in the inertial terms in the momentum conservation equation.

   (c) The variation in density is incorporated only in the buoyancy term in the momentum conservation equation, and a linear approximation, Eq. 10.1.6, is used for the density dependence of the temperature.

3. The Boussinesq equations are valid in gases when the temperature difference between the heated surface and ambient is much smaller than the absolute temperature. In liquids, the Boussinesq approximation is valid when the variation in viscosity and thermal conductivity between the heated surface and ambient is small.

10.2 High Grashof Number Limit

The Grashof number was defined in Section 2.4 (Chapter 2) in by balancing the inertial and buoyancy forces. Here, the Grashof number emerges from the non-dimensionalisation of the momentum and energy equations. The scaled temperature is defined as $T^* = (T - T_\infty)/(T_0 - T_\infty)$. The spatial co-ordinates are scaled by the characteristic length l_c, and the velocity is scaled by the characteristic velocity v_c. The gravitational acceleration vector is non-dimensionalised by its magnitude, $\mathbf{g}^* = (\mathbf{g}/g)$. The dynamical pressure can be non-dimensionalised by either the viscous or inertial terms. Here, we choose to non-dimensionalise the pressure by the fluid inertia, and the appropriate non-dimensional pressure is $p_d^* = (p_d/\rho_\infty v_c^2)$. To summarise, the non-dimensional quantities are,

$$\begin{aligned}
\mathbf{x}^* &= (\mathbf{x}/l_c), & T^* &= (T - T_\infty)/(T_0 - T_\infty), \\
\boldsymbol{v}^* &= (\boldsymbol{v}/v_c), & \boldsymbol{\nabla}^* &= l_c\boldsymbol{\nabla}, \\
\mathbf{g}^* &= (\mathbf{g}/g), & p_d^* &= (p_d/\rho_\infty v_c^2).
\end{aligned} \tag{10.2.1}$$

The mass and momentum conservation equations, Eqs. 10.1.7 and 10.1.8, are

$$\boldsymbol{\nabla}^* \cdot \boldsymbol{v}^* = 0, \tag{10.2.2}$$

$$\frac{\rho_\infty v_c^2}{l_c}\boldsymbol{v}^* \cdot \boldsymbol{\nabla}^*\boldsymbol{v}^* = -\frac{\rho_\infty v_c^2}{l_c}\boldsymbol{\nabla}^*p_d^* + \frac{\mu_\infty v_c}{l_c^2}\boldsymbol{\nabla}^{*2}\boldsymbol{v}^* + \rho_\infty\beta(T_0 - T_\infty)gT^*\mathbf{g}^*. \tag{10.2.3}$$

Since the flow is driven by the last term on the right in Eq. 10.2.3, this has to balance either the inertial or viscous terms. As shown in Examples 2.4.1 and 2.4.2 in Chapter 2, the inertial forces are much larger than the viscous forces in most practical applications, and so the the equation is scaled by the coefficient of the inertial term in Eq. 10.2.3,

$$\boldsymbol{v}^* \cdot \boldsymbol{\nabla}^*\boldsymbol{v}^* = -\boldsymbol{\nabla}^*p_d^* + \frac{\mu_\infty}{\rho_\infty v_c l_c}\boldsymbol{\nabla}^{*2}\boldsymbol{v}^* + \frac{\beta(T_0 - T_\infty)gl_c}{v_c^2}T^*\mathbf{g}^*. \tag{10.2.4}$$

The characteristic velocity is determined by setting the coefficient of the last term on the right to 1, so that the body force and inertial term are comparable,

$$\boxed{v_c = \sqrt{l_c\beta(T_0 - T_\infty)g}.} \tag{10.2.5}$$

The non-dimensional momentum conservation equation is,

$$\boldsymbol{v}^* \cdot \boldsymbol{\nabla}^*\boldsymbol{v}^* = -\boldsymbol{\nabla}^*p_d^* + \frac{1}{\mathrm{Gr}^{1/2}}\boldsymbol{\nabla}^{*2}\boldsymbol{v}^* + T^*\mathbf{g}^*, \tag{10.2.6}$$

where the Grashof number is,

$$\boxed{\mathrm{Gr} = \frac{\rho_\infty^2\beta(T_0 - T_\infty)l_c^3 g}{\mu_\infty^2} = \frac{\beta(T_0 - T_\infty)l_c^3 g}{\nu_\infty^2},} \tag{10.2.7}$$

where $\nu_\infty = (\mu_\infty/\rho_\infty)$ is the kinematic viscosity of the ambient fluid. From Eqs. 10.2.5 and 10.2.7, the square root of the Grashof number is the ratio of inertial and viscous terms, $\mathrm{Gr}^{1/2} = (\rho_\infty v_c l_c/\mu_\infty)$. Another dimensionless number used in natural convection problems is the Rayleigh number,

$$\boxed{\mathrm{Ra} = \frac{\rho_\infty^2 C_p\beta(T_0 - T_\infty)l_c^3 g}{\mu_\infty k_\infty} = \frac{\beta(T_0 - T_\infty)l_c^3 g}{\nu_\infty\alpha_\infty},} \tag{10.2.8}$$

where α_∞ is the thermal diffusivity in the ambient fluid.

The divergence of the velocity is set equal to zero on the left side of Eq. 10.1.9, because the density is considered a constant, and the values of the density, viscosity and specific heat in the ambient are substituted in the equation. The scaled energy equation, Eq. 10.1.9, is

$$v^* \cdot \nabla^* T^* = \frac{1}{\Pr \mathrm{Gr}^{1/2}} \nabla^{*2} T^*, \qquad (10.2.9)$$

where the Prandtl number,

$$\Pr = \frac{C_p \mu_\infty}{k_\infty}, \qquad (10.2.10)$$

is the ratio of the momentum and thermal diffusivity. It can be verified that the prefactor $(1/\Pr \mathrm{Gr}^{1/2})$ on the right side of Eq. 10.2.9 is $(\alpha_\infty/v_c l_c)$, the inverse of the Peclet number based on the characteristic velocity v_c and the thermal diffusivity $\alpha_\infty = (k_\infty/\rho_\infty C_p)$.

The limit $\mathrm{Gr}^{1/2} \gg 1$ and $\Pr \mathrm{Gr}^{1/2} \gg 1$ is considered here. This approximation is applicable to many practical situations, as shown in Examples 2.4.1 and 2.4.2 in Chapter 2, where the convection velocity varies from cm/s to m/s, the Grashof number is in the range 10^4–10^{11}, and $\Pr \mathrm{Gr}^{1/2}$ is large even for fluids with low Prandtl number. In this limit, the diffusion terms on the right in Eqs. 10.2.6 and 10.2.9 are small compared to the convection terms on the left if the size of the heated object is considered the characteristic length scale. There is a boundary layer close to the surface of thickness small compared to l_c where convection and diffusion are comparable.

The momentum and energy equations are re-examined in the limit of high Grashof number. We consider an object of arbitrary shape, as shown in Fig. 10.2. In boundary layer theory for forced convection in the previous chapter, we had seen that at high Peclet number, the thickness of the boundary layer is much smaller than the characteristic dimension of the object. There, the analysis was carried out in a local co-ordinate system at the surface of the object. A similar procedure is used here for natural convection. A Cartesian co-ordinate system is used, where the stream-wise x direction is parallel to the surface, and the cross-stream z direction is perpendicular to the surface, as shown in Fig. 10.2. The scaled mass, momentum and energy equations, Eqs. 10.2.2, 10.2.6 and 10.2.9, written in terms of the velocity

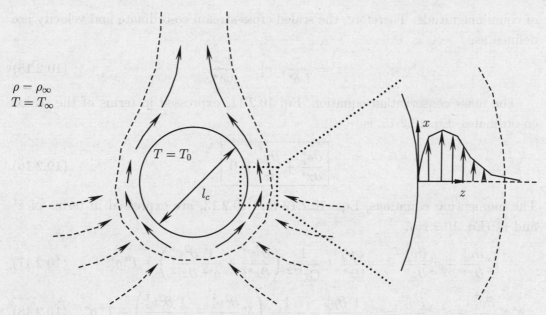

FIGURE 10.2. Co-ordinate system used for analysing the thermal boundary layer flow around a hot object with surface temperature T_0 in an ambient fluid with density ρ_∞ and temperature T_∞ far from the object.

components, are

$$\frac{\partial v_x^*}{\partial x^*} + \frac{\partial v_z^*}{\partial z^*} = 0, \tag{10.2.11}$$

$$v_x^* \frac{\partial v_x^*}{\partial x^*} + v_z^* \frac{\partial v_x^*}{\partial z^*} = -\frac{\partial p_d^*}{\partial x^*} + \frac{1}{\mathrm{Gr}^{1/2}} \left(\frac{\partial^2 v_x^*}{\partial x^{*2}} + \frac{\partial^2 v_x^*}{\partial z^{*2}} \right) + T^* g_x^*, \tag{10.2.12}$$

$$v_x^* \frac{\partial v_z^*}{\partial x^*} + v_z^* \frac{\partial v_z^*}{\partial z^*} = -\frac{\partial p_d^*}{\partial z^*} + \frac{1}{\mathrm{Gr}^{1/2}} \left(\frac{\partial^2 v_z^*}{\partial x^{*2}} + \frac{\partial^2 v_z^*}{\partial z^{*2}} \right) + T^* g_z^*, \tag{10.2.13}$$

$$v_x^* \frac{\partial T^*}{\partial x^*} + v_z^* \frac{\partial T^*}{\partial z^*} = \frac{1}{\mathrm{Pr}\,\mathrm{Gr}^{1/2}} \left(\frac{\partial^2 T^*}{\partial x^{*2}} + \frac{\partial^2 T^*}{\partial z^{*2}} \right). \tag{10.2.14}$$

The cross-stream (z) co-ordinate is rescaled by the boundary layer thickness, which is determined from the condition that convection and diffusion are comparable in the limit Gr $\to \infty$. It is also necessary to rescale the cross-stream velocity v_z so that the two terms in the mass conservation equation, Eq. 10.2.11, are of equal magnitude. The magnitude of the stream-wise velocity v_x is v_c, while the stream-wise length x is scaled by the characteristic dimension of the object, l_c. The cross-stream distance is scaled by the boundary layer thickness $\delta^* l_c$, where the dimensionless parameter $\delta^* \ll 1$. It is evident that the cross-stream velocity v_z has to scale as $(v_c \delta^*)$, so that the two terms in the mass conservation equation, Eq. 10.2.11, are of equal magnitude. Therefore, the scaled cross-stream co-ordinate and velocity are defined as,

$$z^\dagger = \frac{z}{l_c \delta^*}, \quad v_z^\dagger = \frac{v_z}{v_c \delta^*}. \tag{10.2.15}$$

The mass conservation equation, Eq. 10.2.11, expressed in terms of the scaled co-ordinates, Eq. 10.2.15, is

$$\boxed{\frac{\partial v_x^*}{\partial x^*} + \frac{\partial v_z^\dagger}{\partial z^\dagger} = 0.} \tag{10.2.16}$$

The momentum equations, Eqs. 10.2.12 and 10.2.13, are expressed in terms of $z^\dagger$ and $v_z^\dagger$ (Eq. 10.2.15),

$$v_x^* \frac{\partial v_x^*}{\partial x^*} + v_z^\dagger \frac{\partial v_x^*}{\partial z^\dagger} = -\frac{\partial p_d^*}{\partial x^*} + \frac{1}{\mathrm{Gr}^{1/2}} \left(\frac{\partial^2 v_x^*}{\partial x^{*2}} + \frac{1}{\delta^{*2}} \frac{\partial^2 v_x^*}{\partial z^{\dagger 2}} \right) + T^* g_x^*, \tag{10.2.17}$$

$$\delta^* v_x^* \frac{\partial v_z^\dagger}{\partial x^*} + \delta^* v_z^\dagger \frac{\partial v_z^\dagger}{\partial z^\dagger} = -\frac{1}{\delta^*} \frac{\partial p_d^*}{\partial z^\dagger} + \frac{1}{\mathrm{Gr}^{1/2}} \left(\delta^* \frac{\partial^2 v_z^\dagger}{\partial x^{*2}} + \frac{1}{\delta^*} \frac{\partial^2 v_z^\dagger}{\partial z^{\dagger 2}} \right) + T^* g_z^*. \tag{10.2.18}$$

Similarly, the energy conservation equation, Eq. 10.2.14, is expressed in terms of $z^\dagger$ and $v_z^\dagger$,

$$v_x^* \frac{\partial T^*}{\partial x^*} + v_z^\dagger \frac{\partial T^*}{\partial z^\dagger} = \frac{1}{\mathrm{Pr}\,\mathrm{Gr}^{1/2}} \left(\frac{\partial^2 T^*}{\partial x^{*2}} + \frac{1}{\delta^{*2}} \frac{\partial^2 T^*}{\partial z^{\dagger 2}} \right). \tag{10.2.19}$$

In the momentum and energy equations, Eqs. 10.2.17 and 10.2.19, the cross-stream diffusion terms $(1/\delta^{*2})(\partial^2/\partial z^{\dagger 2})$ are the largest terms on the right side. In the limit $\mathrm{Gr} \gg 1$, these terms are comparable to the convective terms on the left for $\delta^* \sim \mathrm{Gr}^{-1/4}$. Without loss of generality, we can define

$$\boxed{\delta^* = \mathrm{Gr}^{-1/4}.} \tag{10.2.20}$$

With this substitution, the momentum and energy equations are,

$$v_x^* \frac{\partial v_x^*}{\partial x^*} + v_z^\dagger \frac{\partial v_x^*}{\partial z^\dagger} = -\frac{\partial p_d^*}{\partial x^*} + \delta^{*2} \frac{\partial^2 v_x^*}{\partial x^{*2}} + \frac{\partial^2 v_x^*}{\partial z^{\dagger 2}} + T^* g_x^*, \tag{10.2.21}$$

$$\delta^* v_x^* \frac{\partial v_z^\dagger}{\partial x^*} + \delta^* v_z^\dagger \frac{\partial v_z^\dagger}{\partial z^\dagger} = -\frac{1}{\delta^*} \frac{\partial p_d^*}{\partial z^\dagger} + \left(\delta^{*3} \frac{\partial^2 v_z^\dagger}{\partial x^{*2}} + \delta^* \frac{\partial^2 v_z^\dagger}{\partial z^{\dagger 2}} \right) + T^* g_z^*, \tag{10.2.22}$$

$$v_x^* \frac{\partial T^*}{\partial x^*} + v_z^\dagger \frac{\partial T^*}{\partial z^\dagger} = \frac{1}{\mathrm{Pr}} \left(\delta^{*2} \frac{\partial^2 T^*}{\partial x^{*2}} + \frac{\partial^2 T^*}{\partial z^{\dagger 2}} \right). \tag{10.2.23}$$

In Eq. 10.2.22, the largest term is the pressure gradient term on the right, which is multiplied by $(1/\delta^*)$. The largest terms are retained in the limit $\delta^* \ll 1$ to obtain the boundary layer momentum and energy equations,

$$v_x^* \frac{\partial v_x^*}{\partial x^*} + v_z^\dagger \frac{\partial v_x^*}{\partial z^\dagger} = -\frac{\partial p_d^*}{\partial x^*} + \frac{\partial^2 v_x^*}{\partial z^{\dagger 2}} + T^* g_x^*, \tag{10.2.24}$$

$$0 = -\frac{\partial p_d^*}{\partial z^\dagger}, \tag{10.2.25}$$

$$v_x^* \frac{\partial T^*}{\partial x^*} + v_z^\dagger \frac{\partial T^*}{\partial z^\dagger} = \frac{1}{\mathrm{Pr}} \frac{\partial^2 T^*}{\partial z^{\dagger 2}}. \tag{10.2.26}$$

From the cross-stream momentum equation, Eq. 10.2.25, the pressure is independent of the cross-stream co-ordinate, and the pressure is only a function of the stream-wise co-ordinate x^*. Since $p_d = 0$ far from the surface where the ambient fluid is stationary, it can be concluded from Eq. 10.2.25 that the dynamic pressure is zero for all $z^\dagger$ in the boundary layer. Therefore, the pressure gradient term in the stream-wise momentum Eq. 10.2.24 is also zero.

The boundary condition for the velocity is the no-slip condition at the surface of the heated object,

$$v_x^* = 0, \;\; v_z^\dagger = 0 \text{ at } z^\dagger = 0, \tag{10.2.27}$$

and the temperature conditions $T = T_0$ at the surface and $T = T_\infty$ in the ambient fluid,

$$T^* = 1 \text{ at } z^\dagger = 0, \;\; T^* = 0 \text{ for } z^\dagger \to \infty. \tag{10.2.28}$$

The simplified conservation equations, Eqs. 10.2.16, 10.2.24, 10.2.25 and 10.2.26, have to be solved numerically for objects of specified shape. However, the magnitude of the Nusselt number can be estimated by realising that the conservation equations, Eqs. 10.2.24–10.2.26, depend only on the Prandtl number and are independent of the Grashof number. The boundary conditions, Eqs. 10.2.27–10.2.28, do not depend on any parameters. Therefore, the solutions for v^* and T^* are also independent of the Grashof number, and depend only on the Prandtl number.

The local heat flux can now be estimated,

$$q_z = -k \left. \frac{\partial T}{\partial z} \right|_{z=0} = -\frac{k(T_0 - T_\infty)}{l_c \delta^*} \left. \frac{\partial T^*}{\partial z^\dagger} \right|_{z^\dagger = 0}. \tag{10.2.29}$$

Here, the scaled temperature gradient $(\partial T^*/\partial z^\dagger)$ at the surface $z^\dagger = 0$ is a function of the Peclet number alone. The average heat flux is obtained by integrating the local heat flux over the surface, and dividing by the surface area,

$$q_{av} = \frac{1}{S} \int dS \, q_z = \frac{C(\mathrm{Pr})\mathrm{Gr}^{1/4} k(T_0 - T_\infty)}{l_c}, \quad (10.2.30)$$

where S is the surface area of the object and $C(\mathrm{Pr})$ is a constant that depends only on the Prandtl number. Here, Eq. 10.2.20 is used to substitute $\delta^* = \mathrm{Gr}^{-1/4}$ in Eq. 10.2.30. The Nusselt number is,

$$\boxed{\mathrm{Nu} = \frac{q_{av}}{k(T_0 - T_\infty)/l_c} = C(\mathrm{Pr})\mathrm{Gr}^{1/4}.} \quad (10.2.31)$$

Thus, the Nusselt number is $\mathrm{Gr}^{1/4}$ times a function of the Prandtl number for natural convection.

Summary (10.2)

1. The characteristic flow velocity generated by the buoyancy forces is given by Eq. 10.2.5.

2. The Grashof number, Eq. 10.2.7, is the square of the Reynolds number based on the characteristic flow velocity. An alternate dimensionless group in natural convection is the Rayleigh number, Eq. 10.2.8.

3. In practical applications, the characteristic velocity varies in the range of cm/s to m/s, and the Grashof number is usually large.

4. In the limit of high Grashof number, momentum and thermal diffusion is restricted to a boundary layer of thickness $\mathrm{Gr}^{-1/4} l_c$, where l_c is the characteristic dimension of the object. The boundary layer thickness is defined in Eq. 10.2.20.

5. When the boundary layer thickness is scaled by the cross-stream co-ordinate, the Boussinesq equations, Eqs. 10.2.16 and 10.2.24–10.2.26, are independent of the Grashof number, and they depend only on the Prandtl number. The boundary conditions, Eqs. 10.2.27–10.2.28, are independent of parameters when expressed in terms of the scaled temperature.

6. The Nusselt number, 10.2.31, is $\mathrm{Gr}^{1/4}$ times a function of Prandtl number.

10.3   Low and High Prandtl Number

In the above analysis, we have taken the limit of large Grashof number, $\text{Gr} \gg 1$, while considering the Prandtl number to be finite. Consequently, the correlation for the Nusselt number depends on $C(\text{Pr})$, which is an unspecified function of the Prandtl number. In this section, a definite form for $C(\text{Pr})$ in Eq. 10.2.31 is obtained in the limits of low Prandtl number, $\text{Pr} \ll 1$, and high Prandtl number, $\text{Pr} \gg 1$.

10.3.1   Low Prandtl number

When the Prandtl number is low, the thermal diffusivity is much higher than the momentum diffusivity. In comparison to momentum, heat diffuses further away from the surface before being convected downstream in the presence of strong convection. Therefore, the momentum boundary layer thickness δ_v^* is much smaller than the thermal boundary layer thickness δ_T^*, as shown in Fig. 10.3.

The boundary layer thickness δ_T^* is used to define the scaled cross-stream co-ordinates in Eq. 10.2.15,

$$z^{\ddagger} = \frac{z}{l_c \delta_T^*}, \quad v_z^{\ddagger} = \frac{v_z^*}{v_c \delta_T^*}. \tag{10.3.1}$$

These are substituted into the energy equation, Eq. 10.2.14,

$$v_x^* \frac{\partial T^*}{\partial x^*} + v_z^{\ddagger} \frac{\partial T^*}{\partial z^{\ddagger}} = \frac{1}{\text{Pr}\,\text{Gr}^{1/2}} \left(\frac{\partial^2 T^*}{\partial x^{*2}} + \frac{1}{\delta_T^{*2}} \frac{\partial^2 T^*}{\partial z^{\ddagger 2}} \right), \tag{10.3.2}$$

and δ_T^* is determined from the condition that convection and diffusion are comparable in the thermal boundary layer for $\text{Pr}\text{Gr}^{1/2} \gg 1$,

$$\boxed{\delta_T^* = \text{Pr}^{-1/2}\text{Gr}^{-1/4}.} \tag{10.3.3}$$

Substituting the above expression for δ_T^* in Eq. 10.3.2, and retaining the largest terms in the limit $\delta_T^* \ll 1$, the energy equation in the thermal boundary layer is,

$$v_x^* \frac{\partial T^*}{\partial x^*} + v_z^{\ddagger} \frac{\partial T^*}{\partial z^{\ddagger}} = \frac{\partial^2 T^*}{\partial z^{\ddagger 2}}. \tag{10.3.4}$$

When Eq. 10.3.3 for the boundary layer thickness is substituted into Eq. 10.2.12 for the stream-wise momentum, we obtain,

$$v_x^* \frac{\partial v_x^*}{\partial x^*} + v_z^{\ddagger} \frac{\partial v_x^*}{\partial z^{\ddagger}} = -\frac{\partial p_d^*}{\partial x^*} + \text{Pr} \frac{\partial^2 v_x^*}{\partial z^{\ddagger 2}} + T^* g_x^*. \tag{10.3.5}$$

The form of the mass conservation equation, Eq. 10.2.11, is unchanged. The cross-stream momentum conservation equation, Eq. 10.2.13, reduces to the zero pressure gradient condition, Eq. 10.2.25.

The diffusion term on the right in Eq. 10.3.5 is $O(\mathrm{Pr})$ smaller than the inertial and body force terms in the limit $\mathrm{Pr} \ll 1$. If this term is neglected in the thermal boundary layer, there is a balance between the inertial and body force terms. In addition, the equation reduces from a second order differential equation to a first order equation in the cross-stream co-ordinate. In order to satisfy the zero velocity boundary condition at the surface, it is necessary to postulate a smaller momentum boundary layer of thickness δ_v^* where momentum convection and diffusion are comparable. However, this calculation is not necessary for determining the correlation for the heat flux at the surface.

If the $O(\mathrm{Pr})$ term is neglected in Eq. 10.3.5, Eqs. 10.3.4–10.3.5 do not contain any parameters. The temperature is scaled so that $T^* = 1$ at the surface and $T^* = 0$ in the ambient. The equations and boundary conditions do not depend on any parameters; therefore, the solution is also independent of parameters. The heat flux

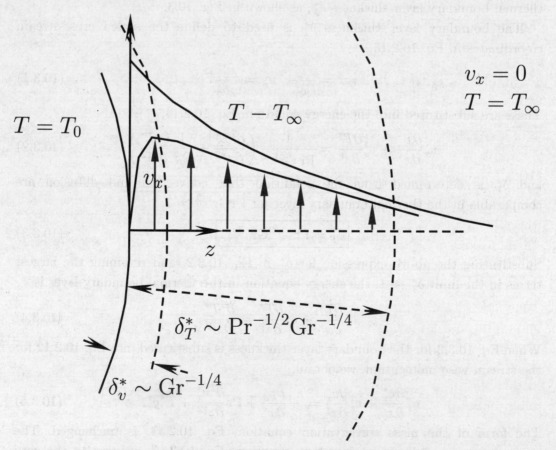

FIGURE 10.3. The momentum and thermal boundary layers in the limit of high Grashof number and low Prandtl number.

at the surface is,

$$q_z = -k \left. \frac{\partial T}{\partial z} \right|_{z=0} = -\frac{k(T_0 - T_\infty)}{l_c \delta_T^*} \left. \frac{\partial T^*}{\partial z^\ddagger} \right|_{z^\ddagger = 0} . \tag{10.3.6}$$

The average heat flux is determined by integrating the local heat flux over the surface,

$$
\begin{aligned}
q_{av} &= \frac{1}{S} \int dS \, q_z = -\frac{k(T_0 - T_\infty)}{l_c \delta_T^*} \frac{1}{S} \int dS \left. \frac{\partial T^*}{\partial z^\ddagger} \right|_{z^\ddagger = 0} \\
&= \frac{k(T_0 - T_\infty) C_{Pr \ll 1} \mathrm{Pr}^{1/2} \mathrm{Gr}^{1/4}}{l_c} .
\end{aligned}
\tag{10.3.7}
$$

Since the solution for T^* is independent of parameters, the surface integral in the first line on the right in Eq. 10.3.7 is a constant, designated $-C_{Pr \ll 1}$. The constant is negative because the heat flux is positive when the temperature decreases with distance from a heated surface. In the last line of Eq. 10.3.7, the right side of Eq. 10.3.3 has been substituted for δ_T^*. The Nusselt number is,

$$\boxed{\mathrm{Nu} = \frac{q_{av}}{k(T_0 - T_\infty)/l_c} = C_{Pr \ll 1} \mathrm{Pr}^{1/2} \mathrm{Gr}^{1/4} .} \tag{10.3.8}$$

Thus, the Nusselt number is proportional to $\mathrm{Gr}^{1/4} \mathrm{Pr}^{1/2}$ in the limit $\mathrm{Pr} \ll 1$ and $\mathrm{Gr}^{1/2} \mathrm{Pr} \gg 1$.

10.3.2 High Prandtl number

In the limit of high Prandtl number, momentum diffusion is much faster than thermal diffusion, and the boundary layer thickness for momentum is much larger than that for heat, as shown in Fig. 10.4. The relative thicknesses of the two boundary layers will be determined shortly from the conservation equations. However, we can identify the dominant terms in the momentum and temperature equations in the thermal boundary layer as follows. There is a balance between momentum convection and diffusion in the momentum boundary layer. Since the momentum boundary layer is much thicker than the thermal boundary layer, it is expected that momentum diffusion will be large compared to momentum convection within the thermal boundary layer. Based on the above discussion, the following physical insights are used in the analysis.

1. Within the thermal boundary layer, there is a balance between momentum diffusion and the body force in the momentum balance equation, and thermal convection and diffusion in the temperature equation.

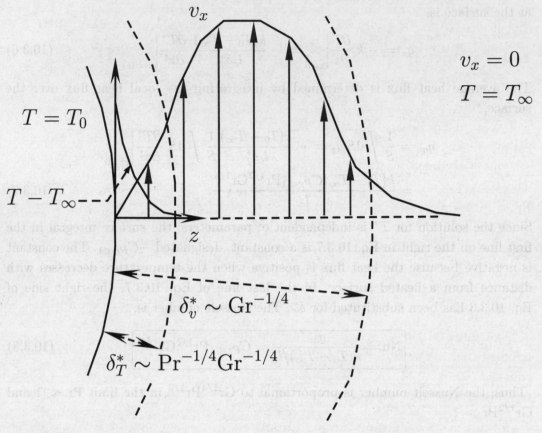

FIGURE 10.4. The momentum and thermal boundary layers in the limit of high Grashof number and high Prandtl number.

2. Due to the no-slip condition at the surface, the velocity in the thermal boundary layer is much smaller than the characteristic velocity v_c in the momentum boundary layer, and a linear approximation can be used for the increase in the velocity in the thermal boundary layer.

Based on the second condition above, the stream-wise velocity in the thermal boundary layer is expressed as,

$$v_x^* = \mathrm{Pr}^{-q} v_x^{\ddagger}, \tag{10.3.9}$$

where the exponent q is positive in the thermal boundary layer where $v_x^* = (v_x/v_c) \ll 1$. The exponent q will be determined form the condition that momentum diffusion and the body force balance in the thermal boundary layer. The definition, Eq. 10.3.1, for the scaled cross-stream distance is used in the momentum and energy equations. From the mass conservation equation, Eq. 10.2.2, the velocity

v_z^* perpendicular to the surface is smaller than the stream-wise velocity by a factor δ_T^*,[1]

$$v_z^* = \mathrm{Pr}^{-q} \delta_T^* v_z^{\ddagger}. \tag{10.3.10}$$

The energy balance equation is

$$\mathrm{Pr}^{-q} \left(v_x^{\ddagger} \frac{\partial T^*}{\partial x^*} + v_z^{\ddagger} \frac{\partial T^*}{\partial z^{\ddagger}} \right) = \frac{1}{\mathrm{Pr}\,\mathrm{Gr}^{1/2}} \left(\frac{\partial^2 T^*}{\partial x^{*2}} + \frac{1}{\delta_T^{*2}} \frac{\partial^2 T^*}{\partial z^{\ddagger 2}} \right). \tag{10.3.11}$$

The stream-wise diffusion term on the right is neglected in the limit $\mathrm{Gr} \gg 1$ and $\mathrm{Pr} \gg 1$, to obtain the simplified energy balance equation.

When Eqs. 10.3.9 and 10.3.10, are substituted into the stream-wise momentum balance equation, Eq. 10.2.12, we obtain

$$\mathrm{Pr}^{-2q} \left(v_x^{\ddagger} \frac{\partial v_x^{\ddagger}}{\partial x^*} + v_z^{\ddagger} \frac{\partial v_x^{\ddagger}}{\partial z^{\ddagger}} \right) = -\frac{\partial p_d^*}{\partial x^*} + \frac{\mathrm{Pr}^{-q}}{\mathrm{Gr}^{1/2}} \left(\frac{\partial^2 v_x^{\ddagger}}{\partial x^{*2}} + \frac{1}{\delta_T^{*2}} \frac{\partial^2 v_x^{\ddagger}}{\partial z^{\ddagger 2}} \right) + T^* g_x^*. \tag{10.3.12}$$

The value of the exponent q is determined from the energy and stream-wise momentum conservation equations, Eqs. 10.3.11 and 10.3.12. In the energy balance equation, Eq. 10.3.11, the convection term on the left and the cross-stream diffusion term on the right are comparable for

$$\delta_T^* \sim \mathrm{Pr}^{-(1-q)/2} \mathrm{Gr}^{-1/4}. \tag{10.3.13}$$

In the stream-wise momentum conservation equation, Eq. 10.3.12, the momentum diffusion is comparable to the body force for

$$\delta_T^* \sim \mathrm{Pr}^{-q/2} \mathrm{Gr}^{-1/4}. \tag{10.3.14}$$

The solutions for δ_T^* from the stream-wise momentum and energy equations are identical for $q = \frac{1}{2}$, and the thermal boundary layer thickness is

$$\boxed{\delta_T^* = \mathrm{Pr}^{-1/4} \mathrm{Gr}^{-1/4}.} \tag{10.3.15}$$

In the stream-wise momentum conservation equation, Eq. 10.3.12, the inertial term on the left is Pr^{-1} smaller than the viscous term on the right.

With the δ_T^* definition, Eq. 10.3.15, the energy equation, Eq. 10.3.11 (with stream-wise diffusion neglected) and stream-wise momentum equation, Eq. 10.3.12

---

[1]Please refer to the arguments leading up to Eq. 10.2.15

(with the inertial and stream-wise diffusion terms neglected) are independent of the Grashof and Prandtl number. The cross-stream momentum equation is given by Eq. 10.2.25, and the mass conservation equation has the same form as Eq. 10.2.16, with $v_x^\ddagger, v_z^\ddagger$ and $z^\ddagger$ replacing $v_x^*, v_z^\dagger$ and $z^\dagger$, respectively. The boundary conditions (Eqs. 10.2.27–10.2.28) are also independent of parameters. Therefore, the solution for the scaled equations depends only on the geometry and is independent of the Prandtl and Grashof numbers.

The Nusselt number correlation is calculated using the same procedure as that in Eqs. 10.3.6–10.3.8, with the definition, Eq. 10.3.15, for the thermal boundary layer thickness. The final expression for the Nusselt number is

$$\boxed{\mathrm{Nu} = C_{Pr \gg 1} \mathrm{Gr}^{1/4} \mathrm{Pr}^{1/4},} \tag{10.3.16}$$

where $C_{Pr \gg 1}$ is a constant.

Summary (10.3)

1. In the limit of low Prandtl number, where thermal diffusivity is much higher than momentum diffusivity,

    (a) The thermal boundary layer thickness is much larger than the momentum boundary layer thickness.

    (b) The viscous terms in the momentum equation are small compared to inertial terms in the thermal boundary layer.

    (c) The thermal boundary layer thickness is $\mathrm{Pr}^{-1/2}\mathrm{Gr}^{-1/4}$ times the characteristic length (Eq. 10.3.3), and the Nusselt number is proportional to $\mathrm{Pr}^{1/2}\mathrm{Gr}^{1/4}$ (Eq. 10.3.8).

2. In the limit of high Prandtl number, where momentum diffusivity is much larger than thermal diffusivity,

    (a) The momentum boundary layer thickness is much larger than the thermal boundary layer thickness.

    (b) The flow in the thermal boundary layer is dominated by viscosity, and the flow velocity in the thermal boundary layer is much smaller than the characteristic flow velocity.

(c) The thermal boundary layer thickness is $Pr^{-1/4}Gr^{-1/4}$ times the characteristic length (Eq. 10.3.15), and the Nusselt number is proportional to $Pr^{1/4}Gr^{1/4}$ (Eq. 10.3.16).

Exercises

EXERCISE 10.1 For a two-dimensional heated cylinder of infinite length along its axis, if the surface temperature is T_0 and the ambient fluid temperature is T_∞, how does the total heat transfer rate per unit length due to natural convection depend on the diameter?

EXERCISE 10.2 For a heated sphere at temperature T_0 in ambient fluid at temperature T_∞, how does the total heat transfer rate from the sphere due to natural convection depend on the diameter?

EXERCISE 10.3 The Nusselt number correlation, Eq. 10.2.31, which is derived assuming steady state, was used in Example 4.8.5 to determine the rate of change of the temperature. The steady-state assumption is valid when the timescale for the boundary layer development is small compared to the characteristic time for the sphere cooling. Determine the condition for the validity of this assumption.

EXERCISE 10.4 Consider natural convection from a heated horizontal tube of diameter $d = 2$ cm into a liquid mercury pool. If the difference in temperature between the tube surface and the ambient mercury is 100°C, what is the characteristic convection velocity, the Grashof and Prandtl number for heat transfer? What is the approximate thickness of the momentum and thermal boundary layer? The properties of liquid mercury are as follows: density $\rho = 1.3 \times 10^4$ kg/m^2, viscosity $\mu = 1.5 \times 10^{-3}$ kg/m/s, specific heat $C_p = 140$ J/kg/°C, thermal conductivity $k = 8.5$ W/m/°C, and thermal expansion coefficient $\beta = 1.8 \times 10^{-4}(°C)^{-1}$.

EXERCISE 10.5 When a suspended salt pill dissolves in water, the water near the surface with higher salt concentration is heavier than the ambient pure water. This causes natural convection because the heavier fluid close to the pill surface descends, to be replaced by ambient water. Determine the characteristic velocity of natural convection and the equivalents of the Grashof and Prandtl number, and the thickness of the momentum and concentration boundary layers for a salt pill of diameter 1 cm. The density of the salt solution can be approximated as $\rho = 10^3(1 + 0.7x)$ kg/m^3, where x is the mass fraction of salt in the solution, and the saturation mass fraction of the salt is 0.26. The viscosity of water is 10^{-3} kg/m/s, and the mass diffusivity of sodium chloride in water is 1.3×10^{-9} m^2/s.

EXERCISE 10.6 Natural convection could be important in the evaporation of water from moist objects. The density of moist air varies with moisture content due to the difference in the molecular weights of water, 18 kg/kmol and air, 29 kg/kmol. Using the ideal gas law, derive an expression for the dependence of the density of moist air on the specific humidity $x = m_w / m_a$, the ratio of the mass of moisture per unit mass of dry air. Simplify the expression for $x \ll 1$, to determine the equivalent of the thermal expansion coefficient.

Determine the Grashof number for the drying of a moist object of characteristic dimension 1 cm in air, where the specific humidity of air at the surface is 0.1, and the air is dry far from the object. The density and viscosity of dry air are 1.25 kg/m^3 and 1.8×10^{-5} kg/m/s, respectively.

Bibliography

[1] R.B. Bird, E.N. Lightfoot, and W.E. Stewart. *Transport Phenomena*. J. Wiley, 2002.

[2] L. G. Leal. *Laminar Flow and Convective Transport Processes*. Butterworth-Heinemann, 1992.

[3] E.L. Cussler. *Diffusion: Mass Transfer in Fluid Systems*. Cambridge Series in Chemical Engineering. Cambridge University Press, 1997.

[4] David J Griffiths. *Introduction to electrodynamics; 4th ed.* Pearson, Boston, MA, 2013.

[5] G. K. Batchelor. *An Introduction to Fluid Dynamics*. Cambridge Mathematical Library. Cambridge University Press, 2000.

[6] R.L. Panton. *Incompressible Flow*. John Wiley & Sons, 1984.

[7] H. Tennekes and J.L. Lumley. *A First Course in Turbulence*. MIT Press, 1978.

[8] L. F. Moody. Friction factors for pipe flow. *Trans. A. S. M. E.*, 66:671–684, 1944.

[9] C. F. Colebrook and White C. M. Experiments with fluid friction in roughened pipes. *Proc. R. Soc. Lond.*, A161:367–381, 1937.

[10] H. Schlichting and K. Gersten. *Boundary-Layer Theory*. Springer, 8th edition, 2000.

[11] M. Van Dyke. *An Album of Fluid Motion*. An Album of Fluid Motion. Parabolic Press, 1982.

[12] S. Taneda. Experimental investigation of the wake behind a sphere at low reynolds numbers. *Journal of the Physical Society of Japan*, 11:1104–1108, 1956.

[13] Z. Naumann and L. Schiller. A drag coefficient correlation. *Z. Verein. Deutsch. Ing.*, 77:318–323, 1935.

[14] F. A. Morrison. *An Introduction to Fluid Mechanics*. Cambridge University Press, 2013.

[15] S. Ergun. Fluid flow through packed columns. *Chem. Eng. Prog.*, 48:89–94, 1952.

[16] J. O. Hinze. Fundamentals of the hydrodynamic mechanism of splitting in dispersion processes. *AIChE Journal*, 1:289–295, 1955.

[17] T. H. Chilton and A. P. Colburn. Mass transfer (absorption) coefficients prediction from data on heat transfer and fluid friction. *Industrial & Engineering Chemistry*, 26:1183–1187, 1934.

[18] E. N. Sieder and G. E. Tate. Heat transfer and pressure drop of liquids in tubes. *Ind. Eng. Chem.*, 28:1429–1435, 1936.

[19] W.E. Ranz and W.R.J. Marshall. Evaporation from drops part i. *Chem. Eng. Prog.*, 48:141–146, 1952.

[20] S. Whitaker. Forced convection heat transfer correlations for flow in pipes, past flat plates, single cylinders, single spheres, and for flow in packed beds and tube bundles. *AIChE J.*, 18:361–371, 1972.

[21]   S. Chapman, T.G. Cowling, D. Burnett, and C. Cercignani. *The Mathematical Theory of Non-uniform Gases: An Account of the Kinetic Theory of Viscosity, Thermal Conduction and Diffusion in Gases.* Cambridge Mathematical Library. Cambridge University Press, 1990.

[22]   A. N. Kolmogorov. The local structure of turbulence in incompressible viscous fluid for very large reynolds numbers. *C. R. Acad. Sci. URSS*, 30:301–305, 1941.

[23]   Th. von Karman. Mechanische ahnlichkeit und turbulenz. *Math. -Phys. Klasse*, pages 58–76, 1930.

Index